# INTERNATIONAL ECONOMIC POLICIES AND THEIR THEORETICAL FOUNDATIONS

## A Source Book

This is a volume in
ECONOMIC THEORY, ECONOMETRICS,
AND MATHEMATICAL ECONOMICS

A Series of Monographs and Textbooks

Consulting Editor: KARL SHELL

A complete list of titles in this series appears at the end of this volume.

# INTERNATIONAL ECONOMIC POLICIES AND THEIR THEORETICAL FOUNDATIONS
## A Source Book

*Edited by*

### JOHN M. LETICHE

*Department of Economics*
*University of California*
*Berkeley, California*

1982

ACADEMIC PRESS

*A Subsidiary of Harcourt Brace Jovanovich, Publishers*

New York   London

Paris   San Diego   San Francisco   São Paulo   Sydney   Tokyo   Toronto

1309532

ACADEMIC PRESS, INC.
111 Fifth Avenue, New York, New York 10003

*United Kingdom Edition published by*
ACADEMIC PRESS, INC. (LONDON) LTD.
24/28 Oval Road, London NW1 7DX

**Library of Congress Cataloging in Publication Data**

Letiche, John M., Date
    International economic policies.

    (Economic theory, econometrics, and mathematical
economics)
    1. Commercial policy--Addresses, essays, lectures.
2. Commerce--Addresses, essays, lectures. 3. International economic relations--Addresses, essays, lectures.
I. Title. II. Series.
HF1411.L454        337        82-1705
ISBN 0-12-444280-3        AACR2

PRINTED IN THE UNITED STATES OF AMERICA

82 83 84 85     9 8 7 6 5 4 3 2 1

# Contents

DEVELOPING COUNTRIES

## V.  INTERNATIONAL CARTELS, COMMODITY AGREEMENTS, AND THE OIL PROBLEM

## VI.  MULTINATIONALS AND INTERNATIONAL INVESTMENT

## VII.  COMMERCIAL POLICIES

DEVELOPING MARKET ECONOMIES

# Preface

Increasing differentiation of international economic policies among the developed and developing market economies, as well as among the centrally planned economies, has been a marked characteristic of contemporary world trade and payments. In selecting readings for this book, I have endeavored to draw upon works that take cognizance of this fact without sacrificing the analysis of fundamental principles of international economics. The objective has been to serve the interests of teachers, students, and government officials for the long term rather than for the short run. Cogency, relevance, and basic principles were, therefore, given high degrees of priority.

An important aim of the book is to provide a complement to textbooks in international economics. Sincere acknowledgments are due to authors and publishers for their generous permission to use materials held in copyright.

I am grateful to a long line of able students who have shared their judgment with me in the selection of articles: Charles Engel, Charles Marston, and Ronald L. Solberg gave me not only highly valued advice but also assistance in checking references and in utilizing libraries. To my wife, Emily Kuyper Letiche, and to my good friend, Howard S. Ellis, I am indebted for literary criticism and for the inspiration that renders every form of scholarly activity a satisfying experience.

John M. Letiche

# INTRODUCTION

JOHN M. LETICHE

This book is composed of nine parts. Part I deals with the evolution of the present international economic order, the recent experience of developing nations, and the deep conflicts involved in the North–South debates, culminating in the call for a new economic order. Part II presents the foundations of terms-of-trade and gains-from-trade theory underlying key policy issues of economic growth and development.

Part III consists of studies in world food production, international trade, and agriculture. In Chapter 5, Schultz shows economic productivity and human well-being to be vitally interrelated in both poor and rich countries. Powerful arguments come to light for rejecting the view that limitations of space, energy, cropland, and other physical properties are the decisive constraints to human betterment. Instead, Schultz argues that the acquired characteristics of the peoples of the earth—their education, experience, skills, and health—are the vital factors in economic progress. Attention is devoted to attributes of traditional agriculture and to allocative efficiency, including its implications for commercial policy throughout the world. A general argument, with supporting evidence, is presented in terms of the increasing value of human time, a formulation that is applicable to both theory and policy pertaining to various terms-of-trade concepts.

Because Chapter 5 is taken from a book, and limitations of space have necessitated severe condensation, the broader framework of its context must be examined.[1] Schultz explains why even the giants of economics such as Adam Smith, David Ricardo, and Thomas Malthus could not have foreseen that the economic development of Western industrial nations would depend primarily on "population quality." A predominant part of their national income (4/5 in the United States) is now derived from earnings and only a small part from property. Drawing on his expertise in agriculture, trade, and development, the author provides fresh insights into the following fundamentals: (1) the process whereby advances in knowledge enhance both physical and human capital; (2) the underlying reasons economists in rich countries find it exceedingly difficult to comprehend the implications of the severe resource constraints on low-income countries; (3) the nature of the performances of poor people that determine their economic choices; and (4) the implications of the ample evidence that poor people in developing countries are no less motivated to work hard, to adapt efficiently to domestic and international market forces (if proper

---

[1]Theodore W. Schultz, *Investing in People: The Economics of Population Quality* (Berkeley: University of California Press, 1981), especially Chapters 2, 3, 6, and 7.

incentives are adopted), and to improve their lot and that of their children, than are those with incomparably greater advantages. Schultz pays particular attention to the consequences of government actions upon national economies and upon their attendant international impacts. The theory developed is robust and applicable to the following issues bearing on this book: (1) to international economic policies on basic research in agriculture, (2) to investment in developing and improving domestic and international market structures, (3) to investment in entrepreneurial ability in developed and developing countries, (4) to economic distortions caused by the International Donor Community, and (5) to trade discrimination by higher-income countries. A major economic implication of this discrimination is that free internal market prices, consistent with international prices in Western European countries, Japan, and some of the other high-income countries, would not only be a boon for consumers in high-income countries, but would contribute substantially to the export opportunities of many low-income countries. "Gains from such trade," Schultz observes, "would probably contribute more to the agricultural development of low-income countries than foreign aid."[2] Systematic research on this issue, particularly as it pertains to contemporaneous uneconomic and obsolete forms of overt and covert "tied aid," is likely to yield high returns to both donor and recipient countries.

Developments in the world food situation is the topic of Chapter 6. Johnson reveals gross error, exaggeration, and oversimplification in numerous current policy approaches and government pronouncements. The author calls attention to unsettled issues and to promising policy research.

Part IV considers the economics of common markets. A key passage from the modern, classic formulation is given, and subsequent chapters survey the more recent literature with reference to developed and developing nations. Although a definitive study on the economic effects of the European common market since its construction has not yet appeared, a scholarly and provocative synthesis of the available evidence to date is provided in Chapter 10.[3] Under the usual assumptions,

---

[2]Ibid., p. 143. Attention has also been directed to the effect of differential growth rates on productivity in food and manufacturing on trade between tropical and industrial countries. This trade has failed dramatically as an engine of economic growth. See W. Arthur Lewis "Aspects of Tropical Trade: 1883–1965. Wicksell Lectures, Stockholm, 1969.

[3]William Wallace, "Economic Divergence in the European Community: Conclusions," Chapter 10 in this publication. For statistical material on the period 1958–1979 (data for 1980 is estimated) demonstrating that the member states of the community experienced an increase during the 1970s in the degree of divergence in terms of inflation rates and exchange rates, in comparison with earlier years, see E. C. Hallett, "Economic Convergence and Divergence in the European Community: A Survey of the Evidence," Chapter 2, in *Economic Divergence in the European Community,* edited by Michael Hodges and William Wallace (London: George Allen and Unwin, 1981), p. 30. While there was no increased disparity in growth rates, there was increased synchronization of cyclical fluctuations and an emergence of a two-tier community caused by a convergence within one group of countries (France, Belgium, The Netherlands, Luxembourg, Denmark, and Germany), "and with current prices and exchange rates apparently forcing a wedge further between the upper and lower tiers" (ibid., p. 30). The reader may also wish to consult the expert studies on *Economic Policies of the Common Market,* edited by Peter Coffey (London: Macmillan, 1979); and Pierre Maillet, *The Construction of a European Community: Achieve-*

it appears that less developed countries (the LDCs) are unlikely to derive economic gains from the formation of common markets. However, econometric projections for the 1980s and 1990s show that the growth rates of world output and trade are more likely to approximate those of the dismal 1970s rather than those of the heady 1960s. In consequence—and because of the failure, to date, to establish a realistic, gradual, and step-by-step process for the evolution of a North–South expansion in specified categories of foreign trade—these factors have led to a resurgence of interest in new forms of economic integration among countries of sub-Saharan Africa, the Pacific Basin region, and the Caribbean. The analysis of the pertinent issues needs to be made more operational, especially as it applies to trade among the more—and the less—rapidly industrializing nations. At present, the industrial countries of Europe are preoccupied with their domestic economic affairs. Consequently, if leadership in international economic policy is to be provided in this field, the responsibility then falls on the government of the United States.

The selections included in Part V pertain to international cartels, commodity agreements, and the oil problem. Emphasis centers on the economic analysis and historical experience of relevant price fluctuations and price trends. The economics of exhaustible resources and of the world oil outlook is briefly but critically examined. This appraisal lends itself to comparison with the Link System simulations for 1981–1990, presented in the concluding chapter of this volume, the results of which furnish a long-term projection of world trade, output, and prices when the real price of oil remains steady, rises gradually, or rises by way of discontinuous shocks.[4]

In Part VI, multinationals and international investment are considered. The chapters provide a synthesis of the roles of direct foreign investment and foreign trade under competitive and monopolistic conditions. Original contributions appear on the "product cycle" and on "uneven development."

Part VII concentrates on commercial policies of developed and developing market economies. Salient facts of industry and trade of seven developing countries have been published by the Organization for Economic Cooperation and Development (OECD).[5] Basic information is further provided in nine volumes of a special

*ments and Prospects for the Future* (New York: Praeger, 1977), originally published in French as *La Construction Européenne* (Paris: Presses Universitoires de France, 1975), especially Chapters 1 and 3 for the effects of the common market on Europe, and Chapter 4 on European macroeconomic adjustments.

[4]See John H. Lichtblau, "Factors Affecting World Oil Prices in the 1980s," Chapter 13 in this publication; and Lawrence R. Klein, "Some Economic Scenarios for the 1980s," Chapter 31. For a more extensive treatment, see L. R. Klein, S. Fardoust, V. Filatov, and V. Su, "Simulations of the World Impact of Oil Price Increases: An exercise in Supply Side Economics," to appear in Lawrence R. Klein, *Supply and Demand Side Economics: Two Blades of the Shears* (Berkeley: University of California Press, 1982.). Excellent discussions on some of the theoretical issues involved, with references to practical problems, are provided by R. M. Solow, "The Economics of Resources or the Resources of Economics," *American Economic Review* 64, no. 2 (May 1974), pp. 1–14; and D. M. G. Newbery, "Oil Prices, Cartels, and the Problem of Dynamic Inconsistency," *Economic Journal* 91 (September 1981), pp. 617–646.

[5]See especially I. M. D. Little, T. Scitovsky, and M. Scott, *Industry and Trade in Some Developing Countries: A Comparative Analysis* (Paris: OECD, 1970).

conference series published by the National Bureau of Economic Research.[6] In addition, two volumes summarize the experience of these foreign trade regimes and their impact on development. One explains the causes for successes and failures of economic liberalization attempts, the other outlines the overall results.[7] Even the summaries of these volumes, however, are either too lengthy, or too closely linked to their context, for feasible inclusion in this book. Nevertheless, for an appraisal of current commercial policies of the developing market economies, a brief statement of the core of this research is indispensable.

The evidence appears to be generally consistent with the following conclusions. The more closed the regime, the lower the growth rate in exports and in GNP. Furthermore, for the middle-income LDCs, the more closed regimes were usually associated with the following characteristics:

1. more comprehensive import-substitution programs;
2. greater ''bias'' against exports, denoted by the greater extent to which relative incentives for domestic production of items not traded and import substitutes were distorted away from international relative prices;
3. higher variance in price differentials among commodity categories;
4. more distorted relative factor prices and less efficient relative factor use, and less efficient finished-product markets;
5. larger investment in capital-intensive sectors;
6. lower domestic value added to capital employed;
7. lower real profits;
8. smaller inflow of governmental and private capital, signifying lower gross investment relative to gross national saving as a percent of GNP;
9. less optimal, and more erratic, composition of imported inventories;
10. greater instability of agricultural and food supplies; and
11. larger excess capacity of plant utilization in manufacturing production.

The first selection in Part VII (taken from a more comprehensive source based on independent research that is complementary to the summary in the preceding), succinctly sets forth the lessons and prospects for the choice of a development strategy.[8]    Balassa first analyzes the results of inward- and outward-oriented development strategies. The lessons of experience—such as the exceedingly high cost incurred by many developing nations as a consequence of their governments' implementing erroneous economic policies—stand out clearly. Chapter 18 by Myint, an important article in the history of economic thought, examines the relevance of

[6]See ''A Special Conference Series on Foreign Trade Regimes and Economic Development,'' resulting from a National Bureau of Economic Research project. The series includes nine country studies: Turkey, Ghana, Israel, Egypt, The Philippines, India, South Korea, Chile, and Colombia, and synthesis volumes noted in the following.

[7]Anne O. Krueger, *Liberalization Attempts and Consequences* (Cambridge, Mass.: Ballinger Publishing Company, 1978); and Jagdish N. Bhagwati, *Anatomy and Consequences of Exchange Control Regimes* (Cambridge, Mass.: Ballinger Publishing Company, 1978).

[8]Bela Balassa, *The Process of Industrial Development and Alternative Development Strategies,* Essays in International Finance (Princeton: Princeton University Press, 1980).

classical international trade theory as a guide to commercial policy and to the economic growth of underdeveloped countries.

Chapter 19 by Prebisch deals with the global system of capitalism in Latin America.[9] Because this chapter is a résumé of a critically important book published in Spanish that is not yet translated into English, the book itself needs to be placed in perspective.

Since World War II, Prebisch has not only raised the key economic issues facing Latin America, and proffered solutions for dealing with them, but has perceptively and provocatively altered some of his fundamental approaches and recommendations in the light of new evidence and superior analysis. Prebisch has had a greater influence on Latin American international development policy than any other economist, and his influence on government officials and academic economists in other newly industrializing countries (NICs) has also been marked. This is understandable, for the fundamental issue facing Latin America—the struggle for orderly development in an epoch of revolutionary change, characterized by massive migrations from rural to urban areas, often aggravating the spread of urban decay and the mounting unemployment in the major cities—has plagued other industrializing nations no less than those of Latin America. With the projections of lowered growth rates in world GNP and international trade for the 1980s, the "interior concept" of economic development has undergone a resurgence (though a moderate one) and the author has once again drawn attention to fundamental, emerging issues.

Four developments in Prebisch's writings readily can be distinguished:

1. the emphasis in the late 1950s on strengthening the Latin-American industrialization process through differential tariff protection, import substitution, and the creation of a common market for trade and investment integration;[10]

2. The recognition by the mid-1960s that the "first," or "easy," stage of substitution of domestic manufacturing production for imports has passed, and that Latin American exports should be promoted through the removal by industrial countries of their trade barriers to imports of manufactures from Latin America;[11]

3. the call in the late 1960s for tariff preferences with respect to Latin American exports of manufactures to the United States and other industrial nations, as well as for more effective economic integration among the major Latin American nations through the use of Special Drawing Rights (SDRs) and the elimination of their overvalued exchange rates[12]; and

[9]Raúl Prebisch, Chapter 19 in this publication.

[10]Raúl Prebisch, "Commercial Policy in the Underdeveloped Countries," *American Economic Review* 49 (May 1959), pp. 251–273.

[11]Raúl Prebisch, *Towards a Dynamic Development Policy for Latin America* (New York: United Nations, 1963); *Towards a New Trade Policy for Development* (New York: United Nations, 1964), pp. 20–25.

[12]Cf. United Nations Economic Commission for Latin America, *Economic Bulletin for Latin America* (New York: United Nations, 1967), vol. 12, no. 1, pp. 35–55; also vol. 12, no. 2, pp. 146–147. Prebisch also expressed similar views at a conference on the 25th anniversary of Bretton Woods held at Queen's University, Kingston, Ontario, Canada, June 2–3, 1967.

4. the synthesis of his views in the early 1980s, the essence of which is an enhanced rate of capital accumulation; this is to be achieved *via* a reduced proportion of GNP devoted to profligate government expenditures and class-conscious conspicuous consumption, and an enlarged proportion (directed by government) devoted to more productive investment through the greater use of private entrepreneurial capacity and improved market structures, supported by preferences on exports of manufactures and selective import substitution.[13]

Prebisch has had much practical experience in guiding and evaluating the process of Latin American economic development. With reference to his earlier views, he believes that after World War II, an import-substitution strategy was essential for Latin America because imports were "undercutting" the development of "nascent" industries. To overcome this handicap, trade barriers had to be erected but they were to be temporary—not permanent. Prebisch also recommended policies for Latin American economic integration, which were to help achieve essential internal and external economies of scale. But, he now maintains, the trade barriers against the outside world were to be moderate, whereas most Latin American countries pushed them to the extreme. Furthermore, he had advocated North–South economic cooperation, which, in his judgment, the industrial countries failed to implement: "The centres," Prebisch writes in Chapter 19 of this volume, "have by no means encouraged this process through changes in their production structure; and by failing to open their doors to manufacturing imports from the periphery, they force the latter to continue with import substitution."[14]

Prebisch appears to concur, however, in the general consensus that has arisen regarding the effects of the tariff and of the quantitative restrictions imposed by most Latin American governments in the 1950s and 1960s. They had inordinately reduced foreign competition. Domestic firms became relatively more inefficient as they expanded production of substandard commodities for highly priced and comparatively small, internal markets. These factors, he points out, contributed to his focusing attention on the markets of the "centre countries"—the United States, Western Europe, and Japan. But Latin American manufacturers could not compete in these markets without preferential tariff treatment. Consequently, the United Nations Commission on Trade and Development (UNCTAD) was formed in 1963 to help Latin American countries compete. As its first secretary general, a post he held for 6 years, he made this organization the focal point of the North–South dialogue. In effect, the concepts developed by UNCTAD induced the United Nations General Assembly in 1974 to pass the unattainable, even if desirable, comprehensive resolutions for a "new international economic order."

The more promising paradigm that Prebisch now advocates appears to have taken account of the serious errors made both by the industrializing and the industrial countries in recent decades. The extreme trade restrictions of the NICs not only

[13]Raul Prebisch, Chapter 19 in this publication.
[14]Ibid.

hindered competition from imports, but also had the effect of discriminating against their exports. Explicitly, this was brought about by high taxes on exportables and implicitly, by overvalued exchange rates that resulted from the cost–price raising tendencies of the immoderate trade restrictions. It was the *combination* of ill-advised economic policies and the movement from the first stage of import substitution to the second stage that resulted in the vast and costly economic distortions. The first stage usually entailed the replacement by domestic production of imports of nondurable consumer goods (e.g., clothing, shoes, and household wares), and of their inputs (e.g., textiles, leather, and wood products). Characteristically, the NICs had a dynamic comparative advantage in these industries. They are usually labor intensive, and do not require highly skilled workers and sophisticated technologies or a network of suppliers of parts and components for the attainment and maintenance of international competitiveness. Their efficient scale of output is relatively low and their expansion tends to generate an increasing supply of entrepreneurial ability, improved labor training, and technology.[15]

The second stage of import substitution involved, however, the replacement of imports of consumer and producer durables (as well as intermediate inputs) by domestic production. The intermediate inputs were composed of goods such as steel products and petrochemicals. The required conditions for efficient production, understandably, were converse to those prevailing in industries referred to as appropriate for the first stage of import substitution. Propelled by exorbitantly high rates of effective tariff protection, these products generated exceedingly high domestic resource costs relative to foreign exchange earnings, because the need to import key materials and machinery virtually always was enlarged. Furthermore, the neglect of intraindustry relationships resulted in enormous economic waste, with the overpricing of industrialization to be matched only by the underpricing of agriculture. The result was an uneconomic relative contraction of output in food and agriculture, aggravating inflationary pressures and external deficits. Since nominal rates of interest usually lagged behind the rise of domestic prices, much wasteful private and public investment was undertaken, and credit rationing by the banks and/or government was imposed.

Not surprisingly, Prebisch regards the immense economic waste in the policies of overextended governments of the NICs and in the new consumption patterns of the elite and middle classes—especially in his native Latin America—as basic factors in their distorted economic structures. Nevertheless, he contends that market forces are not and cannot be ''the supreme regulator of the development of the periphery and its relations with the centre.''[16] Notwithstanding their satisfactory growth rates in the 1960s and 1970s, he points out, income distribution in Latin America remains

[15]For an extensive analysis, see Bela Balassa, *The Process of Industrial Development*, pp. 4–9; and the sources cited therein. This essay retains the format of the Frank D. Graham Memorial Lecture delivered by the author at Princeton University. It eschews footnote references, but a list of the author's publications from which the empirical evidence cited in the essay (and in Chapter 17 of this publication) is derived, is contained in an appendix. Ibid., pp. 29–30.

appallingly skewed: at present, the bottom 40% of the population earns only 8% of the national income.

"Dynamic redistribution," according to Prebisch's present position, is a primary requisite of successful economic development, not through aid from industrial nations, but through harnessing the NICs' own resources that are now misallocated in waste. Prebisch has not abandoned his criticism of what he regards as the "major flaws in centre-periphery relations."[16] Nor does he believe that the economic transformation of Latin America is feasible without the state's playing a critical role in enforcing essential austerity. The free market and authoritarian governments, he argues, have not solved the major problems of Latin American economic development. Recognizing the immensity of the task, Prebisch does not take a dogmatic theoretical stand. At the core of his new analysis remains the division of the world into the "centre" and "the periphery," with the center controlling the world economy and the NICs reduced to a dependency relationship of providing raw materials and markets. If the NICs are to prosper and attain equality, he writes, this form of economic dependency must be broken: "Individual decisions in the market place must be combined with collective decisions outside it which override the interests of the dominant groups. All this, however, calls for a great vision, a vision of change, both in peripheral development and in relations with the centres . . . ."[16]

In effect, Prebisch now formulates his analysis in different terms from those of his previous writings. Accumulation—in its human and physical forms—is the key to economic development. Much of the capital stock required for accelerated growth, he says, already exists in Latin America. But it is wasted in huge government establishments, in ornamental military outlays, and in purchases of apartments abroad and fancy automobiles by the elites, and of smaller luxuries such as color television sets by the middle class. They are indulging in the frenetic imitation of consumption in the center, which is a form of waste they cannot afford. Prebisch therefore suggests that if a larger proportion of GNP were allocated to more productive investment, the growth rate of Latin American NICs would be enhanced and the distribution of income improved.

As an antidote to "frenetic consumption," Prebisch challenges the Latin American governments to "build an interior concept of development" along the lines of Japan, with high savings and austerity dedicated to economic growth. Short of that, he recommends progressive taxation on consumption expenditures—not on income—and an enlarged use of government outlays on productive industrial and agricultural development. The aided industries, he now emphasizes, should be *private*, not government owned. Although the governments of Latin American NICs have to play an active role in the accumulation of capital, Prebisch no longer concentrates his attack on the international and domestic market system as such. The major flaws of the system do not lie in private property itself but in the harmful consequences of the concentration of the means of production.[16] Prebisch acknowledges that the market is an extremely useful mechanism, not only economically, but politically. "If you abolish it," he observes, "the decisions are all made by a few at

[16]Raúl Prebisch, Chapter 19 in this publication.

the top and that is not favorable for democracy.''[17] If the state takes into its own hands the ownership and management of the means of production, this option ''is incompatible with the paramount concept of democracy and the human rights inherent in it.''[18] As compared with his earlier views, Prebisch manifestly accords a more balanced role to government and to import substitution. Exportables, entrepreneurial capacity, freer markets, and private capital formation receive more emphasis in his new and improved paradigms.

When his more comprehensive formulation becomes available in English, it will doubtless require analytical examination and testing. But even cursory examination of comparative evidence suggests that the misallocation of resources in Latin America has been massive. In the 1960s and 1970s, the proportion of gross national saving to GNP, and gross investment to GNP, was not significantly different in Latin America and the Caribbean from the proportion in East Asia and the Pacific.[19] But the rate of growth in gross investment and in total GNP—and particularly in GNP per capita—was substantially lower in Latin America and the Caribbean. However, the new emphasis should mark even more than an improved direction in paradigms. It calls for a recognition that the international economic policies of individual governments are more important to the success of their economic growth and development than the arbitrarily assumed structural economic forces ascribed to various ''centers'' or ''peripheries.''

Attendant upon the oil and other crises of the early 1970s, this recognition has been growing among many leaders of the NICs. They increasingly appreciate that the economic problems of their countries are no less the result of domestic economic distortions than of those emanating from either the capitalist or communist centers. Action based on this recognition is an indispensable element even for a modest improvement in international economic policies. A caveat, however, appears to be in order: Prebisch insists that the route to democracy is through better income distribution. If, by this, he means that a larger proportion of GNP devoted to more productive human and physical capital is likely to enhance the real growth rates per capita, reduce the rate of unemployment, and improve income distribution, the argument appears tenable under the conditions that he now postulates. It may, furthermore, apply to new Latin American arrangements for more effective private entrepreneurship in the management of enterprises now publicly owned and managed—a problem that deserves serious attention. If, however, Prebisch means that the route to democracy is by way of direct income distribution in the shape of wealth distribution and/or enlarged transfer payments, the argument appears much less tenable on economic grounds.[20]

Chapter 20 in Part VII considers internal and external effects of the world

[17]See the discussion of Prebisch's present views by Edward Schumacher, ''North-South by South: A New Axis of Waste,'' *New York Times*, May 31, 1981, p. 19EY. For a more critical view of the North–South dilemma, see W. M. Corden, *The NIEO Proposals: a Cool Look*, Thames Essay No. 21 (London: Trade Policy Research Centre, 1979).

[18]Raúl Prebisch, Chapter 19 in this publication.

[19]The data referred to in the text were calculated from World Bank, *Annual Report*, Washington, D.C., 1978, pp. 118–119, and 1980, pp. 130–131.

economic crisis on the seven East European members of the Council for Mutual Economic Assistance (CMEA or Comecon): Bulgaria, Czechoslovakia, East Germany, Hungary, Poland, Romania, and the Soviet Union. It is an exemplary analysis of the microeconomic and macroeconomic effects and policy responses of trade, investment, and financial flows among developed-market economies and centrally planned economies.[21] The results of the GATT Tokyo Round Negotiations are then presented in Chapter 21 and the section concludes with a brief essay on international assistance policies of the United States.

Part VIII consists of fundamental readings on international payments. Chapter 23 by Meade incisively formulates the meaning of internal and external balance. It is followed by an authoritative statement on the monetary approach to the balance of payments. Appraisals are then given of the theory and practice of the Eurodollar market and of the operating principles and procedures of the European Monetary System. Recent monetary policy under exchange-rate flexibility is analyzed, as is the need for surveillance by the International Monetary Fund. The final selection, Chapter 30, presents in tabular form the kinds of exchange rate regimes now in practice, and describes how members use the Fund's resources to meet balance of payment needs.[22]

[20]Limitations of space have proscribed the inclusion of a paper in this publication dealing specifically with the issue of "unequal exchange." The reader interested in the exploration of literature on this problem may wish to consult the survey article by Alain de Janvry and Frank Kramer, "The Limits of Unequal Exchange," *The Review of Radical Political Economies* 11 (Winter 1979), pp. 3–15, and the sources cited therein.

[21]Richard Portes, "Effects of the World Economic Crisis on the East European Economies," Chapter 21 in this publication. See also the studies in *The Impact of International Economic Disturbances on the Soviet Union and Eastern Europe,* edited by Egon Neuberger and Laura D'Andre Tyson (New York: Pergamon Press, 1981); *The EEC and Eastern Europe,* edited by Avi Shlaim and G. N. Yannopoulos (London: Cambridge University Press, 1978); and *East European Integration and East-West Trade,* edited by Paul Marer and John Michael *Montias* (Bloomington: Indiana University Press, 1980). As regards United States–Soviet commercial relations, in particular, see Hertha W. Heiss, Allen J. Lenz, and Jack Brougher, "United States–Soviet Commercial Relations Since 1972," U.S. Senate, Joint Economic Committee (Washington: U.S. Governemt Printing Office, 1979), pp. 189–207; and *An Assessment of the Afghanistan Sanctions: Implications for Trade and Diplomacy in the 1980's,* Report of the Committee on Foreign Affairs, House of Representatives (Washington, D.C.: U.S. Government Printing Office, 1981).

[22]For excellent background articles on the present international monetary system in historical perspective, on the rapid growth of Eurocurrency markets, and on monetary integration, respectively, see Gottfried Haberler, "The International Monetary System After Jamaica and Manila," *Contemporary Economic Problems* (Washington, D.C.: American Enterprise Institute, 1977), pp. 219–240; John Hewson and Eisuke Sakaibara, "A General Equilibrium Approach to the Eurodollar Market," *Journal of Money, Credit and Banking* 8 (August 1976), pp. 297–323; and W. M. Corden, *Monetary Integration,* Essays in International Finance (Princeton: Princeton University Press, 1972). For provocative background papers on developments in international trade theory as they apply to international economic policies, see Robert E. Baldwin, "Determinants of the Commodity Structure of U. S. Trade," *American Economic Review,* March 1971, vol. LXI, no. 1, pp. 126–146; and Max Corden and Ronald E. Findlay, "Concluding Remarks," on Proceedings of a Nobel Symposium held at Stockholm, edited by Bertil Ohlin, Per-Ove Hesselborn, and Per Magnus Wijkman, *The International Allocation of Economic Activity* (London: Macmillan, 1977), pp. 538–556.

Part IX, which provides perspectives on the present state of the international economy—and its relation to the troubled environment in which international economic policy targets will operate—consists of a Nobel lecture on some economic scenarios for the 1980s.

# Part I

## Evolution and Revision of the International Economic Order

Lewis
Bhagwati

## THE EVOLUTION OF THE
## INTERNATIONAL ECONOMIC ORDER

W. Arthur Lewis

# INTRODUCTION

In international circles the topic of the day is the demand of the Third World for a new international economic order. My topic is the evolution of the existing economic order: how it came into existence not much more than a century ago, and how it has been changing.

The phrase "international economic order" is vague, but nothing would be gained by trying to define it precisely. I will discuss certain elements of the relationship between the developing and the developed countries that the developing countries find particularly irksome. These are:

First, the division of the world into exporters of primary products and exporters of manufactures.

Second, the adverse factoral terms of trade for the products of the developing countries.

Third, the dependence of the developing countries on the developed for finance.

Fourth, the dependence of the developing countries on the developed for their engine of growth.

My purpose in treating these topics is not to make recommendations, but to try to understand how we come to be where we are.

Reprinted from "The Evolution of the Economic Order," pp. 3–25, with permission of Princeton University Press, Princeton, Copyright 1978.

# 2

# THE DIVISION
# OF THE WORLD

How did the world come to be divided into industrial countries and agricultural countries? Did this result from geographical resources, economic forces, military forces, some international conspiracy, or what?

In talking about industrialization, we are talking about very recent times. England has seen many industrial revolutions since the thirteenth century, but the one that changed the world began at the end of the eighteenth century. It crossed rapidly to North America and to Western Europe, but even as late as 1850 it had not matured all that much. In 1850 Britain was the only country in the world where the agricultural population had fallen below 50 percent of the labor force. Today some 30 Third World countries already have agricultural populations equal to less than 50 percent of the labor force—17 in Latin America, 8 in Asia not including Japan, and 5 in Africa not counting South Africa. Thus, except for Britain, even the oldest of the industrial countries were in only the early stages of structural transformation in 1850.

At the end of the eighteenth century, trade between what are now the industrial countries and what is now the Third World was based on geography rather than on structure; indeed India was the leading exporter of fine cotton fabrics. The trade was also trivially small in volume. It consisted of sugar, a few spices, precious metals, and luxury goods. It was then cloaked in much romance, and had caused much bloodshed, but it simply did not amount to much.

In the course of the first half of the nineteenth century industrialization changed the composition of the trade, since Britain captured world trade in iron and in cotton fabrics; but the volume of trade with the Third World continued to be small. Even as late as 1883, the first year for which we have a calculation, total imports into the United States and Western Europe from Asia, Africa, and tropical Latin America came only to about a dollar per head of the population of the exporting countries.★

There are two reasons for this low volume of trade. One is that the leading industrial countries—Britain, the United States, France, and Germany—were, taken together, virtually self-sufficient. The raw materials of the industrial revolution were coal, iron ore, cotton, and wool, and the foodstuff was wheat. Between them, these core countries had all they needed except for wool. Although many writers have stated that the industrial revolution depended on the raw materials of the Third World, this is quite untrue. Not until what is

★ For the sources of this and other statistics used here, and generally for more detailed historical analysis, the reader may consult my book, *Growth and Fluctuations 1870-1913*, Allen and Unwin, London 1978.

sometimes called the second industrial revolution, at the end of the nineteenth century (Schumpeter's Third Kondratiev upswing based on electricity, the motor car and so on), did a big demand for rubber, copper, oil, bauxite, and such materials occur. The Third World's contribution to the industrial revolution of the first half of the nineteenth century was negligible.

The second reason why trade was so small is that the expansion of world trade, which created the international economic order that we are considering, is necessarily an offshoot of the transport revolutions. In this case, the railway was the major element. Before the railway the external trade of Africa or Asia or Latin America was virtually though not completely confined to the seacoasts and rivers; the railway altered this. Although the industrial countries were building railways from 1830 on, the railway did not reach the Third World until the 1860s. The principal reason for this was that, in most countries, railways were financed by borrowing in London—even the North American railways were financed in London—and the Third World did not begin to borrow substantially in London until after 1860. The other revolution in transport was the decline in ocean freights, which followed the substitution of iron for wooden hulls and of steam for sails. Freights began to fall after the middle of the century, but their spectacular downturn came after 1870, when they fell by two-thirds over thirty years.

For all these reasons, the phenomenon we are

exploring—the entry of the tropical countries significantly into world trade—really belongs only to the last quarter of the nineteenth century. It is then that tropical trade began to grow significantly—at about four percent a year in volume. And it is then that the international order that we know today established itself.

Now it is not obvious why the tropics reacted to the industrial revolution by becoming exporters of agricultural products.

As the industrial revolution developed in the leading countries in the first half of the nineteenth century it challenged the rest of the world in two ways. One challenge was to imitate it. The other challenge was to trade. As we have just seen, the trade opportunity was small and was delayed until late in the nineteenth century. But the challenge to imitate and have one's own industrial revolution was immediate. In North America and in Western Europe, a number of countries reacted immediately. Most countries, however, did not, even in Central Europe. This was the point at which the world began to divide into industrial and non-industrial countries.

*The early or late this reaction happened determine* ×××

Why did it happen this way? The example of industrialization would have been easy to follow. The industrial revolution started with the introduction of new technologies in making textiles, mining coal, smelting pig iron, and using steam. The new ideas were ingenious but simple and easy to apply. The capital requirement was remarkably small, except for the cost of building railways, which could be had on loan. There were no great economies of scale, so the skills required for managing a factory or workshop were well within the competence and experience of what we now call the Third World. The technology was available to any country that wanted it, despite feeble British efforts to restrict the export of machinery (which ceased after 1850), and Englishmen and Frenchmen were willing to travel to the ends of the earth to set up and operate the new mills.

Example was reinforced by what we now call "backwash." A number of Third World countries were exporting manufactures in 1800, notably India. Cheap British exports of textiles and of iron destroyed such trade, and provided these countries an incentive to adopt the new British techniques. India built its first modern textile mill in 1853, and by the end of the century was not only self-sufficient in the cheaper cottons, but had also driven British yarn out of many Far Eastern markets. Why then did not the whole world immediately adopt the techniques of the industrial revolution?

The favorite answer to this question is political, but it will not wash. It is true that imperial powers were hostile to industrialization in their colonies. The British tried to stop the cotton industry in India by taxing it. They failed because the Indian cotton industry had the protection of lower wages and of lower transportation costs. But they did succeed in holding off iron and steel production in India till as late as 1912. The hostility of imperial powers to industrialization in their colonies and in the "open door" countries is beyond dispute. But the world was not all colonial in the middle of the nineteenth century. When the coffee industry began to expand rapidly in Brazil around 1850, there was no external political force from Europe or North America that made Brazil develop as a coffee exporter instead of as an industrial nation. Brazil, Argentina, and all the rest of Latin America were free to industrialize, but did not. India, Ceylon, Java, and the Philippines were colonies, but in 1850 there were still no signs of industrialization in Thailand or Japan or China, Indo-China or the rest of the Indonesian archipelago. The partition of Africa did not come until 1880, when the industrial revolution was already a hundred years old. We cannot escape the fact that Eastern and Southern Europe were just as backward in industrializing as South Asia or Latin America. Political independence alone is an insufficient basis for industrialization.

We must therefore turn to economic explanations. The most important of these, and the most neglected, is the dependence of an industrial revolution on a prior or simultaneous agricultural revolution. This argument was already familiar to eighteenth-century economists, including Sir James Steuart and Adam Smith.

In a closed economy, the size of the industrial sector is a function of agricultural productivity. Agriculture has to be capable of producing the surplus food and raw materials consumed in the industrial sector, and it is the affluent state of the farmers that enables them to be a market for industrial products. If the domestic market is too small, it is still possible to support an industrial sector by exporting manufactures and importing food and raw materials. But it is hard to begin industrialization by exporting manufactures. Usually one begins by selling in a familiar and protected home market and moves on to exporting only after one has learnt to make one's costs competitive.

The distinguishing feature of the industrial revolution at the end of the eighteenth century is that it began in the country with the highest agricultural productivity—Great Britain—which therefore already had a large industrial sector. The industrial revolution did not create an industrial sector where none had been before. It transformed an industrial sector that already existed by introducing new ways of making the same old things. The revolution spread rapidly in other countries that were also revolutionizing their agriculture, especially in Western Europe and North America. But countries of low agricultural productivity, such as Central and Southern Europe, or Latin America, or China had rather small industrial sectors, and there it made rather slow progress.

If the smallness of the market was one constraint on industrialization, because of low agricultural productivity, the absence of an investment climate was another. Western Europe had been creating a capitalist environment for at least a century; thus a whole new set of people, ideas and institutions was established that did not exist in Asia or Africa, or even for the most part in Latin America, despite the closer cultural heritage. Power in these countries—as also in Central and Southern Europe—was still concentrated in the hands of landed classes, who benefited from cheap imports and saw no reason to support the emergence of a new industrial class. There was no industrial entrepreneurship. Of course the agricultural countries were just as capable of sprouting an industrial complex of skills, institutions, and ideas, but this would take time. In the meantime it was relatively easy for them to respond to the other opportunity the industrial revolution now opened up, namely to export agricultural products, especially as transport costs came down. There was no lack of traders to travel through the countryside collecting small parcels of produce from thousands of small farmers, or of landowners, domestic or foreign, ready to man plantations with imported Indian or Chinese labor.

And so the world divided: countries that industrialized and exported manufactures, and the other countries that exported agricultural products. The speed of this adjustment, especially in the second half of the nineteenth century, created an illusion. It came to be an article of faith in Western Europe that the tropical countries had a comparative advantage in agriculture. In fact, as Indian textile production soon began to show, between the tropical and temperate countries, the differences in food production per head were much greater than in modern industrial production per head.

Now we come to another problem. I stated earlier that the industrial revolution presented two alternative challenges—an opportunity to industrialize by example and an opportunity to trade. But an opportunity to trade is also an opportunity to industrialize. For trade increases the national income, and therefore increases the domestic market for manufactures. Import substitution becomes possible, and industrialization can start off from there. This for example is what happened to Australia, whose development did not begin until the gold rush of the 1850s, and was then based on exporting primary products. Nevertheless by 1913 the proportion of Australia's labor force in agriculture had fallen to 25 percent, and Australia was producing more manufactures per head than France or Germany. Why did this not happen to all the other agricultural countries?

The absence of industrialization in these countries was not due to any failure of international trade to expand. The volume of trade of the tropical countries increased at a rate of about 4 percent per annum over the thirty years before the first world war. So if trade was the engine of growth of the tropics, and industry the engine of growth of the industrial countries, we can say that the tropical engine was beating as fast as the industrial engine. The relative failure of India tends to overshadow developments elsewhere, but countries such as Ceylon, Thailand, Burma, Brazil, Colombia, Ghana, or Uganda were transformed during these thirty years before the First World War. They built themselves roads, schools, water supplies, and other essential infrastructure. But they did not become industrial nations.

There are several reasons for this, of which the most important is their terms of trade. Thus, we must spend a little time analyzing what determined the terms of trade.

# THE FACTORAL TERMS
# OF TRADE

The development of the agricultural countries in the second half of the nineteenth century was promoted by two vast streams of international migration. About fifty million people left Europe for the temperate settlements, of whom about thirteen million went to what we now call the new countries of temperate settlement: Canada, Argentina, Chile, Australia, New Zealand, and South Africa. About the same number—fifty million people— left India and China to work mainly as indentured laborers in the tropics on plantations, in mines, or in construction projects. The availability of these two streams set the terms of trade for tropical and temperate agricultural commodities, respectively. For temperate commodities the market forces set prices that could attract European migrants, while for tropical commodities they set prices that would sustain indentured Indians. These were very different levels.

A central cause of this difference was the difference in agricultural productivity between Europe and the tropics. In Britain, which was the biggest single source of European migration, the yield of wheat by 1900 was 1,600 lbs. per acre, as against the tropical yield of 700 lbs. of grain per acre. The

European also had better equipment and cultivated more acres per man, so the yield per man must have been six or seven times larger than in tropical regions. Also, in the country to which most of the European migrants went (the United States), the yield differential was even higher, not because of productivity per acre, which was lower than in Europe, but because of greater mechanization. The new temperate settlements could attract and hold European immigrants, in competition with the United States, only by offering income levels higher than prevailed in Northwest Europe. Since Northwest Europe needed first their wool, and then after 1890 their frozen meat, and ultimately after 1900 their wheat, it had to pay for those commodities prices that would yield a higher-than-European standard of living.

In the tropical situation, on the other hand, any prices for tea or rubber or peanuts that would offer a standard of living in excess of the 700 lb. of grain per acre level were an improvement. Farmers would consider devoting idle land or time to producing such crops; as their experience grew, they would even, at somewhat higher prices, reduce their own subsistence production of food in order to specialize in commercial crops. But regardless of how the small farmer reacted, there was an unlimited supply of Indians and Chinese willing to travel anywhere to work on plantations for a shilling a day. This stream of migrants from Asia was as large as the stream from Europe and set the level of tropical prices. In the 1880s the wage of a plantation laborer was one shilling a day, but the wage of an unskilled construction worker in Australia was nine shillings a day. If tea had been a temperate instead of a tropical crop, its price would have been perhaps four times as high as it was. And if wool had been a tropical instead of a temperate crop, it could have been had for perhaps one-fourth of the ruling price.

This analysis clearly turns on the long-run infinite elasticity of the supply of labor to any one activity at prices determined by farm productivity in Europe and Asia, respectively. This is applied to a Ricardian-type comparative cost model with two countries and three goods. The fact that one of these goods, food, is produced by both countries determines the factoral terms of trade, in terms of food. As usual one can elaborate by increasing the number of goods or countries, but the essence remains if food production is common to all.

One important conclusion is that the tropical countries cannot escape from these unfavorable terms of trade by increasing productivity in the commodities they export, since this will simply reduce the prices of such commodities. Indeed we have seen this quite clearly in the two commodities in which productivity has risen most, sugar and rubber. The factoral terms of trade can be improved only by raising tropical productivity in the common commodity, domestic foodstuffs.

There are interesting borderline cases where the two groups of countries compete. Cotton is an example. In the nineteenth century, the United

States was the principal supplier of cotton, but the crop could also grow all over the tropics. The United States maintained its hold on the market despite eager British efforts to promote cotton growing in the British colonies. The U.S. yields per acre were about three times as high as the Indian or African yields, but this alone would not have been enough to discourage tropical production. The United States could not have competed with tropical cotton had southern blacks been free to migrate to the North and to work there at white Northern incomes. It was racial discrimination in the United States that kept the price of cotton so low; or, to turn this around, given the racial discrimination, American blacks earned so little because of the large amount of cotton that would have flowed out of Asia and Africa and Latin America at a higher cotton price.

Cotton was one of a set of commodities where low agricultural productivity excluded tropical competition. The tropics could compete in any commodity where the difference in wages exceeded the difference in productivity. This ruled out not only cotton and tobacco, which fell to the ex-slaves in North America, but also maize, beef, and timber, for which there were buoyant markets, and ground was lost steadily in sugar as beet productivity increased. This left a rather narrow range of agricultural exports and contributed to the over-specialization of each tropical country in one or sometimes two export crops. Low productivity in food set the factoral terms of trade, while relative productivity in other agriculture determined which crops were in and which were out.

Minerals fall into this competing set. Labor could be had very cheaply in the tropical countries, so high productivity yielded high rents. These rents accrued to investors to whom governments had given mining concessions for next to nothing, and the proceeds flowed overseas as dividends. Mineral-bearing lands were not infinitely elastic, but the labor force was. With the arrival of colonial independence over the last two decades, the struggle of the newly independent nations to recapture for the domestic revenues the true value of the minerals in the ground, whether by differential taxation, by differential wages for miners, or by expropriation, has been one of the more bitter aspects of the international confrontation.

Given this difference in the factoral terms of trade, the opportunity that international trade presented to the new temperate settlements was very different from the opportunity presented to the tropics. Trade offered the temperate settlements high income per head, from which would immediately ensue a large demand for manufactures, opportunities for import substitution, and rapid urbanization. Domestic saving per head would be large. Money would be available to spend on schools, at all levels, and soon these countries would have a substantial managerial and administrative elite. These new temperate countries would thus create their own power centers, with money, education, and managerial capacity, independent of and somewhat hostile to the imperial power. Thus, Australia, New Zealand, and Canada ceased to be colonies in any political sense long before they acquired formal rights of sovereignty, and had already set up barriers to imports from Britain. The factoral terms available to them offered them the opportunity for full development in every sense of the word.

The factoral terms available to the tropics, on the other hand, offered the opportunity to stay poor—at any rate until such time as the labor reservoirs of India and China might be exhausted. A farmer in Nigeria might tend his peanuts with as much diligence and skill as a farmer in Australia tended his sheep, but the return would be very different. The just price, to use the medieval term, would have rewarded equal competence with equal earnings. But the market price gave the Nigerian for his peanuts a 700-lbs.-of-grain-per-acre level of living, and the Australian for his wool a 1600-lbs.-per-acre level of living, not because of differences in competence, nor because of marginal utilities or productivities in peanuts or wool, but because these were the respective amounts of food that their cousins could produce on the family farms. This is the fundamental sense in which the leaders of the less developed world denounce the current international economic order as unjust, namely that the factoral terms of trade are based on the market forces of opportunity cost and not on the just principle of equal pay for equal work. And of course nobody understood this mechanism bet-

ter than the working classes in the temperate settlements themselves, and in the United States. The
working classes were always adamant against Indian or Chinese immigration into their countries
because they realized that, if unchecked, it would
drive wages down close to Indian and Chinese
levels.★

★ I have borrowed passages from my paper "The Diffusion
of Development" in Thomas Wilson, Editor, *The Market and
the State*, Oxford University Press, Oxford 1976.

# CUMULATIVE FORCES

Now let me come to more recent developments. I must first make the point that, in spite of the poor factoral terms of trade, the opportunity to trade did substantially raise the national incomes of those tropical countries that participated in trade. This was partly because prices had to be set at levels that would bring the produce out. So, although prices were based on the low productivity in food, they had to be set somewhat higher. Just as wages were higher in Australia and Argentina than in Paris or London, so also wages were higher in Ceylon or Burma than in India or China.

The other reason national incomes of some tropical countries increased was that these countries developed by bringing unused resources into use—both unused land and unused labor—so that to a large extent what they produced for export was additional to what they would otherwise have produced. In particular the tropical countries continued to be self-sufficient in food. The agricultural exports were extra output.

This steady increase in income over some sixty or seventy years, right down to the great depression of 1929, very considerably expanded the demand for manufactures. Imports of textiles and of iron goods mounted, putting domestic handicrafts out of business. Why did not these countries set up their own modern factories to cope with this rising demand?

Some did—especially India, Ceylon, Brazil, and Mexico—but progress was slow. Apart from colonialism, which restricted some but not others, three other factors worked against industrialization.

The first reason is that to a large extent the import and export trades of these countries were controlled by foreign hands. This was where the profits were, in a complex of wholesaling, banking, shipping, and insurance. Railway, plantation, and mining profits were much more volatile. Profits provide a major source of funds for reinvestment. Had trading profits accumulated in domestic hands, there would have been more domestic reinvestment, and almost certainly more interest in domestic manufacturing.

Foreigners participated heavily in the external trade of these countries for a variety of economic, cultural, and political reasons. On the economic side there was advantage in large scale operations because they minimized the usual riskiness of trading and avoided the possibility facing small traders, who could be wiped out by a bad season. On the cultural side Europeans had been running big shipping and trading enterprises since the seventeenth century; in this as also in banking and insurance, they had a considerable lead over Latin Americans and Africans, though not over Indians or Chinese. The political factor was a further complication in that some imperial governments deliberately favored their nationals at the expense both of indigenous and of other foreign competitors. Whatever the reason, the points where profits were greatest (wholesaling, banking, shipping, insurance) tended to be foreign-controlled, and this certainly diminished the availability of funds and enterprise for investment in domestic manufacturing.

A second factor to which some nationalist historians attach much importance is the fact that participation in trade itself whets the appetite for foreign goods, in the process destroying local industry. The consumer learns to prefer wheat to yams and cement to local building materials. This is all right if the country has the raw materials and can acquire the new skills for processing them. Otherwise it reduces the export multiplier—the extent to which the proceeds of exports circulate within the country, stimulating domestic industry, before flowing out again. It is difficult to give this quantitative significance for the nineteenth century, since the products destroyed by imports from Britain were mostly cotton and iron manufactures not essentially different from the imports which replaced them. Some of the difference lay in consumer preference, but most of the difference lay in cost. The situation evolved differently in the twentieth century when brand names established their footing in many consumer markets and proved difficult to dislodge even with domestic products of equal cost and quality.

As long ago as 1841, Friederich List emphasized

that the market forces in an agricultural economy work to keep it agricultural unless special measures are taken to arrest their momentum and change their direction. List's remedy was for the government to protect an infant manufacturing industry with tariffs and quotas. But this presupposes that the industrial forces have already conquered the government and can use it to their advantage. The fact that they had not is the third explanation why the agricultural countries, though becoming more prosperous and consuming more and more manufactures, did not industrialize. Imperial power was of course an obstacle in the colonial countries, but is not a necessary explanation since the same happened in the independent countries. The fact is that the very success of the country in exporting created a vested interest of those who lived by primary production—small farmers no less than big capitalists—and who opposed measures for industrialization, whether because such measures might deflect resources from agriculture and raise factor prices, or because they might result in raising the prices of manufactured goods. The outcome therefore depended on the relative political strengths of the industrial and the agricultural interests.

It is not to be supposed that in this confrontation the entrenched agricultural forces always won. On the contrary, they lost in most European countries and in most of the countries of new settlement. In Latin America at the end of the century the liveliness of Brazilian and Mexican entrepreneurs is notable. Egypt contrasts with India in not producing a single industrialist from its prosperous landowning and merchant classes. To unravel the different responses of countries experiencing apparently similar forces is a source of historical excitement. The contrast between Argentina and Australia is particularly instructive. These two countries began to grow rapidly at the same time, the 1850s, and sold the same commodities—cereals, wool, and meat. In 1913 their incomes per head were among the world's top ten. But Australia industrialized rapidly, and Argentina did not, a failure which cost her dearly after the war when the terms of trade moved against agriculture. Some Argentinian nationalists blame this failure on British interests in Argentina, but the British had even more influence in Australia or Canada, which were industrializing rapidly. The crucial difference between the two countries was that Argentinian politics were dominated by an old, landed aristocracy. Australia had no landed aristocracy. Its politics were dominated by its urban communities, who used their power to protect industrial profits and wages. The slowness with which industrial classes emerged in Latin America, or Central Europe, North Africa or much of Asia is explained as much by internal social and political structure as by the impact of external forces.

# Introduction
## Jagdish N. Bhagwati

North-South economic relations, three decades after the decline of colonial empires and the emergence of new developing countries on the international scene, have come to the forefront of international economics and politics. The concerted demands of the South for a new international economic order (NIEO), and the problems they raise for the North in setting the stage for negotiations on concrete proposals related to the NIEO, now define the agenda, as well as the political climate, of the numerous conferences and intergovernmental negotiating groups on international economic matters.[1]

In assessing North-South relations and their prospects and in suggesting the optimal reforms NIEO demands should and can sensibly (in terms of political feasibility) be directed to, a historical perspective is essential. It is necessary to trace the evolution of the economic and political philosophy of the developing countries that currently animates and conditions their views of the current international economic order and prompts their demands for changes therein.

## Developing Countries: Shifting Postures

In fact, the present postures of the developing countries can be traced to three factors.

1. A substantial shift has occurred in the developing countries' perception of the gains to be had from economic relations with the developed countries under the existing rules of the game; the shift has been toward the gloomier side.

2. At the same time, the developing countries now perceive their own economic and hence political power vis-à-vis the developed countries to be sufficiently substantial to warrant a strategy of effective "trade unionism" to change the rules of the game and thereby to wrest a greater share of the world's wealth and income.

3. Finally, a straightforward political desire to participate more effectively in decision making on international economic matters is now evident: this is the "populist" aspect of the current situation.[2] Participation is thus demanded, not merely to ensure that the developing countries' interests are safeguarded but equally as an assertion of their rights as members of an international community and as a desired feature of a just international order.

Reprinted from "The New International Economic Order: The North–South Debate" (J.N. Bhagwati, ed.) pp. 1–24, with permission of The MIT Press, Cambridge, Massachusetts, Copyright 1977.

J. N. Bhagwati

2

A correct appreciation of each of these striking new aspects of the Southern postures is critical for a proper evaluation of the prospects for improved South-North collaboration on international economic issues.

## Shifts in Perception of Existing International Economic Order[3]

The developing countries are linked to the developed countries through trade, aid, investment, and migration. The central issue for them is whether these links work to their detriment or advantage. Several ideologies compete for attention on this question; the influential policymakers in a number of developing countries have moved from more cheerful to gloomier ideologies as they have progressively made more forceful demands for changes in the world economic order.

The ideology that has traditionally been dominant is aptly characterized as that of "benign neglect"—links with the rich nations create benefits for the poor nations. This view of the world economy parallels the utilitarian economists' view that the invisible hand works to promote universal well-being. In this model, the laissez-faire view that private greed will produce public good translates on the international scene into the notion that, while the different actors in the world economy pursue their own interests, the result will nonetheless be to benefit the developing countries. Thus, while multinational corporations invest in these countries to make profits, they will increase these countries' incomes, diffuse technology, and harness their domestic savings.[4] The exchange of commodities and services in trade will reflect the principle of division of labor and hence bring gains from trade to these countries.[5] The migration of skilled labor, instead of constituting a troublesome brain drain, will help to remove impediments to progress such as inadequate remuneration of the educated elite.[6]

In direct contrast to this classical economic viewpoint, there is the doctrine of "malign neglect" which views the impact of these links between the rich and the poor nations as primarily detrimental to the latter group. In the apt description of Osvaldo Sunkel, integration of the developing countries in the international economy leads to their domestic disintegration. This doctrine also supports the economic notion, used extensively by the Swedish economists Knut Wicksell and Gunnar Myrdal, of growing disequilibrium and exploding sequences, rather than the classical notions of equilibrium. Thus, multinational corporations disrupt domestic salary structures by introducing islands of high-income jobs that cause exorbitant wage demands by others seeking to keep up with the Joneses in the multinationals. International trade leads to the perpetuation of the role of developing countries as producers of primary, unsophisticated products that relegate them to a secondary and inferior position in the international division of labor. Furthermore, in the classic Prebisch thesis, the terms of trade of the primary-product-exporting developing countries have declined and will continue to do so, conferring gains on the developed and inflicting losses on the developing countries.[7] The brain drain to the developed countries deprives the developing countries of scarce skills and the talents that make economic progress possible.[8] The attractions of Western standards of living make domestic setting of priorities and raising of savings difficult if not impossible.

These "malign neglect" views are merely the logical extension of the disenchantment with the "benign neglect" model. This disenchantment initially took the form of complaints that, instead of diffusing development, the links with the international economy were of no consequence to the developing countries. Thus, the early revisionist critics of foreign investment argued that these investments led to enclaves and had little genuine impact on the developing countries: the latter remained in consequence at the periphery of the world economy.[9] As Naipaul remarks wryly in his *Guerrillas*, "Tax holidays had been offered to foreign investors; many had come for the holidays and had then moved on elsewhere."[10] The "malign neglect" school takes this revisionism to its logical extreme and turns the argument on its head by claiming that the trouble with foreign investment is not that it makes no impact on the national economies of the developing countries because of its enclave nature but rather that it does and that this impact is adverse.

Also contrasting with these models are the two major ideological positions that focus not on the impact of the links but rather their intended objectives. Thus, the "benign intent" school of thought, to which the "white man's burden" philosophy belongs, considers the international links and institutions to be designed so as to transmit benefits to the poor nations. Private investment is regarded as motivated by the desire to spread the fruits of modern technology and enterprise to the developing countries. In particular, the foreign aid programs are conceived as humanitarian in origin, reflecting the Western ideals of liberalism and the enlightened objective of sharing the world's resources with the poor countries.

The polar opposite of this model is the "malign intent" view of the world, typically favored by the Marxist and New Left writings on the international economy. Foreign aid is seen as a natural extension of the imperialist designs on the poor nations aimed at creating dependence.[11] Private investments, following the flag in past models, are seen now as precursors of the flag, with brazen colonialism replaced by devious neocolonialism.[12]

Clearly, none of these models in their pure form capture the full complexity of the effects that the links with the outside world have on the developing countries' prospects for economic progress. However, it is clear that policy makers in several developing countries have moved over the three postwar decades from a world view based primarily on the benign neglect and intent models to one characterized more by varying shades of the malign neglect and intent models.

Thus, the early posturing of these countries was based on the view that the existing mechanisms governing trade and investment flows were primarily beneficial. Furthermore, aside from utilizing and expanding trade opportunities and attracting foreign investment funds, the developing countries could appeal to the developed countries on a purely moral plane for the provision of technical assistance and foreign aid for developmental programs. These premises were the basis for the first UNCTAD conference in Geneva in 1964, which led to a permanent creation of the UNCTAD secretariat and its eventual emergence as the principal forum for airing the problems of the developing countries.

UNCTAD I at Geneva and UNCTAD II at New Delhi thus concentrated primarily on defining and underlining aid targets for the developed countries, while laying principal stress on two aspects of trade policy: preferential access by developing countries into the markets of the developed countries, and the principle of nonreciprocity.

The trade efforts were to bear fruit, yielding to the developing countries the satisfaction of having utilized collective action at UNCTAD to some advantage. This advantage, however, was rather small; in retrospect, it is evident that the grant of preferential entry by the EEC and by other developed countries, including the United States in 1975, has been of limited value because of numerous exceptions and because of the importance of nontariff barriers to which it did not extend.[13] As for the principle of nonreciprocity, it is now increasingly obvious that the developing countries probably threw away the main instrument that governments have at their command to lower their own trade barriers—the ability to tell their protected industries that the protection must be reduced as part of the reciprocal bargaining process. Recent studies on the foreign trade regimes of the developing countries have shown[14] that the degree and dispersion of the protection enjoyed by the industries of these countries have been disturbing; a continuation of reciprocity would have been most useful if effectively used by willing governments in the developing countries.[15]

The principal disappointments were to be in the field of international aid flows. The developing countries were faced with the incongruous contrast between the UNCTAD targets on foreign aid and the declining overall flows from the leading aid donors, particularly the United States, once the leader of the enlightened donors. Not merely were nominal aid flows decelerating, but their real worth was falling with inflation. It was increasingly clear that their worth was seriously reduced by practices such as aid-tying, which compelled the aid recipients to buy from the donor countries at artificially high prices. Their worth was further reduced because few of the aid funds were anything but loans to be repaid and hence were substantially less by way of genuine aid transfers than the publicized figures implied.[16]

Aside from the failure to meet the obligations which the developed countries appeared to have endorsed, however reluctantly, at international forums such as the UNCTAD, there was also an emerging sense that the declining efforts at international assistance were a reflection of the steady thaw in superpower relations. Thus, it became increasingly difficult to maintain that humanitarian motives, rather than the political necessities of the Cold War, were the major motivating factors behind the aid programs of the 1950s. These cynical perceptions of the aid efforts of the developed world were only to be reinforced by the misguided attempts at enforcing performance criteria in aid distribution. Typically, the aid donors, following economically wise but politically foolish precepts, insisted on examining and endorsing the entire set of economic policies of the recipient nations to ensure that their meager aid assistance was being utilized to advantage,[17] thereby generating resentments and charges of calculated attempts at imposing ideological solutions in the guise of "scientific" economic prescriptions.[18]

The confirmation of covert political interventions in the developing countries, euphemistically described as destabilization operations, engineered by developed countries from which one expected better behavior, often prompted and encouraged by multinational corporations (such as the ITT in Chile), must have helped in strengthening the radical theses regarding the Northern designs and impact on the South.

The growing sense that the benign intent and impact of the developed countries on the well-being of the developing countries could not be taken as the natural order of things under the existing international arrangements was finally to be accentuated and reinforced from yet another direction. The focus during the 1950s on the "gap" in the incomes, living standards, and wealth of the developing and the developed countries and on the gearing of international targets to narrowing and eventually eliminating such differences, was probably helpful in lending animation to the development decades and the attendant programs for developing countries. But it was also to lead inevitably to frustration—such gaps cannot possibly be narrowed in any significant manner in the foreseeable future despite any optimism as to the prospects of the developing countries' growth rates.[19] Thus, despite the fact that the developing countries, as a group, grew at the historically remarkable rate of 5.5 percent per annum during the first development decade of the 1960s, the awareness grew that these rates of growth could neither help measurably in "catching up" with the developed countries nor could they adequately diffuse the fruits of growth to the poor in the developing countries.[20] Poverty, both absolute at home and relative vis-à-vis the developed countries, thus seemed to be inescapable under the existing economic regimes. As a result, as far as domestic policies are concerned the intellectuals have turned increasingly to distributive implications of their developmental programs: the faith in the "trickle-down" process has been badly shattered. At the international level the implication is again for distribution: it is felt now that the growth rates of the poor countries, no matter how rapid, have to be supplemented by increasing transfers of resources on a simple, progressive argument. It is a question of a *moral* imperative that the world's limited wealth and incomes be shared more equitably. This is only the international counterpart of the sociological fact that as access to affluence diminishes, the resentment of success increases and the stress on redistribution is keener. For example, the greater American mobility surely explains the lack of success of socialist doctrines there whereas the social and economic rigidity of the British society explains the stresses on the social contract that are quite evident in their macroeconomic failures. The erosion of faith since the 1950s in the ability of the developing countries to catch up with the developed countries has surely contributed to their present "trade unionist" demands for greater shares in world income through the creation of a new international economic order.[21]

### Post-OPEC Emphasis on Collective Action: The Rise of "Trade Unionism"

It was against the backdrop of this slow but inevitable shift in many developing countries' world view that the dramatic event of the successful cartelization of oil producers, nearly all members of the Third World, was to materialize. The OPEC had existed for a number of years prior to its dramatic success since 1973, but practically no serious analyst had considered its success probable. Indeed, my colleague Morris Adelman had the singular misfortune of writing a superb analysis of the oil industry, predicated entirely on the assumption that OPEC would not succeed, and having it published just as this basic assumption was being falsified![22]

The Third World's reaction to the nearly sevenfold rise in oil prices and the accentuation of the resource and foreign exchange difficulties of many of the poorer nations among the Third World, was to baffle the rich nations that sought to mobilize the poor against the OPEC. The developing countries refused to condemn, and indeed seemed to take great delight in, the oil price increases. This reaction can only be understood in light of the shifts in their views about the rich nations. Clearly there was a need for prudence vis-à-vis the nouveaux riches to whom the developing countries would have to turn for aid. But, far more than that, the developing countries seemed to feel that finally there was one dramatic instance of a set of primary producers in the Third World who were able to get a "fair share" of the world incomes by their own actions rather than by the unproductive route of morally persuading the rich nations for fairer shares. Even while many of them suffered from the fallout of the oil price increases, many developing countries therefore felt a sense of solidarity, a corps d'esprit with the OPEC countries and the exhilarating sense that they could finally take their economic destiny in their own hands. Thus, the stage was to be set psychologically and politically for the present phase of "trade unionist" militancy. The nascent sense that collective action, as crystallized in the developing countries' Group of 77 and the activities at UNCTAD, could yield some results (such as the schemes for preferential entry), was now to be transformed into an act of faith: solidarity in international bargaining, alternatively termed "collective self-reliance," on a variety of fronts would yield much more than had ever been thought possible.

The OPEC success crystallized the concepts of strength through collective action and "solidarity rather than charity." The developing countries also seemed to infer from the OPEC experience that their commodity exports, which had traditionally been viewed as a sign of weakness, could be turned instead into weapons of collective action. Thus, the notion of "commodity power" emerged and has shaped not merely the politics but also the economics of the demands for NIEO (as typified by the Corea plan for commodities at the UNCTAD).

## Populism
The OPEC example was also to hold the further attraction that, contrary to aid flows (whether bilateral or multilateral), the earning of the new resources through improved terms of trade implied that the OPEC countries retained their national sovereignty in deciding how to spend these resources and also began to qualify as nations that commanded some voice in the management of international monetary affairs and therefore in other deliberations on the world economic regime as well.

Many developing countries, seeking both the assertion of fuller national sovereignty over their economic programs and increased participation in international deliberations on trade, aid, and monetary rules, thus saw the OPEC as an ideal case which they would hope to emulate.

The foregoing analysis underlines the complex nature of the current attitudes and demands of the developing countries while defining the limits within which the amicable evolution of a new international economic order will have to be defined. Several major points must be stressed.

1. Our analysis shows that the developing countries' objectives are economic and political. The economic objective is principally to increase their share of the world's income and wealth. The political objectives are that they should have better control over the use of these and their own resources and that they should also be allowed to participate actively in devising the new rules for managing world trade, aid, and monetary and other matters of global concern. Needless to say all of these objectives may not be in harmony; it may be possible to get more resources transferred if their use is not entirely within the prerogative of the recipient country—a conflict that is quite important in practice. However, these objectives do exist manifestly; the twin political objectives are the new elements on the scene, as compared with the 1950s and 1960s.

2. The developing countries' assessment of their capacity to achieve these objectives is grounded in their assessment of their capacity for collective action. There is some evidence that the early optimism about the use of commodity power has receded: except for bauxite, the results of collective cartelization seem to have borne little fruit. This should, in fact, have been expected from a realistic assessment of commodity markets since oil is a very special case with exceedingly low elasticities, considerable macro effects, and

no real parallel for other commodities. The "commodity power" that exists, outside of oil, is therefore only a short run, disruptive power; it may be currently exercised to some advantage[23] but is certainly self-destructing through high elasticities of substitution and through the use of augmented inventory policies by the developed countries (such as those proposed recently in the United States for certain raw materials).

Interestingly therefore the emphasis has shifted from the proposed use of weak and essentially short run "commodity power" to improve prices to pressuring the consuming developed countries to collaborate in the raising of the prices of these commodities to "fair" levels; this "indexing" idea parallels the domestic U.S. parity program for agriculture.[24] As Kindleberger has remarked, this is tantamount to asking the chicken to help in plucking its own feathers! And yet, the idea is not quite absurd; the developing countries now see their power not as accruing from commodities per se but rather from their capacity to create confrontations and impede agreements on a variety of global concerns such as the Law of the Sea. Thus, commodities today have become a chief vehicle through which the developing countries want resources transferred via increased prices, this increase being forced by the use of political power rather than the use of admittedly small commodity power.

3. It should finally be noted that the new affluence of the OPEC countries has already split the Third World into the Third and Fourth Worlds, the former being almost wholly the OPEC developing countries and the latter the rest. The two worlds have managed to collaborate effectively. Thus, the OPEC nations have extended credits and aid on a massive scale to the developing countries while also championing their cause politically.[25] As Fred Bergsten has noted, the OPEC countries managed to get developmental NIEOtype issues included on the agenda by withdrawing from the first ministerial session of the "energy dialogue" in Paris in April 1975. They also successfully pushed for further liberalization of IMF credits at Kingston in January 1976 by linking the usability of their currencies directly to liberalization of the credit tranches and by prompting the liberalization of the compensatory finance facility through their negotiating strategy in Paris.[26] Thus, far from compromising the collective action potential of the developing countries, the emergence of the more powerful OPEC countries has only served to increase the political potential for collective action by the developing countries.

4. In fact, the specific demands of the developing countries for institutional reform in the international economic order clearly reflect both the new objectives and the new political and economic realities that the oil price increases have imposed on the poor countries. Few of the current demands for specific reforms are entirely novel, but the choice of those that have been highlighted at UNCTAD IV and propagated at other forums is revealing. Two principal proposals have been the *Integrated Programme for Commodities*, christened the Corea Plan after the UNCTAD Secretary-General, and the demand for general *debt relief*.

The debt relief proposal would eliminate the accumulated debt burden, which many of the poorer developing countries find particularly onerous after the terms of trade losses from the increased oil prices. The elimination of this burden would provide for a transfer of resources (measurable as the present discounted value of the repayments canceled) in a form that is free of strings and high on the sovereignty scale. Generalized debt relief would also be politically and psychologically reinforcing to the developing countries because bilateral debt reliefs have usually been accompanied by extensive scrutiny and provisions by the creditor nations. Similarly the Integrated Programme for Commodities would also appear to do extremely well in light of the three objectives that were distinguished. Indexing of commodity prices (implied by the phrase "establishment and maintenance of commodity prices at levels which, in real terms, are . . . remunerative to producers"[27]) would yield transfer of resources by suitably raising them and maintaining them there. Indexing would mean that the developing countries had earned these increased incomes and therefore their national sovereignty over them would have been guaranteed. By participating in the arrangements designed to run these schemes, the developing countries would have earned the right to deliberate in international policy making in this sphere of international economic management.

## Developed Countries: Factors Affecting Response

While these two proposals do not constitute the totality of the demands made by the developing countries nor have they been pushed with continuing vigor by the developing countries since UNCTAD IV—in fact, the demand for a generalized debt relief or moratorium has, if anything, lost ground within the Group of 77 itself and the Group of 24 at the Fund/Bank Manila meetings in the fall of 1976 avoided or even repudiated the subject altogether—they do illustrate the problems that the present aspirations and postures of the developing countries pose for the developed countries as the latter contemplate the nature of their response to the demands for the NIEO.

In particular, there are two dimensions to these proposals that are guaranteed to make the response to them lukewarm, if not hostile, in the United States: the developed country whose consent is critical to orderly adoption of such proposals, as it clearly constitutes a *force majeure* on the international economic scene. These dimensions relate to sovereignty and efficiency, both of which can add up to an ideological confrontation that will have to be cooled, if not circumvented, to usher in reforms in the international economic system.

## Sovereignty
While the developing countries have come to stress the sovereignty over the use of their (and the transferred) resources, the rise of intellectual neoconservatism in the United States has tended to move the aid philosophy

precisely in the opposite direction. To the exaggerated complaints about the misuse and inefficiency of aid programs, one must now add the moral concern that the developing countries suppress civil liberties and oppress their populations and that therefore aid cannot be justified any longer on "humanitarian" or progressive principles; the latter should apply to transfers between individuals, not governments.

Indeed, in the year of Adam Smith's bicentennial, it must be sadly admitted that the invisible hand has yielded to the iron fist in a number of developing countries. However, the neoconservative inference that therefore resources should not be provided to developing countries' governments on progressive principles is a nonsequitur.

The nation state, as an entity that transposes itself between individuals in the developing and the developed countries, cannot be wished away; world order therefore must surely be defined in terms of morality as between nations. Moreover, the freewheeling description of developing countries as dictatorships and tyrannies is an exaggeration and self-righteously ignores the moral lapses of the developed countries themselves.[28] In defining the new international economic order, it would therefore seem perfectly legitimate to apply the progressive tax principles to nation states, none of which is characterized by moral perfection and few of which can ever share a common perception of morality in all its dimensions.

However, the neoconservative arguments do have a superficial appeal, especially in the United States where Vietnam and Watergate have crystallized a psychological need for assertion of moral values in policy making. Thus the argument that "we cannot allow ourselves to be pushed around and lectured into giving aid to an undeserving, corrupt Third World" has several adherents in fashionable intellectual circles in the United States.[29] This attitude of hostility to the developing countries has been reinforced by the subtle but propagandistic caricaturing of the positions of developing countries in regard to the New International Economic Order by conservatives and neoconservatives alike in the United States. *Ignoratio elenchi* is a favorite fallacy of intellectuals who are prominent, rather than eminent; it works very well in its intended purpose, but it must be exposed. In particular, it has been argued that the developing countries wish to establish "Western guilt" for their own underdevelopment; thus developing countries seek transfers of resources as reparations for past and present damage to their economic success. On the other hand, no such guilt can be established since internal institutional changes are critical for development and account for the growth of countries; hence the demands for the NIEO are ill founded and must be rejected. This argument would be laughable were it not so superficially plausible, effective, and pernicious.[30] The argument for progressive redistribution of income and wealth does not rest on whether the rich have hurt the poor in the past or are currently doing so. Nor are the majority of the intellectuals in the developing countries so naive as to assume that the colonial rule was necessarily harmful economically; what many do challenge is the opposite thesis of the imperialist historical school that the colonial rule was necessarily beneficial. Moreover, the critical nature of internal reform for rapid economic advance is logically compatible with the importance of external factors in shaping both the nature and the momentum of domestic development. Moreover, the many developing countries that launched five-year plans designed to direct princi-pally domestic efforts at development and many of their left wing intellec-tuals (who ridiculed Western economists' naivete about the ease with which foreign aid programs would take the developing countries to "self sustained" growth without radical internal reforms) would both (for different reasons) find the present lecturing by these Western intellectuals on the role of domestic reforms to be astonishing, and the inference that the external environment needs no fundamental changes to be a self-serving nonsequitur.

Turning from the conservative and neoconservative arguments, we must note next that the few remaining liberals who favor resource transfers to the developing countries on an increased scale remain untutored by the two important historical lessons of aid giving. First, economics rarely gives unique solutions *ex ante*—when you had six economists including Keynes, a witticism went, you had seven opinions—so that imposing donor-country economic solutions on recipient countries must often require an act of missionary zeal and faith. And second, local constraints on political action, much like in the donor countries, will often require the adoption of nth-best policies, contrary to the desires of the unconstrained policy advisers. They continue to see the application of strict, overall, economywide performance criteria as essential to a foreign aid policy. Experience points to the infeasibility of having such criteria scientifically and to the general inability to find such criteria. Assuming that consensus could be reached on what was the optimal policy to adopt in the first place, it would be impossible to apply such criteria meaningfully in light of local constraints. Thus, witness the following argument by Richard Cooper, an influential international economist of liberal persuasion and now at the center of U.S. policy making:[31]

If we are to justify resource transfers on ethical grounds, then, it must be on the basis of knowledge that via one mechanism or another the transferred resources will benefit those residents of the recipient countries that are clearly worse off than the worst-off "taxed" (including taxes levied implicitly through commodity prices) residents of the donor countries. That is, general transfers must be based on some kind of performance criterion satisfied by the recipient country, or else transfers should be made only in a form that benefits directly those who the ethical arguments suggest should be benefited. But this proposition has profound implications, . . . for it implies that no completely general transfer of resources from country to country can be supported on ethical grounds. This restriction would encompass the organic SDR link, general debt relief, actions to improve (not merely stabilize) the terms of trade of developing countries, and a brain drain tax that automatical-ly remit the revenues to developing countries. Ethically based transfers should discriminate among recipient countries on the basis of performance in improving, directly or indirectly, the well-being of their general population, and/or they should discriminate among uses of the transfers to maximize the flow of benefits to those who are the intended beneficiaries, which generally means concentration on general nutrition, health care, and education in the recipient countries.

While this quote speaks for itself, one might note particularly that emphasis on nutrition, education, and health care leaves open important issues of the type that have traditionally created friction between recipients and donors on the utilization of aid and hence problems regarding national sovereignty. What should be the balance between these three areas? How much in total, by way of current and investment expenditures, should be allocated to the three sectors together? What should be the time profile of benefits provided in these sectors, given the limited volume of resources? Within education, what should be the allocation to education at different levels and in different geographical areas? All of these questions raise both economic and political issues and involve issues of intertemporal allocations of costs and benefits; in none of these cases does an analyst have any ability to proceed without making several value judgments and not simply economic behavioral assumptions. No wonder that the overly sensitive developing countries have occasionally felt that the zeal of the donor-country economist and his value-unfree (in Nobel laureate Gunnar Myrdal's sense) recommendations, which are to be imposed by arm twisting at the aid-consortia meetings, disguise not just economic naivete but also ideological intentions.

Whether one takes the neoconservative or the liberal position, the fact remains that the developing countries' growing insistence on sovereignty in the use of resources conflicts increasingly with the preferred stress of the intellectuals in the United States on evaluation of economic performance, on monitoring, and on the morality of their internal and even international conduct. The dilemma therefore is that, while the developing countries by and large have centralized political regimes and are able to formulate and coordinate their demands for resource transfers, the developed countries generally have democratic, decentralized regimes where resource transfers must be justified to the electorates. These transfers cannot be defended when the neoconservative position assiduously tries to undercut the moral case for such actions and the liberal case is undermined by the insistence on performance criteria that are now sought to be rejected by the developing countries seeking sovereignty. If only the developing countries had the democratic political regimes and the developed countries the centralized ones, the dilemma would disappear!

This dilemma obtains chiefly in the United States, and far less so in the European countries and in Japan. In fact, Erik Lundberg has observed that in Sweden, uneasiness over the political conditions in several developing countries has not, as in the United States, been taken to rationalize a neoconservative position of opting out of aid; rather the Swedish political parties have unanimously agreed that the aid transfers are a moral obligation and the debate between the parties has only been about which developing countries ought to get what share of the total Swedish aid. This is of small consolation, however, since the United States still remains the force majeure in this area.

### Efficiency

Both the generalized debt relief scheme and the Corea Plan for commodities offend the tenets of efficiency as far as the transference of resources is concerned, thus raising serious doubts as to the advisability of these schemes as the principal props of current efforts at negotiating a new international economic order.

1. Generalized debt relief has already been seen as such and has already been dropped from the active shopping list of the developing countries. The main problems with the scheme were that many developing countries (among the medium and higher per capita income groups) had come to borrow increasingly from the private capital markets and presumably must continue doing so for the foreseeable future. A generalized default or moratorium could not but adversely affect the creditworthiness of the developing countries; it would also make it extremely difficult to regulate future borrowings since such largesse would affect expectations as to possible gains from similar largesse later—as Paul Streeten observes (in his comment on Kenen's paper in this volume), in economics, bygones are rarely bygones.

Recent statistical analyses suggest that the major beneficiaries would be the poor developing countries only when the debt relief was biased toward them.[32] Since these are the countries that have been hit badly by the rise in oil prices and their ability to borrow petrodollars has been inversely affected (whereas the not-so-poor developing countries are better off and have been able to borrow more successfully to tide over their current foreign exchange *and* real resource difficulties), the simpler and least cost solution[33] to debt relief seems to be to deal with the debt problems of the poor countries on a case by case basis; this was, in fact, suggested by the United States at UNCTAD IV (and has been carried out for many years for several countries).

2. The commodity schemes have not met quite the same fate. This is because that part of their objectives that deals with price stabilization allows for accommodation among both developed and developing countries, whereas the part that deals with indexing, to secure explicit resource transfer to the primary producers in developing countries, raises no enthusiasm and generates much hostility.

The stabilization part, despite the ambiguities of the UNCTAD documentation (which unhappily does not record the analytical work that underlies the declaration of the proposed objectives and hence unnecessarily creates the impression of being economically untutored)[34] appeals to many policy makers in the developed countries. The extremely wide and rapid shifts in commodity prices during 1972-1975 appear to have baffled and upset both producers and consumers, suggesting that international buffer stocks are not entirely unacceptable. Besides, except to a Chicago economist, the fact that Keynes had also proposed a third international institution, the *Commod Control*, to supplement the bank and the fund precisely for commodity price stabilization after the war appears to have endowed the idea with some legitimacy![35]

Unfortunately, the developing countries have set the scheme back by muddling it up, deliberately no doubt, with the indexing idea. As most sensible economists will agree, indexing is crude, simplistic, inefficient, inequitable (among developing countries, exactly like oil price increases), and virtually impracticable (administratively perhaps, and certainly politically to developed countries).

It is apparent that the developing countries have chosen to focus on particularly ill-designed measures to translate their objectives into reality. We have also identified the problem that the developing countries' demand for sovereignty in use of resources is likely to raise in the developed countries, a problem that must somehow be circumvented if the resources are to be transferred on anything like a significant scale via one feasible scheme or another. On the other hand, there are also favorable signs that the conflicts may turn into constructive cooperation. The generalized debt relief scheme has been dropped thus pleasing those who believe that the failure to perceive failure is the chief impediment to progress. And the North-South dialog continues at UNCTAD, in Paris, and elsewhere, with developing countries participating as they wished, and on more harmonious terms than the early "Algerian-style" rhetoric and the "Moynihan-style" ripostes (both of which were petty but popular in their own spheres of influence), thus pleasing those who believe instead that the failure to perceive progress is the chief impediment to progress.

## Desirable and Feasible Dimensions of the New International Economic Order

In a sense there is already a transitional international economic order. The developing countries are already active participants in negotiations where they were previously either ignored or regarded peripherally. Recall that the Committee of 20 grew out of the Committee of 10 in negotiations on international monetary reform which, until that recent shift, were regarded as of immediate concern only to the developed countries. "Populism" therefore is here to stay. However, the developing countries have learned the virtues of avoiding unwieldiness and, within the Group of 77, of biasing participation toward those who are most affected by the decisions being deliberated. It is in fact arguable that such participatory democracy in international economic negotiations has served to cool the inflammatory rhetoric and to reduce the confrontational content of North-South relations: Lord Acton has observed that absolute power corrupts absolutely, but lack of power corrupts equally.[36]

The interesting issues from an economist's viewpoint concern the desirable dimensions, in the form of concrete reform proposals, that the ongoing North-South negotiations should impart to the new international economic order. The record of the developing countries in finding and backing imaginative proposals, as is evident in the foregoing analysis, is not exactly exciting, nor have the developed countries put sufficient energy and initiative into developing a coherent and imaginative response to the specific demands of the developed countries.[37] In short, an overall categorization of desirable targets for the architects of the new international economic order is still to be evolved.

In my view, such a delineation must rest on a judicious combination of two major principles of reform: (1) the developing countries must receive an increased share of resources under the NIEO; and (2) bargains must be struck which are mutually profitable and which therefore appeal also to the developed countries' interests in areas of trade policy, regulation of multi-nationals, transfer of technology, migration, and food policy. By focusing on the transfer-of-resources proposals that can raise substantial resources *while circumventing the objections in the developed countries to the autonomous use of such resources by the developing countries*, (as can be done), and by appealing to the interests of all concerned parties through the mutual gain proposals (that can indeed be developed) in several different areas of international economic policy, the NIEO can be made a reality that accommodates the major political preferences and economic objectives of the developing and the developed countries. These two principles may now be developed briefly by spelling out the major reform proposals that might be entertained under them.

### Resource Transfers
There are two proposals that can qualify as raising fairly substantial revenues for developing countries that are least likely to be perfect substitutes for normal aid flows because of the nature of the rationales on which they are justified and also because of the incidence of the revenues within the developed countries; these proposals are also least likely to raise the moral questions that are attached to uncontrolled transfers of resources from the developed to the developing countries.

The first and major proposal relates to the grant of a share in the profits of seabed mining to the developing countries and extending the arrangement further to oceans generally thereby bringing into the net the possible revenues from regulation of "overfishing" as well. Politically, such proposals are less likely to give rise to objections and therefore be more acceptable to developed countries because rights over these resources are still the subject of international negotiation; thus it is difficult to conceive of the allocation of some of the profits or revenues from seabed mining and licensing of overfished fisheries to the developing countries as a simple grant *from* the developed *to* the developing countries. Furthermore, since the regulation of overfishing is an efficiency-improving business (whereas the taxing of seabed mining would be tantamount to taxing rents unless the returns to innovation and risk were also taxed away), the gut reaction of neoclassical economists is to welcome such methods of yielding resources to the developing countries; in such a case one has that rare example of transfers that may even improve welfare (for fishing) and, at worst, not hurt it (for seabed mining).

The Law of the Sea Conference, which has just terminated its fifth session without reaching any agreement, is the forum for transacting such a resource transfer to the developing countries. The developing countries have, to date, held out for far more substantial control of the ocean resources while resisting lucrative offers of revenues as proposed here.[38] In the end, since the major developed countries are poised to mine the seabeds and will probably effectively threaten to continue unilaterally to do so, it is certain that a bargain will be struck and a substantial source of revenues for the developing countries will emerge. Richard Cooper has put together estimates of the sums that could be raised from the oceans and, while they are necessarily rough and ready, they do indicate the orders of magnitude that one is dealing with: by 1985 a full capture of rents on fish at 1974 levels of output could yield $2.2 billion, a 50 percent profits tax on offshore oil coming from waters more than 200 meters deep would yield $2.2 billion, and a 50 percent profits tax on manganese nodules would yield $0.1 billion. This is a total of $4.5 billion in revenues from the oceans.[39] Thus, even if a half of this revenue was allocated to the developing countries in an international oceans bargain, the developing countries would receive over $2 billion worth of revenues as of 1985. These would be characterized by autonomy in use; because they would be untied grants, they would be worth at least twice as much in grant-equivalent terms as an equal aid flow under the average terms would generally imply.

Yet another example of a proposal that would raise resources/revenues for the developing countries is the suggestion of a brain drain tax. The proposal is to levy a supplementary tax on the incomes earned in the developed countries by skilled migrants from the developing countries. The tax would be levied by the developing countries (for legal-constitutional reasons), collected by the developed countries (for administrative reasons), and the revenues would be transferred *en bloc* to the developing countries via the United Nations (the UNDP) for developmental spending.[40] Given the entry restrictions of the developed countries, the tax may be seen as essentially a tax on the rents generated by such quantitative interferences with the international mobility of labor; it would again be consistent with human rights for the identical reason that the countries that create such interferences on human mobility are the developed countries with their immigration quotas. A moral rationale for taxing the immigrants exists, quite aside from the case for taxing rents, and has been spelled out by me elsewhere:[41]

The rationale behind the tax implementation would consist of two arguments; in order of their importance: (1) Firstly, one would assert the moral principle that, in a world of imperfect mobility, those few who manage to get from LDCs into DCs to practice their professions at substantially-improved incomes ought to be asked to contribute a fraction of their gains for the improved welfare of those left behind in the LDCs; this would effectively be extending the usual principle of progressive taxation across national borders. (2) Moreover, since there is a widely-held presumption, based on several sound arguments and embodied in numerous international resolutions, that the brain drain creates difficulties for the LDCs, it would also constitute a simple and rough-and-ready way for the emigrating professionals to compensate the LDCs for these losses. In fact, the moral obligation to share one's gains with those who are unable to share in these gains would be reinforced if these others were also hurt by one's emigration.

The revenues from imposing the brain drain tax have been estimated with some thoroughness and are to be found in Koichi Hamada's contribution to this volume. They add up to half a billion U.S. dollars for 1976. Allowing for inflation at 5 percent to arrive at a figure comparable to Cooper's 1985 oceans-revenue estimate of over $2 billion (at half of the $4.5 billion overall revenue figure, as that allotted to the developing countries), the brain drain tax revenues would exceed a significant sum of $1 billion.

The brain drain tax proposal is based on well-defined moral principles and ought also to appeal to the developed country populations which often feel that the skilled immigrants from the developing countries also ought to make their contribution to their part of the world;[42] the human-rights objections to it are easily dismissed once the proposal is examined in any depth. In addition, the developed country policy makers should realize that the contributions will come from the skilled immigrants as a supplementary tax; hence the revenue is hardly a matter of significant concern for the overall fiscal policy of the developed countries any more than voluntary contributions to foreign countries under the existing U.S. laws are for example. Given the great concern expressed over the brain drain in many developing countries, and Mr. Kissinger's declaration of willingness to do something in the area at UNCTAD IV[43] ("Finally, the United States proposes that appropriate incentives and measures be devised to curb the emigration of highly trained manpower from developing countries."), it seems logical to place a measure such as the brain drain tax on the agenda for the new international economic order.[44]

### Mutual-Gain Bargains
While these two examples of significant resource-flow proposals appeal primarily to the gains for the developing countries in a manner that does not hurt the economic interests of the developed countries or offend their political sensibilities on the issue of the sovereignty of the developing countries in the use of these resources, there are several possibilities of mutual-gain bargains in other areas of international economic reform. These

possibilities should prove to be more readily negotiable and are discussed in the concluding remarks of this paper.

In the field of trade policy it is clear that mutual gains can be had in two major spheres of policymaking. First, The Corea Plan needs to be explicitly modified to remove any traces of indexing: it should be clearly focused on stabilization. Further, the latter should explicitly distinguish among commodities that have supply problems and those that suffer from demand fluctuations if there is to be any clarity in the kinds of objectives such a stabilization is going to address itself meaningfully to; this also implies that the catch phrase "Integrated Programme" may be fetching but is certainly poor economics. Second, the developing countries have a clear interest in market access for the *sale* of manufactured exports that have run into nontariff barriers and often interrupted entry on the pretext of market disruption. At the same time, the developed countries (since the flourishing of commodity power by the OPEC countries) have become concerned with assuring market access for their *purchases* of raw materials. Here again therefore there is a prospect for a mutual-gain joint deal whereby the developed countries agree to new rules for regulating the use of trade barriers to handle market disruption problems in exchange for an agreement from developing countries (as well as the developed ones) to a new set of rules for regulating their use of export restrictions. Detailed proposals for such new rules, aimed at reforming the GATT on both counts, have been proposed recently and await policy makers' attention.[45] There is a clear case here for a joint bargain as against the piecemeal, issue-by-issue negotiating strategy that the developing countries seem to be following to little advantage, with little overall vision, in their current efforts at GATT and elsewhere.[46]

In the area of food policy the situation again seems to call for mutual-gain deals. The events of 1972-1974, leading up to the crisis-ridden Rome Conference of November 1974, are probably too unique to be repeated. The combined failures of the Soviet, Chinese, and Indian harvests, the depletion of the food surpluses in the United States, the unwillingness of the food exporting countries to part with grain because of inflation fed by the dramatic oil price increases, the foreign exchange difficulties that combined with high prices of fertilizers (related to the high oil price increases) to produce shortages of fertilizer availability in the developing countries where the Green Revolution had been important and could not be sustained without the fertilizers, the additional factor of the tragic drought in Suheil and Saharan Africa—all of these were to produce unprecedented panic in the grain-importing developing countries and confuse thinking on three separate issues that the world food problem raises for solution on a long-term basis.[47] These issues include (1) the need to have buffer stocks, exactly as with other commodities in the Corea Plan, to stabilize wild swings in prices such as happened most dramatically in 1972-1974; (2) the need to insulate the drought-affected developing countries from fearsome consequences by providing them with the assistance, not necessarily but most conveniently, in the form of food; and (3) the need to have an aid program for transferring food to the needy developing countries whose developmental programs cannot be sustained without such aid inflows and attendant food imports (which serve as the wage-goods, in the classical sense, supporting the needed investments for growth).

There is clearly nothing to be gained by the developed countries from having unplanned calls upon their food stocks for emergency relief in situations of drought; nor is it possible for them, on a long-term basis, to deny food aid to needy countries—after all, food has that special quality where conscience does hurt! For both these objectives, therefore, systematic planning ahead will be necessary. As for buffer stocks to handle stabilization, there are now sensible plans afloat that show both the relatively low levels of stocks that such an international operation will have to carry, as well as the feasibility of building up such stocks. The developing countries (such as India) are currently holding very substantial stocks of grain, largely because of successive good harvests but in part because of the immediacy of the lessons learned from the crisis of 1972-1974. Thus, the need for international stockholding seems to be correspondingly reduced; and this is also partly because the possibility of entering the grain market seems much more feasible now that the oil-price-hike-induced inflation has decelerated and harvests in the West, especially the United States, have been good.[48] Clearly, therefore, the time has come for an initiative on a world buffer stock for food program, or a substitute that also sounds like an excellent scheme: the grain insurance scheme of Gale Johnson who has proposed that the United States (and any other interested developed countries) join in a scheme under which they would assure, and build stocks to deliver on such an assurance, the developing countries that join such an insurance scheme that any shortfalls in their trend production larger than a prespecified percentage would be made up by the scheme.[49]

Finally, there is now clearly scope for a code on multinationals (MNCs) and their activities in the developing countries. Until now the developing countries that worried about the MNCs were considered to be somewhat bizarre, if not depraved and corrupted by socialist doctrines. Nothing works to cure one of illusions faster than to be proved naive by unpleasant revelations. The evidence of destabilization efforts directed against foreign regimes, the conviction that the oil companies collaborated in enforcing the Arab oil boycott, the exposé about the corporate bribing of foreign officials that has already ruined a prince and a prime minister,[50] are developments that have come just as the developing countries realized that they do need the technology and management expertise that MNCs can bring. Thus the awareness seems to have grown in influential circles in both the developing and the developed countries that MNCs are a good thing but need to have their international conduct regulated by explicit codes and legal sanctions, including the extension of trust-busting legislation to external operations in the social interest.

In trade, food, MNCs, brain drain, oceans, and a host of other smaller impact areas of international policy making, there seems to be scope for acceptable resource transfers and for mutual-gain bargains. Whether these initiatives will be taken constructively and the current momentum for creating a new international economic order will be directed toward these constructive channels will depend critically on the imaginativeness, empathy, and tolerance that the developed nations show toward the sentiments of the developing nations. For, it is only then that it will be possible to shift the developing countries toward the preferred reforms, as opposed to the ill-conceived proposals now current, and to collaborate with them in constructing a desirable and desired international economic order.

In conclusion, I might address my fellow economists who worry, quite naturally, about efficiency in international economic arrangements, and note explicitly that the only feasible way of guaranteeing such an outcome to the present debate on the new international economic order is to have the world system respond, in some significant manner, to the demands for distributive justice; the alternative politically is disorder from disenchanted developing countries, and that can hardly be expected to lead to conditions for international efficiency. Moreover, the program of reforms set out here fully meets these requirements, while ensuring a sense of fair play and justice to the developing countries and while moving substantially in the direction of assuring them of their sovereignty—newly discovered and therefore much valued. Thus, the brain drain tax, preferably matched by equal contribution from general taxation, would provide some sense of recompense to the developing countries for their present sense of loss of skills thereby making it easier for them to maintain freer flow of such skilled people. The evolution of new rules at the GATT to regulate the use of trade barriers and export controls for so-called market disruption and the holding back of supplies would guarantee freer trade. Commodity price stabilization, aimed at eliminating wild fluctuations in prices of the type that Keynes noted and that were acutely observed in 1972-1975, would supplement, not supplant, the market and would make it more efficient. And, by assuring that resources were transferred through the taxation of disexternalities (such as overfishing) and of rents (such as those from seabed mining and others from skilled migration from developing countries into highly-restricted developed countries), the developing countries would be assured of substantial, automatic, and continuing claims on revenues.

Such a program for the new international economic order can be judiciously embedded in an overall program of aid flows, restored to the higher (aid to GNP) levels enjoyed during the late 1950s and early 1960s, refocused on the poorer Fourth World, and disbursed with diplomatic surveillance, performance criteria, and scrutiny that would make these more substantial flows acceptable to the public opinion and parliamentary institutions of the donor countries.

The full agenda for the new international economic order, as developed and proposed here, is surely fully deserving of economists' support—support that is critical for its adoption by the developed countries.[51]

## Notes

1 Among the major, recent conferences, UNCTAD IV at Nairobi in May 1976 and the ongoing Paris Conference on International Economic Cooperation (CIEC) are the principal ones. The 1976 ILO World Employment Conference and the earlier, 1975 UNIDO Lima Conference should also be mentioned.

2 The characterization as "populist" comes from C. P. Kindleberger, "World Populism," *Atlantic Economic Journal*, 3:2 (November 1975).

3 The alternative ideologies noted in this subsection were distinguished earlier in J. Bhagwati, "The Developing Countries," in *The Great Ideas Today*, (Chicago: Encyclopedia Britannica, Inc., 1976).

4 For a lucid statement of this viewpoint, see Raymond Vernon, *Sovereignty at Bay* (New York: Basic Books), 1972.

5 This is, of course, the central conclusion of the conventional theory of international trade and welfare.

6 Harry Johnson, among others, has noted several positive effects of the brain drain on developing countries. For a review, see J. Bhagwati and M. Partington (eds.), *Taxing the Brain Drain: A Proposal* (Amsterdam: North-Holland Publishing Company, 1976), Chapter 1, Appendix.

7 This view appears to be factually erroneous, though one could construct theoretical models to explain it. Ian Little has argued:

UNCTAD was founded on the mistaken view, which it has enshrined by constant repetition into the myth, that there is a trend in the terms of trade against developing countries as a result of an adverse trend in the terms of trade between manufactures and commodities. The mistake was originated by a League of Nations publication in 1945, and repeated by an early U.N. publication. Some more recent work suggests an *improvement* in the manufactures/commodities terms of trade for nearly a century before 1952-5, when there was a highly favourable and unsustainable peak. in developing countries terms of trade associated with the Korean War boom. Thereafter for at least seven years they worsened, but then improved for a decade, even excluding oil. Any reasonably objective observer would have been saying for many years now that the evidence cannot possibly be held to give grounds for maintaining that there is a trend in the terms of trade against developing countries. Theories have been invented to explain this non-existent trend: they are treated with respect even though they explain what does not exist.

Cf. Little, "Economic Relations with the Third World—Old Myths and New Prospects," *Scottish Journal of Political Economy*, November 1975, p. 227.

8 For several models of such adverse impact, see J. Bhagwati (ed.), *The Brain Drain and Taxation: Theory and Empirical Analysis* (Amsterdam: North-Holland Publishing Company, 1976); note the contributions of Koichi Hamada and J. Bhagwati.

9 The writings of Hans Singer developed this theme; see, for example, *The Strategy of International Development* (New York: International Arts and Sciences Press, 1975).

10 V. S. Naipaul, *Guerrillas* (New York: Alfred S. Knopf, 1975), p. 5.

11 An excellent statement of this type of viewpoint can be found in Thomas Weisskopf, "Capitalism, Underdevelopment and the Future of the Poor Countries," in Bhagwati (ed.), *Economics and World Order* (New York: Macmillan, 1972).

12 See Kwame Nkrumah, *Neocolonialism: The Last Stage of Imperialism* (New York: International Publishers, 1965).

13 See the paper by Harald Malmgren in this volume.

14 For a review of the principal findings on this and other related issues in research by Little-Scitovsky-Scott for the OECD, Bhagwati-Krueger for the NBER, and Balassa for the IBRD, see J. Bhagwati, "Protection, Industrialization, Export Performance and Economic Development," paper prepared for the UNCTAD IV conference in Nairobi, May 1976.

15 One must concede, however, that for the 1950s and most of the 1960s, many of these governments were of the view that their protectionist policies were desirable.

16 See J. Bhagwati, *Amount and Sharing of Aid*, (Washington, D.C.: Overseas Development Council, 1970), Chapters I and II. Several different estimates are reviewed here and it is reported that the net worth of foreign aid to recipients was reduced to less than half of the alleged amounts and, in some cases, to little more than a third.

17 That aid may be misused unless the whole economic program of the recipient country is examined is a lesson that was learned by economists during the Marshall Plan. Needless to say, politically it seems outrageous to sovereign nations receiving such assistance to have their entire economic process be subjected to scrutiny and control by donors who contribute, in general, no more than 1-2 percent of the overall resources in the recipient countries.

18 Such resentments are inevitable as economic policies inevitably reflect ideological preferences and value judgments. An articulate expression of these resentments from a frustrated aid-recipient negotiator can be found in I. G. Patel's contribution in Barbara Ward (ed.), *The Widening Gap: Development in the 1970's* (New York: Columbia University Press, 1971).

19 The consequent need to redefine the goals of international efforts at development in the developing countries has been stressed by several writers. See, for example, the contributions by Pitambar Pant, Göran Ohlin, and J. Bhagwati (ed.) in *Economics and World Order*.

20 One can only agree with Little, however, in the judgment that the so-called statistics on income distribution for developing countries, which suggest *absolute* impoverishment of millions during the 1960s, are totally unreliable. See his critique of the Adelman-Morris claims to this effect in the *Journal of Development Economics*, Vol. 2 (1975).

21 An alternative view is that the demands for the NIEO have resulted from the "revolution of rising expectations" following the high rates of growth in the developing countries. However, this explanation is hardly plausible except for perhaps Brazil.

22 See M. A. Adelman, *The World Petroleum Market* (Baltimore: Johns Hopkins University Press, 1972).

23 For example, Japan as well as Western Europe (as noted by Bergsten in his paper for this volume) have raw material "dependence"—ratios that are of concern to their govern-

ments: the short run disruption to their economies from interruptions of supplies could be quite substantial.

24 For example, the 1972 Algiers demands and the resolutions of the special 1975 United Nations Conference on Raw Materials seem far more optimistic in regard to unilateral initiatives on commodities than does the February 1976 Manila Charter of the Group of 77 which sought to define the concerted position of these countries at the May 1976 Nairobi Conference of the UNCTAD. The Manila Charter demands the indexing of prices of exports of primary commodities by the developing countries to the prices of manufactured goods imported from the developed countries; this indexing is clearly to be achieved by cooperation toward that goal by both producing and consuming nations.

25 See Table 2.2 in the Edelman-Chenery paper in this volume, for the statistics on this.

26 See Bergsten, *Panel Discussion*, this volume.

27 Secretary-General of UNCTAD, 1975, *An Integrated Programme for Commodities: Specific Proposals for Decision and Action by Governments*, Report TD/B/C.1/193, 28 October 1975. For further details on the Integrated Programme, see the paper by Harry Johnson in this volume.

28 Indeed, it is arguable that some of these lapses (for example, the Vietnam war) were a greater affront to one's moral sense than the lapses that the developing countries are accused of.

29 The *locus classicus* of the neoconservative philosophical attitudes and their application to social policy issues is the magazine *The Public Interest*.

30 Variations on this basic theme have appeared in articles by Daniel Moynihan, Peter Bauer, and others, in American magazines such as *Commentary*. Even the *New York Times Sunday Magazine* (November 7, 1976) carried an article on the theme of Western guilt and its untenability written by two staffers of the Hudson Institute.

31 R. N. Cooper, *Panel Discussion*, this volume.

32 See Kenen's paper in this volume for the magnitudes involved.

33 This would be equally true for the not-so-poor developing countries (such as Brazil and Mexico) which fear being tainted by a generalized debt write-off or moratorium.

34 It is worth noting that Mr. Corea, the Secretary-General of UNCTAD, obtained a First in the Economics Tripos at Cambridge University, and that the UNCTAD draws in many consultants with considerable professional expertise and of much distinction. Unfortunately, it has become customary in certain professional circles (in this connection, read Harry Johnson in this volume and elsewhere) to equate the UNCTAD Secretariat with economic illiteracy—a charge that cannot be sustained, certainly relative to other international secretariats (such as the GATT) that also operate under governmental control rather than as autonomous university departments of economics!

35 Keynes' memorandum on the subject was discovered by Dr. Lal Jayawardene, then on the Committee of Twenty; I hastened to publish it in the *Journal of International Economics*, Vol. 4, no. 3 (August 1974).

36 One might add with regard to the subset of dictators, that lack of absolute power corrupts absolutely as well!

37 Thus, the *Tinbergen Report on Reshaping the International Order* (New York: E. P. Dutton & Co., 1976), concludes (p. 54):

It became apparent at the Seventh Special Session that, with few exceptions, the *Western European* Market economy countries and *Japan* although they have potentially more to lose than the U.S. from a failure to forge new international structures, were not only reluctant to take the initiative in redirecting the process of change, but were quite prepared to take refuge behind the United States' position when discussions became serious. That they are unwilling or unable to take serious initiatives is witnessed by the results of the first meeting of EEC Ministers for Development Cooperation held after the Seventh Special Session (3). It ended in complete failure; no agreement was reached on any important agenda point. *Despite a considerable effort, the 'nine' also failed to formulate a common position for UNCTAD IV* (italics added for emphasis).

38 It appears that the desire for full control stems again from residual notions of commodity power: it is feared that the access to seabeds would enable developed countries to reduce the potency of the developing countries' major weapon. Note that manganese is among the commodities that are included in the developing countries' list of commodities for exercise of commodity power!

39 R. N. Cooper, "Oceans as a Source of Revenue," in this volume.

40 For details, see Bhagwati and Partington (ed.), *Taxing the Brain Drain: A Proposal*, Amsterdam: North-Holland Publishing Co., 1976.

41 Bhagwati and Partington, p. 22. Also quoted in Koichi Hamada, "Taxing the Brain Drain: A Global Point of View," in this volume.

42 See the sociological evidence on this point in Partington's contribution in Bhagwati and Partington, *Taxing the Brain Drain.*

43 Note the speech delivered on May 5, 1976, in Nairobi; text available from Department of State, Press Release No. 224, p. 14.

44 While these two proposals do generate substantial magnitudes of resource flows to the developing countries in a manner that is consistent with the new political objectives and constraints spelled out earlier, they are to be regarded as supplements to the traditional aid programs. Indeed, aid flows need to be raised to more substantial levels; this is particularly the case for the poorest countries, many of which have been most seriously affected by the oil price increases. This group of "most affected countries" in the Fourth World is badly in need of substantially augmented resource transfers to maintain preOPEC growth performance, as is evident in the Edelman-Chenery paper in this volume. Furthermore, Edelman and Chenery underline the need to focus on these countries in deciding on the allocation of any given aid funds: the distribution of aid could be improved by giving more weight to poverty and need criteria than to bilateral, strategic and political criteria. Whether the overall aid flows and their distribution could be substantially moved in the directions that are so required is a matter about which one might be skeptical.

45 Proposals related to market disruption (addressed to developed countries but now awaiting the initiative of the developing countries) have been delineated in my paper in this volume. On the other hand, Bergsten's paper develops, at equal length, the proposals for GATT reform to regulate the use of export controls.

46 On the latter point, see Gardner Patterson's observations in this volume.

47 These issues have been distinguished, and their consequences for policy neatly analyzed, in the Sarris-Taylor paper in this volume.

48 Calculations of buffer stock requirements on alternative assumptions have been made recently by Lance Taylor, Alexander Sarris and Philip Abbot at MIT; others no doubt are in progress.

49 See Gale Johnson's paper in this volume.

50 Here, as elsewhere, one should distinguish between demand-determined and supply-determined bribes, however.

51 Thanks are due to Fred Bergsten, Joe Nye, and Hans Singer for helpful comments.

# Part II

## Terms of Trade and
## Gains from Trade

Meier
Letiche, Chambers, and Schmitz

# Terms of Trade

Gerald M. Meier

1. Despite all the ambiguities obscuring their use, the terms of trade still receive considerable attention in discussions of economic development. This is so not only because the terms of trade have sizeable quantitative significance for most poor countries, but also because they are a convenient indication of the net result of many diverse forces, and may have important welfare implications. We shall therefore analyze in this chapter the determinants of secular changes in the terms of trade and attempt to assess the influence of these changes on the development of a poor country.[1]

2. Several different concepts of the terms of trade may be distinguished: the gross barter, net barter or commodity, income,

---

[1] The short-run problem of cyclical fluctuations in the terms of trade and export earnings is analyzed in Chapter 9.

Reprinted from ''The International Development of Economics: Theory and Policy,'' pp. 41–65, with permission of Harper and Row, Copyright 1968.

single-factoral, double-factoral, real cost, and utility terms of trade.[2] These several concepts fall into three groups: (1) those that relate to the ratio of exchange between commodities—the gross barter, net barter, and income terms of trade; (2) those that relate to the interchange between productive resources—the single-factoral and double-factoral terms of trade; and (3) those that interpret the gains from trade in terms of utility analysis— the real cost and utility terms of trade.

In considering the barter terms of trade, Taussig introduced the distinction between "net" and "gross" barter terms.[3] The commodity or net barter terms of trade ($N$) are expressed as $N = P_x/P_m$, where $P_x$ and $P_m$ are price index numbers of exports and imports, respectively. A rise in $N$ indicates that a larger volume of imports could be received, on the basis of price relations only, in exchange for a given volume of exports. According to Taussig, however, the net barter terms are relevant only when nothing enters into the trade between countries except sales and purchases of merchandise.

If the balance of payments includes unilateral payments, so that there is an excess in money value of either exports or imports, then the relevant concept is the gross barter terms ($G$). This measures the rate of exchange between the whole of a country's physical imports as compared with the whole of its exports, and is expressed as $G = Q_m/Q_x$, where $Q_m$ and $Q_x$ are vol-

[2] Jacob Viner, *Studies in the Theory of International Trade*, Harper & Row, 1937, pp. 558–564; W. W. Rostow, "The Terms of Trade in Theory and Practice," *Economic History Review*, Second Series, Vol. III, No. 1, 1950, pp. 1–20; R. G. D. Allen and J. E. Ely, eds., *International Trade Statistics*, John Wiley, 1953, pp. 207–209; Gottfried Haberler, *A Survey of International Trade Theory*, International Finance Section, Princeton University, rev. ed., 1961, pp. 24–29.

[3] F. W. Taussig, *International Trade*, Macmillan Co., 1927, pp. 113, 117, 248–249.

ume index numbers for imports and exports, respectively. A rise in $G$ represents a "favorable" change in the sense that more imports are received for a given volume of exports than in the base year. Since $G = N$ only if the value of imports and value of exports are equal,[4] $G$ and $N$ diverge when there are unilateral transactions. But one must distinguish among the different types of unilateral transactions that cause changes in $G$. It is then more meaningful to consider the significance of various unilateral transactions directly, instead of incorporating them in the terms of trade index.[5]

Since it is especially important for a poor country to take changes in its volume of exports into account, we may want to correct the movements in $N$ for changes in export volume. The income terms of trade $(I)$ do this, and are expressed as $I = N \cdot Q_x$, where $Q_x$ is the export volume index.[6] A rise in $I$ indicates that the country can obtain a larger volume of imports from the sale of its exports; its "capacity to import"—based on exports—has increased. The export-based capacity to import should be distinguished, of course, from the total capacity to import, which depends not only on exports but also capital inflow and other invisible exchange receipts. Nor should a change in

[4] If $V_m$ *and* $V_x$ are index numbers of values of imports and exports, respectively, $\dfrac{G}{N} = \dfrac{Q_m}{Q_x} \cdot \dfrac{P_m}{P_x} = \dfrac{V_m}{V_x}$ .

[5] Cf. Gottfried Haberler, *The Theory of International Trade*, William Hodge & Co., 1936, pp. 164–165; Viner, *op. cit.*, p. 563; Erick Schiff, "Direct Investments, Terms of Trade, and Balance of Payments," *Quarterly Journal of Economics*, February, 1942, pp. 310–316.

[6] G. S. Dorrance, "The Income Terms of Trade," *Review of Economic Studies*, 1948–49, pp. 50–56. The income terms of trade have also been referred to as "the export gain from trade"; A. H. Imlah, "The Terms of Trade of the United Kingdom, 1798–1913," *Journal of Economic History*, November, 1950, p. 176.

the income terms of trade be interpreted as a measure of the gain from trade or an indicator of welfare; it should be used simply as a measure of the quantity of imports bought by exports.

It is significant that, according to the direction and magnitude of the changes in $P_x$ and $Q_x$, the changes in $I$ and $N$ may be in opposite directions. If, for example, with unchanged import prices, export prices have fallen, but export quantities ($Q_x$) have increased by a greater percentage than the decrease in $P_x$, the income terms of trade will have improved despite a deterioration in the commodity terms of trade.

Changes in productivity are obviously also of prime significance in considering development, and one may therefore want to refer to the factoral terms of trade. The single-factoral terms ($S$) correct the commodity terms for changes in productivity in producing exports, and may be expressed as $S = N \cdot Z_x$, where $Z_x$ is an export productivity index. A rise in $S$ is a favorable movement in the sense that a greater quantity of imports can be obtained per unit of factor-input used in the production of exportables.

If $N$ is corrected for changes in productivity in producing imports as well as exports, the result is the double-factoral terms of trade ($D$), expressed as $D = N \cdot Z_x/Z_m$, where $Z_m$ is an import productivity index. A rise in $D$ shows that one unit of home factors embodied in exports now exchanges for more units of the foreign factors embodied in imports. $D$ will diverge from $S$ when there is a change in the factor cost of producing imports, but this has no welfare significance for the importing country, even though it indicates a change in productivity in the other country from which commodities are imported. What matters to the importing country is whether it receives more goods per unit of its "exported factor-input" (an improvement in $S$)—not whether these imports contain more or less foreign inputs than before.

It may also be noted that $N$ will equal $D$ when constant returns to scale prevail, and there are no historical changes in costs and no transport costs. But if costs are variable with respect to output or time, or there are transport costs, $N$ and $D$ will diverge. Although this divergence is analytically significant, it is difficult to measure as long as a productivity index remains an elusive concept. In the offer curve analysis of the preceding chapter, the terms of trade as determined at the positions of equilibrium in Fig. 3 are the commodity terms. If, however, we had followed Marshall, and considered on each axis "representative bundles" or "bales" of commodities that contained a constant quantity of "productive resources," the terms of trade would have been the double-factoral terms.

Proceeding more directly to the level of welfare analysis, we may define in utility terms the total amount of gain from trade as the excess of the total utility accruing from imports over the total sacrifice of utility involved in the surrender of exports.[7] To consider the amount of disutility involved in the production of exports, we may correct the single-factoral terms of trade index by multiplying $S$ by the reciprocal of an index of the amount of disutility per unit of productive resources used in producing exports.[8] The resultant index would be a real cost terms of trade index $(R)$. If $R$ rises as a result of a change in the methods of producing exports, or a change in the factor proportions used in exports, this would indicate that the amount of imports obtained per unit of real cost was greater.

On the side of demand, we may want to allow for changes in the relative desirability of the imports and the domestic commodities whose home consumption is foregone because of the use of resources in export production. It is then necessary to in-

[7] Viner, *op. cit.*, p. 557.
[8] *Ibid.*, p. 559.

corporate into $R$ an index of the relative average utility per unit of imports and of foregone domestic commodities. The resultant index is the utility terms of trade ($U$), equal to $R$ multiplied by an index of the relative utility of imports and foregone commodities.[9]

The difficulty with the use of $R$ and $U$ is, of course, that of calculating the disutility involved in export production, or the relative average utility of various commodities. The welfare significance of changes in the terms of trade must therefore be considered only indirectly, along the lines suggested below in section 5, and not directly through any measurement of $R$ or $U$.

Having minimized the significance of changes in $G$, $D$, $R$, and $U$, we are thus left with $N$, $S$, and $I$ as the most relevant concepts of the terms of trade for poor countries. Movements in $N$, $S$, and $I$ may diverge, however, and these divergences are not merely technical but are due to fundamentally different circumstances. Accordingly, they have different consequences for the country's development. To assess the significance for a poor country of an alteration in its commodity terms of trade—the most frequently cited change—we must therefore analyze the determinants of this change and also the attendant movements in the income and single-factoral terms of trade.

3. Over the short period, the terms of trade may vary as a consequence of changes in commercial policy, exchange rate variations, unilateral transfer payments, or cyclical fluctuations. Over the long period, however, the determinants of changes in the terms of trade are associated with more fundamental structural variations in production and consumption that may be examined in the light of the offer curve analysis of the preceding chapter.

[9] *Ibid.*, pp. 560–561.

As already noted, the shifts in the offer curves will cause movements in the terms of trade, and the various possible shifts in the offer curves can be attributed, in turn, to different types of development.[10]

Assuming that development occurs only in country $E$, so that $G$'s offer curve remains fixed while $E$'s offer curve shifts, we can summarize the various changes in $E$'s commodity terms of trade, according to the different total biases in development, as follows:

| Type of Total Bias in Development | Direction of Change in Commodity Terms of Trade |
|:---:|:---:|
| $N$ | $(-)$ |
| $X$ | $(-)$ |
| $M$ | $(-)$ |
| $UM$ | $(+)$ |
| $UX$ | $(-)$ |

| | |
|---|---|
| $N$ = neutral | $UX$ = ultra-export-bias |
| $X$ = export-bias | $(-)$ = deterioration |
| $M$ = import-bias | $(+)$ = improvement |
| $UM$ = ultra-import-bias | |

When development occurs only in $E$, and $G$'s offer curve is not infinitely elastic, the terms of trade for $E$ deteriorate for each

[10] For other analyses of the effects on international trade of shifts in reciprocal demand schedules of different elasticities, see Murray C. Kemp, "The Relation between Changes in International Demand and the Terms of Trade," *Econometrica*, January, 1956, pp. 41–46; W. R. Allen, "The Effects on Trade of Shifting Reciprocal Demand Schedules," *American Economic Review*, March, 1952, pp. 135–140; J. Bhagwati and H. G. Johnson, "Notes on Some Controversies in the Theory of International Trade," *Economic Journal*, March, 1960, pp. 84–93; Frederic L. Pryor, "Economic Growth and the Terms of Trade," *Oxford Economic Papers*, March, 1966, pp. 45–57.

type of total bias except an ultra-import-bias. As indicated previously (Fig. 3), the deterioration for a given increase in total output is least, however, when there is an import-bias, and the demand for imports increases less than proportionately to the expansion in total output. The deterioration is greatest when there is an ultra-export-bias, and the absolute demand for imports increases more than total output. In general, the rate of deterioration in $E$'s commodity terms of trade will be greater under the following conditions: the larger is the degree of export-bias on balance in $E$; the higher is the rate of increase in $E$'s total output; the lower is $E$'s elasticity of demand for $G$'s goods; and the lower is $G$'s elasticity of demand for imports from $E$.

4. When development occurs in both $E$ and $G$, the movement of the terms of trade depends on the rate of increase in each country's demand for imports from the other country—in other words, on the relative shifts of the offer curves as determined by the type of total bias and the rate of development in each country.

If the total bias in the development of each country is neutral, each country's offer curve shifts outwards, with the extent of the shift depending on the rate of development. The terms of trade will therefore deteriorate for the country that has the higher rate of development.

If in each country development is ultra-export-biased, or export-biased, or import-biased, each country's offer curve again shifts outwards. The terms of trade would then remain constant only if the types and degrees of bias and rates of development had the same total effect on the growth of demand for imports in each country. In the general case, the terms of trade deteriorate for that country which has the greater rate of growth of demand for imports as determined by its degree of bias, as well as rate of development.

If the over-all effects are export-biased in $E$, but import-biased in $G$, then, assuming the rate of development is the same in each country, the terms of trade will deteriorate for $E$. If, however, the rate of development in $E$ is sufficiently lower than in $G$, the terms of trade will improve for $E$, even though its development is export-biased.

The relative rates of development in the two countries may, in many cases, be significant in offsetting the different degrees or types of bias. If, however, the development is ultra-import-biased in only one of the countries, the terms of trade will improve for that country regardless of the type of bias in the other country and the relative rates of development.

From these diverse cases it is apparent that there is no invariant relationship between a country's development and movements in its commodity terms of trade. Depending on the type and degree of bias and the rate of development in each country, the terms of trade may either improve or deteriorate.

5. The connection between changes in the terms of trade and economic welfare is an especially difficult problem: In what sense may a movement in a country's terms of trade be accepted as an index of the trend in economic welfare? Considerable care must be exercised to avoid the fallacy of equating a change in any of the various terms of trade with a variation in the amount or even direction of change in the gains from trade. Such an equation cannot be adduced until we determine the underlying forces associated with the change in the terms of trade, and until we connect the terms of trade, relating to a unit of trade, with the volume of trade.

The welfare implications of a change in the commodity

terms of trade are most directly seen in the effect on real national income. When a country's commodity terms of trade improve, its real income rises faster than output, since the purchasing power of a unit of its exports rises. This increase in real income will supplement the benefit that the country derives from its own development.[11] If, however, a country experiences a deterioration in its terms of trade as it develops, part of the benefit from an expansion in its own output is thereby cancelled.

Insofar as a slower rate of development might allow a country's commodity terms of trade to improve, whereas a higher rate would cause a deterioration, it is possible that the gain from the improvement in the terms of trade might be more than sufficient to compensate for the output foregone by the slower expansion in home output. In the case of an ultra-import-bias, however, a lower rate of development would not tend to augment the improvement in the commodity terms of trade. On the contrary, unlike the other cases, a higher rate of development in this situation will not only increase domestic output further, but will also cause a greater improvement in the commodity terms of trade.

As an extreme case, it is possible that the type and rate of development may cause so severe a deterioration in the terms of trade that the gain from the growth in output is more than offset by the loss from adverse terms of trade, so that the country ends up with a lower real income after growth. This theoretical possibility has been demonstrated by Professor Bhagwati, who de-

---

[11] An improvement in the commodity terms of trade might facilitate an expansion in domestic output by permitting the release of resources from export production to domestic production. If the improvement is due to a rise in export prices, this may contribute to an increase in public saving through export taxes, income taxes, or a rise in the profits of governmental marketing boards.

scribes it as a case of "immiserizing growth."[12] For example, an increase in factor supply or technical progress would raise real income by the amount of the change in output at constant prices, but if the factor accumulation or "factor-saving" is so export-biased that the terms of trade worsen, the negative income effect of the actual deterioration in the terms of trade may then be greater than the positive effect of the expansion in output.

Although analytically interesting, the practical bearing of this possibility is very limited. The conditions necessary for immiserization to result are highly restrictive. In the case of incomplete specialization, the possibility can arise only if the increased quantity of the factor is allocated to export industries, and either the foreign demand for the growing country's exports is inelastic or the country's expansion actually reduces the domestic production of importables.[13] But if external demand is so unfavorable, then additional resources will not flow into the export sector when the situation is such that the very growth of factor supplies may actually have to be induced by the existence of profitable

[12] Jagdish Bhagwati, "Immiserizing Growth: A Geometrical Note," *Review of Economic Studies*, June, 1958, pp. 201–205; "International Trade and Economic Expansion," *American Economic Review*, December, 1958, pp. 941–953; "Growth, Terms of Trade and Comparative Advantage," *Economia Internazionale*, August, 1959, pp. 395–398.
  Some classical and neoclassical economists also recognized this possibility when they considered the impact of technological change upon the terms of trade. See J. S. Mill, *Principles of Political Economy*, Longmans, Green, 1848, Book III, chap. XVIII, sec. 5; C. F. Bastable, *The Theory of International Trade*, Macmillan Co., 1903, appendix C, pp. 185–187; F. Y. Edgeworth, "The Theory of International Values, I," *Economic Journal*, March, 1894, pp. 40–42.
[13] Bhagwati, "International Trade and Economic Expansion," *op. cit.*, pp. 949–952. In the case of complete specialization, it is necessary that both the foreign demand for exports and the domestic demand for imports be inelastic. This proposition is demonstrated by Bhagwati and Johnson, *op. cit.*, pp. 80–81.

openings for the employment of these additional factors. More-over, even if there is an autonomous increase in factors, there is still no basis for "immiserizing growth," inasmuch as increments in factor supplies are as a rule mobile and the economy has some capacity for transforming its structure of output. Factor incre-ments, therefore, need not flow into the export sector in accord-ance with a predetermined pattern of production.[14] To be valid, the "immiserizing growth" argument depends on highly restric-tive conditions with respect to elasticities of demand and supply —conditions which are unlikely to apply when an economy has some flexibility in its structure of output and some capacity for adapting to changed circumstances. It should also be realized that, even if the necessary conditions do exist, the country can still institute offsetting policies and impose taxes on its trade suf-ficient to gain some of the benefits of the expanded production.[15]

If we examine the welfare implications of a change in the terms of trade more broadly, we can readily identify circum-stances under which a country need not be worse off, even though its commodity terms deteriorate. When the deterioration results from a shift only in the foreign offer curve, with the country's own offer curve unchanged, the resultant deterioration in the country's terms of trade is clearly unfavorable. If, how-ever, the domestic offer curve also shifts, then it is necessary to consider the causes of this shift and also the possible changes in the factoral and income terms of trade.

[14] Ragnar Nurkse, *Patterns of Trade and Development*, Wicksell Lec-tures, Almqvist & Wiksell, 1959, pp. 56, 58–59 (reprinted in *Equilibrium and Growth in the World Economy. Economic Essays by Ragnar Nurkse*, Harvard Univ. Press, 1961, pp. 332–334).

[15] R. A. Mundell, "The Pure Theory of International Trade," *Ameri-can Economic Review*, March, 1960, p. 85; Bo Södersten, *A Study of Eco-nomic Growth and International Trade*, Almqvist & Wiksell, 1964, pp. 53–54.

For instance, development may occur in both countries $E$ and $G$, but the rate of growth of demand for imports may be greater in $E$ than in $G$, so that $E$'s commodity terms of trade deteriorate. Nonetheless, $E$ may still be better off than before if the deterioration in its commodity terms is due to export-biased increases in productivity. In this case the single-factoral terms of trade improve, and the deterioration in the commodity terms is only a reflection of the increased productivity in $E$'s export industries. As long as productivity in $E$'s export sector is rising faster than the prices of its exports are falling, its real income rises despite the deterioration in its commodity terms of trade. If the prices of exports in terms of imports fall by a smaller percentage than the percentage increase in productivity, the country clearly benefits from its ability to obtain a greater quantity of imports per unit of factors embodied in its exports.

Classical and neoclassical economists recognized this possibility and attempted to go behind the quantities of exports and imports to consider what, as Pigou remarked, "underlie the exports, namely a given quantity of labor and service of capital." It may then be that although the commodity terms of trade deteriorate when the production costs of exports fall, the country may receive more imports than previously for what "underlies its exports." A divergence between the commodity terms and the factoral terms was meant to be avoided by J. S. Mill's conception of "cost," Bastable's "unit of productive power," and Marshall's "representative bales of commodities," each of which contains a constant quantity of "productive resources." But as already noted, if we allow for more than two commodities, transportation costs, or variable costs of production, the commodity terms and the factoral terms of trade may diverge.

It is also relevant that even if productivity is not rising in the export sector, and the commodity terms of trade are deteriorating, it is still possible for the real income of the factors to rise. This may occur under conditions of a "dual economy" in which factors are initially employed in the backward domestic sector with lower productivity than exists in the advanced export sector. If export production should then expand and attract these factors into the export sector, the factors will gain to the extent that their marginal productivity in the export sector remains above their marginal productivity in the sector from which they withdraw. At the same time, the real prices of export products may be falling, and the commodity terms of trade may be worsening.[16]

A high degree of export-bias on the side of consumption may also cause a deterioration in $E$'s commodity terms of trade. But if this export-bias is due to a change in tastes or a redistribution of income, it is difficult to reach any welfare conclusion. For the intervening change in the preference system makes it impossible to conclude that the later result is inferior to the previous situation merely because the commodity terms have deteriorated. If the terms worsen because demand increases for imports, it may not be true that from the criterion of "utility" a loss is incurred. What must be considered is not the utility of the import alone, but also its utility relative to that of the domestic commodities whose domestic consumption is precluded by allocation of resources to production for export. Were it measurable, the utility terms of trade index would be appropriate for this type of change.

We should also realize that it is possible for the country's income terms of trade to improve despite, or sometimes even because of, a deterioration in the country's commodity terms of

[16] Theodore Morgan, "The Long-Run Terms of Trade Between Agriculture and Manufacturing," *Economic Development and Cultural Change*, October, 1959, pp. 17–18.

trade. If the foreign offer curve is elastic,[17] or if the foreign offer curve shifts out sufficiently, the volume of exports may increase enough to improve the income terms of trade despite the deterioration in the commodity terms. The country's capacity to import is then greater, and this can be of decided significance for a developing country. Such an improvement in the capacity to import is especially important for a poor country which has a high average propensity to import. It would, of course, be even better for the country if its greater volume of exports could be traded at unchanged prices. But this involves a comparison with a hypothetical situation, whereas the relevant consideration is the effect of the actual change between the previous and present situations.

In contrast, a country's development program may be handicapped, despite an improvement in its commodity terms, if its capacity to import is reduced because of a fall in the volume of exports that is not offset sufficiently by the improved commodity terms. If, for example, a country's development is ultra-import-biased so that its commodity terms improve, but the foreign offer curve is not inelastic, or it shifts inwards relatively more than does the domestic offer curve, the country's income terms will deteriorate. Regardless of its more favorable commodity terms, the country's capacity to import is then reduced, and this may hamper the country's developmental efforts if the growth in output has not been sufficiently import-saving.

These examples illustrate that the mere knowledge of a change in the commodity terms of trade does not in itself allow a firm conclusion as to the effect on the country's economic welfare. It is essential to proceed beyond this superficial level and consider whether the change has been caused by a shift only in

[17] When the offer curve is of the normal "elastic" sort, more imports are demanded and more exports are supplied as the price of imports falls.

the foreign offer curve or by a shift in the domestic offer curve. If
by the latter, then the cause of the shift becomes relevant and
may deserve more emphasis than the fact of the change itself.
Attention to the underlying cause is especially needed for recognizing movements in the single-factoral terms as well as commodity terms of trade, and for determining possible changes in
the pattern of demand. Finally, changes in the volume of trade
must always be considered along with price variations.

6. With the foregoing general considerations in mind, we may
now examine the validity of the often-repeated contention that
the poor countries have suffered a secular deterioration in their
commodity terms of trade.[18] On the basis of inferences from the
United Kingdom's commodity terms of trade, proponents of this
view claim that "from the latter part of the nineteenth century
to the eve of the second world war . . . there was a secular
downward trend in the prices of primary goods relative to the
prices of manufactured goods. On an average, a given quantity
of primary exports would pay, at the end of this period, for only
60 percent of the quantity of manufactured goods which it could
buy at the beginning of the period."[19]

[18] This allegation appears in several reports of the United Nations and
in various writings by Raúl Prebisch, Hans Singer, W. A. Lewis, and
Gunnar Myrdal, among others. It is noteworthy that this view is completely at variance with that commonly held by classical economists who
believed that the operation of diminishing returns in primary production
would cause the prices of primary products to rise relatively to prices of
manufactures. Keynes restated the classical view in his "Reply to Sir William Beveridge," *Economic Journal*, December, 1923, pp. 476–488; also,
D. H. Robertson, *A Study of Industrial Fluctuation*, P. S. King & Son,
1915, p. 169.
[19] United Nations, Department of Economic Affairs, *Relative Prices
of Exports and Imports of Underdeveloped Countries*, 1949, p. 72. The

The causes of this deterioration are supposedly associated with differences in the distribution of the gains from increased productivity, diverse cyclical movements of primary product and industrial prices, and disparities in the rates of increase in demand for imports between the industrial and primary producing countries. Since technical progress has been greater in industry than in the primary production of poor countries, it is suggested that if prices had been reduced in proportion to increasing productivity, the reduction should then have been less for primary products than for manufactures, so that as the disparity between productivities increased, the price relationship between the two should have improved in favor of the poor countries. It is alleged, however, that the opposite occurred: In respect to manufactured commodities produced in more developed countries, it is contended that the gains from increased productivity have been distributed in the form of higher wages and profits rather than lower prices, whereas in the case of food and raw material production in the underdeveloped countries the gains in productivity, although smaller, have been distributed in the form of price reductions.[20]

---

indices used are based on Werner Schlote, *Entwicklung und Struktur-wandlungen des englischen Aussenhandels von 1700 bis zur Gegenwart.* Probleme der Weltwirtschaft, No. 62, Jena, 1938. Other indices constructed by Professors Imlah and Kindleberger do not show as marked an improvement for Britain as do Schlote's; A. H. Imlah, *Economic Elements in the Pax Britannica,* Harvard Univ. Press, 1958, chap. IV, Table 8; C. P. Kindleberger, *The Terms of Trade, A European Case Study,* John Wiley, 1956, pp. 53 ff.

W. A. Lewis' consideration of the prices of primary products and manufactures also relies heavily on Schlote's data; Lewis, "World Production, Prices and Trade, 1870–1960," *Manchester School of Economic and Social Studies,* May, 1952, Table II.

[20] United Nations, Department of Economic Affairs, *The Economic Development of Latin America and Its Principal Problems,* 1950, pp. 8–14;

This contrasting behavior of prices in industrial and primary producing countries is also attributed to the different movements of primary product prices and industrial prices over successive business cycles and to the greater number of monopoly elements in industrial markets.[21] According to this reasoning, the prices of primary products have risen sharply in prosperous periods, but have subsequently lost their gain in the downswing of the trade cycle. In contrast, it is asserted that although manufacturing prices have risen less in the upswing, they have not fallen as far in depression as they have risen in prosperity, because of the rigidity of industrial wages and price inflexibility in the more monopolistic industrial markets. It is therefore concluded that over successive cycles the gap between the prices of the two groups of commodities has widened, and the primary producing areas have suffered an unfavorable movement in their terms of trade.

Proponents of the secular deterioration hypothesis also argue that the differential price movements between poor and rich countries have been accentuated by a relative decrease in the demand for primary products and a relative increase in the demand for industrial products. This is attributed to the operation of Engel's law, and also, in the case of raw materials, to technical progress in manufacturing, which reduces the amount of raw ma-

---

*Relative Prices of Exports and Imports of Underdeveloped Countries, op. cit.*, pp. 13–24, 126; H. W. Singer, "The Distribution of Gains Between Investing and Borrowing Countries," *American Economic Review, Papers and Proceedings*, May, 1950, pp. 477–479; W. A. Lewis, "Economic Development with Unlimited Supplies of Labour," *Manchester School of Economic and Social Studies*, May, 1954, pp. 183–184; F. Mehta, "The Effects of Adverse Income Terms of Trade on the Secular Growth of Underdeveloped Countries," *Indian Economic Journal*, July, 1956, pp. 9–21.

[21] *The Economic Development of Latin America and Its Principal Problems, op. cit.*, pp. 12–14.

terials used per unit of output.[22] The low income elasticity of demand and the structural changes result in a secular decline in the demand for primary products. In other words, the consumption effect of development in the poor country is export-biased (pro-trade-biased), whereas in the rich country it is import-biased (anti-trade-biased).

If the alleged secular deterioration in the terms of trade of poor countries were true it would mean that there has been an international transfer of income away from the poor countries, and this decrease in purchasing power would be significant in reducing their capacity for development. The thesis is, however, highly impressionistic and conjectural. When its content is examined more rigorously, the argument appears weak—both statistically and analytically.[23]

Although the relevant long-run data for individual poor countries are not readily available, the substitution of the "inverse" of the United Kingdom's terms of trade is merely an expedient and does not provide a sufficiently strong statistical foundation for any adequate generalization about the terms of

---

[22] Singer, *op. cit.*, p. 479; Raúl Prebisch, "Commercial Policy in Underdeveloped Countries," *American Economic Review, Papers and Proceedings,* May, 1959, pp. 261–264. For a quantitative approach to some of the factors considered by Singer and Prebisch, see M. K. Atallah, *The Terms of Trade Between Agricultural and Industrial Products*, Netherlands Economic Institute, 1958.

[23] The most systematic and thorough-going critiques of the argument have been presented by Gottfried Haberler, "Terms of Trade and Economic Development," Howard S. Ellis, ed., *Economic Development for Latin America*, St. Martin's Press, 1961, pp. 275–297; M. June Flanders, "Prebisch on Protectionism: An Evaluation," *Economic Journal*, June, 1964, pp. 309–316. On the basis of several objections, largely similar to those we discuss below, both of these papers conclude that the reasons which have been advanced for the alleged trend are either fallacious or are entirely inadequate in their explanation.

trade of poor countries.[24] The import-price index is a mixed bag, concealing the heterogeneous price movements within and among the broad categories of foodstuffs, raw materials, and minerals. An aggregation of primary products cannot be representative of the wide variety of primary products exported by poor countries. Nor, of course, is it legitimate to identify all exporters of primary products as poor countries. Some primary producing countries are also importers of primary products. Moreover, the composition of exports from other industrial countries differs markedly from the United Kingdom's, making it unlikely that the United Kingdom's terms of trade can be truly representative for other industrial countries. It has been shown that the terms of trade for other industrial countries have behaved quite differently from those of the United Kingdom.[25]

Even if we were willing to use the British terms of trade as indirect evidence for the terms of trade between industrial and nonindustrial countries, we should still have to be extremely skeptical about the reliability of the British data. Apart from all

[24] Morgan, *op. cit.*, pp. 6–20. From a consideration of six countries other than the United Kingdom, Professor Morgan concludes that the highly diverse demand and supply experience for particular commodities of the different countries emphasizes the importance of refraining from generalizing about the experience of other countries by using the experience of the United Kingdom. Particular supply influences, and particular demand changes, for different commodities, countries, and times, have dominated the historical picture (p. 20).

Also see Morgan, "Trends in Terms of Trade and Their Repercussions on Primary Producers," R. F. Harrod and D. C. Hague, eds., *International Trade Theory in a Developing World*, Macmillan Co., 1963, pp. 57–59; Robert E. Lipsey, *Price and Quantity Trends in the Foreign Trade of the United States*, Princeton Univ. Press, 1963, pp. 8–24, 76; Harry G. Johnson, *Economic Policies toward Less Developed Countries*, Brookings Institution, 1967, appendix A.

[25] Kindleberger, *op. cit.*, pp. 53 ff., 233.

the statistical pitfalls connected with the construction of import and export price indices, there are strong biases in the United Kingdom series that make the terms of trade appear less favorable to poor countries than they actually were.[26] No allowance is made for changes in the quality of exports and imports; nor is there adequate coverage for the introduction of new commodities. Insofar as the improvements in quality and the introduction of new commodities have undoubtedly been more pronounced for industrial products than for primary products, a simple inversion of the United Kingdom's terms of trade would thus overstate any unfavorable movement for countries exporting primary products to the United Kingdom and importing industrial products from it.

Furthermore, there is no allowance for the fact that transportation costs were falling, making it invalid to infer from the British data what the terms of trade were for the primary producing countries trading with Britain. If the recorded terms of trade were corrected for the decline in transportation costs that occurred, the improvement in the United Kingdom's terms would appear substantially less. This is because British exports are valued at the port of exit, while the value of imports includes shipping costs. A large part of the decline in British import prices, however, was caused by the fall in ocean freights, and if Britain's export price index were corrected for transportation costs it would show a greater decline than does the recorded British export price index.[27] A proper consideration of transportation costs

[26] Morgan, "The Long-Run Terms of Trade Between Agriculture and Manufacturing," *op. cit.*, pp. 4–6; R. E. Baldwin, "Secular Movements in the Terms of Trade," *American Economic Review, Papers and Proceedings*, May, 1955, pp. 267–268.

[27] Statistical confirmation is given by L. Isserlis, "Tramp Shipping Cargoes and Freights," *Journal of Royal Statistical Society*, 1938, p. 122;

makes the terms of trade of primary producers appear less unfavorable.

These statistical imperfections do not allow much support for the hypothesis of a secular deterioration in the terms of trade for poor countries. It might even be maintained that their terms of trade improved because of quality improvements in their imports, access to a wider range of imports, and the great relative decline in transportation costs as compared with the prices of the commodities transported.

If the empirical evidence does not bear close scrutiny, still less does the analytical explanation. The validity of the appeal to monopolistic elements in the industrial countries depends on the existence of monopoly in not only factor markets but also product markets,[28] so that the increasing productivity could be distributed in the form of rising money wages and profits, with stable or rising prices. It is an open question whether trade unions and firms actually possessed and exercised sufficient monopoly powers. But even if they did, the existence of such monopoly elements would at most explain movements in the absolute domestic price level and not changes in relative world prices of manufactures and primary products. World price levels depend on world conditions of supply and demand, and a country with a relatively high domestic price level may simply find itself priced out of international markets unless it makes some adjustment in its domestic prices or exchange rate.

Further, allowing for the neglected influence of transport

---

Kindleberger, *op. cit.*, pp. 20–21, 336–339; C. M. Wright, "Convertibility and Triangular Trade as Safeguards against Economic Depression," *Economic Journal*, September, 1955, pp. 424–426; P. T. Ellsworth, "The Terms of Trade Between Primary Producing and Industrial Countries," *Inter-American Economic Affairs*, Summer, 1956, pp. 47–65.

[28] Kindleberger, *op. cit.*, pp. 246–247, 304.

costs over the cycle, we may also note many instances in which during a recession the prices of primary products declined in the United Kingdom, while actually rising at the ports of shipment in the primary producing countries.[29] Nor is the pre-1914 evidence on the purchasing power of primary products consistent with the cyclical explanation: Britain's terms of trade actually deteriorated during most depressions before 1914; Britain's food import prices fluctuated less in most trade cycles before 1914 than did British export prices; and a substantial number of primary products—especially foodstuffs—actually gained in purchasing power during many pre-1914 depressions.[30]

As for the appeal to disparities in the rates of increase in the demand for imports, it is true that, *ceteris paribus*, different Engel curves could cause a deterioration in the terms of trade. It is, however, essential to consider also the rates of development and changes in supply conditions, as has been stressed in the analysis of shifts in the offer curves. For even though the percentage of expenditure on a given import might be a decreasing function of income, the absolute demand for the import may still be greater as development proceeds. In addition, shifts of the long-term supply elasticities within industrial countries may be such as to prevent the domestic output of importables from keeping up with demand, so that the import requirements may rise relatively to income growth in the industrial countries. It should also be remembered that Engel's law applies only to foodstuffs—not to industrial raw materials or minerals. And even if an income

[29] Wright, *op. cit.*, pp. 425–426.

[30] K. Martin and F. G. Thackeray, "The Terms of Trade of Selected Countries, 1870–1938," *Bulletin of the Oxford University Institute of Statistics*, November, 1948, pp. 380–382; W. W. Rostow, "The Historical Analysis of the Terms of Trade," *Economic History Review*, Second Series, Vol. IV, No. 1, 1951, pp. 69–71.

elasticity of demand of less than unity is accepted as reasonable for primary products, what is significant for a specific primary producing country is not this over-all elasticity but the expansion in demand for its own exports.

Finally, aside from its statistical and analytical weaknesses, the entire argument has been unduly restricted to only the commodity terms of trade. Also significant are changes in the income terms of trade and especially the single-factoral terms. It is clearly possible, as already noted, that a country's income terms and single-factoral terms might improve at the same time as its commodity terms deteriorate. Since the exports from poor countries have grown so considerably, and productivity in export production has increased, the income terms and single-factoral terms have undoubtedly improved for poor countries. This is actually implicit in the secular deterioration argument, insofar as it relies on productivity increasing in both primary producing and industrial countries, but at a higher rate in the latter. Although their double-factoral terms of trade may have deteriorated, this did not affect the welfare of poor countries; they were better off when their own single-factoral terms improved and they received more imports per unit of their "exported factors," regardless of whether the single-factoral terms also improved for other countries exporting to them. Their capacity to import and their imports per unit of productive resources exported have increased—regardless of any changes in the relative prices for their products.

The most favorable situation, of course, would be an improvement in the commodity terms of trade as well as in the single-factoral and income terms. But the ruling conditions may frequently be incompatible with such a simultaneous improvement. Nonetheless, to look only at changes in the commodity terms is to neglect the favorable effects of the greater capacity to import through improvement in the income terms and the benefits from the improvement in the single-factoral terms. When it is assessed within this wider analysis, a change in the commodity terms of trade may prove to be of small moment for a developing economy in comparison with the more fundamental changes that have occurred at the same time.

# THE DEVELOPMENT OF GAINS FROM TRADE THEORY: CLASSICAL TO MODERN LITERATURE

John M. Letiche
*Professor of Economics, University of California, Berkely*

Robert G. Chambers
*Assistant Professor of Agricultural Economics, Ohio State University*

Andrew Schmitz
*Professor of Agricultural and Resource Economics, University of California, Berkeley*

## CONTENTS

Even though classical, neoclassical, and every school of post-classical economists have long emphasized that nations can gain from an expanding volume of world trade, in the last years of the 1970s, many sectors of national economies were once again becoming more protectionist. In view of that fact, this paper will examine the arguments supporting the free trade doctrine in the context of major theoretical developments in the gains from trade literature. The classical writings on the issue are voluminous. They have been excellently apparised, however, making it possible to present the essential foundations with comprehensive brevity.[1] Most of the focus will be on the normative aspects of neoclassical trade theory, and only the standard results will be considered.[2]

Reprinted from "Economic Perspectives: An Annual Survey of
Economics" (M.B. Ballabon, ed.), Vol. 1, pp. 119–149, with
permission of Harwood, Copyright 1979.

## THE EARLY WRITINGS

The classical theory of international trade was formulated primarily to provide guidance on important questions of national policy. It originated in Britain, where a major issue revolved around the potential gains to England from free trade, as well as the distribution of gains from trade between England and the rest of the world. The classical economists utilized three methods in dealing with the question of the gains from trade: (1) the doctrine of comparative costs; (2) the increase in income as a criterion of gain; and (3) the terms of trade as an index of the gains from trade and its distribution.

The free trade doctrine, as presented by Adam Smith was based on the apparent advantage to a country of importing, in exchange for domestic products, those goods which either could not be produced at home or could be produced only at costs absolutely greater than those at which they could be produced abroad. Accordingly, the producer of exportables had to have an absolute advantage; and this advantage was generally measured in terms of the labor costs of production. Under free trade all products entering international commerce would be produced in those countries where the real costs were lowest, and from this doctrine Adam Smith drew laissez-faire conclusions. As he wrote over 200 years ago:

> If a foreign country can supply us with a commodity cheaper than we ourselves can make it, better buy it from them with some part of the produce of our own industry, employed in a way in which we have some advantage. The general industry of the country . . . will not thereby be diminished . . . but only left to find out the way in which it can be employed with the greatest advantage.[3]

David Ricardo and Robert Torrens corrected Smith's theorizing to show that a country can gain from trade even if it has no absolute real cost advantage in the production of any commodity. The early theory of comparative cost was formulated in terms of a single factor of production (labor) and assumed that, in the absence of trade, goods were exchanged internally at their relative unit labor cost. In the case of free trade, gains from trade resulted when countries specialized in the production of those commodities in which they had a comparative advantage, denoted by the respective ratio of productivity of the resources between the countries concerned.[4]

The case for a large volume of profitable trade, it was held, depends on the width of these gaps in comparative advantage of different countries for the production of different goods. If a country has an absolute advantage in agriculture, this does not stop it from specializing in something in which it has a greater advantage still. The comparative-cost approach, in effect, emphasized that trade minimizes the aggregate real cost at which a given level of real income can be obtained or maximizes the aggregate level of real income obtainable from a given (full employment) utilization of resources. Theoretical contributions made by J. S. Mill, Alfred Marshall, F. Y. Edgeworth, and F. W. Taussig demonstrated that comparative costs are an essential element in the gains from trade, but that reciprocal demand functions have to be incorporated into the analysis insofar as they are welfare functions representing net income.[5]

In this connection Mill contributed a brilliant discussion on the relationship between reciprocal demand and the commodity terms of trade. Marshall in turn generalized the analysis by an ingenious fusion of "reciprocal demand-and-supply"—a simultaneous incorporation of macroeconomic and microeconomic theory which, for its time, achieved not only a remarkable degree of theoretical "generality" but also a constructive link with reality. In this theoretical framework, production is carried on under diminishing marginal value productivity, resulting from the combined operation of the law of diminishing returns and the law of diminishing utility. Hence it is assumed that output per unit of input of labor and/or capital diminishes after a point. Further, as output of some commodities increases, the exchange value per unit of this commodity declines; i.e., its terms of trade for other commodities diminishes. Therefore, if all commodities are produced under increasing costs, each country produces some of each good and, with minor exceptions, imports or exports the difference between its domestic demand and supply, respectively. By implication at least, for developed countries partial specialization is the rule, its extent depending upon how rapidly relative costs rise.

The classical school believed that in the medium term manufactures were produced at constant or decreasing costs per unit of output, with the check to specialization occurring via a decline in the relative value of the commodity as its supply increases. Agricultural production, however, was assumed to occur under increasing costs. But even to classical economists this formulation leads logically to free trade conclusions, in some cases, only under these assumptions:

1. Free competition exists in each country entering into international trade.

2. Labor and capital migrate within the same country more easily and freely than among different countries.

3. Costs and prices are appreciably more variable than employment; hence full employment conditions are taken for granted in the assumed long-term analysis.

4. Costs incurred during the process of occupational mobility and during production adjustment are abstracted from, and transportation costs usually are assumed to be zero, internally and internationally.

5. Entrepeneurs expect current prices to continue for that part of the future that is relevant to their decisions.

6. Maximization of national real income or aggregate world real income is considered to be the most important economic objective, with some recognition (but not of practical importance) of the possible conflict between countries as to the relative level of real income or its distribution.

7. Multilateral convertibility prevails in the modern sense of the term plus approximately stable exchange rates; monetary and credit systems work either automatically or under central bank management in such a way as to bring the monetary and real factors easily and readily toward domestic and international equilibrium.

Since free trade makes imports cheaper and exports dearer, the wages of labor employed in import fields would tend to fall and in export fields to rise. Distribution of income would therefore be affected. With this important exception, on the basis of their assumptions, there appear to be no qualifications of consequence to the free trade position. Even changes in the distribution of income resulting from free trade could be compensated for by changes in taxation.

There was, however, a logical error in this early free trade doctrine. The argument was expressed in terms of free trade versus no trade, whereas the real choice usually was between free trade and tariff protection. Implicitly, some of the writers might have assumed that the movement from protection to free trade would not alter a country's terms of trade. But in time this issue was explicitly faced and it brought about a rift in the ranks of free traders. Torrens, Mill, Marshall, Edgeworth, and Taussig conceded the theoretical error. They realized that by imposing a tariff a country would tend to reduce the volume of its imports; the price of its exportables would tend to be raised and that of importables lowered, improving its terms of trade. Such gains were likely to be offset by losses from resource misallocation. The gains from protection were only to be derived up to a point, Edgeworth noted, and they were not accurately ascertainable. Even if a country gave careful consideration to the national interest, rather than to pressure groups, all these writers believed that it would still probably err in the direction of excess protection. Torrens qualified his argument by advocating reciprocal rather than unilateral tariff reductions. The gains from trade to be derived by tariff protection, however, were regarded as national—not cosmopolitan—gains, for a gain in terms of trade to one country is a loss to another, while the misallocation of resources is a general loss.

Even to a single country, if the terms of trade argument were to hold, the reciprocal demand of the outside world should have a low elasticity and its own reciprocal demand a high elasticity. Not only should the tariffs be administered with skill and integrity, but there should be no retaliation—an important issue reconsidered in the later literature on the "second-best." Understandably, therefore, the classical writers before World War I believed that in the long run the scope for nationally profitable protection was in practice narrowly limited. From a cosmopolitan point of view, they believed the free trade doctrine to have remained intact. Long-term protection, they held, may injure all who practice it, and can benefit none except at the cost of greater injury to others. The only exception would entail the effective use of tariff protection by a poor country that would benefit at the expense of its rich trading partners. The resulting reduction in aggregate world income would hence justify the use of income-offsets from the rich countries to the poor in order to maximize world welfare.

Although the classical economists considered the long-term arguments for free trade as substantive arguments against protection, they conceded the theoretical validity of the short-term "young country" or "infant industry" arguments for protection. In its crude form the argument was often stated in the seventeenth century. The classical economists, however, formulated the argument in specific terms.[6] Gains from trade could therefore be derived through the use of temporary protection by speeding the establishment and development of potentially profitable industries. The argument was sometimes rejected, however, on these practical grounds: (1) the selection of industries was likely to be arbitrary or irrational; (2) it would open the door to promiscuous protection and "ad hoccery"; (3) it would stifle or delay genuine progress through the spread of industrial inertia, inefficiency, and restrictive monopoly; (4) temporary protection would tend to become permanent when market forces revealed the incapacity to operate without artificial support; (5) the extensive use of infant industry protection for the economic development of a young country would tend to increase costs as a whole and thereby hinder its specialization in domestic industries and export fields by way of comparative advantage; (6) the widespread use of temporary protection tends to make commodities too expensive for the budget of many consumers and thereby hinders the development of mass production and distribution; (7) when justified, infant industries could be more efficiently protected through the use of subsidies rather than tariffs.[7] For these reasons plus noneconomic arguments for free trade, the classical writers for the most part did not regard tariff protection as an intelligent form of economic planning, but a perversion to laissez-faire.

Many of the conclusions concerning the gains from trade derived from later theoretical developments are similar to those obtained from the models presented by these early writers. To investigate one of these models more fully, consider Figure 1. Let $S^X$, $D^X$ represent the supply and demand situation in the exporting country and $S^M$, $D^M$ represent the supply and demand situation in the importing country. Without trade, price in the importing country is $P^M$, in the exporting country, $P^X$. With trade, in the absence of transportation costs and other impediments, price $P^W$ obtains in both countries. The movement to free trade results in a net welfare gain of the two crosshatched areas. For example, in the exporting country the introduction of trade results in a loss of consumers' surplus of $BP^W P^X C$ while producers' surplus is increased by the area of $AP^W P^X C$.[8]

The early writers also used this framework to examine the welfare impacts of restrictive trade policies, especially tariffs. In Figure 1, consider the introduction of a tariff, $T$. The effect on the importing country, for example, is a loss in consumer surplus equal to the area $GIEP^W$, a gain in producer surplus equal to the area $GHFP^W$, and the tariff revenue obtained is $HIKJ$. Interestingly, modern facets of the optimal tariff were developed within this framework.[9] Thus, for the importing country to gain from the imposition of a tariff, the sum of the tariff revenue and the gain in producers' surplus must exceed the loss in consumers' surplus. This will always be the case for the optimal tariff when the foreign excess supply curve is upward sloping.

## POST-WORLD WAR I LITERATURE

The main theme of the normative aspects of modern trade theory supports the claim of the classical economists that, with "appropriate compensation," trade results in an improvement in economic welfare. In pur-

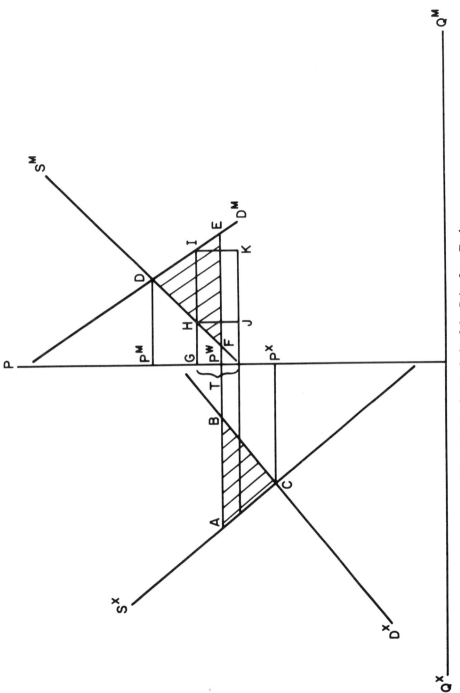

**Figure 1.** Surplus Analysis of the Gains from Trade.

suing the normative aspects of trade theory, Leontief, in a classic paper, demonstrated how standard consumer theory can be extended to the analysis of international trade.[10] Let $TT$ in Figure 2 represent the transformation curve of the home country derived under conditions of increasing costs, and let $U_i$ represent community indifference curves. Autarkic equilibrium occurs where community curve $U_0$ is tangent to $TT$ at $\alpha$. The home country produces and consumes $OF$ of $A$ and $OD$ of $B$. Let $p^0p^0$ represent the free trade terms of trade.[11] In equilibrium, the home country now produces $OE$ of $A$ but consumes $OC$ of $A$ and exports $EC$ of $A$ for $XY$ of $B$. Clearly, there are gains from trade since utility level $U_0$ is below utility level $U_1$.[12]

A more rigorous proof of the existence of either actual or potential gains from trade was provided by Samuelson.[13] Assuming perfect competition, the existence of monotonic ordinal preferences for each individual [i.e., $x^* \geqslant x$ implies $u(x^*) \geqslant u(x)$] –profit maximization and utility maximization–denote commodities consumed by the vector $x = (x_1 \ldots x_n)$, commodities produced by the vector $\overline{x} = (\overline{x}_1 \ldots \overline{x}_n)$, productive services by the vector $a = (a_1 \ldots a_n)$, and prices of commodities and productive services by the respective vectors $p = (p_1 \ldots p_n)$ and $w = (w_1 \ldots w_n)$. Let superscript $o$ denote autarkic quantities and prices. In autarkic equilibrium it is clear that $\overline{x}^o = x^o$.

Consider the opening of trade. Denote free trade commodity prices and factor prices, respectively, by $p'$ and $w'$. Under these assumptions the optimal bundles of commodities produced and productive services utilized will maximize the difference betwen total commodity value and total productive service value for any feasible production set. That is,

$$p'\overline{x}' - w'a' \geqslant p'\overline{x} - w'a \quad Vx \neq x', w \neq w' \tag{1}$$

and specifically

$$p'\overline{x}' - w'a' \geqslant p'\overline{x}^o - w'a^o$$

where $p'\overline{x}'$ is understood to be the inner product. Then assuming balanced trade, i.e., $p'(\overline{x}' - x') = 0$, (1) can be rewritten as

$$p'x' - w'a' \geqslant p'x^o - w'a^o \tag{1'}$$

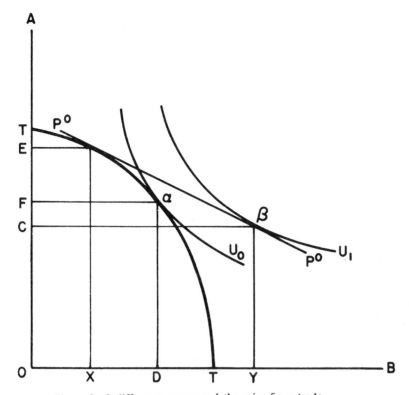

**Figure 2.** Indifference curves and the gains from trade.

Based on expression (1'), Samuelson derived the following theorem on the assumption that all members of the economy are identical in every respect.[14]

*The introduction of outside (relative) prices differing from those which would be established in our economy in isolation will result in some trade, and as a result every individual will be better off than he would be at the prices which prevailed in the isolated state.*[15]

When individuals are not alike (given that the introduction of outside relative prices generates trade), trade may make some people actually worse off. However, there are *potential* gains to be had from trade since the trade consumption bundle can be redistributed by means of lump-sum income transfers such that everyone could be made better off.

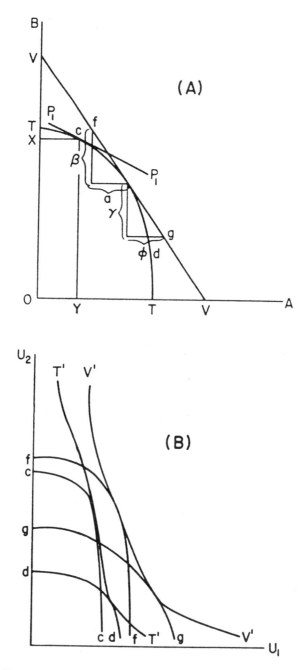

Figure 3. The new welfare economics and the gains from trade.

In other words, if a unanimous decision were required in order for trade to be permitted, it would always be possible for those who desired trade to buy off those opposed to trade, with the result that all could be made better off.[16]

## The New Welfare Economics and the Gains from Trade

In a later article, Samuelson showed in a broader framework that nations can potentially gain from international trade.[17] Let $TT$ in Figure 3A represent the transformation curve in the home country. $VV$ represents the free trade consumption possibility frontier.[18] Assume $c$ represents the no trade equilibrium with relative prices $P_1P_1$. At $c$, $OX$ of $B$ and $OY$ of $A$ are produced and consumed. The utility possibility frontier corresponding to this bundle is represented by $cc$ in Figure 3B. Consider any other point, $d$, on the transformation curve. The corresponding utility possibility frontier for $d$ is $dd$ in Figure 3B. The envelope of all such frontiers is known as the grand utility possibility frontier under no trade ($T'T'$).[19] Now, consider point $f$ which is a bundle obtainable through international trade, i.e., $\alpha$ of $A$ is exported for $\beta$ of $B$. The utility possibility frontier for $f$ is $ff$ in Figure 3B. Likewise, consider point $g$ where $\gamma$ of $B$ is traded for $\phi$ of $A$. The utility possibility frontier for $g$ is $gg$ in Figure 3B. The envelope of all such frontiers is $V'V'$. It follows that, for all positive trade, everyone can be made better off, with proper income redistribution, than in the no trade case because $V'V'$ lies everwhere outside of $T'T'$ (at the limit $V'V'$ may touch $T'T'$).

This argument can be seen in another way. Let $TT$ in Figure 4 represent the home country's transformation curve. Suppose $c$ represents autarkic equilibrium and free trade prices are given by the slope of $VV$. Evaluating $c$ at the free trade prices ($V'V'$), it is clear that national income is greater in the trading case than in the no trade case.

An alternative geometric proof for the proposition that free trade is always potentially superior to autarky has been provided by Peter B. Kenen, who developed the commodity space analogue to the utility possibility frontier known as the Paretian contract curve.[20]

The classical economists conjectured that, typically, the more the terms of trade for a country improved, the larger the gains from trade. Samuelson has posed the question, "Is it possible to state that the more prices 'deviate' from those of the isolated state the better off all individuals will be?"[21] In this regard it has been demonstrated that if the terms of trade improve it is possible for society to consume at least as much of each commodity without employing more of any factor.[22] Furthermore,

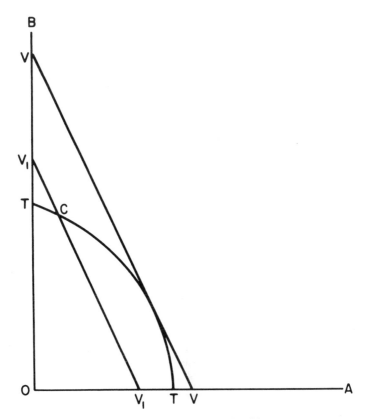

**Figure 4.** Trade and national income.

for two commodities produced with fixed factor supplies, it has been shown that the more the terms of trade improve the greater the gains from trade.[23] This result, however, does not apply to a country engaged in tariff-ridden trade.[24] In such a case, as illustrated in Figure 5, if the export good is inferior in consumption it is possible that an improvement in terms of trade will leave everyone worse off. Suppose the home country always produces a fixed amount of $A$, say $a$, which it exports for $B$. The original free trade terms of trade are given by the slope of $ab$. Free trade quilibrium is at $c$ where a community indifference curve (not drawn) is tangent to $ab$. Now suppose that a duty is levied at $B$ at a constant ad valorem rate. Equilibrium will then obtain at $c'$ where $ab$ intersects a community indifference curve $U_1$ at the point where the indifference curve has the same slope as the internal price ratio (represented by

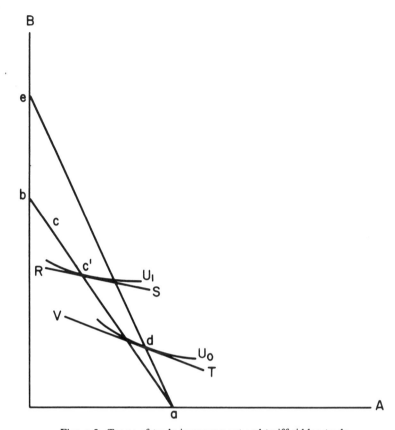

**Figure 5.** Terms of trade improvement and tariff ridden trade.

the slope of *RS*). If the terms of trade improve to, say, the slope of *ae*, the new equilibrium can be found anywhere along *ae*. Figure 5 illustrates the possibility of a deterioration of the home country's welfare—a deterioration possible only when the export good is inferior: A rise in the domestic price of exportables (the new internal price ratio, given by the slope of *VT*, is greater than that of *RS*) is associated with an increase of their domestic consumption.

## Tariffs and the Gains from Trade

As already noted, a considerable amount of attention was given in the early writings to the welfare effects of tariffs. In the neoclassical literature many authors have rigorously demonstrated, in a general equilibrium framework,

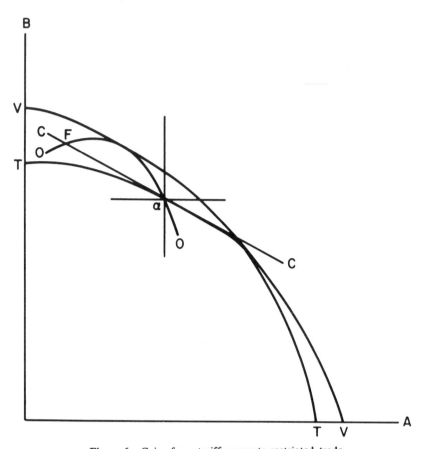

**Figure 6.** Gains from tariff or quota restricted trade.

the conditions under which tariffs can bring about a higher level of national welfare than free trade.

In a well known article Robert E. Baldwin used the "envelope curve" to show how a country potentially can gain from restricting trade by imposing a tariff or a quota.[25] Assuming nonintersecting community indifference curves, let *TT* in Figure 6 represent the home country's transformation curve and *OO* the foreign country's offer curve. The envelope, *VV*, can be derived by sliding the foreign country's offer curve around *TT*, always keeping its axis parallel to the axis of the transformation curve. Suppose that the free trade terms of trade, determined by the intersection of the home country's offer curve and *OO*, is given by the slope of *cc*. The free trade point, *F*, will always lie within the envelope *VV*. Hence the grand utility possibility frontier of the bundles represented by the envelope will lie everywhere to the northeast of the grand utility possibility frontier derived from the free trade bundles.[26] Therefore, by imposing tariffs or quotas—in the absence of retaliation—there is a potential gain from restricting trade, since points on the envelope denoting higher levels of community welfare can be reached by these means. This result follows from the fact that the imposition of tariffs or quotas by countries having monopoly power on imports (and, symmetrically, of taxes and quotas on exports) can in the postulated conditions improve their terms of trade.[27]

In one of his early articles on optimum tariffs and retaliation, Harry G. Johnson refuted the conclusion of earlier writers that tariff retaliation would necessarily leave both (all) countries worse off than under free trade. Using a myopic Cournot-type of retaliation mechanism such that each country in turn imposed on optimal tariff, Johnson demonstrated that the country initially imposing an optimum tariff could be made better off than under free trade. But in his later writings he reversed the policy implications of this *curiosum,* reducing the case for tariffs to a *second-best* category in the presence of all distortions *other* than that of the traditional case implied by the presence of monopoly power in foreign trade. In a related note, he provided yet another argument against protection: if a small country grew subject to a distortionary tariff, the country could experience immiserization with no change in either domestic or international terms of trade. Furthermore, if technical change in the protected industry was the source of growth, the likelihood of immiserating growth was even greater, since the primary gain from growth (measured as optimal policy) could now be more outweighed by the accentuation of the loss from the distortion imposed by the tariff following growth. This argument reinforced the importance of measuring growth at world rather than domestic prices in trade-dis torted economies, and furnished an important explanation of why import-substituting LDCs have as a rule done worse than the export-promoting LDCs.

Not only have these contributions stimulated scholars to examine whether tariff-induced capital flows would necessarily be welfare-improving, but they led Johnson and others to theoretic-empirical work on the cost of protection, on the gains or losses to a country of joining a customs union, on the so-called "scientific tariff," and on a more generalized theory of distortions and welfare.[28]

As the objectives of tariff policy became more extensive, new concepts were developed to analyze and appraise them. Listing a large number of commonly recognized arguments of modern tariff protection, Johnson applied the "scientific tariff" for an examination of the following objectives: (1) to promote national self-sufficiency, independence, diversification, agriculturalization, or industrialization; (2) to increase military preparedness; and (3) to strengthen a country's international bargaining position. With regard to gains from trade, he formulated alternative maximization policies for a developing country's exports of primary products, in terms of both export and import tariffs designed to maximize government tariff revenue.

In comparing the results of tariff protection with other commercial policies, the problem has been further analyzed as to whether a higher tariff is necessarily preferable to a lower one. In Figure 7, let $TT$ be the transformation curve, $A$ the exportable good and $ab$ the international terms of trade. Manifestly, an appropriate tariff on $B$ could lead to production at $a$ and consumption at $b$. A higher tariff would lead to production at $c$ and consumption at $d$, where $d$ is on a higher indifference curve than $b$. In this illustration given by Bhagwati, a higher tariff provides superior results to a lower one, refuting the assertion that a higher tariff is necessarily inferior to a lower one. It will be noted, however, that in this illustration the export good is inferior in consumption: evaluated at the original tariff-level price, $d$ represents an increase in real income, but consumption of $A$ has declined. In a controversy on this issue, Kemp criticized some of the welfare comparisons which were made on the basis of this particular example. He showed that the maximum feasible utility one individual can attain—given all other utility levels as fixed—is greater in a comparatively low-tariff situation.[29]

Having unified a considerable body of literature on the theory of distortions and welfare, Bhagwati derived "duality" relationships between the analysis of policy rankings under various market imperfections and policy rankings to achieve noneconomic objectives. The seven areas treated, and key conclusions reached, point to the strength and weakness of this form of analysis. Postulating the existence of four important market imperfections—for example, in factor markets a wage differential between sectors, in product markets a production externality, in consumption a consumption externality, and in trade monopoly power—the suboptimality of laissez-faire is demonstrated. In other words, under such conditions laissez-faire would not be the optimal policy. Providing examples of immiserization—economic growth bringing about a deterioration in the country's terms of trade; an optimum tariff (before growth)

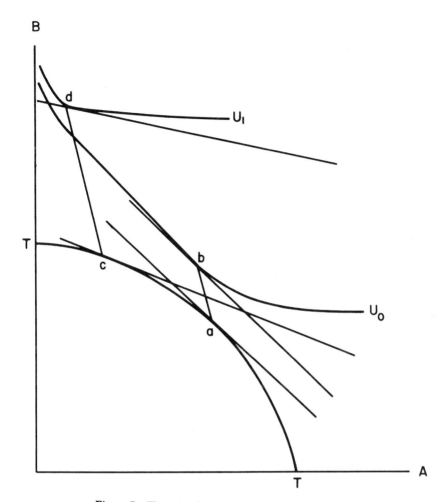

**Figure 7.** The gains from high vs. low tariff protection.

becoming suboptimal after growth; a distortionary tariff used for the purpose of import-substitution; and cases where a wage differential exists in the factor market—it is shown that all these distortions may result in the country becoming worse off after growth. Ranking alternative policies under these market imperfections, Bhagwati demonstrates the optimal policy intervention in regard to them. Under the heading of "ranking of tariffs," policies constituting impediments to the attainment of optimality are evaluated. This is the category in which Kemp's theorem can be illustrated—that a country without monopoly power in trade (and no other imperfection), but with no inferior goods, would be worse off with a higher tariff than a lower one. Similarly, tariffs around the optimal tariff for a country with monopoly power in trade may be evaluated. Introducing various kinds of market imperfections (e.g., Hagen's case of wage differentials or Haberler's case of production externalities), or the existence of commodity taxes, it is feasible to rank conditions of free trade and autarky demonstrating that free trade is not necessarily superior to self-sufficiency. Finally, noneconomic objectives are considered with the attendant ranking of policies. This analysis comprises optimal policy intervention when the values of certain variables are constrained so that full optimality is unattainable.

After considering the theory of distortions in terms of four "pathologies," representing cases where the economic system would not satisfy conditions for an economic maximum, viz., where the equalities for the marginal rate of transformation in domestic production, the marginal foreign rate of transformation, and the marginal rate of substitution in domestic consumption are not met, Bhagwati analyzes the distortions arising potentially from endogenous (existing market imperfections), policy-imposed (concrete government acts), and autonomous (accidental) distortions. With regard to gains or losses from trade, these are the noteworthy conclusions. Where market imperfections exist and noneconomic objectives are postulated, the required policy directly attacks the source of the distortion. Hence, when distortions are introduced into the economy, because the values of certain variables (e.g., production or employment of a factor in an activity) have to be constrained, the least-cost method of achieving this objective is to choose the policy intervention affecting directly the constrained variable. Specifically, a trade-level noneconomic objective is achieved at least cost by introducing a policy-imposed distortion via a trade tariff or subsidy; a production noneconomic objective by introducing a policy-imposed distortion via a production tax-cum-subsidy; a consumption noneconomic objective by introducing a policy-imposed distortion via a consumption tax-cum-subsidy; and a factor-employment (in a sector) noneconomic objective by introducing a policy-imposed distortion via a factor tax-cum-subsidy.[30]

Bhagwati has shown that although each distortion, whether endogenous, policy-imposed, or autonomous in origin, may be welfare-ranked and symmetrical with a "corresponding" class of imposed distortions (e.g., the ranking of policies for a production externality is identical with the ranking of policies when production is constrained as a noneconomic objective), the policies can be ranked as optimal or second-best, etc.—as well as when they may be superior to free trade—but they cannot be ranked uniquely vis-à-vis one another. When several distortions exist, successive reductions in a tariff will not necessarily improve welfare steadily, if at all, and the gains from trade may turn into losses. The theorems on the possible inferiority of free trade (i.e., zero tariff) to no trade (i.e., prohibitive tariff) when there is a production externality or a wage differential are thus only special cases of the general theorem that reductions in the "degree" of a distortion will not necessarily be welfare-increasing if there is another distoriton in the system. Similarly, if distortions exist under free trade and under autarky, the free trade situation may be immiserizing (therefore, free trade inferior to autarky) if the loss from the distortion is accentuated and outweighs the primary gain from the shift to (free) trade itself. The conclusion that deserves particular attention for policy formation is that distortions cannot be ranked vis-à-vis one another, which applies to all classes of distortions.

This literature on distortions and welfare, with its related emphasis on potential gains or losses from trade, grew primarily out of the conditions and ferment of ideas in Delhi, India, during 1963-68. In many ways, with differences being no smaller than similarities, the problems of growth, development, and trade under extreme uncertainty of the industrial countries of North America, Western Europe, Australasia, and Japan have brought about an illuminating literature with references to the "gains from trade." Inter alia, these writings have dealt with customs union theory, optimum-currency areas, various exchange rate regimes, as well as with import-substitution, effective tariffs, and the relation of these phenomena to geographical and national income distribution.

A considerable emphasis has been placed on the notion of "effective tariffs." The concept of the effective rate of protection was developed within the traditional framework of tariff theory. It is a tariff measure that takes into account tariffs on both intermediate and final products. In the simplest case of one imported input, the effective rate of protection is:

$$ERP = \frac{t_c - A_{yc}\, t_y}{V_c}$$

where $t_c$ is the nominal tariff on the output, $t_y$ is the nominal rate on the imported input, $A_{yc}$ is the share of the input in the total value of the output, and $V_c$ is the value added.[31] The literature on effective tariffs has endeavored to show the impact of various restricted trade policies upon growth and development. Understandably, much attention has been given to the implications for the LDCs of reduced gains from trade resulting from high effective tariffs of developed countries on imports of processed goods and semifinished manufactures. To date the results of this literature have been highly controversial, and the theoretical concepts are still being integrated into the corpus of modern trade and development theory.[32]

## GAINS FROM TRADE: SOME RECENT MATHEMATICAL CONTRIBUTIONS

With the emergence of a new economic era in the 1970s—deceleration in the rates of economic growth and volume of foreign trade among the major capitalist countries—not surprisingly there appeared a series of articles providing a reexamination of the gains from trade under nonautarkic versus autarkic conditions. These works were written in mathematical terms and their authors either confirmed or derived the following results. Free trade is *potentially noninferior* to autarky. Trade restriction via tariffs, or quotas, is *potentially noninferior* to autarky. If a small country is in trade equilibrium and experiences an autonomous change in prices that is potentially beneficial, a further change in prices in the same direction is potentially unharmful. Similarly, if a large country experiences this phenomenon, an expansion of foreign trade is potentially unharmful.[33]

Proofs have been provided for the existence of a world "Pareto-optimal" competitive equilibrium, as well as for "equilibrium" under conditions of autarky. A general theorem demonstrates the existence of a competitive world equilibrium which is potentially noninferior for the consumers of each nation as compared with "equilibrium" under autarky.[34] The relationships between welfare criteria in utility space and in commodity space have been rigorously formalized.[35] The concept of a social utility frontier has been used to prove a variant of the proposition that free trade is potentially noninferior to autarky.[36]

Furthermore, the gains from trade have been investigated for situations where the trading economy is faced with nonmarket externalities and the conditions formulated under which trade would still be superior to autarky.[37] A unified analysis of many of these problems shows that conditions can be postulated under which economic growth with free trade, or growth with tariffs, is potentially noninferior to trade under autarky; trade with subsidies—solely financed by tariff revenues—is not worse than no trade; and a generalization has been provided for the relationship between price divergence and terms-of-trade improvement.[38] It has also been shown that, if there exists an economy, say $A$, such that every agent in $A$ benefits from the movement from autarky to trade, an economy $A'$ comprised of agents sufficiently similar to those of $A$ will benefit from the opening of trade.[39]

This mathematical literature has had the beneficial effect not only of defining the precise conditions under which restricted trade can better all members of society as compared with autarky, but indirectly of providing the foundation for empirical investigations of the feasibility of achieving greater gains from trade by reducing serious fluctuations in the terms of trade.

## TRADE, DEVELOPMENT AND UNCERTAINTY

In the last two decades, full-length analytical studies have appeared on trade, development and uncertainty. Basically this work was founded on a two-sector growth model with trade. It has also been utilized to analyze such problems as the impact of slow-export-demand growth upon the growth of per capita income in developing countries. Bardhan has shown that, if a less-developed country faces an externally determined rate of growth of demand for its exports and if its domestic production requires the use of an imported intermediate product, the long-run rate of growth for this country will be influenced by export demand factors. The long-run rate of growth will be a weighted average of the rate of growth of the primary factors (as in standard neoclassical models) *and* the rate of growth of export demand. If exports grow at a sluggish rate so that the latter rate is lower than the former, long-run capital accumulation for such a country will be slower than the "natural" growth rate.[40] More generally, several articles have reconsidered the problems of factor proportions and comparative advantage within the framework of long-run growth and development. Important results have been obtained: the basic conclusion of the static Heckscher-Ohlin theory, which maintains that a labor- (capital-) abundant country will have a comparative advantage in producing labor- (capital-) intensive goods, also applies to the long run when an economy experiencing population growth and capital accumulation has reached a steady growth path.[41] Moreover, it has been demonstrated that in the long run a higher savings ratio will inevitably lead to relatively higher output of capital-intensive goods. This analysis was carried out for a small country which, by means of labor and capital, produces three goods, one of which is a nontraded capital good. As is the case with most such analyses, the results may not be generalizable but the basic conclusions appear to be of prime importance to the theory and practice of the gains from trade. In the long run, the *form* of comparative advantage will differ from the initial situation; per se, however, dynamic considerations do not contravene the results of standard comparative advantage theory—arguments to the contrary of many critics notwithstanding.

The gains from trade via import-substitution have also been reconsidered. It has been shown convincingly that a heavy-industry emphasis does not necessarily lead to the most rapid growth process. Clearly this conclusion has an important bearing on the trade and development policies of LDCs.

Attempts have also been made to relate monetary and exchange rate arrangements to the gains from trade. Specifically, in analyzing the controversy between monetarists and structuralists on the issue of inflation in Latin American countries, Findlay has designated market imperfections and lack of factor mobility for special consideration. This has opened the way for the incorporation of these conditions into standard international trade theory via well-developed economic techniques. Similarly, it has been shown that certain exchange-rate regimes have important distribution effects on production and consumption.[42] An overvalued exchange rate regime benefits importers and consumers of internationally traded goods and penalizes producers; it restricts the inflow of foreign captial. *Per contra*, devaluation shifts income from consumers to producers of these goods. Although many country-studies have appeared indicating the distorting effects of inflexible—usually overvalued—exchange rate regimes on the gains from trade and its distribution, further theoretical-quantitative research on this issue is long overdue.

An important and growing literature has devoted attention to the more general incorporation of uncertainty into gains from trade theorizing. One author has investigated the case of a small trading country experiencing large fluctuations in terms of trade as a result of uncertain transaction costs and/or of erratic movements in spot exchange rates. Under otherwise customary assumptions, the following nonautarky theorem was derived: Given trade uncertainty (and excluding "pathological" distortions), autarky will not be optimal regardless of variations in the terms of trade. Similarly, by introducing forward markets into the analysis, it was shown that in the long run autarky cannot be optimal regardless of the variations in terms of trade and whatever the level of forward prices.[43] In a jointly authored paper, Batra and Russell examined the effects of increasing uncertainty of world prices on the social welfare of a trading nation and demonstrated that it would bring about a decline of expected utility. Under conditions of uncertainty, therefore, free trade may not be an optimal policy. To reduce the effects of uncertainty, and to increase the potential gains from trade, the authors considered various governmental policies designed to minimize the cost to consumers resulting from variations in actual as compared with expected terms of trade.[44]

Batra further examined these issues by adopting the well known approach of introducing a random variable in the production function and proceeding with the assumption that the expected utility of the producers is to be maximized. With this change in assumptions the theory of international trade was reconsidered along standard lines.[45] The analysis was confined to the narrow world of two commodities, two factors, and two countries. It also utilized a special form of the production function, for uncertainty appears as a multiplicative factor grafted on to a standard production function with full certainty. The conclusions therefore depend not only on the producers being risk averters but upon the precise nature of their risk-aversion. Although the theorems derived are narrow in scope, they provide insights for the development of gains from trade theory and practice, especially in regard to the "commodity problem" and the stabilization of terms of trade.

In effect, the introduction of uncertainty is shown to have important consequences for standard trade theory. Uncertainty cannot be assumed away as a minor modification. Contrary to the standard Heckscher-Ohlin conclusion, given constant commodity prices and the customary assumptions regarding homogeneity—with uncertainty, changes in factor endowments do affect relative factor prices. Still, the following related theorems remain robust: (1) the Stopler-Samuelson theorem, which demonstrates that a tariff increases the return to factors used intensively in the import-competing industry—although Kemp has pointed out that the Stolper-Samuelson theorem may not hold in the many-commodity, many-factor case;[46] (2) the Samuelson theorem on the one-to-one correspondence between international commodity-price ratios and factor-price ratios;[47] and (3) the celebrated Rybczynski theorem: At constant commodity prices, accumulation of a factor increases the output of the commodity that uses that factor intensively and reduces the output of the other commodity.[48] Even the Heckscher-Ohlin theorem, it is shown, can be rescued for conditions of uncertainty if we define factor abundance in physical terms. However, under certain specifications of risk aversion, complete factor-price equalization is ruled out. But the weak factor-equalization theorem, i.e., the existence of a *tendency* toward factor-price equalization, remains. The volume of trade and the gains from trade, it is conclusively demonstrated, are smaller under uncertainty than under certainty.

The Ricardian model of international trade also has been reworked and this conclusion reached: Expected gains from trade for a risk-averse country which, under certainty, would wish to trade may, under postulated assumptions with price uncertainty, become negative, causing it to cease trading.[49] Several authors have formulated a trading model in which both price uncertainty and storage activities were included. They showed that with nonincreasing risk aversion but increased price uncertainty, both importers and exporters of the stored commodity tend to reduce the volume of their trade and, at the limit, would be better off not trading.[50] A framework has been developed for the analysis of trade under conditions of substantial fluctuations in prices, given alternative specifications of risk by the governments of respective trading partners.[51]

Although models based on the analysis of "economic surplus" cannot provide general conclusions on the potential gains or losses from trade, a number of papers have extended such models by comparing various restricted trade situations rather than confining the analysis to the comparison of autarkic and free trade equilibria.[52] The sources of instability, it has been noted, have different effects under different trading regimes.[53] Marketing boards engaged in foreign trade are likely to stabilize producers' prices (although in developing countries they have been notorious for destabilizing producers' incomes) whereas private firms engaged in "marketing" are more likely to destabilize producers' prices. By relaxing the linearity assumption made in previous studies, the conclusion has been reached that producers of an export good generally would be better off with price instability, while consumers and importers would be better off with price stability. In a cosmopolitan framework, however, they would jointly be better off with price stability.[54]

Theoretically, the presence of uncertainty in the foreign trade of goods and assets can, to a degree, be treated as an additional or joint "commodity," denoted as "risk." In the absence of satisfactory, or feasible, risk-sharing arrangements—reflected by the fact that this "commodity" is not internationally traded—the prices of these risk elements in production and distribution are manifestly different in different countries. As a consequence, some of the basic properties of international trade theory may not hold. Without such minimal risk-sharing arrangements, analysis of the effects of substantial uncertainty on the basic properties of trade theory (e.g., comparative advantage) has little predictive value for international specialization and the pattern of trade. Nevertheless, studies which have considered a planning model in which a social welfare function (expected utility) is maximized subject to the country's technological constraint, can be interpreted as models which include *domestic* stock markets and/or *domestic* Arrow-Debreu contingent commodity markets.[55] For effective gains-from-trade analysis under uncertainty, the incorporation into these models of international trade in firms' equities and/or other national and international risk-sharing arrangements appears to be a primary requisite.

## CONCLUDING REMARKS

The classical literature first compared the effects of free trade versus no trade; then, those of free trade versus protection. The neoclassical literature, in turn, increasingly compared the general effects of protection versus quantitative trade restrictions; then, those of various forms of trade restrictions versus autarky. By examining the precise conditions under which restricted trade can improve the welfare of society as compared to autarky, the recent literature on trade and welfare also provided the foundation for empirical research on optimizing the gains-from-trade by reducing extreme fluctuations in the terms of trade. These theoretical developments reflected both the growing interdependence of the world economy and the increasing pressures for protection—particularly in times of deep and prolonged recession—resulting from this interdependence. Unless the standard theory accords these factors their due importance, a pendular reaction to free-trade doctrine may have the effect of institutionalizing new and excessive forms of protection unwarranted by either modern gains-from-trade theory or by relevant statistical evidence.

## NOTES

1. The *locus classicus* of international trade theory before Adam Smith is Jacob Viner, *Studies in the Theory of International Trade* (New York: Harper & Bros., 1937), Ch. 1 and 2. For evidence on increasing protectionism in the latter 1970s, see International Monetary Fund, *Annual Report on Exchange Restrictions*, e.g., 1977; Jan Tinbergen, ed., *R. I. O. Reshaping the International Order* (New York: Dutton, 1976), Ch. 3; and Harold B. Malmgren, "Trade Policies of the Developed Countries for the Next Decade," in Jagdish Bhagwati, *The New International Economic Order: The North-South Debates* (Cambridge: M.I.T. Press, 1977), Ch. 8.
2. The many paradoxes in modern trade theory are discussed by Stephen P. Magee, "Twenty Paradoxes in International Trade Theory," Chapter 5, ed. Jimmye S. Hillman and Andrew Schmitz, *International Trade and Agriculture: Theory and Policy* (Boulder: Westview Press, 1979). Some of the material presented in this

paper is based on an earlier work by Chambers, Letiche, and Schmitz, *ibid.*, Chapter 4.

3. Adam Smith, *The Wealth of Nations*, eds. R. H. Campbell and A. S. Skinner (Oxford: Clarendon Press, 1976), vol. 1, p. 457.

4. David Ricardo is deservedly given credit for the first clear and reasonably comprehensive exposition of the comparative cost doctrine. See *The Works and Correspondence of David Ricardo*, ed. Piero Sraffa with M. H. Dobb (New York: Cambridge University Press, 1951), vol. 1, *On the Principles of Political Economy and Taxation* (first published in 1817), especially pp. 133-49; and Robert Torrens, *An Essay on the External Corn Trade* (London: J. Hatchard, 1815), especially pp. 264-66.

5. The most precise exposition of their views is to be found in the following original sources. John Stuart Mill, *Essays on Some Unsettled Questions of Political Economy* (London: Parker West Strand, 1844), pp. 47-74; Alfred Marshall, "The Pure Theory of Foreign Trade," in *Pure Theory (Foreign Trade-Domestic Values)* (London: London School of Economics, 1949), pp. 1-28; Francis Y. Edgeworth, "The Theory of International Values," *Economic Journal* 4 (March 1894):34-50; and F. W. Taussig, whose earlier writings are incorporated in *International Trade* (New York, 1927).

6. See Francis Y. Edgeworth, *Papers Relating to Political Economy* (London: Macmillan and Co., 1925); and Philip E. Sorensen, "Edgeworth on Monopoly, Taxation and International Trade" (Ph.D. dissertation, University of California, Berkeley, 1966), pp. 127-63.

7. The detailed evolution of these views is discussed by Viner in his *Studies in the Theory of International Trade* (Chs. 8 and 9) and *International Trade and Economic Development* (Glencoe: Free Press, 1952), Ch. 3.

8. Enrico Barone used this form of back-to-back diagram to reach conclusions with respect to gains from trade. He borrowed the diagram from H. Cunynghame, *A Geometric Political Economy* (Oxford, 1904), Figure 51, p. 98. Cf. Enrico Barone, *Grundzüge der theoretischen Nationalökonomie* (German trans. by Hans Staehle of the original Italian ed. of 1908; Bonn, 1927), Figure 30, p. 102, and Figure 32, p. 105. Barone, however, makes no reference to Cunynghame. On the formidable errors of using such Marshallian domestic-trade demand and supply curves in terms of monetary forces for gains-of-trade theory, see Viner, *Studies in the Theory of International Trade,* pp. 589-93; and John M. Letiche, *Balance of Payments and Economic Growth* (New York: Harper & Bros., 1958), pp. 71-73. See also the survey paper by John M. Currie, John A. Murphy, and Andrew Schmitz, "The Concept of Economic Surplus and its Use in Economic Analysis," *Economic Journal* 81 (December 1971):741-99.

9. See, e.g., Steven Enke, "The Monopsony Case for Tariffs," *Quarterly Journal of Economics* 58 (February 1944):229-44.

10. Wassily W. Leontief, "The Use of Indifference Curves in the Analysis of Foreign Trade," *Quarterly Journal of Economics* 48 (May 1933):493-503.

11. The terms of trade $p^o p^o$ can be obtained under the small-country assumption or derived from the standard offer curve analysis for large-country models. For an excellent exposition of offer curves, with illustrative diagrams, see James E. Meade, *A Geometry of International Trade* (London: George Allen, 1952).

12. An important assumption made by Leontief is that community indifference curves do not intersect.

13. Paul A. Samuelson, "The Gains from International Trade," *Canadian Journal of Economics and Political Science* 5 (May 1939):195-205, reprinted in *Readings in the Theory of International Trade*, eds. Howard S. Ellis and Lloyd A. Metzler (Philadelphia: Blakiston Co., 1949), pp. 239-52.

14. This means that the same ordinal preference schedule relating commodities and productive services is assumed for every individual as well as the same ownership in the means of production. It does not mean, however, that the utilities of different individuals are comparable. For as Samuelson notes, since all individuals are identical, if one is bettered (in an ordinal sense) by the introduction of trade, all will be bettered, and this makes it unnecessary to make any welfare comparisons between individuals.

15. Samuelson, pp. 245-46 (italics in original). Although Samuelson's original, rigorous proof on the gains from trade used the small-country assumption, Murray C. Kemp demonstrated that the proof holds for countries of any size in "The Gains from International Trade," *Economic Journal* 72 (December 1962):803-19.

16. Samuelson, p. 251. Samuelson suggests that the reader see "Professor Viner's interesting remarks in his *Studies in the Theory of International Trade*, pp. 532-4." An arithmetic illustration, which Viner used in his international trade course at Chicago, may be of interest to some readers: *Before trade,* with customary free competitive assumptions, France's output was: 20 units of labor produced 200 cotton; the remaining 20 units of labor produced 50 wine. Therefore 4 cotton exchanged for 1 wine. *After trade,* assuming France had a comparative advantage in wine, France could produce with its 40 units of labor 100 wine. The rate of interchange would be 1 wine = 4(+) cotton. Therefore, for 50 wine France could obtain 200(+) cotton. Hence France could always have at least 50 wine and more than the original amount of 200 cotton; or, 50(+) wine and 200 cotton.

17. Paul A. Samuelson, "The Gains from International Trade Once Again," *Economic Journal* 72 (December 1962):820-29.

18. Slope of $VV$ represents the free trade price line (i.e., the terms of trade). Profit maximization insures production at a point like $U$. Free trade allows consumption along the $VV$ frontier; the volume of trade depends upon consumer preferences.

19. This terminology was used by Francis M. Bator, "The Simple Analytics of Welfare Maximization," *American Economic Review* 47 (1957):22-59.

20. See Peter B. Kenen, "On the Geometry of Welfare Economics," *Quarterly Journal of Economics* 71 (August 1957):426-47.

21. Samuelson, "The Gains from International Trade," p. 250.

22. Kemp, "The Gains from International Trade," passim.

23. A. O. Krueger and H. Sonneschein, "The Terms of Trade, the Gains from Trade and Price Divergence," *International Economic Review* 8 (February 1967):pp. 121-27.

24. Murray C. Kemp, "Some Issues in the Analysis of the Gains from Trade," *Oxford Economic Papers* 20 (July 1968): 149-61.

25. Robert E. Baldwin, "The New Welfare Economics and Gains in International Trade," *Quarterly Journal of Economics* 66 (February 1952):91-101. A key assumption in the analysis on the gains from trade is that factors of production are internationally immobile. For example, see the work by Robert Mundell, "International Trade and Factor Mobility," *American Economic Review* 47 (June 1957): 321-35; and A. Schmitz and P. Helmberger, "Factor Mobility and International Trade: The Case of Complementarity," *American Economic Review* LX (September 1970):761-67, where this assumption is relaxed.

26. The envelope $VV$ and the free trade possibility frontier will coincide at the no-trade point. Correspondingly, the grand utility possibility frontier for the $VV$ envelope may be tangent to the trade grand utility possibility frontier.

27. On the basis of this proposition, an expression can be obtained for the optimal tariff. It is simply: $1/(\epsilon - 1)$, where $\epsilon$ is the elasticity of the foreign offer curve. For a graphic and algebraic proof, see Richard E. Caves and Ronald W. Jones, *World Trade and Payments* (Boston: Little Brown, 1973), pp. 239-40. Among

outstanding contributions on the subject of tariffs and quotas are: Wolfgang F. Stolper and Paul A. Samuelson, "Protection and Real Wages," *Review of Economic Studies* 9 (November 1941):58-73; Tibor Scitovsky, "A Reconsideration of the Theory of Tariffs," *Review of Economic Studies* 9 (Summer 1942):89-100; Lloyd A. Metzler, "Tariffs, the Terms of Trade and the Distribution of National Income," *Journal of Political Economy* 62 (February 1949):2-29, reprinted in *Readings in International Economics*, eds. Richard E. Caves and Harry G. Johnson (Homewood, Ill.: Richard D. Irwin, 1968), pp. 24-57; and A. P. Lerner, "The Symmetry between Import and Export Taxes," *Economica* 3 (August 1936):306-13, reprinted in *Readings in International Economics*, rev., pp. 197-203.

28. Harry G. Johnson, "Optimum Tariffs and Retaliation," *Review of Economic Studies* 21 (1953-54):142-53; idem, "Optimal Trade Intervention in the Presence of Domestic Distortions," in R. E. Baldwin et al., *Trade, Growth and the Balance of Payments*, Essays in Honor of Gottfried Haberler (Chicago: Rand McNally, 1965), pp. 3-34; and idem, "The Possibility of Income Losses from Increased Efficiency or Factor Accumulation in the Presence of Tariffs," *Economic Journal* 77 (March 1967):151-54. This brief paper also induced Richard Brecher and Carlos Diaz-Alejandro to show that distortionary tariff-induced capital inflows will necessarily be immiserizing if the protected industry is capital intensive; see R. Brecher and Carlos Diaz-Alejandro, "Tariffs, Foreign Capital and Immiserizing Growth," *Journal of International Economics* 7 (November 1977), 317-22. For further work on the theory of tariffs, see Harry G. Johnson, "The Cost of Protection and the Scientific Tariff," *Journal of Political Economy* 68 (August 1960):327-45; and "The Alternative Maximization Policies for Developing Country Exports of Primary Products," *Journal of Political Economy* 43 (May/June 1968): 489-93.

29. Kemp, "Gains from International Trade," pp. 803-19, and "Some Issues in the Analysis of the Gains from Trade," pp. 149-61; Jagdish Bhagwati, "The Gains from Trade Once Again," *Oxford Economic Papers* 28 (July 1960):137-48. One of the issues dealt with here is the question of ranking policies that themselves constitute impediments to the attainment of optimality: e.g., for a country *without* monopoly power in trade (but no other imperfections), the question is whether a higher tariff is worse than a lower tariff. In addition to the results given, *supra*, Bhagwati and Kemp have analyzed the problem for tariffs around the optimal tariff for a country *with* monopoly power in trade. See J. Bhagwati and M. C. Kemp, "Ranking of Tariffs under Monopoly Power in Trade," *Quarterly Journal of Economics* 83 (May 1969): The Kemp and Bhagwati-Kemp theorems dealing with these issues have deduced the proposition that reductions in the degree of *an only* distortion are successively welfare-increasing until the distortion is fully eliminated. But cases relating to this proposition require the exclusion of inferior goods and attendant multiple equilibria if the possibility of the competitive system "choosing" an inferior-welfare equilibrium under the lower degree of distortion is to be ruled out. See, J. N. Bhagwati, "The Generalized Theory of Distortions and Welfare," in *Trade, Balance of Payments and Growth*, Papers in International Economics in Honor of Charles P. Kindleberger, ed. Jagdish Bhagwati, Ronald W. Jones, Robert A. Mundell, and Jaroslav Vanek (New York: North-Holland Publishing Co., 1971), p. 84. Bhagwati demonstrates, moreover, that Kemp's theorem of the superiority of tariff-restricted trade over no trade does not extend to cases in which trade is restricted instead by policies such as consumption and production tax-cum-subsidies. In this instance two distortions are being compared: (1) a consumption tax-cum-subsidy with a situation of autarky and (2) a production tax-cum-subsidy with a situation of autarky. Unique

ranking is also impossible between autarky and restricted trade where trade restriction occurs via use of a factor tax-cum-subsidy. In these cases, a more general proposition applied: "Distortions cannot be ranked (uniquely) vis-à-vis one another" (p. 88).

30. Bhagwati, "Generalized Theory of Distortions and Welfare," p. 78. See also Gottfried Haberler, "Some Problems in the Pure Theory of International Trade," *Economic Journal* 60 (June 1950):223-40; W. M. Corden, "Tariffs, Subsidies and the Terms of Trade," *Economica* 24 (August 1957), and E. Hagen, "An Economic Justification of Protectionism," *Quarterly Journal of Economics* 72 (November 1958).

31. Although aspects of this problem were referred to in the pre-World War II literature on commercial treaties and tariff protection, the modern literature gained much impetus from the writings by C. L. Barber, W. M. Corden, G. Bascvi, B. Belassa, and J. H. Young. An extensive bibliography is to be found in Herbert G. Grubel and Harry G. Johnson, eds., *Effective Tariff Protection* (Geneva, 1971), pp. 299-305. (However, the reference on p. 305 to J. H. Young, *Canadian Commercial Policy*, is misdated; it should read 1957 rather than 1967.) See especially Clarence L. Barber, "Canadian Tariff Policy," *Canadian Journal of Economics and Political Science* 21 (November 1955):513-30; W. M. Corden, "Effective Protective Rates in a General Equilibrium Model: A Geometric Note," *Oxford Economic Papers* 21 (July 1969), *The Theory of Protectionism* (New Jersey: Oxford University Press, 1971), and *Trade Policy and Economic Welfare* (Oxford: Cloverdon Press, 1974).

32. Cf. Herbert G. Grubel "Effective Tariff Protection: A Non-specialist Introduction to the Theory, Policy Implications and Controversies," *Effective Tariff Protection*, pp. 1-15; Wilfred K. Ethier, "General Equilibrium Theory and the Concept of the Effective Rate of Protection," *Effective Tariff Protection*, pp. 17-43; and the critical analysis by V. K. Ramaswami and T. N. Srinivasan, "Tariff Structure and Resource Allocation in the Presence of Factor Subsiitution," in *Trade, Balance of Payments and Growth*, ed. Bhagwati, et al., in which the authors "conclude that substitution effects will often be significant and that, therefore, effective protective rates are likely to be a poor guide to the resource movements resulting from the levy of a tariff. . . . Domestic distortions, externalities, and noneconomic objectives are appropriately handled through taxes and subsidies on domestic economic variables. Thus in many situations the optimal policy with regard to foreign trade is nonintervention. When the levy of trade taxes is justified, the optimal tariff must be determined with reference to a specified objective: and the effective protective rates entailed by this tariff are of no relevance" (p. 298). Regarding American agriculture, empirical estimates of effects of these tariffs were made by Rachel Dardis and J. Dennisson, "The Welfare Costs of Alternative Methods of Protecting Raw Wool in the United States," *American Journal of Agricultural Economics* 51 (May 1969):303; and by Larry J. Wipf, "Tariffs, Non-tariff Distortions and Effective Protection in U.S. Agriculture," *American Journal of Agricultural Economics* 53 (August 1971):423-30.

33. Murray Kemp and Henry Y. Wan, "The Gains from Free Trade," *International Economic Review* 13 (October 1972):509-22.

34. J. M. Grandmont and Daniel McFadden, "A Technical Note on Classical Gains from Trade," *Journal of International Economics* 2 (May 1972):109-28.

35. Yoshihiko Otani, "Gains from Trade Revisited," *Journal of International Economics* 2 (May 1972):127-56.

36. John S. Chipman and James C. Moore, "Social Utility and the Gains from Trade," *Journal of International Economics* 2 (May 1972):157-72.

37. Henry Y. Wan, "A Note on Trading Gains and Externalities," *Journal of Interna-*

*tional Economics* 2 (May 1972):173-80. For an early, classical contribution to this problem, see Gottfried Haberler, "Some Problems in the Pure Theory of International Trade," *Economic Journal* 60 (June 1950):223-40.

38. See the report by Akira Takayama on Ohyama's work in *International Trade: An Approach to the Theory* (New York: Holt, Rinehart & Winston, 1972), Ch. 17.

39. Donald J. Roberts, "Continuity in the Gains from Trade with Similar Consumers," *Journal of International Economics* 4 (February 1974):25-36.

40. Pranab K. Bardhan, *Economic Growth, Development and Foreign Trade: A Study in Pure Theory* (New York: John Wiley & Sons, 1970), Ch. 4; also, H. Oniki and H. Uzawa, "Patterns of Trade and Investment in a Dynamic Model of International Trade," *Review of Economic Studies* XXXII (January 1965):15-38.

41. Ronald E. Findlay, "Factor Proportions and Comparative Advantage in the Long-Run," *Journal of Political Economy* (February 1970), republished in *International Trade and Development Theory* (1973), Ch. 7 (rev.), and Ch. 2 and 11, to which the immediate discussion refers.

42. Cf. Carlos Diaz-Alejandro, *Exchange Rate Devaluation in a Semi-Industrialized Country* (Cambridge, Mass.: M.I.T. Press, 1965) and *Essays on the Economic History of the Argentine Republic* (New Haven: Yale University Press, 1970), pp. 362-90 and Ian Little, Tibor Scitovsky, and Maurice Scott, *Industry and Trade in Some Developing Countries* (New York: Oxford University Press, 1970), especially pp. 67-69; see, also, I. A. McDougall and R. H. Snape, eds., *Studies in International Economics* (Amsterdam: North-Holland Publishing Co., 1970); Jagdish Bhagwati, *Anatomy and Consequences of Exchange Control Regimes*, Vol. 11 of the National Bureau of Economic Research, Inc., *Foreign Trade Regimes and Economic Development* (Cambridge, Massachusetts: Ballinger Publishing Company, 1978), Chs. 2-7; Ann O. Krueger, *Liberalization Attempts and Consequences*, Vol. 10, *ibid.*, Chs. 10-12; and Bent Hansen and Jagdish Bhagwati, "Should Growth Rates be Evaluated at International Prices?" eds. J. H. Baghwati and R. Eckaus, *Development and Planning*, essays in honor of P. Rosenstein-Rodan (London: George Allen & Unwin, 1973), pp. 53-68.

43. R. J. Ruffin, "International Trade under Uncertainty," *Journal of International Economics* 4 (August 1974):243-59.

44. Raveendra N. Batra and William R. Russell, "Gains from Trade under Uncertainty," *American Economic Review* 64 (December 1974):1040-48.

45. Raveendra N. Batra, *The Pure Theory of International Trade under Uncertainty* (New York: John Wiley & Sons, 1975), especially Chs. 3-7.

46. Murray C. Kemp, *Three Topics in the Theory of International Trade: Distribution, Welfare, and Uncertainty* (Amsterdam: North Holland Publishing Co., 1976). Chs. 4 and 5 analyze an "n x m" model, which includes intermediate goods. Elaborating on the essence of the Stolper-Samuelson theorem, viz., that an increase in any commodity price must lead to a decrease in the price of *at least one* other factor, Kemp shows that this result holds for the more realistic conditions examined. However, Chs. 1, 2, and 3 examine whether a small increase in any product price, all other product prices remaining the same, is associated with an *unambiguous* increase in the real reward of the factor used relatively intensively in this sector and with an unambiguous decline in the real reward of any other factor or in the real reward of the factor used relatively unintensively in this sector. Kemp shows that this requires setting extreme restrictions on the technology in the "n x n" model which makes it very much like the "2 x 2" model. Cf. Assaf Razin's careful review in *Journal of Political Economy* 86 (February 1978):162-65.

47. Paul A. Samuelson, "International Trade and the Equalization of Factor Prices," *Economic Journal* 58 (June 1948):163-84. "I learn from Professor Lionel Robbins," Samuelson has written, "that A. P. Lerner, while a student at L.S.E.,

dealt with this problem. I have had a chance to look over Lerner's mimeographed report, dated December 1933, and it is a masterly, definitive treatment of the question, difficulties and all." Paul A. Samuelson, "International Factor-Price Equalization Once Again," *Economic Journal* 59 (June 1949):181.

48. T. M. Rybczynski, "Factor Endowments and Relative Commodity Prices, *Economica* (November 1955):336-41, republished in *Readings in International Economics* (1968), pp. 72-77. The figure I has been redrawn to meet a criticism of E. J. Mishan, "Factor Endowment and Relative Commodity Prices: A Comment," *Economica* 23 (November 1956):352-59.

49. Stephen J. Turnovsky, "Technological and Price Uncertainty in a Ricardian Model of International Trade." *Review of Economic Studies* 61 (April 1974): 201-17. This conclusion, it will be noted, is different from that derived by Ruffin, under somewhat different assumptions.

50. Gershon Feder, Richard E. Just, and Andrew Schmitz, "Storage with Price Uncertainty in International Trade," *International Economic Review* 18 (October 1977):553-68.

51. James Anderson and J. Riley, "The International Trade with Fluctuating Prices," *International Economic Review* 17 (February 1976):76-97.

52. See especially Kemp, *Three Topics in the Theory of International Trade,* Part 2; and Jagdish Bhagwati, *The Theory and Practice of Commercial Policy: Departures From Unified Exchange Rates* (International Finance Section, Princeton University, 1968), pp. 11-66, and "The Generalized Theory of Distortions and Welfare," pp. 78ff.

53. C. F. Darrell Hueth and Andrew Schmitz, "International Trade in Intermediate and Final Goods: Some Welfare Implications," *Quarterly Journal of Economics* 86 (August 1972):351-65; Jurg Bieri and Andrew Schmitz, "Market Intermediaries and Price Instability," *American Journal of Agricultural Economics* 56 (May 1974):280-85. The work on the welfare aspects of market intermediaries in trade has recently been extended by R. Just, A. Schmitz, and D. Zilberman, "Price Controls and Optimal Export Policies Under Alternative Market Structures," *American Economic Review* (forthcoming).

54. Richard E. Just, Ernst Lutz, Andrew Schmitz, and Stephen J. Turnovsky, "The Distribution of Welfare Gains from International Price Stabilization under Distortions," *American Journal of Agricultural Economics* 59 (November 1977): 652-61.

55. See K. J. Arrow, "Le rôle des valeurs boursières pour la répartition la meilleure des risques," *Econométric* (Centre National de la Recherche Scientifique, Paris, 1953), pp. 41-48; idem, *Aspects of a Theory of Risk-Bearing* (Helsinki: Yrjo Jahnsson Lectures, 1965); Gérard Debreu, "Une économie de l'incertain," (Electricité de France, 1953); idem, *Theory of Value* (New Haven: Yale University Press, 1959), Ch. 7; E. Helpman and A. Razin, "Uncertainty and International Trade in the Presence of Stock Markets," *Review of Economic Studies* (forthcoming); and *idem,* "Welfare Aspects of International Trade in Goods and Securities," *Quarterly Journal of Economics* XCII (August 1978):489-508.

# Part III

## World Food, International Trade, and Agriculture

Schultz
Johnson

# THE ALLOCATIVE EFFICIENCY
# OF TRADITIONAL AGRICULTURE

Theodore W. Schultz

The economic acumen of people in poor agricultural communities is generally maligned. It is widely held that they save and invest too little of their income in view of what capital earns, that they pay no heed to changes in prices, and that they disregard normal economic incentives at every turn. For 'these and other reasons, it is frequently said they do badly in using the factors they have. But is this true? The aim of this chapter is to examine the efficiency with which farmers within traditional agriculture allocate the factors at their disposal.

## THE ECONOMIC EFFICIENCY HYPOTHESIS

There is, as has already been noted, a large class of poor agricultural communities in which people have been doing the same things for generations. Changes in products and factors have not crowded in on them. For them neither consumption nor production is studded with new gadgets.

Reprinted from "Transforming Traditional Agriculture," pp. 36–52, with permission of Yale University Press, New Haven, Copyright 1964.

The factors of production on which they depend are known through long experience and are in this sense "traditional." While the communities in this class differ appreciably one from another in the quantity of factors they possess, in what they grow, in the arts of cultivation, and culturally, they have one fundamental attribute in common: they have for years not experienced any significant alterations in the state of the arts. This means simply that farmers of this class continue year after year to cultivate the same type of land, sow the same crops, use the same techniques of production, and bring the same skills to bear in agricultural production. To examine the allocative behavior of these farmers, the following hypothesis is proposed:

> *There are comparatively few significant inefficiencies in the allocation of the factors of production in traditional agriculture.*

The factors of production under these circumstances consist of traditional factors, and the hypothesis is restricted to those factors at the disposal of the people of a particular community. It should be made clear that not all poor agricultural communities have the economic attributes of traditional agriculture. Some are excluded on the ground that they have been subject to change. Any community that has experienced a significant alteration to which it has not had time to adjust fully is excluded. When a new road or railroad is built, as a rule it takes some years for the communities affected to adapt to it. The economic routine of the affected communities is also disturbed by a new large dam, irrigation canals, structures to control floods and to reduce soil erosion. A serious adversity of nature—a flood or a drought followed by famine—can be a source of disequilibrium. Some poor agricultural communities must be excluded because they have been subject to large political changes, for example by partition, by recruitment of many men into the armed services, or by the destruction of both human and nonhuman resources by war. Large changes in relative prices of products because of outside developments affecting the terms of trade can also upset the quiet economic life of particular communities. In modern times, the most pervasive force disturbing the equilibrium of agricultural communities is the advance in knowledge useful in agricultural production. Any poor agricultural community that is adjusting its production to one or more of these circumstances is excluded from traditional agriculture to which the *efficient but poor hypothesis* applies. The fact that particular communities are excluded because they are making major adjustments in production does not imply that they are inherently inefficient in making the adjustments. The test in that case, however, is different.

Whether one wishes to test or to examine the implications of the proposed hypothesis, it will be necessary to distinguish between an efficient allocation of the stock of factors devoted to current production and an optimum rate of investment to increase the stock of such factors. It will be convenient at this stage, in working with this hypothesis, to assume that the rate of return to investment is given and, whether the rate is low or high, that the total stock of factors can be increased only a little per year. Accordingly, the rate of return can be either low or high, or, if one prefers, the price of additional income streams can be either dear or cheap. The hypothesis at this point pertains only to the allocation of the existing factors in current agricultural production, with the prevailing rate of return to investment given. The question of investment will be considered later.

It may be helpful to mention a few of the implications of

the hypothesis. The principal implication is of course that no appreciable increase in agricultural production is to be had by reallocating the factors at the disposal of farmers who are bound by traditional agriculture. It follows, therefore, that the combination of crops grown, the number of times and depth of cultivation, the time of planting, watering, and harvesting, the combination of hand tools, ditches to carry water to the fields, draft animals and simple equipment—are all made with a fine regard for marginal costs and returns. Implied also is that significant indivisibilities will not show their ugly heads. Product and factor prices will reveal themselves as flexible. Another implication is that an outside expert, however skilled he may be in farm management, will not discover any major inefficiency in the allocation of factors. To the extent that any of these implications are contrary to the observable and relevant facts, the hypothesis here proposed would be under a cloud of doubt.

Mindful of what an outside expert can usefully do that goes beyond a reallocation of existing factors, it must be underscored that in testing this hypothesis it is not permissible to alter the technical properties of the factors of production at the disposal of the community. Nor is it permissible to provide new useful knowledge about superior factors that exist in other communities, that is, provide such knowledge at a cost that would be less than it was formerly. Doing so would alter the costs and the return to the search for information pertaining to alternative economic opportunities. Obviously, the introduction of better varieties of seeds and other technically superior inputs by the expert is precluded in making this test. If the outside expert were successful in these respects, he would alter the established equilibrium that may otherwise have characterized the economic activities of the community being investigated.

# Transforming Traditional Agriculture

Still another implication of this hypothesis is that no productive factor remains unemployed. Each parcel of land is used that can make a net contribution to production, given the existing state of the arts and other available factors. So are irrigation ditches, draft animals, and other reproducible forms of capital. Also, each laborer who wishes and who is capable of doing some useful work is employed. It is of course possible to conceive of exotic technical conditions in agriculture that preclude "full" employment. Workers conceivably could become so numerous as to be in each other's way. There could be indivisibilities in factors of production. But these seem to be paper tigers, for they are not found in this class of agricultural communities. The recent doctrine that agricultural production activities are often such that capable workers contribute nothing to production at the margin—that is, that a part of the agricultural labor force has a marginal productivity of zero value—will be examined in the following chapter. The efficient but poor hypothesis does not imply that the real earnings (production) of labor are not meager. Earnings less than subsistence are not inconsistent with this hypothesis provided there are other sources of income, whether from other factors belonging to workers or from transfers within the family or among families in the community.

In turning to the real world to test the hypothesis here advanced, the main difficulty is the paucity of usable data. The propensity to take any estimates, however weak they may be, and force them into a Cobb-Douglas type of production function, is as a rule a sheer waste of time. Fortunately, some social anthropologists studying particular communities of this type for extended periods have diligently recorded product and factor prices, costs and returns of the major economic activities, and the institutional framework in which

# Allocative Efficiency

production, consumption, savings, and investment occur. Two of these studies—one pertaining to a Guatemalan Indian community and another to an agricultural community in India—are especially useful and relevant. These two studies will now be examined in relation to the proposed hypothesis.

## PANAJACHEL, GUATEMALA: VERY POOR BUT EFFICIENT

A classic study by Sol Tax, *Penny Capitalism,*[1] opens with these words: it is "a society which is 'capitalist' on a microscopic scale. There are no machines, no factories, no co-ops or corporations. Every man is his own firm and works ruggedly for himself. Money there is, in small denominations; trade there is, with what men carry on their backs; free entrepreneurs, the impersonal market place, competition—these are in the rural economy." Tax leaves no doubt that this community is very poor, that it is under strong competitive behavior, and that its 800 people are making the most of the factors and techniques of production at their command.[2]

No one ought to be surprised that the people are very poor. Tax puts their poverty this way: they "live without medical aid or drugs, in dirt-floored huts with hardly any furniture, the light only of the fire that smokes up the room, or of a pitch-pine torch or a little tin kerosene lamp; the

1. Originally published by the Smithsonian Institution, Institute of Social Anthropology, Publication No. 16 (Washington, U.S. Government Printing Office, 1953). Reprinted by the University of Chicago Press, 1963.

2. This study is keyed to data covering the period from 1936 to 1941. Professor Tax lived in the community on and off from the autumn of 1935 to the spring of 1941 (see his preface).

# Transforming Traditional Agriculture

mortality rate is high; the diet is meager and most people cannot afford more than a half-pound of meat a week. . . . Schools are almost nonexistent; the children cannot be spared from work in the field. . . . Life is mostly hard work."[3] Tax presents many data measuring the consumer goods and the level and cost of living to support this poignant testimony on the poverty of the community.

Competition is present everywhere in the way products and factors are priced. "All household utensils—pottery, grinding stones, baskets, gourds, china, and so on—and practically all household furnishings such as tables and chairs and mats, must be brought in from other towns. So must many articles of wearing apparel, such as material for skirts and cloaks, hats, sandals, blankets, and carrying bags, as well as cotton and thread for weaving the other things. So must most of the essential foodstuffs: the greater part of the corn, all lime, salt and spices, most of the chile, and most of the meat. . . . To get the money they depend upon the sale of agricultural produce . . . . Onions and garlic, a number of fruits, and coffee are the chief commodities produced for sale."[4] Prices are in every respect highly flexible.

Tax goes on to document the fact that the Indian is "above all else an entrepreneur, a business man," always looking for new means of turning a penny. He buys the goods he can afford with a close regard for price in various markets, he calculates with care the value of his labor in producing crops for sale or for home consumption against his working for hire, and he acts accordingly. He rents and pawns parcels of land with a shrewd eye to the return, and he does likewise in acquiring the few producer goods that he buys from others. All of this business, "may be characterized as *a money econ-*

3. Tax, p. 28.
4. Ibid., pp. 11–12.

*omy organized in single households as both consumption and production units, with a strongly developed market which tends to be perfectly competitive."*[5]

The economy has been geared to a stable, virtually stationary, routine pattern. Not that the Indian is not always looking for new ways to improve his lot. Tax notes that "he is on the lookout for new and better seeds, fertilizer, ways of planting." But such improvements come along infrequently, and their effects upon production are exceedingly small. There was a growing demand by "foreigners" for some shore land along Lake Atitlan but this development was having very little effect upon the land Indians used for producing crops and on which they built their huts. Some buses and trucks had become available for transport to more distant towns and they were being used to go to and from markets in these towns because it was "cheaper" than walking and carrying the goods. There were more tourists in and about the lake but these too were having little or no discernible influence on the community.[6]

All the evidence revealed in the careful documentation of the behavior of the people in *Penny Capitalism* and in the many tables showing prices, costs, and returns strongly supports the inference that the people are remarkably efficient in allocating the factors at their disposal in current production. There are no significant indivisibilities in methods of production, none in factors, and none in products. There is no disguised unemployment, no underemployment of either men, women, or children old enough to work, and

5. Ibid., p. 13. Italics are from Tax.
6. Professor Tax had occasion to revisit this Indian community in Guatemala after a lapse of 20 years. The overwhelming impression of the brief visit was that life and the economy had remained virtually unchanged.

for the least of them there is no such thing as a zero marginal product. Because even very young children can contribute something of value by working in the field, they cannot be spared the time to go to school. Product and factor prices are flexible. People respond to profit. For them every penny counts.

SENAPUR, INDIA: POOR BUT EFFICIENT

A study by W. David Hopper, "The Economic Organization of a Village in North Central India,"[7] portrays an economy in another part of the world performing as if it too were highly efficient in using the factors at hand. Hopper, like Tax, entered upon his study of this village as an anthropologist. Like Tax, after having lived in the community for a period and having observed its cultural, social, and other characteristics, he decided to concentrate heavily on the economy of the village.

For students of anthropology, there are undoubtedly important cultural and social differences between Senapur, India, and Panajachel, Guatemala.[8] There are also some

7. An unpublished Ph.D. thesis presented at Cornell University, June 1957. The village of Senapur is located on the Ganges Plain. At the time of the study, it comprised 1,046 acres and had a population of about 2,100. Hopper resided in Senapur from October 1953 to February 1955.

8. Senapur, for instance, has a long-established caste system, whereas the Guatemalan community has singular flexibility in the movement of families up and down its social status scale. In Senapur the families of the privileged castes, mainly the Thakur, have perpetuated their wealth, privileges, and social status for many generations. In the Guatemalan community, Tax found marriages cutting across wealth lines and much mobility on the economic and social scale; moreover, the social and economic gap separating the top and the other families has not been substantial. Thus even the varying winds of fortune make for much mobility from one generation to the next as one sees these families in *Penny Capitalism*.

differences in the level of production and consumption. Senapur is not as poor as Panajachel, but by Western standards it is nevertheless poor. Senapur has a school with grades 1 to 5 which until very recently served mainly the more privileged castes. The number of "productive" animals is unbelievably large: 270 milch cows and buffaloes; 480 bullocks to work in the fields. The stock of capital includes irrigation wells, ditches, storage ponds, digging tools, plows, chaff-cutters, and some small equipment. There is more specialization in Senapur than in the Guatemalan community: well-diggers, potters, carpenters, brick-makers, a blacksmith, and others. But for all that Senapur is poor.

Hopper examines with care the factors of production over which the people of Senapur have command. There is a fine set of natural resources characteristic of that part of India, and there is a substantial set of reproducible resources both within this community and outside that also serves its production and consumption activities. He then traces the behavior of the competitive forces as these are revealed through the established product and factor markets.

Hopper summarizes an important part of this study thusly: "An observer in Senapur cannot help but be impressed with the way the village uses its physical resources. The age-old techniques have been refined and sharpened by countless years of experience, and each generation seems to have had its experimenters who added a bit here and changed a practice there, and thus improved the community lore. Rotations, tillage and cultivation practices, seed rates, irrigation techniques, and the ability of the blacksmith and potter to work under handicaps of little power and inferior materials, all attest to a cultural heritage that is richly endowed with empirical wisdom." Hopper then puts this question to himself: "Are the people of Senapur realizing the full economic po-

tential of their physical resources? . . . From the point of
view of the villagers the answer must be 'Yes' for in general
each man comes close to doing the best that he can with his
knowledge and cultural background."[9]

Fortunately, the data that Hopper collected have permitted him to make a rigorous test of the allocative hypothesis under consideration.[10] He made such a test by determining the set of relative prices of products and factors
implicit in the allocation decisions revealed in the data. In
determining the allocative efficiency of the farmers from the
prices implicit in their production activities, Hopper used
the price of barley as the numeraire. The implicit prices in
terms of barley for each product estimated from factor allocations, with barley at 1.00, are wheat, 1.325; pea, .943;
and gram, .828. The implicit price estimates for each factor
based on its production use at the average product prices
and their standard errors are as follows.[11]

| | Barley | Wheat | Pea | Gram | Average |
|---|---|---|---|---|---|
| Average Price | 1.00 | 1.325 | .943 | .828 | |
| Price of: | Used in Production of: | | | | |
| Land (acres) | 4.416 (1.056) | 4.029 (.855) | 4.405 (1.185) | 4.845 (.857) | 4.424 |
| Bullock time (hours) | .0696 (.0116) | .0716 (.0098) | .0820 (.0180) | .0834 (.0156) | .0774 |
| Labour (hours) | .0086 (.0026) | .0097 (.0037) | .0087 (.0021) | .0076 (.0030) | .0086 |
| Irrigation Water (750 gals.) | .0355 (.0122) | .0326 (.0078) | .0305 (.0111) | .0315 (.0234) | .0325 |

9. Hopper, p. 161.
10. W. David Hopper, "Resource Allocation on a Sample of Indian
Farms," University of Chicago, Office of Agricultural Economics Research, Paper No. 6104 (April 21, 1961, mimeo.).
11. W. David Hopper, "Allocation Efficiency in Traditional Indian
Agriculture," *Journal of Farm Economics* (forthcoming).

# Allocative Efficiency

From these data and his test, Hopper infers that "there is a remarkably close correspondence between the various price estimates. It would appear that the average allocations made by the sample of farms were efficient within the context of the prevailing technical relationships. There is no evidence that an improvement in economic output could be obtained by altering the present allocations as long as the village relies on traditional resources and technology."

The implicit prices also match closely the market prices of products and factors for which there were market prices. These prices follow:

| Product or factor | Relative barley price | Adjusted to the barley price (in rupees) | Actual market price (in rupees) |
|---|---|---|---|
| Barley (md.) | 1.00 | 9.85 | 9.85 |
| Wheat (md.) | 1.325 | 13.05 | 14.20 |
| Pea (md.) | .943 | 9.29 | 10.40 |
| Gram (md.) | .828 | 8.16 | 10.85 |
| Land (acres) | 4.424 | 43.57 | 8.00 to 30.00 (cash rent only) |
| Bullock time (hrs.) | .0774 | .762 | not available |
| Labor (hrs.) | .0086 | .085 | .068 (cash and kind only) |
| Irrigation water (750 gals.) | .0325 | .321 | not available |

The implicit prices of these products, except for gram, match closely the actual market prices. Hopper observes that in the case of gram there is a lagged response under way to a strengthening market for gram; the relative price of gram had been rising for three years prior to the date of Hopper's study. The findings in this important study show that there is a "close approximation between market and implicit prices." The factors of production available to the people of Senapur were allocated efficiently, and the test therefore strongly supports the hypothesis here proposed.

### INFERENCES AND IMPLICATIONS

The data pertaining to the allocation of factors for current production in Panajachel and Senapur are consistent with the hypothesis proposed at the outset of this chapter. It is important to note, however, in drawing the inference that there are no significant inefficiencies in the allocation of factors in these two communities, that the concept of factors includes more than land, labor, and capital as these are commonly defined. It also includes the state of the arts, or the techniques of production, that are an integral part of the material capital, skills, and technical knowledge of a people. In other words, factors are not treated by abstracting from the state of the arts. By this all-inclusive concept of factors, the community is poor because the factors on which the economy is dependent are not capable of producing more under existing circumstances. Conversely, under these simplified conditions, the observed poverty is not a consequence of any significant inefficiencies in factor allocation.

Although it is not feasible to show that these two communities are typical of a large class of poor agricultural communities, the assumption that they are seems highly plausible. Moreover, this plausibility is supported by the fact that the hypothesis under consideration appears to be consistent with a wide array of other empirical studies of such communities. The well-known studies of the farm economy of China by

Buck[12] lend support as do the many examples cited by Bauer and Yamey.[13] A comprehensive examination of all such data is, however, beyond the scope of this study.

The economic premises on which the hypothesis rests, and the support it receives empirically, warrant treating it as a proposition likely to be widely useful. As such, it has a number of implications, some already mentioned.

What does illiteracy imply? The fact that people are illiterate does not mean that they are therefore insensitive to the standards set by marginal costs and returns in allocating the factors they have at their disposal. What it does indicate is that the human agent has fewer capabilities than he would have if he had acquired the skills and useful knowledge associated with schooling. Although schooling may increase greatly the productivity of the human agent, it is not a prerequisite to an efficient allocation of the existing stock of factors. The notion that these poor agricultural communities do not have enough competent entrepreneurs to do a satisfactory job in using the factors at hand is in all probability mistaken. In some cases these entrepreneurs may be subject to political or social restraints that give rise to allocative inefficiencies, but the adverse production effects of such restraints are quite another matter.

There is another inference that is contrary to a widely held view, namely that farmers in these communities do not respond to developments that alter the stock of factors at their disposal. This view holds that the farmers do not adjust to changes in relative prices of products and factors.

12. John Lossing Buck, *Chinese Farm Economy* (Chicago, University of Chicago Press, 1930).

13. Peter T. Bauer and Basil S. Yamey, *The Economics of Under-Developed Countries,* a Cambridge Economic Handbook (Chicago, University of Chicago Press, 1957), Ch. VI.

If this is true, it is inconceivable that the community could ever become essentially efficient in factor allocation, except by sheer accident. Both Hopper and Tax, however, are explicit in noting that these farmers do respond. The question may be formulated thusly: if an irrigation canal is constructed or a new and better variety of a particular crop becomes available, do they respond? A pioneering study by Raj Krishna[14] of the supply responses of farmers in the Punjab during the twenties and thirties indicates that the lag in adjustment in producing cotton was about the same as it has been for cotton farmers in the United States. Quite aside from the rate at which they adjust to alterations in economic conditions, however, the important fact at this juncture of the analysis is that they do respond. It therefore follows that whenever such a community has for decades been living a quiet, routine economic life it has long since achieved an essentially efficient allocation of factors at its disposal.

There is one set of estimates based on cross-sectional Cobb-Douglas type production functions which includes six classes of farms in India that appear to show extraordinary inefficiencies in factor allocations. Heady includes these six Indian sets in a list of 32 that covers locations in various parts of the world.[15] The six sets covering farms in India are based presumably on data for the middle 1950s. In these the marginal returns to labor range from .03 to 1.78 for each

14. Raj Krishna, "Farm Supply Response in the Punjab (India-Pakistan): A Case Study of Cotton," (unpublished Ph.D. dissertation, University of Chicago, 1961).

15. Earl O. Heady, "Techniques of Production, Size of Productive Units, and Factor Supply Conditions," Paper presented at the Social Science Research Council Conference on Relations between Agriculture and Economic Growth, Stanford University, Stanford, California, November 11–12, 1960.

1.00 (unit) of labor costs.[16] For land, the range of the marginal returns to costs is even wider, with the lowest at .05 and the highest at 3.60 for 1.00 (unit) of land rental. The most extreme results, however, are those reported for reproducible material capital. For these the marginal returns range from –.85 to 6.97 per 1.00 (unit) of capital costs.

Although Heady mentions the possible limitations that qualify the usefulness of these estimates,[17] they are nevertheless treated as if they could be taken seriously. Hopper's careful examination of the data problem in Senapur, India, makes it abundantly clear that such "monthly wage rates" and "rental returns" to land are most inaccurate. In the case of capital, Heady reveals that the "selection of interest rate for capital is itself a problem,"[18] because it ranges "from 6 to 200 percent." It is no wonder that the results of working with such data are so meaningless. If these agricultural communities in India had been experiencing rapid economic development and were therefore confronted by large changes in factor and product prices to which they had not as yet had time enough to adjust, there would be a logical basis for some inequalities in marginal returns relative to the costs of factors. But no such major developments had taken place in India at that time. It is noteworthy that no logical explanation of the extreme ranges in the estimates cited for

16. In the case of labor, no estimate is shown for wheat farming in Uttar Pradesh, India, undoubtedly the most absurd of the lot, for it is this set that shows a marginal return to land of 2.22 and to capital of 6.97 for each unit (1.00) of input costs.

17. Among these possible limitations, Heady lists "(a) specification bias, (b) aggregation, (c) algebraic form, (d) sampling, and (e) other facets of statistical inference. However, we believe that the data do, even though they represent a small stratum of national agriculture, provide some qualitative types comparisons." P. 35.

18. Heady, "Techniques of Production," p. 35.

the six sets of farming in India is offered. Had one been attempted, the untenable nature of the results would have become apparent.

Still another implication stemming from the proposition that a large class of poor agricultural communities shows comparatively few significant inefficiencies in factor allocation is that competent farm managers, whether national or foreign, cannot show the farmers how to allocate better the existing factors of production. Once again it must be stressed that this implication holds provided these competent experts are restricted in advising farmers to the existing factors, which means that they do not alter the opportunity to increase production by introducing other factors, including knowledge about the availability of such other factors.

Lastly, then, there is the implication that no part of the labor force working in agriculture in these communities has a marginal productivity of zero. But since this particular implication runs counter to a well-established doctrine, the next chapter is devoted to an examination of the basis of this doctrine and the reasons why it is a misleading conception of the economics of labor productivity in poor agricultural communities.

# THE ECONOMICS OF THE VALUE OF HUMAN TIME[1]

THEODORE W. SCHULTZ

The difference in the economic value of human time between low- and high-income countries is very large At the time when the foundations of classical economics were established, however, the value of human time throughout Western Europe was exceedingly low. In view of economic changes since then, are corresponding improvements possible for low-income countries? While it is all too convenient to believe that it can be accomplished by law supported by rhetoric, it is clearly not possible to achieve this objective on command by government. Increases in the earnings of labor depend basically on achieving increases in the value productivity of labor. Investment in population quality is one of the important means of doing so. The economic dynamics are, however, exceedingly complex, as is evident in accounting for the increases in the value of human time that have occurred over time in high-income countries.

In the United States, for instance, real earnings per hour of work have risen fivefold since 1900. The upward trends in real wages in industry in France, Germany, Sweden, and the United Kingdom are much like that in the United States, with some notable differences. France and the United Kingdom show no increase between 1900 and 1910, and, as of 1925, the increases show Sweden and the United States substantially ahead of the other three countries (was this a consequence mainly of differences in the effects of the war and its aftermath). Sweden and the United States maintain their advantage over the others up to 1960, with the United Kingdom losing ground in relative terms. Finally, at the end of the 1960s, France and Germany join Sweden and the United States in showing a fourfold and more relative increase in real annual wages in industry over the period from 1900 to 1970, whereas the relative increase for the United Kingdom is slightly less than threefold.

The trend of the deflated natural resource commodity prices over this period was slightly downward, compared to the more than fivefold rise in real hourly wages. In agriculture, the deflated prices of crops declined about one-third, despite various government price supports during parts of this period; the index for livestock closed at the level at which it began. In general, the costs of producing livestock products have been affected more by the increase in the price of human time than have the costs of producing crops. The deflated prices of mineral fuels indicate that whereas the deflated price index for all mineral fuels was about one-fourth less at the end of this period compared to 1900, the price of bituminous coal rose and that of petroleum fell. It is undoubtedly true that the rise in real wages accounts for a good deal of the increase in coal prices.

---

[1]This is a condensation entirely in the words of the original; all changes were made with author approval.

Reprinted (with author approved deletions) from "Investing in People: The Economics of Population Quality," pp. 59–84, with permission of California University Press, Berkeley, Copyright 1981.

Without a useful theory, there can be no satisfactory analysis of the determinants that account for the changes in relative prices. Since "growth" implies changes over time, the theory that is required could be referred to as a theory of economic growth. But it is fair to say that, as yet, there is no growth theory that is sufficiently comprehensive in specifying the factors and events that determine the changes in relative prices and stocks of resources that occur as a consequence of observable economic behavior, and that are in turn consistent with that behavior.

A simple supply-and-demand approach helps to clarify matters. It is the intercept of the supply of and the demand for human time that reveals the price we observe. Shifts in the supply and demand schedules then account for the recorded increases in this price over time. The key to this pricing problem is in the factors that determine such shifts. We know a good deal about the factors that increase the *supply,* both in terms of the size of the labor force and of the quality attributes of the workers. But this is, at best, a partial picture of the price changes that occur. The nub of the unresolved problem is that we know very little about the factors that shift the demand upward over time so strongly.

In devising an approach to get at the factors that explain the shifts in these two schedules, an all-inclusive concept of capital formation is necessary. In using this concept, it is essential to see the heterogeneity of the various old and new forms of capital and to specify them in sufficient detail to determine not only the substitutions but also the interacting complementarity between these forms of capital. Inasmuch as captial formation entails investment, it is important not to conceal the changes over time in incentives—that is, the anticipated rates of return to be had from alternative investment.

Changes in investment opportunities, events, and human behavior alter the scale of value and the composition of the stock of capital. Alterations that enhance the scope of choices are favorable developments. The various forms of capital differ significantly in their attributes. Natural resources are not reproducible, but structures, equipment, and inventories of commodities and goods are. Human beings are productive agents with the attributes of human capital, and they are also the optimizing agents. In a fundamental sense, human preferences determine what use is made of the various forms of capital. It is noteworthy that in high-income countries the rate at which human capital increases exceeds that of nonhuman capital.

To specify the heterogeneity of capital, it is not sufficient to classify the capital forms as natural resources, reproducible material forms, and human capital, because of the important role that new forms of capital within each of these classes play in altering relative prices (returns) and in shifting supply-and-demand schedules. It will be necessary to make room in this approach to growth and changes in relative prices for the following propositions:

1. The Ricardian principle that an increasing share of national income accrues to land rent (natural resources) needs to be replaced by the proposition that this share tends to decline as a consequence of man-made substitutes for land. A notable example is the creation of hybrid corn, which may be viewed either as a substitute or

as a new input augmenting the yield from land. Plastics and aluminum become substitutes for various metals and wood, and nuclear energy becomes a substitute for fossil fuels. The economics of producing such substitutes (research and development) is still in its infancy, and the prospective output of this sector is subject to the same uncertainty as are other advances in useful knowledge.

2. Some new forms of capital complement other forms of capital in production. A consequence of such complementarity is that particular new forms of material capital increase the demand for particular human skills (a subclass of human capital). These complementary forms of capital need to be identified and included in the analytical model.

3. Making room in economic growth models for changes in relative prices over time is a return to the approach of early classical economics. Since modern macro-growth models tend to take prices as given (usually fixed), the inclusion of relative prices and their function is a radical analytical proposition. Be that as it may, relative prices, which include the alternative rates of return on investment, are the mainspring that drives the economic system. If this mainspring did not exist, we would have to invent it by appealing to shadow prices.

The shifts in demand in favor of productive services of labor that contribute to increases in the price of human time are, in large part, a consequence of the complementarity proposed in the second of these propositions. But the state of the art of economics does not as yet permit us to identify and determine the effects of this complementarity on the demand for labor.

The price and income effects of increases in the value of human time include enlargement of institutional protection of the rights of workers, favoring human capital relative to property rights; increases in the value added by labor, relative to that added by materials in production; a decline in hours worked; increases in labor's share of national income; a decline in fertility; and the high rate at which human capital increases. The human agent becomes ever more a capitalist by virtue of his personal human capital, and he seeks political support to protect the value of that capital. The rise in the value of human time makes new demands on institutions. Some political and legal institutions are especially subject to these demands. What we observe is that these institutions respond in many ways. The legal rights of labor are enlarged and in the process some of of the rights of property are curtailed. The legal rights of tenants are also enhanced. Seniority and safety at work receive increasing protection. The history of national income by type indicates clearly that large changes have occurred over time that parallel, and are associated with, a rise in the real earnings of workers.

By 1970, about ¾ of the official United States national income by type consisted of employee compensation. The remaining one-fourth is classified as proprietors' income, rental income, net interest, and corporate profits. These four classes of "property" income include considerable earnings that accrue to human agents for the productive time they devote to self-employed work and to the management of their property assets. A conservative estimate of the aggregate contribution of human agents in 1970, measured by employee compensation, plus self-employment earnings and management of assets within the domain of the market sector, was fully $\frac{4}{5}$ of the value of the production accounted for in national income.

The price and income effects of hourly earnings explain a wide array of changes in the allocation of time. When expected future earnings from more education rise, the response of youth is to postpone work-for-pay in order to devote more years to education. The advantage of youth in acquiring additional education is twofold: the wages foregone are lower than they would be later, and there are more years ahead to cash in on the anticipated higher earnings and satisfactions. As wages increase, people who earn their income by working can afford to retire at an earlier age because of the larger retirement income that they are able to accumulate. This is counterbalanced by the improvement in health that is purchased, which extends the years that individuals may opt to work. The rise in the value of the time of women is an incentive to substitute various forms of physical capital in household production, and, inasmuch as children are labor-intensive for women, the demand for children is reduced, and an increasing part of women's time is allocated to the labor market.

Economic theory has in recent years been extended to explain the accumulation of human capital, and the price and income effects of this form of capital. The theory has led to important new approaches in bringing economics to bear on human behavior in both developed and developing countries. Throughout most of the world, labor still earns a pittance. In a few countries, however, the value of the time of working people is exceedingly high. The high price of human time that characterizes these exceptional countries is, from the viewpoint of economic history, a recent development. In these countries, the increases in real wages and salaries represent gains in economic welfare that are the most significant achievement of their economic growth. Much less time is allocated to work-for-pay. Most of the work is no longer hard physically. Ever more skills are demanded, and the supply response of skills is strong and clear. But the increases in demand are still concealed in the complementarity between the various new forms of capital.

The historical fact is that, despite the vast accumulation of capital, the real rate of return on investment has not diminished over time. There has been much aimless wandering in analyzing growth that could have been avoided had the perceptions of Alfred Marshall been heeded.

> Capital consists in a great part of knowledge and organization: knowledge is the most powerful engine of production. . . . The distinction between public and private property in knowledge and organization is of great and growing importance: in some respects of more importance than that between public and private property in material things.[1]

Public and private investment in human capital and in useful knowledge are a large part of the story in accounting for the increases in the value of human time.

[1]Alfred Marshall, *Principles of Economics*, 8th ed. (New York: Macmillan, 1920), bk. 4, pp. 138–39.

# THE WORLD FOOD SITUATION: DEVELOPMENTS DURING THE 1970s AND PROSPECTS FOR THE 1980s[1]

D. GALE JOHNSON

## INTRODUCTION

The lead sentence in an Associated Press story of November 11, 1979, was "the Third World is moving toward a massive food shortage that could result in 'economic disaster' within 20 years, a United Nations report says."[2] This gloomy projection was a reporter's interpretation of a conference report by the director-general of the Food and Agriculture Organization (FAO) of the United Nations. The report expressed the fear that developing countries would require ever-increasing grain imports and that the cost of these imports would become a burden that many low-income countries could not bear.

Has the world food situation deteriorated during the 1970s? Have the prices of basic foods increased over time? Are the majority of the world's poor eating less and less as time goes by? If the dire predictions concerning the world's food situation, which were so common in 1973 through 1975, had proven to be valid, the answer to each of these questions would be in the affirmative.

But such is not the case. The world food situation did not deteriorate during the 1970s. The prices of basic foodstuffs are low by historical standards. There is no evidence that the world's poor people are eating less well now than they did a decade ago. In fact, on each point the evidence is to the contrary. The 1970s, as did each of the two prior decades, saw an improvement in per capita food supply for the world and in the low-income countries.

In *World Food Problems and Prospects,* published in early 1975, I concluded, "I am cautiously optimistic that the food supply situation of the developing countries will continue to improve over the coming decades. If I had as much confidence in

---

[1]This is a condensation entirely in the words of the original; all deletions were made with the author's approval.

[2]*Sunday Sun-Times* (Chicago), November 11, 1979. My reading of this United Nations report entitled *Agriculture: Toward 2000* leaves me with a quite different impression than the newspaper article gives. First, the report makes no projections of future events; instead it presents what is likely to occur if recent trends persist and what could be achieved if substantial increases were made in investments in agricultural inputs and research and if appropriate incentives were provided for farmers. To quote from the introduction: "in the study no attempt has been made to forecast or to predict what is likely to happen by the end of the century." Second, there is very little emphasis given to energy. Of course, it is possible that discussion of the report at the conference did emphasize the bleakest possible picture of the future. *Agriculture: Toward 2000* was prepared by the Food and Agriculture Organization (FAO) of the United Nations and presented at the twentieth FAO Conference held in Rome, November 10–29, 1979.

Reprinted from "Contemporary Economic Problems 1980," pp. 301–339, with permission of American Enterprise Institute for Public Policy Research 1980, Washington, D.C., Copyright 1980.

the political process in both the industrial and developing countries as I do in the farmers of the world, I would drop the qualification *cautiously*.''[3] For the 1970s the cautious optimism was justified. Per capita food supplies in the developing countries improved. The improvement, however, was modest, and it was far from uniform. In fact, in Africa, per capita food production declined during the decade, but there were substantial differences in growth of food production within Africa that cannot be attributed primarily to natural conditions.

## FOOD SUPPLY AND DEMAND DURING THE 1970s

In the mid-1970s the predominant view was that demand was likely to grow more rapidly than supply and that the real prices of food would quite probably increase rather than decrease from the high levels of 1973 and 1974. A number of reasons were given for the expectation that the food supply situation would deteriorate after the early 1970s. These included the assumption that higher energy prices would greatly increase the price of farm inputs, especially fertilizer; that expanding the cultivated area was no longer a possibility as a significant means of increasing food production; and that the growing use of grain as feed would reduce the food supplies of low-income countries.

It is appropriate to look at what actually happened to per capita food supplies during the 1970s and, especially, after 1972. For all low-income countries (except centrally planned) per capita food production in 1972 was at the 1961–1965 level, a decline from the 1970 index of 106 (1961–1965 = 100). The recovery from the low 1972 level was rapid, and by 1975 the index of per capita food production stood at 107. Further progress occurred through 1978, by which time the index had increased to 110. Preliminary indications, however, indicate a sharp reduction to 106 for 1979. The average index for the last 4 years of the 1970s was 108.2 compared with 101.2 for the same period a decade earlier. The annual growth in per capita food production for the period 1966–1969 to 1976–1979 was approximately 0.7%. This may seem low, but it was higher than the growth rate for the 1960s.

Viewed in terms of the performance of agriculture, the growth of food production during the 1970s in the low-income countries reflects a remarkable performance. For the decade, food production grew at 3.2% annually, compared to 2.0% for the high-income countries. The modest rise in the per capita food production figure was due to the rapid growth in population.

### Imports and Food Supply

The consumption of food in a given year is generally not the same as the amount of food produced. The supply actually consumed consists of food produced, changes in stocks, and net trade. We have inadequate data on changes in stocks for the low-income countries. If we are interested, however, in average food consumption for a number of years, changes in stocks have only a limited effect. The major

---

[3]D. Gale Johnson, *World Food Problems and Prospects* (Washington, D.C.: American Enterprise Institute, 1975), p. 81.

source of a difference between food production and food consumption is net trade.

As a group, the low-income countries are net importers of grain. During the 1970s their grain imports increased by 25 million tons to 48 million tons net in 1979–1980. The increase in grain imports would provide about 40 calories per day for more than 2 billion people. On the basis of an average daily per capita intake of 2000 calories, the increase in grain imports added 2% to the available calories. This is not to imply that the increase in grain imports made a major contribution to the improvement of food supplies in the low-income countries but rather to make it clear that the increase in per capita food supplies was equal to or greater than the increase in per capita food production during the 1970s.

Increased imports of food are often assumed to be a necessity caused by failure of agricultural production to increase at an appropriate pace. An FAO study, for example, recently noted that the developing countries increased their grain imports by more than 50% between 1975 and 1979, and stated that this increase portends both increasing dependence upon high-income countries (especially the United States) and increasing difficulties in paying for a rising level of food imports.[4] But for many developing countries, increasing food imports may be an efficient use of their resources and a response to rising real incomes—not a signal of imminent or eventual disaster. Associated with the notion that significant food imports are evidence of agricultural failure is the view that most, if not all, developing countries should be self-sufficient in food.[5]

The fact that the vast majority of developing countries are net exporters of agricultural products is often lost sight of in the discussion of food self-sufficiency. In 1977 the developing market economies had an excess of $17 billion of agricultural exports over agricultural imports; in terms of percentage, the excess of exports over imports was almost 60%. If the oil exporters are excluded, the value of agricultural exports was nearly double the value of agricultural imports in 1977 for the developing market economies.[6] The best course for some nations may be to produce crops for export and to import a significant fraction of their food.

It is interesting to note that although concern has been expressed about the

---

[4]In a discussion of recent trends in world trade in agricultural products in the United Nations report, *Agriculture: Toward 2000*, the following was stated: "every developing region . . . shared in both the downward share of world agricultural exports and the rising share of world agricultural imports. . . . A major feature of this deterioration—from the point of view of the developing countries—of trade in agricultural products was the stubborn upward trend in their cereals imports" (p. 12). It was noted that the degree of cereal self-sufficiency declined from 96% in 1963 to 92% in 1975.

[5]*Agriculture: Toward 2000* is somewhat ambivalent about the merits of self-sufficiency in food and export expansion. For example, "indeed, the pursuit of the objective of greater self-sufficiency in food is not necessarily inconsistent with a drive toward commodity development aimed at enhancing export earnings. Import substitution for food saves foreign exchange and production for foreign exchange earns foreign exchange. With limited land resources and with the need for foreign exchange earnings—for promoting industrialization and other development efforts—the most profitable use of agricultural resources would depend upon the relative gains from import substitution and export promotion. The individual countries would need to select the right mix of foodstuff production and export crop production, taking into account the relative costs and returns" (p. 189).

[6]*FAO Commodity Review and Outlook: 1977–79* (Rome: FAO, 1979), p. 11.

increase in grain and other food imports by the developing countries, little attention has been given to the even larger increase in the value of agricultural exports by the same group of countries. Although food imports of the non-oil-producing developing countries increased at an annual rate of $1.4 billion between 1970 and 1977, the value of their agricultural exports increased by $2.9 billion per year.[6]

**Grain Prices**

Another measure of the supplies of food available to the low-income countries is the price of grain in international markets. If demand had been growing more rapidly than supply, as many predicted would occur during the latter half of the 1970s, the real or deflated prices of grain would have increased. But the United States export prices of wheat and corn, in 1967 dollars per ton for the period from 1910 to date, reveal a long-term downward trend for the real prices of both products. In fact, prices during the late 1970s have been below those that prevailed from 1930–1934, during the Great Depression, and significantly below the prices of the late 1930s. Prices in recent years are also lower than during the late 1950s and very near to the low prices of the late 1960s.

These price data indicate that for the grains the supply to the international market has increased somewhat more than demand during the last seven decades. In fact, compared with the late 1920s, it is an understatement to say that supply has increased at a "somewhat" faster rate than demand. The real prices of both wheat and corn have fallen by approximately one-third in a half century. During this time output has grown substantially as, of course, has demand. It may be noted that what were considered to be the very high grain prices in 1974 were less than one-tenth higher than average prices for 1925–1929 and below the prices that prevailed after World War II. Within 3 years after 1974, the real export price for wheat fell by almost half, and the corn price fell by one-third. Clearly the world market for grains was not under strong demand pressure during the late 1970s. The prices prevailed even though the total world trade in grains doubled during the 1970s as did the volume of grain imports by the low-income market economies.

**MALNUTRITION—HOW MUCH?**

I have argued that the availability of food in the low-income countries improved during the 1970s, as it had during the previous two decades. I have confidence that the 1980s will see further improvement. However, the extent of malnutrition or hunger that exists in the world has not been considered. Unfortunately our knowledge of the number of percentage of people inadequately fed, either in the past or today, is limited.

Estimates for recent years of the number of people who are malnourished vary enormously. The Presidential Commission on World Hunger, in its preliminary report issued in December 1979, presented no new estimates of the extent of hunger and malnutrition. It relied upon previous estimates made by FAO and the World Bank, yet one finds in the opening paragraph of the preface the following:

Widespread hunger is a cruel fact of our time. In 1974, after poor harvests and oil price increases disrupted the international food system, the World Food Conference called on all governments to accept the goal that in ten years' time no child would go to bed hungry, no family would fear for its next day's bread, and no human being's future and capacities would be stunted by malnutrition. *Today, however, the world is even farther from that goal than it was then.* While the good harvests of recent years have prevented widespread famine, the next world food crisis will find the world not much better prepared than it was in 1974. This need not be the case [Emphasis added].[7]

It is true that in the resolutions of the World Food Conference there was a call for the elimination of hunger within a decade. The elimination of hunger, as defined in the quotation, is an impossibility; it is impossible for the world to ensure that there will always be enough food for everyone all of the time. Nonetheless, it is a worthy goal to move toward a reduction of hunger and malnutrition. The sentence in the preceding quotation shown in italics is an astounding statement to have been produced by a commission appointed by the President of the United States. It is stated as a fact that the world is farther from the goal of reducing hunger and malnutrition than it was in 1974. Not the slightest bit of evidence is presented to back up the statement. It is not even clear what the sentence actually means. Is it that half of the 10 years has passed and there has been little improvement? Or is it that conditions are now actually worse, with more children going to bed hungry in 1979 than in 1974 and with more families fearing for their next day's bread? Whatever may be intended, most readers will interpret the sentence to say that in 1979 the food situation was even more serious for the world's poor people than it was in 1974 with its poor harvests.

It is not my intention to engage in a battle of numbers or to claim that the world's poorer poeple do not face serious problems in obtaining adequate amounts of nutritious food.[8] Instead my objective is to try to measure recent progress or change in per capita food availability and to assess the prospects for improvements in the years ahead. Nor, in emphasizing food production and food prices and costs, am I denying that the primary reason for inadequate nutrition is poverty. In the lower-income countries of the world, limited availability of food for purchase is not the reason poor people have inadequate diets. The reason for inadequate diets is insufficient income to buy enough food. But I hasten to add that the majority, perhaps as many as $\frac{3}{4}$, of the poor people in the low-income countries live in rural areas. Thus, for many of these people there are intimate relationships among income, food, and agricultural production. Thus, while the primary source of food inadequacy is low income, an important factor in the low incomes for hundreds of millions is the limited productivity of agriculture. In this situation, many of the measures that

[7]Presidential Commission on World Hunger, *Preliminary Report of the Presidential Commission on World Hunger,* Washington, D.C., 1979, preface.

[8]In a section headed "Problems of Nutrition" in *Agriculture: Toward 2000,* the obvious truth is stated: "This is an area of relative ignorance, with limited clinical and anthropometrical data" (p. 147). Yet estimates of the number of malnourished continue to be made.

result in increased productivity in agriculture will also increase incomes and the adequacy of diets.

## PRICE INSTABILITY

Recent grain export prices are low by comparison with the past. The data for the 1970s, however, indicate a high degree of price instability for the decade. Although there have been some positive changes affecting world food supply and demand during the 1970s, there has been little change with respect to the potential for price instability for grains in international markets and in countries where prices are permitted to vary with international market prices. Was the much greater grain price instability of the 1970s compared with the 1960s due to greater variability in grain production? The evidence clearly supports the conclusion that the source of the increased price instability was not nature, but man.

Much of the increase in grain prices from 1972 to 1974, and much of the subsequent decline, was due to governmental policies rather than to variations in production.[9] The policies responsible for a major fraction of the upward movement of prices (and the subsequent reductions) were those designed to achieve domestic price stability through varying net international trade. A large fraction of the world's consumption of wheat and other grains occurs in countries that divorce their domestic prices from international prices. In other words, numerous countries solve their own problems of instability by imposing that instability on the rest of the world by varying net trade.

The policies of the European Community, the Soviet Union, the Eastern European countries, China, and Japan did not change during the 1970s. Each still maintains a high level of domestic price stability by imposing its internal instability on the rest of the world. The most notorious example is the Soviet Union, which has highly variable grain production but has attempted to stabilize domestic availability of grain by varying net imports. The Soviet Union follows a policy of stable prices for livestock products; a production shortfall does not result in higher retail prices in the state stores, but rather in longer queues at the retail markets. The policy decision to maintain retail meat prices constant in current rubles puts a high premium on providing the feed supplies required to keep meat production growing at a reasonably constant rate. To meet this objective when grain production declines requires grain imports.

The shortfalls in Soviet grain production are sometimes greater than can be made up by imports. There is some evidence that part of the shortfall not met by imports is met by stock changes. Since the size of grain stocks is regarded as a state secret, changes in grain stocks can only be inferred. For example, the 1975 grain crop was 55 million tons below the 1974 crop (all references to tons are in metric tons); in 1975–1976 Soviet net grain imports were 25 million tons larger than during the prior year; net grain imports in 1979–1980 in the absence of the United States

[9]Johnson, *World Food Problems and Prospects,* op. cit. p. 33–34.

suspension of grain exports would have been approximately 34 million tons as an offset to a production decline of 58 million tons.[10] To some degree the amount of grain imported is influenced by the capacities of the Soviet ports and transport facilities to move the grain away from the ports. It is probable that the intended 1979–1980 level of imports was at or near that capacity, and this may also have been true in 1975–1976.

Since countries that consume well over half of the world's grain stabilize internal grain prices, sharp increases in international market prices such as occurred in 1973 and 1974, cannot be ruled out. In fact, it is probable that a similar pattern of price increases could occur during the 1980s. A rather modest decline in world grain production below trend levels could create the potential for similar price changes. Until or unless domestic price policies are changed in many countries, the countries that permit their domestic prices to vary with international market prices will be subject to significant price instability.

## FOOD SECURITY, POVERTY, AND MALNUTRITION

As already noted, the primary reason for inadequate food consumption and malnutrition in the low-income countries is poverty rather than any lack of food availability. Barring war and civil insurrection, food is available almost anywhere in the world today if there is income to purchase it. Such was not always the case. It has only been in the past two centuries that famine was not a potential danger facing the majority of the world's population. Less than a century ago, famine was a threat to life for millions in most of Asia and Russia.

The twentieth century has brought a revolutionary change in world food security. Improvements in communication and transportation have brought all but a tiny minority of the world's population into a world food community. The world's food supply is now available to virtually anyone anywhere in the world who has the money to pay for food, except when governments block such purchases. The purchase price now approaches the lowest at which the primary food products (grains) have been available during the past century. The statement that food is now available to anyone who has the money to pay for it is not a cynical description of the alternatives available to the poor people of the world. It is a statement that realistically applies to a large and increasing fraction of the world's population, including the majority of the poorest people.

Unfortunately, the great deal of attention given to food problems during the mid-1970s has resulted in almost no improvement in the factual base for improving our understanding of malnutrition, its causes, and its extent. True, we know that most malnutrition is associated with low incomes, but we also know that some very poor people are adequately fed while others with equal or higher incomes have less than adequate diets. If we better understood why such differences exist in consump-

[10]U.S. Department of Agriculture, Foreign Agricultural Service, *Foreign Agriculture Circular, Grains,* FG-4-80, January 15, 1980.

tion patterns and what health effects, if any, such differences induce, we would be better able to assist those in the greatest need.

## PROSPECTS FOR THE 1980s

Per capita food production in the world and in the developing countries increased during the 1970s. The improvement in the developing countries was modest overall and in some areas, particularly Africa, per capita food production declined. But the increase achieved in per capita food production during the 1970s represented a continuation of past trends.

There is no significant reason to expect that the 1980s will show less improvement in the per capita food production of the low-income countries than each of the last two decades. The potential for greater improvement clearly exists, though a cautious and realistic view is that during the 1980s, as during the previous two decades, the realization will fall short of the potential. However, there are some positive factors that may result in a higher rate of growth of per capita food production and consumption in the low-income countries during the 1980s than during the 1970s.

Three reasons have been emphasized to support a pessimistic view of the outcome of the race between world demand and supply of food: higher energy prices, the near disappearance of uncultivated arable land, and the competition between livestock and people for grain. Of these three issues, only higher energy prices merits discussion. There remains much land that can be brought under cultivation.[11] The presumed competition for grain between livestock and people has always been a false issue. It remains a false issue because the amount of grain produced in the world is as much a function of demand as of supply. If less grain were demanded, less grain would be produced. It is as simple as that. And because it is so simple, this point is frequently misunderstood or neglected.

### Energy

Energy prices will continue to increase in real terms, and the real prices of farm inputs using significant amounts of energy will increase with the higher energy prices. However, the experience of the 1970s shows there is no simple, one-to-one relationship between the price of energy and the price of an input, such as fertilizer, that has a high energy component. Whereas the index of fertilizer prices paid by United States farmers has increased less since 1970 than the index of prices for all farm production items, the index of prices of fuels and oil has increased by about 35% more than all production items.[12] The share of fertilizer, oil, and fuel expenditures in total farm production expenses increased between 1970 and 1978, but probably by less than most think. Such expenditures on energy products accounted

[11]*Agriculture: Toward 2000* indicates that arable land in the low-income countries (excluding China) could increase from 730 million ha in 1975 to 830 million ha in 1990 and to 930 million ha by 2000 (p. 30).

[12]U.S. Department of Agriculture, Economics, Statistics, and Cooperatives Service, Crop Reporting Board, *Agricultural Prices,* various issues.

for 9.1% of all production expenditures in 1970 and 11.1% in 1978. Preliminary estimates indicate that approximately 11.7% of all 1979 farm production expenses were devoted to these energy products.[13]

I do not intend to minimize the importance of energy prices in affecting the costs of producing agricultural products in the United States and other high-income countries. Food and other agricultural products would be priced lower today and the output of food would be somewhat greater if energy prices were at their pre-OPEC real levels. But we should not exaggerate the effect of energy prices on agriculture nor should we assume that farmers can make no adjustment to the higher real prices.

One additional point should be made about energy prices. Energy derived from fossil fuels is a much smaller component of farm production costs in low-income countries than in the United States or Western Europe.[14] Consequently, increased energy prices have had less effect on agriculture in the poor countries than in the United States. This difference will persist in the years ahead. And we have seen that the effect of higher energy costs on United States agriculture has been quite small so far. One indication of this is that real grain prices are now approximately the same as they were a decade ago, when energy prices were much lower.

## The Green Revolution

For many, a look ahead at what the 1980s holds for food supplies in the low-income countries is influenced by their answer to this question: "What happened to the Green Revolution?" The question often implies that the benefits of research leading to the new grain varieties and the cultivation practices that together were called the *Green Revolution* were less than had been anticipated. One reason the question is asked was the unfortunate designation of the new high-yielding varieties of rice and wheat as miracle varieties. The use of the term *Green Revolution* also held out the hope of major, indeed revolutionary, changes in food supplies. The development of new varieties was a remarkable achievement, but it was unfortunate that they were tagged as revolutionary. One lesson must be learned about research and agriculture—a single revolution is never enough. What is required is a stream of revolutions or changes, each perhaps with a rather modest effect but with significant cumulative influence.

For several reasons the new high-yielding varieties did not have the effects on output expected by their most enthusiastic supporters. First, the new varieties were well-adapted to only part of the areas producing rice and wheat; second, the yield differentials between the new and old varieties were significantly smaller on the farm than under experimental conditions; and, third, farmers had to learn how to use

[13]U.S. Department of Agriculture, Economics, Statistics, and Cooperatives Service, *Farm Income Statistics*, Statistical Bulletin No. 627, October 1979, pp. 42, 46. These data do not include expenditures on electricity. The 1979 estimate is made by the author.

[14]It is estimated that in the low-income countries, agriculture produced approximately 24% of the total output of the economies but used only 3.1% of all fossil energy consumed in 1975 (*Agriculture: Toward 2000*, p. 233). In the high-income countries, agriculture consumes approximately the same amount of fossil energy per unit of output as the economy as a whole.

the new varieties effectively. Specifically in the case of the new varieties, they could be used only under irrigation that permitted effective control of water depths. Many of the irrigation systems in South and Southeast Asia do not provide effective control of water levels. On the other hand, since most wheat irrigation is done with tube wells in South Asia, there has been much wider adoption of the new varieties of wheat than of rice. In India in 1976–1977, 35% of all rice was sown with the new varieties, whereas more than 70% of the wheat area was sown with new varieties. The doubling of wheat yields and the trebling of wheat output in India would have been impossible without the new wheat varieties.

## Hybrid Corn

When hybrid corn was first introduced in the 1930s, the general methods of cultivating and producing corn had been unchanged for several decades. Little or no fertilizer was used on corn. In this setting the yield advantage of hybrid corn over the traditional (open pollinated) varieties was approximately 15%. The average corn yield in the United States prior to the introduction of hybrids in years of average weather was 1.5 tons/ha (25 bushels/acre). Thus hybrid corn would have increased average yield by about 0.22 tons/ha (less than 4 bushels/acre). In the Corn Belt, where yields were higher than the national average and hybrids were first introduced, the absolute yield advantage of the hybrids was of the order of 0.3 tons/ha (5 bushels/acre). Corn yields are now 6.2 tons/ha (100 bushels/acre) with average weather. Hybrid corn has been an important contributor to the quadrupling of yield over the past four or five decades. But there were many other changes that made the recent yields possible—fertilizer, herbicides, insecticides, more plants per unit of land. Increasingly, corn hybrids are adapted to the climatic and soil and moisture conditions of quite small areas, and some part of the yield increases in the last decade or two are a result of such increased adaptation of hybrid seeds.

## Grain Yield Trends

The 1970s saw repeated references to the possibility, if not the probability, that the growth rate of grain yields was declining in the world. The feeling of unease was buttressed for many by United States grain yields, especially corn, that were below prior peak levels in 1975, 1976, and, to a lesser degree, in 1977. Thus, some concluded that grain yields in the United States were slowing down and perhaps had reached their peak.

However, research undertaken at the University of Illinois has shown that the tapering off of corn yield growth after 1974, at least in Illinois, was due to climatic factors. After account was taken of the climatic effects, there was no evidence of a slowing down of the growth in yields of corn or soybeans. In 1978 and 1979, weather variables influencing corn production were more favorable than in 1975 and 1976, and United States corn yields increased substantially, reaching an average yield of 109 bushels per acre (6.7 ton/ha) in 1979. This was between 9% and 10% in excess of the trend level of corn yields for that year.

Prior to 1975 most developing countries held down the price of at least one important food crop and protected farm input-producing industries, such as fertilizer. Although there remain numerous examples of exploitation of farmers for the presumed benefit of consumers or input-producing sectors, many developing countries now have support and purchase prices for major farm crops that are in excess of world market prices. Agricultural exports are still often burdened by an overvalued currency, as in Argentina and, until very recently, in Brazil. The probable improvement in the structure of incentives for farmers in the low-income countries will have a positive impact on the growth of agricultural output.[15]

## Agricultural Research

During the past two decades there has been a major growth in the expenditures for agricultural research on the problems of increasing farm production in the low-income countries. In the 15 years between 1959 and 1974, agricultural research in the major developing regions (Latin America, Africa, and Asia) increased from $228 million to $957 million (in 1971 constant United States dollars). This increase occurred both in the international agricultural research centers and in the research institutions of the low-income countries. As a share of an increasing world expenditure on agricultural research, the expenditures in the three low-income regions increased from 17% in 1959 to 25% in 1974.[16] Since there is a lag of 5–10 years between research investment and actual application on farms, most of the effects of the increased investments made after 1973 have still not been felt.

## Irrigation

Earlier it was noted that during the past decade there has been a substantial increase in the irrigated area in low-income countries. Much of the increase in irrigation has occurred in the densely populated regions of the world, principally Asia. Current rates of investment in irrigation in South and Southeast Asia are high by historical standards. Irrigation, whether it comes about through new projects or improvement of existing systems, is capital-intensive. This is an investment area where the availability of foreign capital may be important in improving per capita food availability. A report of the Trilateral Commission indicated that it would require more than $50 billion (in 1975 purchasing power) of investment to provide the irrigation expansion and improvements in South and Southeast Asia required for a doubling of rice production by the early 1990s.[17] Approximately one-third of the required investment could come from sources within the region.

---

[15]For a discussion of the effects of incentives and of distortions of those incentives upon agricultural development, see Theodore W. Schultz (ed.), *Distortions of Agricultural Incentives* (Bloomington: Indiana University Press, 1978).

[16]Robert Evenson, "The Organization of Research to Improve Crops and Animals in Low-Income Countries," ibid., p. 224. The reader interested in further exploration of the potential contributions of agricultural research to improving food production and nutrition may wish to consult the following: Steering Committee of the National Academy of Sciences, *World Food and Nutrition Study: The Potential Contributions of Research* (Washington, D.C.: National Academy of Sciences, 1977). See also, Sterling Wortman and Ralph W. Cummings, Jr., *To Feed This World: The Challenge and the Strategy* (Baltimore: Johns Hopkins University Press, 1978). At the end of each chapter there is an excellent classified bibliography.

[17]Umberto Colombo, D. Gale Johnson, and Toshio Shishido, *Reducing Malnutrition in Developing Countries: Increasing Rice Production in South and Southeast Asia* (New York: The Trilateral Commission, 1978).

## Income Growth

Two other factors will have a positive effect on per capita food consumption by the end of the 1980s. One of these is the continued growth of per capita income in the developing countries. The World Bank projects the annual growth of per capita gross domestic product of the low-income (per capita income of less than $300) developing countries during the 1980s at 2.7%.[18] The middle-income developing economies are projected to have a significantly higher growth rate of 3.4%. The population (as of 1976) of the low-income developing countries was 1193 million and of the middle-income countries, 1037 million. An income growth rate of 2.7% annually means a 30.5% growth in income level in a decade. Such a growth of income would, on the average, result in at least a 15% increase per capita in real food expenditures for low-income persons. If this growth in demand for food is reasonably well distributed among all segments of the population, the growth will go a substantial distance toward reducing both severe and moderate malnutrition in the low-income countries.

## Slower Population Growth

The other factor that will have a positive effect on per capita food consumption is the likely decline in the rate of population growth in many low-income countries. The decline in the rate of population growth will not significantly alleviate the world's food deficiencies during the 1980s. A slowing down of population growth has two effects on per capita food supply—a larger fraction of the population is in the working ages during the transition period to a lower rate of population growth, and there are fewer people eating from a given food supply after some period of time. These effects will not be large during most of the 1980s, though they will be of modest significance. The difference in a decade between 2.3% and 2.1% annual growth rates is a little less than 2% for the total population. The two growth rates are those for India for the 1960s and from 1970 to 1977. If there is a further decline in the population growth rate for the 1980s to 1.9%, a further 2% lower total population would result by 1990. These differences appear to be small, but compared with annual per capita increases in food consumption of 0.5% or 0.6%, the effect on per capita food consumption of a 2% difference in population size would increase the growth of consumption by one-third. It is by small and additive measures that the nutrition of the world's poorest people will be improved during the rest of this century.

## CONCLUDING COMMENTS

It is common in the discussion of the potential growth of food production to emphasize the restraints imposed by nature—the finite limits of our natural resources and the vagaries of climate. It is easy to paint a gloomy picture of the future of the world, if you accept the view that our supplies of fossil energy will be exhausted in some finite period, that erosion will destroy a large part of the land

---

[18]*World Development Report, 1979* (Washington, D.C.: The World Bank, August 1979), p. 13. The population data and growth estimates exclude China.

used for crops, and that future scientific discoveries increasingly will be less relevant to increasing food production.

But if there is an enemy that will prevent improving the nutrition of the world's poor people, it is man and not nature. Man, not nature, will be the primary factor in determining the rate of growth of food production in the rest of this century. Nature is often niggardly and at times terribly cruel. But we now know enough about what is required to expand food production not to blame our failures upon nature.

When I say that man, not nature, will impose the greatest barriers to the potential expansion of food production, I am not referring to the consequences of the acts of farmers, traders, processors, or suppliers, but to man as he functions in the political process. If policies are adopted that give appropriate incentives for farmers, provide them access to supplies at reasonable prices, and permit farmers to sell at the best possible prices, production growth will increase. And if policies are followed that encourage research on problems of the agriculture of the low-income countries, all will gain. The mistakes of the past have been especially costly for the world's poor; we can only hope that fewer mistakes will be made in the next decade.

# Part IV

## Common Markets
### Developed Countries

Viner
Lipsey
Wallace

### Developing Countries

Mikesell

# THE ECONOMICS OF CUSTOMS UNIONS

JACOB VINER

## CUSTOMS UNION AS AN APPROACH TO FREE TRADE

The literature on customs unions in general, whether written by economists or noneconomists, by free-traders or protectionists, is almost universally favorable to them, and only here and there is a skeptical note to be encountered, usually by an economist with free-trade tendencies. It is a strange phenomenon that unites free-traders and protectionists in the field of commercial policy, and its strangeness suggests that there is something peculiar in the apparent economics of customs unions. The customs union problem is entangled in the whole free-trade–protection issue, and it has never yet been properly disentangled.

● ● ●

A customs union is more likely to operate in the free-trade direction, whether appraisal is in terms of its consequence for the customs union area alone or for the world as a whole:

1. the larger the economic area of the customs union and therefore the greater the potential scope for internal division of labor;

2. the lower the "average" tariff level on imports from outside the customs union area as compared to what that level would be in the absence of customs union;

3. the greater the correspondence in kind of products of the range of high-cost industries as between the different parts of the customs union that were protected by tariffs in both of the member countries before customs union was established, that is, *less* the degree of complementarity—or the *greater* the degree of rivalry—of the member countries with respect to *protected* industries, prior to customs union;[1]

4. the greater the differences in unit costs for protected industries of the same kind as between the different parts of the customs union, and therefore the greater the economies to be derived from free trade with respect to these industries within the customs union area.

5. the higher the tariff levels in potential export markets outside the customs union area with respect to commodities in whose production the member countries of the customs union would have a comparative advantage under free trade, and therefore the less the injury resulting from reducing the degree of specialization in production as between the customs union area and the outside world;

6. the greater the range of protected industries for which an enlargement of the

---

[1] In the literature on customs union, it is almost invariably taken for granted that rivalry is a disadvantage and complementarity is an advantage in the formation of customs unions. See *infra*, pp. 73 ff., with reference to the Benelux and Franco-Italian projects.

Reprinted from "The Customs Union Issue," (taken from) pp. 41–56, with permission of Carnegie Endowment for International Peace, New York, Copyright 1950.

market would result in unit costs lower than those at which the commodities concerned could be imported from outside the customs union area;

7. the smaller the range of protected industries for which an enlargement of the market would not result in unit costs lower than those at which the commodities concerned could be imported from outside the customs union area but which would nevertheless expand under customs union.

Confident judgment as to what the overall balance between these conflicting considerations would be, it should be obvious, cannot be made for customs unions in general and in the abstract, but must be confined to particular projects and be based on economic survey thorough enough to justify reasonably reliable estimates as to weights to be given in the particular circumstances to the respective elements in the problem. Customs unions are, from the free-trade point of view, neither necessarily good nor necessarily bad; the circumstances discussed in the preceding are the determining factors.

● ● ●

## CUSTOMS UNION AND THE "TERMS OF TRADE"

There is a possibility, so far not mentioned, of economic benefit from a tariff to the tariff-levying country, which countries may be able to exploit more effectively combined in customs union than if they operated as separate tariff areas. This benefit to the customs area, however, carries with it a corresponding injury to the outside world. A tariff does not merely divert consumption from imported to domestically produced commodities—this is, from the free-trade point of view, the economic disadvantage of a tariff for the tariff-levying country and one of its disadvantages for the rest of the world—but it also alters in favor of the tariff-levying country the rate at which its exports exchange for the imports that survive the tariff, or its "terms of trade," and within limits— which may be narrow and which can never be determined accurately—an improvement in the national "terms of trade" carries with it an increase in the national total benefit from trade. The greater the economic area of the tariff-levying unit, the greater is likely to be, other things being equal, the improvement in its terms of trade with the outside world resulting from its tariff.[2] A customs union, by increasing the extent of the territory that operates under a single tariff, thus tends to increase the efficacy of the tariff in improving the terms of trade of that area vis-à-vis the rest of the world.

---

[2]The greater the economic area of the tariff unit, other things equal, the greater is likely to be the elasticity of its "reciprocal demand" for outside products and the less is likely to be the elasticity of the "reciprocal demand" of the outside world for its products, and consequently the greater the possibility of improvement in its terms of trade through unilateral manipulation of its tariff.

# THE THEORY OF CUSTOMS UNIONS:
## A GENERAL SURVEY [1]

### Richard G. Lipsey

THIS paper is devoted mainly to a survey of the development of customs-union theory from Viner to date; since, however, the theory must be meant at least as an aid in interpreting real-world data, some space is devoted to a summary of empirical evidence relating to the gains from European Economic Union. It is necessary first to define customs-union theory. In general, the tariff system of any country may discriminate between commodities and/or between countries. Commodity discrimination occurs when different rates of duty are levied on different commodities, while country discrimination occurs when the same commodity is subject to different rates of duty, the rate varying according to the country of origin. The theory of customs unions may be defined as that branch of tariff theory which deals with the effects of geographically discriminatory changes in trade barriers.

Next we must turn our attention to the scope of the existing theory. The theory has been confined mainly to a study of the effects of customs unions on welfare rather than, for example, on the level of economic activity, the balance of payments or the rate of inflation. These welfare gains and losses, which are the subject of the theory, may arise from a number of different sources: (1) the specialisation of production according to comparative advantage which is the basis of the classical case for the gains from trade; (2) economies of scale; [2] (3) changes in the terms of trade; (4) forced changes in efficiency due to increased foreign competition; and (5) a change in the rate of economic growth. The theory of customs unions has been almost completely confined to an investigation of (1) above, with some slight attention to (2) and (3), (5) not being dealt with at all, while (4) is ruled out of traditional theory by the assumption (often contradicted by the facts) that production is carried out by processes which are technically efficient.

Throughout the development of the theory of customs unions we will find an oscillation between the belief that it is possible to produce a general conclusion of the sort: " Customs unions will always, or nearly always, raise welfare," and the belief that, depending on the particular circumstances present, a customs union may have any imaginable effect on welfare. The

---

[1] An earlier version of this paper was read before the Conference of the Association of University Teachers of Economics at Southampton, January 1959. I am indebted for comments and suggestions to G. C. Archibald, K. Klappholz and Professor L. Robbins.

[2] Points (1) and (2) are clearly related, for the existence of (1) is a *necessary* condition for (2), but they are more conveniently treated as separate points, since (1) is not a *sufficient* condition for the existence of (2).

earliest customs-union theory was largely embodied in the oral tradition, for it hardly seemed worthwhile to state it explicitly, and was an example of an attempt to produce the former sort of conclusion. It may be summarised quite briefly. Free trade maximises world welfare; a customs union reduces tariffs and is therefore a movement towards free trade; a customs union will, therefore, *increase* world welfare even if it does not lead to a world-welfare *maximum*.

Viner showed this argument to be incorrect. He introduced the now familiar concepts of trade creation and trade diversion [1] which are probably best recalled in terms of an example. Consider the figures in the following Table:

TABLE I

*Money Prices (at Existing Exchange Rates) of a Single Commodity (X) in Three Countries*

| Country | A | B | C |
|---|---|---|---|
| Price | 35s. | 26s. | 20s. |

A tariff of 100% levied by country A [2] will be sufficient to protect A's domestic industry producing commodity X. If A forms a customs union with either country B or country C she will be better off; if the union is with B she will get a unit of commodity X at an opportunity cost of 26 shillingsworth of exports instead of at the cost of 35 shillingsworth of other goods entailed by domestic production.[3] This is an example of trade creation. If A had been levying a somewhat lower tariff, a 50% tariff, for example, she would already have been buying X from abroad before the formation of any customs union. If A is buying a commodity from abroad, and if her tariff is non-discriminatory, then she will be buying it from the lowest-cost source—in this case country C. Now consider a customs union with country B. B's X, now exempt from the tariff, sells for 26s., while C's X, which must still pay the 50% tariff, must be sold for 30s. A will now buy X from B at a price, in terms of the value of exports, of 26s., whereas she was formerly buying it from C at a price of only 20s. This is a case of Viner's trade diversion, and since it entails a movement from lower to higher real cost sources of supply, it represents a movement from a more to a less efficient allocation of resources.

[1] Jacob Viner, *The Customs Union Issue* (New York: Carnegie Endowment for International Peace, 1950). See the whole of Chapter 4, especially pp. 43–4.
[2] In everything that follows the "home country" will be labelled A, the "union partner" B and the rest of the world C.
[3] This argument presumes that relative prices in each country reflect real rates of transformation. It follows that the resources used to produce a unit of X in country A could produce any other good to the value of 35s. and, since a unit of X can be had from B by exporting goods to the value of only 26s., there will be a surplus of goods valued at 9s. accruing to A from the transfer of resources out of X when trade is opened with country B.

This analysis is an example of what Mr. Lancaster and I have called " The General Theory of Second Best ": [1] if it is impossible to satisfy *all* the optimum conditions (in this case to make all relative prices equal to all rates of transformation in production), then a change which brings about the satisfaction of *some* of the optimum conditions (in this case making some relative prices equal to some rates of transformation in production) may make things better or worse.[2]

Viner's analysis leads to the following classification of the possibilities that arise from a customs union between two countries, A and B:

1. Neither A nor B may be producing a given commodity. In this case they will both be importing this commodity from some third country, and the removal of tariffs on trade between A and B can cause no change in the pattern of trade in this commodity; both countries will continue to import it from the cheapest possible source outside of the union.

2. One of the two countries may be producing the commodity inefficiently under tariff protection while the second country is a non-producer. If country A is producing commodity X under tariff protection this means that her tariff is sufficient to eliminate competition from the cheapest possible source. Thus if A's tariff on X is adopted by the union the tariff will be high enough to secure B's market for A's inefficient industry.

3. Both countries may be producing the commodity inefficiently under tariff protection. In this case the customs union removes tariffs between country A and B and ensures that the least inefficient of the two will capture the union market.[3]

In case 2 above any change must be a trade-diverting one, while in case 3 any change must be a trade-creating one. If one wishes to predict the welfare effects of a customs union it is necessary to predict the relative strengths of the forces causing trade creation and trade diversion.

This analysis leads to the conclusion that customs unions are likely to cause losses when the countries involved are complementary *in the range of commodities that are protected by tariffs.* Consider the class of commodities produced under tariff protection in each of the two countries. If these classes overlap to a large extent, then the most efficient of the two countries will capture the union market and there will be a re-allocation of resources in a more efficient direction. If these two classes do not overlap to any

---

[1] R. G. Lipsey and K. J. Lancaster, " The General Theory of Second Best," *Review of Economic Studies,* Vol. XXIV (1), No. 63, 1956–57.

[2] The point may be made slightly more formally as follows: the conditions necessary for the maximising of *any* function do not, in general, provide conditions sufficient for an increase in the value of the function when the maximum value is not to be obtained by the change.

[3] One of the two countries might be an efficient producer of this commodity needing no tariff protection, in which case, *a fortiori*, there is gain.

great extent, then the protected industry in one country is likely to capture the whole of the union market when the union is formed, and there is likely to be a re-allocation of resources in a less-efficient direction. This point of Viner's has often been misunderstood and read to say that, in some general sense, the economies of the two countries should be competitive and not complementary. A precise way of making the point is to say that the customs union is more likely to bring gain, the greater is the degree of overlapping between the class of commodities produced under tariff protection in the two countries.

A subsequent analysis of the conditions affecting the gains from union through trade creation and trade diversion was made by Drs. Makower and Morton.[1] They pointed out that, *given that trade creation was going to occur,* the gains would be larger the more dissimilar were the cost ratios in the two countries. (Clearly if two countries have almost identical cost ratios the gains from trade will be small.) They then defined competitive economies to be ones with similar cost ratios and complementary economies to be ones with dissimilar ratios, and were able to conclude that unions between complementary economies would, if they brought gain at all, bring large gains. The conclusions of Viner and Makower and Morton are in no sense contradictory. Stated in the simplest possible language, Viner showed that gains will arise from unions if both countries are producing the same commodity; Makower and Morton showed that these gains will be larger the larger is the difference between the costs at which the same commodity is produced in the two countries.[2]

We now come to the second major development in customs-union theory —the analysis of the welfare effects of *the substitution between commodities* resulting from the changes in relative prices which necessarily accompany a customs union. Viner's analysis implicitly assumed that commodities are consumed in some fixed proportion which is independent of the structure of relative prices. Having ruled out substitution between commodities, he was left to analyse only bodily shifts of trade from one country to another. The way in which Viner's conclusion that trade diversion necessarily lowers welfare depends on his implicit demand assumption is illustrated in Fig. 1. Consider the case of a small country, A, specialised in the production of a single commodity, Y, and importing one commodity, X, at terms of trade independent of any taxes or tariffs levied in A. The fixed proportion in

---

[1] H. Makower and G. Morton, "A Contribution Towards a Theory of Customs Unions," ECONOMIC JOURNAL, Vol. LXII, No. 249, March 1953, pp. 33–49.

[2] Care must be taken to distinguish between complementarity and competitiveness in costs and in tastes, both being possible. In the Makower–Morton model these relations exist only on the cost side. An example of the confusion which may arise when this distinction is not made can be seen in F. V. Meyer's article, "Complementarity and the Lowering of Tariffs," *The American Economic Review*, Vol. XLVI, No. 3, June 1956. Meyer's definitions, if they are to mean anything, must refer to the demand side. Hence he is not entitled to contrast his results with those of Makower and Morton, or of Viner, all of whom were concerned with cost complementarity and competitiveness.

which commodities are consumed is shown by the slope of the line $OZ$, which is the income- and price-consumption line for all (finite) prices and incomes.   $OA$ indicates country A's total production of commodity Y, and the slope of the line $AC$ shows the terms of trade offered by country C, the lowest cost producer of X.   Under conditions of free trade, country A's equilibrium will be at $e$, the point of intersection between $OZ$ and $AC$.   A will consume $Og$ of Y, exporting $Ag$ in return for $ge$ of X.   Now a tariff which does not affect A's terms of trade and is not high enough to protect a domestic industry producing Y will leave her equilibrium position un-changed at $e$.[1]   The tariff changes relative prices, but consumers' purchases

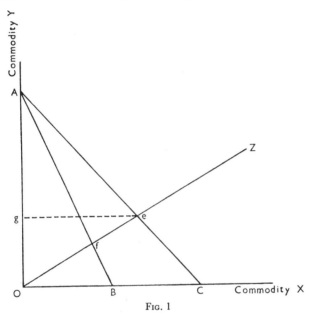

Fɪɢ. 1

are completely insensitive to this change and, if foreign trade continues at terms indicated by the slope of the line $AC$, the community must remain in equilibrium at $e$.   Now consider a case where country A forms a trade-diverting customs union with country B.   This means that A must buy her imports of X at a price in terms of Y higher than she was paying before the union was formed.   An example of this is shown in Fig. 1 by the line $AB$. A's equilibrium is now at $f$, the point of intersection between $AB$ and $OZ$; less of both commodities are consumed, and A's welfare has unambiguously diminished.   We conclude therefore that, under the assumed demand conditions, trade diversion (which necessarily entails a deterioration in A's terms of trade) *necessarily* lowers A's welfare.

   [1] It is assumed throughout all the subsequent analysis that the tariff revenue collected by the Government is either returned to individuals by means of lump-sum subsidies or spent by the Government on the same bundle of goods that consumers would have purchased.

Viner's implicit assumption that commodities are consumed in fixed proportions independent of the structure of relative prices is indeed a very special one.    A customs union necessarily changes relative prices and, in general, we should expect this to lead to some substitution between commodities, there being a tendency to change the volume of already existing trade with more of the now cheaper goods being bought and less of the now more expensive.    This would tend to increase the volume of imports from a country's union partner and to diminish both the volume of imports obtained from the outside world and the consumption of home-produced commodities.    The importance of this substitution effect in consumption seems to have been discovered independently by at least three people, Professor Meade,[1] Professor Gehrels [2] and myself.[3]

In order to show the importance of the effects of substitutions in consumption we merely drop the assumption that commodities are consumed in fixed proportions.    I shall take Mr. Gehrels' presentation of this analysis because it illustrates a number of important factors.    In Fig. 2 $OA$ is again country A's total production of Y, and the slope of the line $AC$ indicates the terms of trade between X and Y when A is trading with country C.    The free-trade equilibrium position is again at $e$, where an indifference curve is tangent to $AC$.    In this case, however, the imposition of a tariff on imports of X, even if it does not shift the source of country A's imports, will cause a reduction in the quantity of these imports and an increase in the consumption of the domestic commodity Y.    A tariff which changes the relative price in A's domestic market to, say, that indicated by the slope of the line $A'C'$ will move A's equilibrium position to point $h$.    At this point an indifference curve cuts $AC$ with a slope equal to the line $A'C'$; consumers are thus adjusting their purchases to the market rate of transformation and the tariff has had the effect of reducing imports of X and increasing consumption of the home good Y.    In these circumstances it is clearly possible for country A to form a trade-diverting customs union and yet gain an increase in its welfare.    To show this, construct a line through $A$ tangent to the indifference curve $I''$ to cut the $X$ axis at some point $B$.    If A forms a trade-diverting customs union with country B and buys her imports of X from B at terms of trade indicated by the slope of the line $AB$, her welfare will be unchanged.    If, therefore, the terms of trade with B are worse than those given by C but better than those indicated by the slope of the line $AB$, A's welfare will be increased by the trade-diverting customs union.    A's

---

[1] J. E. Meade, *The Theory of Customs Unions* (Amsterdam: North Holland Publishing Company, 1956).

[2] F. Gehrels, "Customs Unions from a Single Country Viewpoint," *Review of Economic Studies*, Vol. XXIV (1), No. 63, 1956–57.

[3] R. G. Lipsey, "The Theory of Customs Unions: Trade Diversion and Welfare," *Economica*, Vol. XXIV, No. 93, February 1957.    My own paper was first written in 1954 as a criticism of the assumption of fixed ratios in consumption made by Dr. Ozga in his thesis (S. A. Ozga, *The Theory of Tariff Systems*, University of London Ph.D. thesis, unpublished).

welfare will be diminished by this trade-diverting union with B only if B's terms of trade are worse than those indicated by the slope of *AB*.

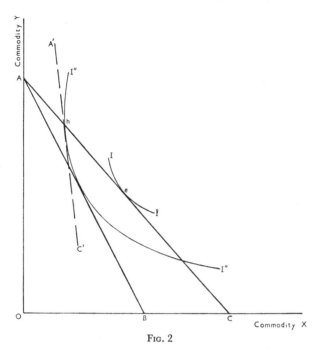

FIG. 2

The common-sense reason for this conclusion may be stated as follows:

"The possibility stems from the fact that whenever imports are subject to a tariff, the position of equilibrium must be one where an indifference curve [surface or hyper-surface as the case may be] cuts (*not* is tangent to) the international price line. From this it follows that there will exist an area where indifference curves higher than the one achieved at equilibrium lie below the international price line. In Fig. 2 this is the area above *I''* but below *AC*. As long as the final equilibrium position lies within this area, trade carried on in the absence of tariffs, at terms of trade worse than those indicated by *AC*, will increase welfare. In a verbal statement this possibility may be explained by referring to the two opposing effects of a trade-diverting customs union. First, A shifts her purchases from a lower to a higher cost source of supply. It now becomes necessary to export a larger quantity of goods in order to obtain any given quantity of imports. Secondly, the divergence between domestic and international prices is eliminated when the union is formed. The removal of the tariff has the effect of allowing . . . consumer[s] in A to adjust . . . purchases to a domestic price ratio which now is equal to the rate at which . . . [Y] can be transformed into . . . [X] by means of international trade. The final welfare effect of the trade-diverting customs union must be the net effect of these two opposing tendencies; the first working to lower welfare and the second to raise it." [1]

On this much there is general agreement.   Professor Gehrels, however, concluded that his analysis established a general presumption in favour of gains from union rather than losses.   He argued that " to examine customs unions in the light only of *production* effects, as Viner does, will give a biased judgment of their effect on countries joining them," [2] and he went on to say that the analysis given above established a general presumption in favour of gains from union.   Now we seemed to be back in the pre-Viner world, where economic analysis established a general case in favour of customs unions.   In my article " Mr. Gehrels on Customs Union " [3] I attempted to point out the mistake involved.   The key is that Gehrels' model contains only two commodities: one domestic good and one import.   There is thus only one optimum condition for consumption: that the relative price between X and Y equals the real rate of transformation (in domestic production or international trade, whichever is relevant) between these two commodities.   The general problems raised by customs unions must, however, be analysed in a model containing a minimum of three types of commodities: domestic commodities (A), imports from the union partner (B) and imports from the outside world (C).   When this change is made Gehrels' general presumption for gain from union disappears.   Table II

TABLE II

| Free trade (col. 1) | Uniform *ad valorem* tariff on all imports (col. 2) | Customs union with country B (col. 3) |
|---|---|---|
| $\dfrac{P_{Ad}}{P_{Bd}} = \dfrac{P_{Ai}}{P_{Bi}}$ | $\dfrac{P_{Ad}}{P_{Bd}} < \dfrac{P_{Ai}}{P_{Bi}}$ | $\dfrac{P_{Ad}}{P_{Bd}} = \dfrac{P_{Ai}}{P_{Bi}}$ |
| $\dfrac{P_{Ad}}{P_{Cd}} = \dfrac{P_{Ai}}{P_{Ci}}$ | $\dfrac{P_{Ad}}{P_{Cd}} < \dfrac{P_{Ai}}{P_{Ci}}$ | $\dfrac{P_{Ad}}{P_{Cd}} < \dfrac{P_{Ai}}{P_{Ci}}$ |
| $\dfrac{P_{Bd}}{P_{Cd}} = \dfrac{P_{Bi}}{P_{Ci}}$ | $\dfrac{P_{Bd}}{P_{Cd}} = \dfrac{P_{Bi}}{P_{Ci}}$ | $\dfrac{P_{Bd}}{P_{Cd}} < \dfrac{P_{Bi}}{P_{Ci}}$ |

Subscripts $A$, $B$ and $C$ refer to countries of origin, $d$ to prices in A's domestic market, and $i$ to prices in the international market.

shows the three optimum conditions that domestic prices and international prices should bear the same relationship to each other for the three groups of commodities, A, B and C.[4]   In free trade all three optimum conditions

[1] R. G. Lipsey, " Trade Diversion and Welfare," *op. cit.*, pp. 43–4.   The changes made in the quotation are minor ones necessary to make the notation in the example comparable to the one used in the present text.

[2] Gehrels, *op. cit.*, p. 61.

[3] R. G. Lipsey, " Mr. Gehrels on Customs Unions," *Review of Economic Studies*, Vol. XXIV (3), No. 65, 1956–57, pp. 211–14.

[4] If we assume that consumers adjust their purchases to the relative prices ruling in their domestic markets, then the optimum conditions that rates of substitution in consumption should equal rates of transformation in trade can be stated in terms of equality between relative prices ruling in the domestic markets and those ruling in the international market.

No. 279.—VOL. LXX.   L L

will be fulfilled.  If a uniform tariff is placed on both imports, then the relations shown in column 2 will obtain, for the price of goods from both B and C will be higher in A's domestic market than in the international market.  When a customs union is formed, however, the prices of imports from the union partner, B, are reduced so that the first optimum condition is fulfilled, but the tariff remains on imports from abroad (C) so that the third optimum condition is no longer satisfied.  The customs union thus moves country A from one non-optimal position to another, and in general it is impossible to say whether welfare will increase or diminish as a result. We are thus back to a position where the theory tells us that welfare may rise or fall, and a much more detailed study is necessary in order to establish the conditions under which one or the other result might obtain.

The above analysis has lead both Mr. Gehrels and myself [1] to distinguish between *production effects* and *consumption effects* of customs unions.  The reason for attempting this is not hard to find.  Viner's analysis rules out substitution in consumption and looks to shifts in the location of production as the cause of welfare changes in customs unions.  The analysis just completed emphasises the effects of substitution in consumption.  The distinction on this basis, however, is not fully satisfactory, for consumption effects will themselves cause changes in production.  A more satisfactory distinction would seem to be one between *inter-country substitution* and *inter-commodity substitution*. Inter-country substitution would be Viner's trade creation and trade diversion, when one country is substituted for another as the source of supply for some commodity.  Inter-commodity substitution occurs when one commodity is substituted, at least at the margin, for some other commodity as a result of a relative price shift.  This is the type of substitution we have just been analysing.  In general, either of these changes will cause shifts in both consumption and production.

Now we come to Professor Meade's analysis.  His approach is taxonomic in that he attempts to classify a large number of possible cases, showing the factors which would tend to cause welfare to increase when a union is formed and to isolate these from the factors which would tend to cause welfare to diminish. [2]  Fig. 3 (i) shows a demand and a supply curve for any imported commodity.  Meade observes that a tariff, like any tax, shifts the supply curve to the left (to $S'S'$ in Fig. 3) and raises the price of the imported commodity.  At the new equilibrium the demand price differs from the supply price by the amount of the tariff.  If the supply price indicates the utility of the commodity to the suppliers and the demand price its utility to the purchasers, it follows that the utility of the taxed import is higher to purchasers than to suppliers, and the money value of this difference in utility is

[1] Gehrels, *op. cit.*, p. 61, and Lipsey, " Trade Diversion and Welfare," *op. cit.*, pp. 40–1.

[2] The point of his taxonomy or of any taxonomy of this sort, it seems to me, must be merely to illustrate how the model works.  Once one has mastered the analysis it is possible to work through any particular case that may arise, and there would seem to be no need to work out all possible cases beforehand.

the value of the tariff. Now assume that the marginal utility of money is the same for buyers and for sellers. It follows that, if one more *unit of expenditure* were devoted to the purchase of this commodity, there would be a net gain to society equal to the proportion of the selling price of the commodity composed of the tariff. In Fig. 3 the rate of tariff is $\frac{cb}{ab}\%$, the supply price is *ab* and the demand price is *ac*, so that the money value of the " gain " (" loss ") to society resulting from a marginal increase (decrease) in expenditure on this commodity is *bc*.

Now assume that the same *ad valorem* rate of tariff is imposed on all imports so that the tariff will be the same proportion of the market price of

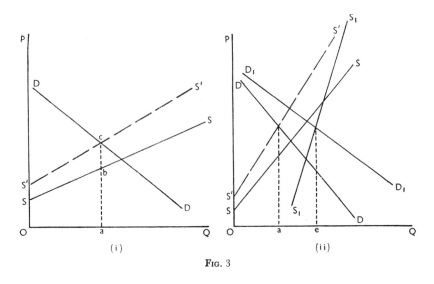

Fig. 3

each import. Then the gain to society from a marginal increase in expenditure (say one more " dollar " is spent) on any import is the same for all imports, and this gain is equal to the loss resulting from a marginal reduction in expenditure (one less " dollar " spent) on any import. Now consider *a marginal reduction* in the tariff on one commodity. This will cause a readjustment of expenditure, in the various possible ways analysed by Meade, so that in general more of some imports and less of others will be purchased. Since, *at the margin*, the gain from devoting one more unit of expenditure to the purchase of any import is equal to the loss from devoting one less unit of expenditure to the purchase of any import, the welfare consequences of this discriminatory tariff reduction may be calculated by comparing the increase in the volume of imports (trade expansion) with the decrease in the volume of other imports (trade contraction). If there is a net increase in the volume of trade the customs union will have raised economic welfare. A study of the welfare consequences of customs unions can, therefore, be devoted to the factors which will increase or decrease the volume of international trade. If the influences which tend to cause trade expansion are found to predominate it may be predicted that a customs union will raise welfare. The main body of Meade's analysis is in fact devoted to a study of those factors which would tend to increase, and to those which would tend to decrease, the volume of trade. Complications can, of course, be introduced, but they do not affect the main drift of the argument.[1]

Meade's analysis, which makes use of demand and supply curves, suffers from one very serious, possibly crippling, limitation. It will be noted that we were careful to consider only *marginal reductions* in tariffs. For such changes Meade's analysis is undoubtedly correct. When, however, there are *large* changes in many tariffs, as there will be with most of the customs unions in which we are likely to be interested, it can no longer be assumed that the demand and supply curves will remain fixed; the *ceteris paribus* assumptions on which they are based will no longer hold, so that both demand and supply curves are likely to shift. When this happens it is no longer obvious how much welfare weight should be given to any particular change in the volume of trade (even if we are prepared to make all of the other assumptions necessary for the use of this type of classical welfare analysis). In Fig. 3 (ii), for example, if the demand curve shifts to $D_1D_1$ and the supply curve to $S_1S_1$, what are we to say about the welfare gains or losses when trade changes from $Oa$ to $Oe$?

There is not time to go through a great deal of Professor Meade's or my own analysis which attempts to discover the particular circumstances in which it is likely that a geographically discriminatory reduction in tariffs will raise welfare. I shall, therefore, take two of the general conclusions that emerge from various analyses and present these in order to illustrate the type of generalisation that it is possible to make in customs-union theory.

The first generalisation is one that emerges from Professor Meade's analysis and from my own. I choose it, first, because there seems to be general agreement on it and, second, although Professor Meade does not make this point, because it is an absolutely general proposition in the theory of second best; it applies to all sub-optimal positions, and customs-union theory only provides a particular example of its application. Stated in terms of customs unions, this generalisation runs as follows: when only some

---

[1] For example, the same rate of tariff might not be charged on all imports. In this case it is only necessary to weight each dollar's increase or decrease in trade by the proportion of this value that is made up by tariff—the greater is the rate of tariff the greater is the gain or loss. It is also possible, if one wishes to make inter-country comparisons, to weight a dollar's trade in one direction by a different amount than a dollar's trade in some other direction. These complications, however, do not affect the essence of Meade's analysis, which is to make a *small change* in some tariffs and then to observe that the welfare consequences depend on the net change in the volume of trade and to continue the study in order to discover in what circumstances an increase or a decrease in the net volume of trade is likely.

tariffs are to be changed, welfare is more likely to be raised if these tariffs are merely *reduced* than if they are completely *removed*. Proofs of this theorem can be found in both Meade [1] and Lipsey and Lancaster,[2] and we shall content ourselves here with an intuitive argument for the theorem in its most general context. Assume that there exist many taxes, subsidies, monopolies, etc., which prevent the satisfaction of optimum conditions. Further assume that all but one of these, say one tax, are fixed, and inquire into the second-best level for the tax that is allowed to vary. Finally, assume that there exists a unique second-best level for this tax.[3]  Now a change in this one tax will either move the economy towards or away from a second-best optimum position. If it moves the economy away from a second-best position, then, no matter how large is the change in the tax, welfare will be lowered. If it moves the economy in the direction of the second-best optimum it may move it part of the way, all of the way or past it. If the economy is moved sufficiently far past the second-best optimum welfare will be lowered by the change. From this it follows that, if there is a unique second-best level for the tax being varied, a small variation is more likely to raise welfare than is a large variation.[4]

The next generalisation concerns the size of expenditure on the three classes of goods—those purchased domestically, from the union partner, and from the outside world—and is related to the gains from inter-commodity substitution. This generalisation follows from the analysis in my own thesis [5] and does not seem to have been stated in any of the existing customs-union literature. Consider what happens to the optimum conditions, which we discussed earlier, when the customs union is formed (see Table II). On the one hand, the tariff is taken off imports from the country's union partner, and the relative price between these imports and domestic goods is brought into conformity with the real rates of transformation. This, by itself, tends to increase welfare. On the other hand, the relative price between imports from the union partner and imports from the outside world are moved away from equality with real rates of transformation. This by itself tends to reduce welfare. Now consider both of these changes. As far as the prices of the goods from a country's union partner are concerned, they are brought

---

[1] *Op. cit.*, pp. 50–1.

[2] *Op. cit.*, Section V.

[3] A unique second-best level (*i.e.*, the level which maximises welfare subject to the existence and invariability of all the other taxes, tariffs, etc.) for any one variable factor can be shown to exist in a large number of cases (see, for example, Lipsey and Lancaster, *op. cit.*, Sections V and VI), but cannot be proved to exist in general (*ibid.*, Section VIII).

[4] This may be given a more formal statement. Consider the direction of the change—towards or away from the second-best optimum position—caused by the change in the tax. Moving away from the second-best optimum is a *sufficient*, but not a necessary, condition for a reduction in welfare. Moving towards the second-best optimum is a *necessary*, but not a sufficient, condition for an increase in welfare.

[5] R. G. Lipsey, *The Theory of Customs Unions: A General Equilibrium Analysis*, University of London Ph.D. thesis, unpublished, pp. 97–9, and Mathematical Appendix to Chapter VI. This thesis was subsequently published (1973) by Weidenfeld and Nicholson.

into equality with rates of transformation *vis à vis* domestic goods, but they are moved away from equality with rates of transformation *vis à vis* imports from the outside world. These imports from the union partner are thus involved in both a gain and a loss and their size is *per se* unimportant; what matters is the relation between imports from the outside world and expenditure on domestic commodities: the larger are purchases of domestic commodities and the smaller are purchases from the outside world, the more likely is it that the union will bring gain. Consider a simple example in which a country purchases from its union partner only eggs while it purchases from the outside world only shoes, all other commodities being produced and consumed at home. Now when the union is formed the " correct " price ratio (*i.e.*, the one which conforms with the real rate of transformation) between eggs and shoes will be disturbed, but, on the other hand, eggs will be brought into the " correct " price relationship with all other commodities —bacon, butter, cheese, meat, etc., and in these circumstances a customs union is very likely to bring gain, for the loss in distorting the price ratio between eggs and shoes will be small relative to the gain in establishing the correct price ratio between eggs and all other commodities. Now, however, let us reverse the position of domestic trade and imports from the outside world, making shoes the only commodity produced and consumed at home, eggs still being imported from the union partner, while everything else is now bought from the outside world. In these circumstances the customs union is most likely to bring a loss; the gains in establishing the correct price ratio between eggs and shoes are indeed likely to be very small compared with the losses of distorting the price ratio between eggs and all other commodities. If, to take a third example, eggs are produced at home, shoes imported from the outside world, while everything else is obtained from the union partner, the union may bring neither gain nor loss; for the union disturbs the " correct " ratio between shoes and everything else except eggs, and establishes the " correct " one between eggs and everything else except shoes. This example serves to show that the size of trade with a union partner is not the important variable; it is the relation between imports from the outside world and purchases of domestic goods that matters.

This argument gives rise to two general conclusions, one of them appealing immediately to common sense, one of them slightly surprising. The first is that, *given a country's volume of international trade*, a customs union is more likely to raise welfare the higher is the proportion of trade with the country's union partner and the lower the proportion with the outside world. The second is that a customs union is more likely to raise welfare the lower is the total volume of foreign trade, for the lower is foreign trade, the lower must be purchases from the outside world relative to purchases of domestic commodities. This means that the sort of countries who ought to form customs unions are those doing a high proportion of their foreign trade with their union partner, and making a high proportion of their total expenditure on domestic trade. Countries which are likely to lose from a customs union, on the other hand, are those countries in which a low proportion of total trade is domestic, especially if the customs union does not include a high proportion of their foreign trade.

We may now pass to a very brief consideration of some of the empirical work. Undoubtedly a serious attempt to predict and measure the possible effects of a customs union is a very difficult task. Making all allowances for this, however, a surprisingly large proportion of the voluminous literature on the subject is devoted to guess and suspicion, and a very small proportion to serious attempts to measure. Let us consider what empirical work has been done on the European Common Market and the Free Trade Area, looking first at attempts to measure possible gains from specialisation. The theoretical analysis underlying these measurements is of the sort developed by Professor Meade and outlined previously.

The first study which we will mention is that made by the Dutch economist Verdoorn, subsequently quoted and used by Scitovsky.[1] The analysis assumes an elasticity of substitution between domestic goods and imports of minus one-half, and an elasticity of substitution between different imports of minus two. These estimates are based on some empirical measurements of an aggregate sort and the extremely radical assumption is made that the same elasticities apply to all commodities. The general assumption, then, is that one import is fairly easily substituted for another, while imports and domestic commodities are not particularly good substitutes for each other.[2]

Using this assumption, an estimate was made of the changes in trade when tariffs are reduced between the six Common Market countries, the United Kingdom and Scandinavia. The estimate is that intra-European trade will increase by approximately 17%, and, when this increase is weighted by the proportion of the purchase price of each commodity that is made up of tariff and estimates for the reduction in trade in other directions are also made, the final figure for the gains from trade to the European countries is equal to about one-twentieth of one per cent of their annual incomes. In considering this figure, the crude estimate of elasticities of substitution must cause some concern. The estimate of an increase in European trade of 17% is possibly rather small in the face of the known fact that Benelux trade increased by approximately 50% after the formation of that customs union. A possible check on the accuracy of the Verdoorn method would have been to apply it to the pre-customs union situation in the Benelux countries, to use the method to predict what would happen to

---

[1] T. de Scitovsky, *Economic Theory and Western European Integration* (Allen and Unwin, 1958), pp. 64–78.

[2] Note also that everything is assumed to be a substitute for everything else; there are no relations of complementarity.

Benelux trade and then to compare the prediction with what we actually know to have happened. Whatever allowances are made, however, Scitovsky's conclusion is not likely to be seriously challenged:

> " The most surprising feature of these estimates is their smallness.
> . . . As estimates of the total increase in intra-European trade contingent upon economic union, Verdoorn's figures are probably underestimates; but if, by way of correction, we should raise them five- or even twenty-five-fold, that would still leave unchanged our basic conclusion that the gain from increased intra-European specialisation is likely to be insignificant." [1]

A second empirical investigation into the possible gains from trade, this time relating only to the United Kingdom, has been made by Professor Johnson.[2] Johnson bases his study on the estimates made by *The Economist* Intelligence Unit of the increases in the value of British trade which would result by 1970, first, if there were only the Common Market and, second, if there were the Common Market and the Free Trade Area. Professor Johnson then asks what will be the size of the possible gains to Britain of participation in the Free Trade Area? His theory is slightly different from that of Professor Meade, but since it arrives at the same answer, namely that the gain is equal to the increased quantity of trade times the proportion of the purchase price made up of tariff, we do not need to consider the details. From these estimates Johnson arrives at the answer that the possible gain to Britain from joining the Free Trade Area would be, *as an absolute maximum*, 1% of the national income of the United Kingdom.

Most people seem to be surprised at the size of these estimates, finding them smaller than expected. This leads us to ask: might there not be some inherent bias in this sort of estimate? and, might not a totally different approach yield quite different answers? One possible approach is to consider the proportion of British factors of production engaged in foreign trade. This can be taken to be roughly the percentage contribution made by trade to the value of the national product, which can be estimated to be roughly the value of total trade as a proportion of G.N.P., first subtracting the import content from the G.N.P. This produces a rough estimate of 18% of Britain's total resources engaged in foreign trade. The next step would be to ask how much increase in efficiency of utilisation for these resources could we expect: (1) as a result of their re-allocation in the direction of their comparative advantage, and (2) as a result of a re-allocation among possible consumers of the commodities produced by these resources. Here is an outline for a possible study, but, in the absence of such a study, what would we guess? Would a 10% increase in efficiency not be a rather conservative

[1] Scitovsky, *op. cit.*, p. 67.
[2] H. G. Johnson, " The Gains from Free Trade with Europe: An Estimate," *Manchester School*, Vol. XXVI, September 1958.

estimate? Such a gain in efficiency would give a net increase in the national income of 1·8%. If the resources had a 20% increase in efficiency, then an increase in the national income of 3·6% would be possible. At this stage these figures can give nothing more than a common-sense check on the more detailed estimates of economists such as Verdoorn and Johnson. Until further detailed work has been done, it must be accepted that the best present estimates give figures of the net gain from trade amounting to something less than 1% of the national income (although we may not, of course, have a very high degree of confidence in these estimates).

When we move on from the possible gains from new trade to the question of the economic benefits arising from other causes, such as economies of scale or enforced efficiency, we leave behind even such halting attempts at measurement as we have just considered. Some economists see considerable economies of scale emerging from European union. Others are sceptical. In what what follows, I will confine my attention mainly to the arguments advanced by Professor H. G. Johnson.[1] His first argument runs as follows:

"It is extremely difficult to believe that British industry offers substantial potential savings in cost which cannot be exploited in a densely-populated market of 51 million people with a G.N.P. of £18 billion, especially when account is taken of the much larger markets abroad in which British industry, in spite of restrictions of various kinds, has been able to sell its products.[2]

Let us make only two points about Professor Johnson's observation. First, many markets will be very much less than the total population. What, for example, can we say about a product sold mainly to upper middle-class males living more than 20 miles away from an urban centre? Might there not be economies of scale remaining in the production of a commodity for such a market? Secondly, in the absence of some theory that tells us the statement is true for 51 and, say, 31, but not 21, million people, the argument must remain nothing more than an unsupported personal opinion. As another argument, Professor Johnson asks, "Why are these economies of scale, if they do exist, not already being exploited?".[3] It is, of course, well known that unexhausted economies of scale are incompatible with the existence of perfect competition, but it is equally well known that unexhausted economies of scale are compatible with the existence of imperfect competition as long as long-run marginal cost is declining faster than

---

[1] In singling out Professor Johnson, I do not wish to imply that he is alone in practising the sort of economics which I am criticising. On the contrary, he is typical of a very large number of economists who have attempted to obtain quantitative conclusions from qualitative arguments.

[2] H. G. Johnson, "The Criteria of Economic Advantage," *Bulletin of the Oxford University Institute of Statistics*, Vol. 19, February 1957, p. 35. See also "The Economic Gains from Free Trade with Europe," *Three Banks Review*, September 1958, for a similar argument.

[3] Johnson, "Economic Gains," *op. cit.*, p. 10, and "Economic Advantage," *op. cit.*, p 35.

marginal revenue.   Here it is worthwhile making a distinction, mentioned by Scitovsky,[1] between the long-run marginal cost of producing more goods, to which the economist is usually referring when he speaks of scale effects, and the marginal cost of making and selling more goods (which must include selling costs).   This leads to a distinction between increasing sales when the whole market is expanding and increasing sales when the market is static, and thus increasing them at the expense of one's competitors.   The former is undoubtedly very much easier than the latter.   It is quite possible for the marginal costs of *production* to be declining while the marginal costs of *selling* in a static market are rising steeply.   This would mean that production economies would not be exploited by the firms competing in the market, but that if the market were to expand so that *all* firms in a given industry could grow, then these economies would be realised.

Let us also consider an argument put forward in favour of economies of scale.   Writing in 1955, Gehrels and Johnson argue that very large gains from economies of scale can be expected.[2]   In evidence of this they quote the following facts: American productivity (*i.e.*, output per man) is higher than United Kingdom productivity for most commodities; the differential is, however, greatest in those industries which use mass-production methods.   From this they conclude that there are unexploited economies of mass production in the United Kingdom.   Now this may well be so, but, before accepting the conclusion, we should be careful in interpreting this meagre piece of evidence. What else might it mean?   Might it not mean, for example, that the ratios of capital to labour differed in the two countries so that, if we calculate the productivity of a factor by dividing total production by the quantity of one factor employed, we will necessarily find these differences?   Secondly, would we not be very surprised if we did not find such differences in comparative costs between the two countries?   Are we surprised when we find America's comparative advantage centred in the mass-producing industries, and, if this is the case, must we conclude that vast economies of mass production exist for Europe?

Finally, we come to the possible gains through forced efficiency.   Business firms may not be adopting methods known to be technically more efficient than those now in use due to inertia, a dislike of risk-taking, a willingness to be content with moderate profits, or a whole host of other reasons.   If these firms are thrown into competition with a number of firms in other countries who are not adopting this conservative policy, then the efficiency of the use of resources may increase because technically more efficient production methods are forced on the business-man now facing fierce foreign competition.   Here no evidence has as yet been gathered, and, rather than report the opinions of others, I will close by recording the personal guess that this

[1] Scitovsky, *op. cit.*, pp. 42 ff.
[2] Gehrels and Johnson, "The Economic Gains from European Integration," *Journal of Political Economy*, August 1955.

is a very large potential source of gain, that an increase in competition with foreign countries who are prepared to adopt new methods might have a most salutary effect on the efficiency of a very large number of British and European manufacturing concerns.[1]

R. G. Lipsey

*London School of Economics.*

[1] Milton Friedman's argument that survival of the fittest proves profit maximisation notwithstanding (see *Essays in Positive Economics*, Chicago: University of Chicago Press, 1953). What seems to me to be a conclusive refutation of the Friedman argument is to be found in G. C. Archibald, " The State of Economic Science," *British Journal of the Philosophy of Science*, June 1959.

# Conclusions

## William Wallace

---

The problem of economic divergence came to preoccupy the Community at the end of the 1970s, in the wake of the economic, monetary and energy crises which coincided with the first enlargement, and in anticipation of the difficulties which the second round of enlargement was likely to pose. Yet − as previous chapters have argued − the underlying issues have a much longer history, and a much longer-term significance. The analysis of Daniel Jones's chapter makes it clear that the underlying trends of divergence in structure and performance are secular phenomena. In the case of Britain and Germany, it is possible to trace the divergence in performance and the distinctive strengths and weaknesses which underlay this to the early years of the twentieth century; though the pattern for France and Italy is less unilinear. The Community may have failed to reverse these trends, but it can in no way be held responsible for bringing them about. The crises of 1973−5 and the consequent sharpened divergence in monetary parities, inflation rates and patterns of economic recovery re-emphasised these differences, perhaps even exaggerated them; but, again, they cannot be said to have created the problem.

It should also be clear that the convergence issue carries echoes of arguments about the Community's objectives and proper role which go back through the 1960s to the Treaty of Rome and before. Some members of the Spaak Committee were well aware of the awkward questions of economic policy co-ordination and of regional imbalance which lay beyond the initial stages of the removal of barriers and the establishment of the customs union; but they recognised that this was a political minefield, and prudently left it for later resolution within a Community which they hoped would by then have become sufficiently integrated to withstand the strains which open debate of these issues would impose. It is hardly surprising, therefore, that Article 2 of the Rome Treaty is open to different interpretations, and that the only specific reference to 'reducing the

Reprinted from "Economic Divergence in the European Community" (M. Hodges and W. Wallace, eds.), pp. 205–223, with permission of George Allen & Unwin, London, Copyright 1981.

differences existing between the various regions and the backwardness of the less favoured regions' is in the Preamble. To have confronted the differences between French and German conceptions of economic policy, or to have considered directly how far the Community should assume the responsibility for assisting the development of the Italian Mezzogiorno, would have threatened the successful conclusion of the negotiations.

The Treaty was very thin on the objectives and instruments of economic policy. Article 103 commits member states to 'regard their conjunctural [short-term economic] policies as a matter of common concern'; Article 105 states that member states 'shall co-ordinate their economic policies', but only in pursuit of the objectives of maintaining balance-of-payments equilibrium and 'confidence' in their currencies. As so often repeated in later debates and proposals, it was easier to agree on monetary collaboration than on the other dimensions of economic policy. The Treaty therefore established a Monetary Committee, and indicated its objectives, but postponed for later discussion the institutions and objectives of other aspects of economic co-operation.

There were, it is true, some limited attempts to broach these difficult issues during the Community's first decade. A Conjunctural Policy Committee was set up in March 1960, in pursuance of Article 103. In April and May 1964 the Council of Ministers agreed to establish three further committees of national officials to promote the co-ordination of economic policies – a Medium-Term Policy Committee, a Budgetary Policy Committee and a Committee of Central Bank Governors to strengthen co-operation in monetary policy and in managing the exchanges (this last reflecting the extent to which the Monetary Committee had already come to concentrate on the co-ordination of policies towards international monetary co-operation). The Commission's Proposals of 31 March 1965 on financing the Common Agricultural Policy and on independent revenue for the Community included (in Article 5) the requirement that 'If...the Commission provides for payments to the Member States, it shall take into account the economic and social situation in the different regions of the Community and the need to ensure that burdens are equitably shared within the Community'; the accompanying explanatory memorandum dared even to refer to 'redistribution'.[1] But other parts of these ambitious proposals sparked off the dispute which culminated in the Luxembourg Compromise, in which such advanced ideas were put on one side.

In spite of this check to the too easily assumed momentum of economic integration, 'the extremely favourable conditions for the creation of the customs union in the 1960s created an erroneous impression about the role of unconstrained choice of objectives and of political will exercised in pursuit of them by the Community'.[2] Thus, in the aftermath of the retirement of President de Gaulle and the devaluation of the French franc in 1969, the Commission and national governments chose to ignore the increasingly unsettled international environment and to use as the vehicle for launching the second stage of economic integration the 'technicians' scheme' put forward by the Werner Committee. The hope, again, was that early experience of successful collaboration in the monetary sphere would provide sufficient momentum to carry the Six on towards Economic Union over the hard choices about conflicting priorities in economic policy, about the distribution of resources, and above all about the transfer of authority away from national governments, which the Werner Committee had lightly touched on and left for later decision. The politicians and experts who participated in the whole attempt to launch Economic and Monetary Union were, of course, distantly aware of the strains which progress might impose upon the Community's weaker economies. But it was easy to accept the monetarist prescription as the way forward, to assume that convergence of policies would bring about convergence of performance, and to trust in the apparent dynamism of the Italian economy – and to conclude that such problems as would occur would be temporary and minor.

The problem of divergence similarly loomed in the background of the negotiations for the first enlargement, without ever fully emerging into the light. The British government chose not to complicate the difficulties of entry by raising fundamental issues; the Irish government put its faith in the likely benefits of the CAP. The issue of the Regional Development Fund was thus played tactically, as the *quid pro quo* for Britain's contribution to the CAP, rather than as an essential part of the Community's commitment to 'strengthen the unity' of its member states' economies 'and to ensure their harmonious development' – the phrase in the Treaty Preamble which its negotiators took to quoting six years later.[3] The Commission, as guardian of the Community interest, registered its concern about the imbalance in the pace of expansion between the richer and the poorer regions, and bluntly stated that 'rapid progress towards Economic and Monetary Union would be arrested if . . . excessive divergencies between the economies of Member States' were not avoided by positive measures of policy.[4] But the Commission was only a minor actor in the bargaining between Britain, Germany and France over the size and shape of the RDF.

The impact of the 1973 Middle East War and its after-math, however, transformed the discussion of economic policies and priorities within the Community, and brought the issue of divergence – or, at least, some aspects of the issue of divergence – tó the fore. Movements in exchange rates were worrying in themselves, reflecting and reinforcing widening differences in inflation rates as member states struggled with varying success to absorb the increased cost of imported energy and to pay for that energy through increasing exports. The stronger European economies were thus immediately concerned with the threat of imported inflation, and with the implications of continuing divergence in exchange rates for the CAP and for the Community as a whole. Less immediately apparent were the fundamental strains which the necessity for more rapid economic and industrial adjustment placed on the Community's weaker economies, as the pressures of international recession, increased raw material prices and heightened competition for international markets weighed down the sluggish British and Italian economies with problems which the established dynamism of the German economy was sufficient to overcome.

The sober and gloomy conclusions of the Marjolin Report, from a study group commissioned to examine progress towards 'Economic and Monetary Union 1980' in the adverse conditions of 1974, explicitly pulled together for the first time most of the central themes of the convergence/divergence debate. It sharply criticised 'the centrifugal movement which characterises national policies' – and which had come to characterise national policies in response to increasing international economic and monetary turbulence even before the shocks of 1973. It noted the absence of 'Community solidarity' in the aftermath of the 1973 shocks, as undermining the political assumptions on which the commitment to economic union had been made. It noted also the progressive 'concentration of wealth to the detriment of certain peripheral regions', accepting the arguments of regional policy exponents that the workings of a common market tend to reinforce existing imbalances rather than to spread benefits equally or promote convergence. Though it laid much stress on the reassertion of 'political will' as the key to overcoming this succession of setbacks, it listed among the 'necessary decisions to bring nearer the time when the creation of an Economic and Monetary Union might be seriously envisaged ... the establishment of a Community budget on such a scale that the important transfers which the maintenance of Economic and Monetary Union will require can take place and be financed out of Community taxation'. 'The need for a large-scale regional policy', it added, 'is urgent'; venturing beyond a simple regional fund to consider investment incentives for disadvantaged regions and to touch on such delicate and difficult issues as Community support for unemployment benefits and social security transfers.[5] These arguments have since been taken further, in the MacDougall Report on *The Role of Public Finance in European Integration*,[6] and in the series of published and unpublished papers which accompanied the formulation of the European Monetary System − though the evident absence of sufficient political will within the member states to support anything more than modest additional burdens in conditions of economic recession has led to increased preoccupation within successive Commission proposals with minimalist schemes which might somehow combine acceptability with effectiveness.

The arguments of 1978−80, in the context of the establishment of the EMS and the Budget controversy, have however helped to clarify the terms of the debate and to force all member governments to consider the underlying problems. A report from the European Parliament's Economic and Monetary Affairs Committee on progress with the EMS (the Ruffolo Report), in April 1980, still found it necessary to criticise 'the ambiguity as to the real meaning of "convergence"' and of 'parallel measures'; noting that the weaker economies had allowed the issue to be reduced to a struggle over the reallocation of Budget contributions and receipts, while the wealthier member states still identified convergence with economic stabilisation and the reduction of inflation.[7] But the 'Conclusions' of successive European Councils from Bremen on reflect a gradual acceptance of a wider definition. The carefully negotiated language of the Conclusions of the Paris European Council of March 1979, in phrases hard-fought between the British and German delegations, commit the Community explicitly to a concern with 'increased convergence of the economic policies *and performances* of the Member States' (my italics), and charge the Council of Ministers (Economic and Financial Affairs) and the Commission 'to examine in depth how the Community could make a greater contribution, by means of all its policies taken as a whole, to achieving greater convergence of the economies of the Member States *and to reduce the disparities between them*'. The Conclusions of the Dublin Council of November 1979 'expressed its determination . . . to reinforce those policies most likely to favour the harmonious growth of the economies of the Member States and to reduce the disparities between these economies. They further declared the need, particularly with a view to the enlargement of the Community and necessary provisions for Mediterranean agriculture, to strengthen Community action in the *structural field*.'

## What Lies Behind the Convergence Debate?

The debate about convergence and divergence has, of course, been for many of its participants fundamentally about the necessary conditions for economic union. Divergences in economic policy and performance derailed the first, ill-thought-out, attempt at Economic and Monetary Union. Successive Commission study groups were therefore detailed to diagnose the problem more deeply and to prescribe more subtle remedies. The Community remains in principle and in rhetoric committed to European Union, a concept less precise even than that of convergence but which necessarily implies economic union as an integral aim. Explicitly to abandon that aim would be to throw into question the whole ideological underpinning of the Community. The Commission, as institutionally charged with the promotion and protection of the Community interest, must therefore address itself to the removal of obstacles to economic union, and strive to persuade member governments to agree to the actions necessary. In so far as the EMS may be seen as a cautious step forward towards EMU – which was certainly the Commission's perspective – the arguments over parallel measures were thus a direct continuation of the discussions in 1970 about the economic dimensions of the Werner Plan and the controversy in 1972–4 over regional policy.

Commitment to the eventual aim of economic union overlapped with the much more immediate concern, both within the Commission and within many member governments, that the Community's established common policies, the *acquis communautaire*, would be put at risk if trends within different national economies remained so disparate. Part of the impetus for the initial attempt at Economic and Monetary Union, after all, lay in the fears of the French government and others that divergence in exchange rates would threaten the CAP. By the late 1970s, there were understandable fears – as Stephen Woolcock notes – that the difficulties of the weaker economies and the increasing imbalance between the strong and the weak would progressively undermine the common market itself. Anthony Crosland, in his 'presidential' speech to the European Parliament on 12 January 1977, remarked that divergences in economic performance had now increased 'to an extent that in practice rules out major measures of integration'. His successor as Foreign Secretary and President of the Council of Ministers, David Owen, asserted more strongly that 'were the present trends of economic divergence to become firmly established, they would present a serious threat to the cohesion of the Community'.[8] Few in other governments disputed his analysis – even though they resisted the British proposals to alleviate the problem. The debate has thus in another sense been about preventing the Community's disintegration, and about considering what changes in national and Community policies might be necessary to prevent the re-erection of barriers to trade.

From another perspective, the debate of 1978–80 was about the Community Budget and its distribution of costs and benefits. Since of the three 'weaker economies' Britain was by far the most dis-advantaged in terms of receipts, this appeared to be primarily a British problem – all the more so because the British defined it in those terms, and thus failed to create and sustain an alliance with their less prosperous partners. During much of the discussion over the EMS in 1978 Ireland and Italy worked in informal alliance, successfully gaining some useful concessions on interest rates and EIB loans as a compensation for joining the scheme, with Britain offering only wavering support. Thereafter the Irish government, as a massive beneficiary from the CAP, was satisfied, though the Italians continued to maintain their concern with the broad issues of convergence – as they had in related negotiations ever since the drafting of the Rome Treaty. The British government rested its case for adjustment of its budgetary receipts and contributions primarily on grounds of equity, using the Treaty's commitment to promote harmonious and balanced development as a supporting argument. But in so doing it reduced the scope of the debate about convergence and divergence into a battle over figures and financial transactions, instead of a wider concern with the impact of Community policies as a whole. The dominance of the Budget issue in arguments over convergence in this period is demonstrated by the repeated appearance, in agendas for European Councils and official drafts, of the combined heading 'Convergence and Budgetary Questions'.

For the French, as Jonathan Story points out, and less directly for other member states, the problem of convergence and divergence is also the problem of Germany. The rapidity of the German adjustment to the changes in energy costs in 1973–4, its success in containing the inflationary surge which followed, the continuing strength of its exports through repeated revaluations of the Deutsche Mark (which in themselves posed problems of balance for the other member states), all impressed and concerned Germany's less buoyant partners. A report from a 'Group of Experts' to the Commission in 1979 on 'Changes in industrial structure in the European economies since the oil crisis, 1973–8' (the Maldague Report) spelt out the extent to which the German economy had now distanced itself from the rest of the Community. 'The Federal Republic of Germany's industry is well ahead of that of its main Community partners', partly because it had become 'highly specialised . . . and . . . concentrated on products with a high technology input, requiring highly-qualified labour'. 'The Federal Republic of Germany is the only European Community country in as favourable a position as the United States and Japan' in terms of high skill and capital input into goods traded, and adjustments out of sectors in which the newly industrialising countries were becoming competitive; 'overall it is clearly a case apart from its EEC partners'. 'The position of France is less certain', while both Italy and Britain were clearly highly vulnerable to competition from faster-growing countries, and ill-supplied with new products to replace declining industries.[9] As Daniel Jones argues in his chapter, these all represent the continuation of trends established well before the shocks of 1973–5; but the effect of those shocks was to increase the pressures upon the Community's economies, and thus to increase the pace of divergence.

Divergence is not therefore a simple picture of a Community developing into two separate tiers, but of a range of divergent developments, in which only Germany (and to an extent also the Netherlands, as its closest economic associate) could afford to be concerned more about problems in other members' economies than in its own. The picture is complicated further when Greece, Spain and Portugal are included. Loukas Tsoukalis and Geoffrey Denton note the scale of the challenges which the second round of enlargement poses for the Community. The Commission's approach to enlargement, from its controversial *Opinion* on the Greek application (in January 1976) on, has been to emphasise the difficulties which it would pose for the existing balance of Community policies, and the consequent need for the Council to face up to the need for some major changes in that balance. This was, for example, the theme of Roy Jenkins's speech to the first Council of Ministers discussion on the implications of enlargement, in October 1977, which specifically raised the issues of reducing disparities and promoting convergence − arguing that 'the problem of the economic gap, which has long weakened the process of integration, is exacerbated by and central to the question of enlargement'.[10]

Initial reactions within the German SPD to the need to reinforce newly democratic governments with economic assistance led to discussions about a 'European Solidarity Fund', a Marshall Aid type of programme for the three applicants, to be financed by the more prosperous European countries. This emerged as a budget authorisation of DM500m., authorised by the Bundestag Finance Committee in January 1978; but it was blocked by the Finance Ministry, and the proposal lost in the gathering arguments over the Community Budget and more specific demands for economic assistance to Turkey. In 1979–80 the argument over the Budget and convergence paid remarkably little attention to the linked problems raised by Greek entry in January 1981 and by the progress of the Spanish and Portuguese negotiations; the agreement on Britain's contributions to the Budget for 1980–2, finally reached at the end of May 1980, appeared not to have taken into account that Greece would be a member of the Community for two of the three years covered. But the complications of enlargement were at the back of the minds of most of those involved in the whole convergence debate from 1977 onwards.

### The Implications for the Community

As we have seen, the Treaty of Rome committed its signatories to the principle of economic convergence, but refrained from spelling out any of the implications of such a commitment. The economies of four of the original six members – the Benelux and Federal Germany – were already closely linked; the French were prepared, in the end, to stake their hopes on a successful trade-off between agricultural and industrial free trade. The problem of the sixth member, Italy, was recognised and alleviated by the inclusion of the European Social Fund and the European Investment Bank. But the general expectation that the Community would generate more rapid growth, and its realisation under favourable international circumstances in the early 1960s, pushed the issue of how balanced that growth might be into the background. The assumptions of economic integration theory that market integration brought benefits for all – in effect, brought about an automatic process of complementarity – were thus easy to accept. They enabled governments to avoid acrimonious argument on issues of economic and budgetary policy which would have sharply divided them.[11] The absence of any substantial economic critique of the assumptions of market integration also made it easier to accept such an optimistic approach; with few exceptions, economic studies of regional imbalance and of the possibility that market integration reinforces imbalance did not emerge until after the common market had been created.

By 1980 most of these favourable conditions had ceased to apply. The end of the long period of rapid economic growth and its replacement by prolonged international recession unavoidably redirected attention away from the size of the cake and how fast it was rising to the distribution of the slices. The evident dynamism of the German economy in contrast to those of Britain and Italy (and south-western France), the phenomenon of the 'Golden Triangle' with its concentration of industry and services, have undermined the faith of all but the most dedicated free traders in the equitable effects of untrammelled market integration. The French government has proved itself much more successful than most had expected in 1958 at competing in industrial goods; but by the end of the 1970s it was becoming much less certain that the CAP operated to its advantage, rather than to that of Germany.[12] A Community of Nine was already notably different in the balance of economic interests from the original Six. A Community of Twelve, stretching from Salonika to Lisbon and from the Straits of Gibraltar to the Skaggerak, would be fundamentally different in the spread and diversity of economic interests from the original grouping, of which all but Italy shared a common heartland in the Rhine valley and delta and northern France.

Increased concern about the problem of divergence and the role which Community policies play – and could play – in exacerbating or alleviating it is therefore understandable. Unfortunately, it is equally understandable that opinions on the appropriate Community response are diverse. Economists of different schools and in different member states are further away from consensus on the instruments and objectives of economic policy than they were twenty years ago; the advice available to the Commission and to member governments is therefore various, at times even contradictory. Important national and sectional interests are at stake: that of German industry in preventing its dynamism being inhibited by restrictive measures to aid industrial development in the weaker countries, those of the national finance ministries which would be asked to contribute more to an expanded budget and of those other finance ministries who would hope to receive more, regional interests both in the less prosperous and the more prosperous states, and so on.

Since the Community is clearly not primarily responsible for the problem of divergence, it is open to argument how far it should now assume the responsibility for its correction − or how far it *can* assume that responsibility. The Conclusions of the Paris European Council in March 1979 emphasise, in language which owes much to the efforts and arguments of the German government, that 'achievement of the convergence of economic performances requires *measures for which the Member States concerned are primarily responsible*'; the Italian and British governments were largely responsible for the balancing phrase 'but in respect of which *Community policies can and must play a supporting role* within the framework of increased solidarity' (my italics). Self-evidently, the social, political and economic contexts within which British and Italian policies are framed are important factors in their long-term tendency towards economic weakness. These, their more successful partners have argued in private, require national action to alter, to bring down the rate of inflation and to raise the level of investment, to change the attitudes of management and labour. To ask the more prosperous countries to burden their own economies and possibly to restrict their own industrial and economic growth, through financial transfers, crisis cartels and industrial location policies, without requiring the less prosperous to look to their own failings, would result in wasted resources and a general lowering of the Community's economic capacities. Politicians and economists from the stronger economies have therefore naturally tended to hold to the benign view of market integration, to emphasise the importance of political action to deal with structural problems by the less prosperous states themselves, and to insist that Community action must be ancillary to this.[13] Theirs is an argument with much plausibility when the subject is Italy or Britain − but hardly applicable to the situation of Greece, Portugal, or Spain when they enter the Community.

A further and major conceptual problem in considering the

appropriate Community response is the scepticism of economists and the uncertainty of politicians about how best to correct regional imbalances and promote convergence. The record of regional policies within the member states is mixed, in spite of a range of instruments and a scale of financial transfers far more extensive than the Community could hope to aspire to. Financial transfers to disadvantaged regions may simply enable the postponement of economic adjustments which are needed, instead of encouraging change. Automatic inter-regional transfers through social security systems, pension payments and fiscal equalisation schemes are characteristics of unified states and of developed federations with large federal budgets, extremely difficult to envisage on a Community basis within the foreseeable future. Radical solutions such as a Community input into long-term subsidies for employment costs in disadvantaged regions, to encourage inward investment and discourage outward migration, are faced with formidable barriers in economic and financial orthodoxy.[14]

Yet the Community cannot avoid some sort of response. Leaving aside the exact interpretation of the opening phrases of the Rome Treaty, there are three reasons why the Community has to be concerned to reduce divergence and promote convergence: the distributive and redistributive effects of its current Budget and common policies, the expectations of its weaker members and the renewed commitment to make the Community 'an area of monetary stability' (in the words of the Presidency Conclusions from the Strasbourg European Council in June 1979) through the establishment and further development of the EMS.

In the course of the controversy over the British contribution to the Budget, in 1978–80, a great deal of evidence was collected on the distributive effects of the Community Budget and on the impact of common policies on the terms of intra-Community trade. Their incontestable conclusions were that the Community was already redistributing resources, primarily through the CAP, in a perverse manner. The effects of agricultural trade and transfers were strongly to the advantage of the Netherlands and Denmark, two of the most prosperous members of the Community, and to the disadvantage of Britain – and would be as strongly to the disadvantage of Portugal, as a substantial agricultural exporter, on its accession. The CAP's effects were also regressive *within* member states, raising the income of prosperous as well as marginal farmers at the expense of consumers. The redistributive effects of Community policies were therefore to reinforce divergence rather than reduce it: at once at odds with the Community's declared objectives and politically unacceptable to the disadvantaged states. Those who benefited from the current *acquis* still maintained the argument that this was simply a secondary effect of the Community's common policies, and that deliberate redistribution of resources was not an objective of the Treaty; but this was an increasingly difficult position to maintain in the face of the evidence and the protests of the disadvantaged.[15]

Whether or not the promotion of convergence was an aim of the Community on its establishment, the expectation that membership would have 'dynamic effects' on their economies has been shared by all applicants – except perhaps for Denmark, which was already heavily dependent on its economic ties with Britain and Germany. The anticipation of direct and indirect economic benefits from entry was fulfilled in the case of Ireland, largely because of the operations of the CAP. It was disappointed in Britain, partly because successive British governments had underestimated the severity of the weaknesses of their domestic economy, partly because Britain's entry coincided with international economic crisis and the end of the era of fast growth. An understandable sense of grievance and disappointment therefore adversely affected Britain's political relations with its partners in the Community. Greece, Spain and Portugal in their turn all approached the Community in the expectation of gaining direct and indirect economic benefits from membership – expectations not necessarily any better founded than those of Britain. If the other member states attach political importance to their adhesion and satisfied membership – and the commitment to enlargement, in both the first and the second rounds, was fundamentally political – the predictable outcome if these hopes were disappointed would be to undermine, perhaps even destroy, the political objectives of entry.

The close link between convergence and moves towards economic union has been noted above. What needs to be questioned further is how seriously the Community and its members – above all its more prosperous members – are committed to economic union. This in turn takes us back to the underlying question of the Community's objectives and its member governments' perceptions of the purpose of membership. If the Community was fundamentally a political creation, of which the economic expression was a convenient symbol and path forward, then the continued reality of shared political aims and interests might be sufficient to hold it together despite economic divergence and imbalance. The rhetoric of economic union might still be rehearsed, in obeisance to the stated objectives of the founders; but the unwillingness of member governments to breathe life into the concept of economic union, while demonstrating their commitment to co-operation in other fields, would point to the underlying realities of a limited but valued political community.

The argument can be advanced further along these lines that divergence is thus not a vital problem for the future of the Community. The European Community has, after all, survived a considerable period of economic divergence without collapse and without more than temporary checks to its political progress. Certainly, it is important that the workings of the Community should help rather than hinder the economic prospects of its weaker members; the British problem (and potentially the problems presented by the new Mediterranean members) did therefore present real difficulties. But the Community was held together by sets of expectations and interests other than simply economic advantage. Shared political commitments, to democratic governments within Western Europe and to an open and stable international order outside, shared concerns about the Atlantic alliance and American leadership and about the Soviet threat, shared advantages in acting as a caucus in international negotiations both on economic and on security issues, now provided the real cement of European co-operation.

This argument cannot be dismissed lightly. The motives which led the original Six to negotiate and accept the Treaties of Paris and Rome were at least as much political as economic; so were those which led Britain and Ireland (though not Denmark) to apply, and which led in their turn Greece, Portugal and Spain to follow. The political dimension of European co-operation seemed all the more important in 1980, in an international environment dominated by an uncertain United States and an expansionist Soviet Union, and all the more firmly established with the democratic authority of the directly elected European Parliament and the regular and close relations between governments from the European Council downwards. But it ignores the close and mutually dependent relationship between political and economic factors, which has been a central characteristic of the European Community since the creation of the ECSC. For the German public and their political and industrial leaders, the Community has been the framework through which they have regained international acceptance and influence. But it has also been the framework for the sustained expansion of their economy – and it is the coincidence of these which has made the German commitment to the Community so strong. For successive British governments, the strongest arguments for joining and remaining in the Community were political; the economic case was always doubtful. Dissatisfaction with the perceived economic disadvantages, however, spilled over into political attitudes to the Community, among both leaders and public. Whether the Community is at bottom primarily a political or an economic undertaking is in this sense immaterial; it has to be *both*, to provide demonstrable benefits for its members both in the economic and the political sphere if it is to maintain its cohesion.

The relationship between political influence and economic strength is extraordinarily difficult to describe; but that there is a relationship, that governments see such a relationship, is clear. The French government's preoccupation with the German economy and its attempts to keep abreast of it were at once political and economic, fearing that to fall behind a dominant German economy would force it also to accept German political influence. The hostility of the British government, in particular, to suggestions about a 'two-tier' Community owed much to the fear that acceptance of a second-class status in economic terms would unavoidably carry political implications. The growing German self-confidence in foreign policy and in political relations with its partners during the 1970s owed much to the consciousness of economic success, just as the increasing British hesitation owed much to the consciousness of economic failure.

The most direct link between political and economic influence is of course provided by the leverage economic instruments supply for political ends. European aid to Turkey, in response to Turkish pleas and American pressure in conditions of rising East–West tension in 1979–80, meant above all German aid; the British, Irish and Italian governments contributed far less, and accepted that their influence was thus also lessened. It is not only in external policy, however, that economic leverage carries political implications. Within the Community, as within the IMF, creditor nations expect to exert influence over the policies of debtor nations, and resist attempts to influence their own policies in return. The issue at stake in the argument among member governments over a parity grid or a currency basket basis for the European Monetary System was about whether the weaker economies could avoid being forced to take the full strain of policy adjustment within a fixed rate or crawling peg system. Their demand for 'symmetrical rights and obligations' was essentially political: an attempt to frame the rules of the EMS so that strong economies (most of all Germany) would also have to accept the obligation to adjust domestic policies when their currency diverged from those of their partners. It was a measure of the political advantage which economic strength provides that the 'divergence indicator' which resulted bound the German government to respect the 'obligations' of a strong economy to its weaker partners much less than the majority of member governments had initially desired. The issue of conditionality in Community loans and funding is of similar character, reflecting the understandable attitude of those who provided the largest contributions to such transfers that they should have some say in the way such funds are used and the policies they support.[16] But such a claim to 'Mitspracherrecht' is naturally less welcome to the recipient countries, who see in such economic leverage an interference in their domestic policy-making which has no counterpart for them in their efforts to influence the policies of the more prosperous.

We are thus talking about an essentially political process, in which the different member governments (and the parties, groups and lobbies which influence them) pursue economic interest and advantage, using the political and economic instruments available to them. The rules, policies and stated objectives of the Community are a significant resource for the actors in this political process. Thus arguments over the intentions of the founding fathers, attempts to define – and redefine – the objectives of economic co-operation, appeals to 'Community solidarity' or the *acquis communautaire*, are part and parcel of the process of political bargaining among the member states. Motives and concerns are almost always mixed, for all governments; bending the objectives and rules of the Community to support immediate interests, but at the same time bearing in mind their commitment to maintaining the Community, and the necessity therefore of playing the game within the rules. It is characteristic of this process that the Luxembourg European Council in April 1980 found itself debating a draft on the 'fundamental principles' of the Community as an integral part of its efforts to reach a compromise on the budgetary dispute. It is characteristic, too, that the French representatives attempted to substitute for the Presidency's draft a longer statement which included commitments to exports of food and the maintenance of family farming concerns among this list of principles.[17]

The Treaty is sufficiently explicit about the economic aims of European co-operation to support the arguments of the less prosperous countries – half the membership in a Community of Twelve, accepting the definition of 'less prosperous' hammered out by the Economic Policy Committee in the context of the EMS negotiations – that positive action is required to reduce divergence. Their case that Community policies and financial transfers should, at the least, not have the effect of promoting divergence is particularly difficult for the more prosperous to resist. A minimalist interpretation of what response is required of the Community would thus be that the structural reform of the Budget, agreed in principle in the settlement of the dispute over Britian's contribution in May 1980 but left for later definition in detail, should lead to a pattern of contributions and receipts which bore some relation to the capacities and needs of member states; that regional policies should be further developed; and that the Community should pay more attention in its competition policies, rules on state aids and industrial adjustment measures, and in its trade policy, to the needs of its less prosperous members. Such a response could be accommodated within the current framework of Community co-operation, without raising major issues of institutional change and the transfer of authority.

A more ambitious interpretation of the response needed would however raise major political issues. If, for example, a recognition among the more prosperous that the advantages of monetary stability, or the preservation of the common market, justified financial and policy concessions were combined with a concerted effort by the less prosperous to promote more integrated policies, then the promotion of convergence might become a more operational objective. The issues skirted and suppressed during the Spaak Committee, the Werner Committee and the whole debate over Economic and Monetary Union would then come to the surface: effective co-ordination of economic policies, as required by the more prosperous, a larger financial and policy-making role for the Community, industrial policies broadly defined, and – if the demands of the less prosperous were to be met – acceptance of a degree of discrimination in Community policies to redress the balance of advantages which an open market offers to the centrally situated and already successful. All of these presuppose a substantial degree of consensus on the objectives and priorities of economic policy and on the political values which they serve. To spell them out is to require a much closer definition of the 'decision-taking centre' which the Werner Report warned would be needed, 'the creation or transformation of a certain number of Community organs to which powers until then exercised by the national authorities will have to be transferred'.[18] Closer integration of economic policies thus implies closer political integration; the two go together.

### Economics and Politics, Theory and Practice

In the confident years of the 1950s and 1960s economists claimed to be able to predict the consequences of pursuing different combinations of economic policies. Political scientists aimed to emulate them by careful examination of trends and the construction of theoretical frameworks within which they might be interpreted. The corpus of integration theories built up during the early years of European co-operation aspired to prediction, postulating a cumulative process through economic integration, sector by sector, towards a gradual transfer of political loyalties, and thus eventual political union. A degree of initial political commitment was of course necessary; but once an economic entity was established, it could be argued that 'the progression from politically inspired common market to an economic union to a political union amongst states is automatic'.[19] Economic causes, political consequences; political integration followed from economic integration. As with many economic theories, this perspective carried with it implicit political assumptions; and, again as with many economic theories, it became embedded in the rhetoric of several governments, though without being fully accepted or understood.

After the shocks and disappointments of the early 1970s, 'the idea which has been the basis for the past twenty years of the views of many Europeans, namely, that European political unity, particularly in the economic and monetary field, will come about in an almost imperceptible way', came to be widely questioned. 'It is clear that experience up to now shows nothing that supports the validity of this idea. One may legitimately wonder today if what may be required in order to create the conditions for an economic and monetary union is not perhaps on the contrary a radical and almost instantaneous transformation . . . giving rise at a precise point in time to European political institutions.'[20] What was needed, according to this alternative approach, was the reassertion of 'political will', rather than the accumulation of economic interests. It was a political commitment which had created the European Communities, overcoming a number of economic hesitations and obstacles; a political commitment to over-ride immediate economic concerns was the necessary condition for the transformation of European co-operation into union. From this perspective, economics follows politics: first the commitment, then the discovery of mutual economic benefit.

Looking at the process of European co-operation in 1980, one is struck far more by the number and variety of different forces at work. Some economic interests pull the Community members together, while others pull them apart. External threats and uncertainties provide powerful incentives for closer collaboration; but they also impose considerable strains on the uncompleted structure, leading to near-disintegration in conditions of crisis. Domestic constraints in the different states vary, as governments and elections come and go and the political context changes, but provide continuing limits to Community co-operation. Political leadership can in exceptional circumstances over-ride these limits, building new coalitions of interests, redefining priorities, easing the path to consensus; but changing political leadership can also make co-operation more difficult, as politicians pay domestic electoral debts and as they bring non-consensual attitudes into the intimate atmosphere of inter-governmental consultation.

The Community has made some significant political advances during the late 1970s, in the further development of Political Co-operation, in the move to a directly elected European Parliament, in the closer interlinking of its member governments through the whole network of meetings of which the European Councils provide the summit. It has made some significant economic advances as well, most notably in the establishment of the European Monetary System. But there have also been failures, both in economic and in political collaboration; it has proved extraordinarily difficult to adjust Community policies, except under crisis conditions, and the pace of

collaboration has been painfully slow. It is hard to discern steady
trends, either in the development of a political community or in the
Community's capacities to manage economic integration. It is
arguable that the strength provided by the close and extensive network
of political consultation and by the commitment of the more
prosperous countries to the maintenance of the Community's
cohesion is now sufficient to offset the strains exerted by economic
divergence. But it is evident from the preceding chapters that
economic divergence *does* exert considerable strains on the political
cohesion of the Community, which might well prove more than the
limited achievements of political integration can bear.

There is no way the political scientist can predict how the
Community will respond to the political problems posed by economic
divergence. This should not be a matter for despair; in the uncertain
conditions of the late 1970s, most economists too became less
confident of their predictive powers, more cautious about the
relationship between different trends. Politics is an intricate process,
with a very large number of variables. All one can say with confidence
is that the problem has now been placed firmly on the Community's
political agenda; and that the prospect of enlargement will make it all
the more sensitive an issue for a Community of twelve.

### Notes: Chapter 11

1 Supplement no. 5 to *Bulletin of the EEC*, 1965.
2 Andrew Shonfield, 'The aims of the Community in the 1970s', *Report of the Study Group 'Economic and Monetary Union 1980'*, Vol. II (Commission of the European Communities, Brussels, 1975).
3 Helen Wallace, 'The establishment of the Regional Development Fund: common policy or pork barrel?', in Helen Wallace *et al.* (eds.), *Policy-Making in the European Communities* (Wiley, Chichester, 1977).
4 'Regional Problems in the Enlarged Community', *Bulletin of the European Communities*, Supplement 8/73.
5 *Report of the Study Group 'Economic and Monetary Union 1980'*, Vol. I (Commission of the European Communities, Brussels, 1975), pp. 2, 6, 15 and 34.
6 *Report of the Study Group on the Role of Public Finance in European Integration* (Commission of the European Communities, Brussels, 1977).
7 *The EMS as an Aspect of the International Monetary System*, Report of the Economic and Monetary Affairs Committee of the European Parliament (Ruffolo Report), Working Document No. 1/63/80, 1980.
8 House of Commons, 1 March 1977.
9 This report appeared as a special issue of *European Economy*, Commission of the European Communities, November 1979; the quotations are from pp. 43 and 79.
10 *Agence Europe*, 28 October 1977.
11 Jacques Pelkmans, 'Economic theories of integration revisited', *Journal of Common Market Studies*, June 1980, pp. 333.
12 Yao-Su Hu, 'German agricultural power: the impact on France and Britain', *The World Today*, November 1979; *L'Europe: les vingt prochaines années* (Commissariat Général du Plan, Paris, 1980).
13 See, for example, Carsten Thoroe, 'The transfer of resources: comment', in

William Wallace (ed.), *Britain in Europe* (Heinemann, London, 1980), pp. 153–7.

14  Wynne Godley, 'The United Kingdom and the Community Budget', in Wallace, op. cit., pp. 83–4, argues for such a Community 'Regional Employment Premium'.

15  See Geoffrey Denton's contribution on the Budget issue, above, Wynne Godley's chapter in Wallace, op. cit., and Helen Wallace, *Budgetary Politics: The Finances of the European Communities* (Allen & Unwin, London, 1980), particularly ch. 2.

16  Helen Wallace, op. cit., pp. 47–8.

17  *Agence Europe*, 28–29 April 1980.

18  Supplement no. 11 *Bulletin of the EEC*, 1970, p. 24. See also William Wallace, 'The administrative implications of Economic and Monetary Union within the European Community', *Journal of Common Market Studies*, 1974, pp. 410–45.

19  Ernst Haas, ' "The Uniting of Europe" and "The Uniting of Latin America" ', *Journal of Common Market Studies*, June 1967, p. 315.

20  Marjolin Report (note 5), p. 5.

# THE THEORY OF COMMON MARKETS AS APPLIED TO REGIONAL ARRANGEMENTS AMONG DEVELOPING COUNTRIES

BY

### R. F. MIKESELL
University of Oregon, Eugene, Oregon

## I. INTRODUCTION

B y and large the theory of customs unions has been confined to considerations of welfare gains or losses arising from a disturbance of the existing pattern of trade which is assumed to reflect comparative advantages in the commodities traded as determined by existing factor endowments. Some attention has been paid to the realization of gains from the economies of scale and from increased competition. However, the effects of the creation of regional markets on the more fundamental problems of developing countries such as increasing opportunities for profitable foreign and domestic investment, broadening the export base, achieving balance of payments equilibrium, mobilizing unemployed resources and avoiding economic dualism, have been largely neglected. Some of these problems, which are concerned with the dynamics of economic growth, are of interest to under-developed and to industrially advanced countries alike. I doubt, for example, if the most significant gains from the creation of the European Economic Community are to be discovered through a comparison of trade-diverting and trade-creating effects on welfare, even if we could measure them. Rather, the major impact will occur as a consequence of the effects on entrepreneurial decisions arising out of the new market structure and out of the acute awareness of the continual generation of new products, new processes and new methods of distribution on the part of competitors within the broad regional market. In other words, broadening of the area of unfettered activity of competitive enterprise creates opportunities for innovations and forces changes in investment patterns which constitute the dynamic elements of growth. These intangible factors, which are basic to business

Reprinted from "International Trade Theory in a Developing World" (R. Harrod and D. Hague, eds.), pp. 205–229, with permission of Macmillan, London, Copyright 1964.

decisions and expectations, often lie outside the economist's analytical framework.

But the fact that analytical work on common markets has been largely directed to problems of welfare under somewhat static assumptions which permit the employment of the analytical tools at our disposal, does not mean that the conclusions reached have no relevance for economic growth or for developing countries generally. I believe, however, that the theoretical analysis of customs unions or of regional preference arrangements generally should be directed more towards the problem of their impact on the direction of investment in the developing countries for future output rather than limited to an analysis of the welfare implications of shifting existing trade patterns. There are two general reasons for this conclusion, the first of which also has applicability for regional markets among industrially advanced countries. One is that plans for the creation of a customs union or free trade area usually involve relatively long time-periods for fruition so that the initial impact, and perhaps the most important one, is on expectations regarding future market opportunities rather than on existing trade patterns arising directly out of changes in intra-regional trade restrictions. Thus what is most relevant are the effects on investments which will determine trade and production patterns a decade in the future, as compared with what they might have been in the absence of the creation of the regional trading arrangements. The second factor, which is related to the first, is that developing countries are undergoing rapid and far-reaching changes in the structure of their production and trade. Very often there is relatively little trade among the members of regional trading blocs to begin with and virtually no exports of manufactures either between members or to the rest of the world. Hence, while the European Common Market and the European Free Trade Area are striving to achieve an expansion of intra-regional trade within the framework of an existing economic structure, developing countries, such as the members of the Latin American Free Trade Area, are seeking to bring about within the next decade or two a fundamental change in the structure of their production and trade and have sought to fashion a regional trade mechanism which will help to orient their economies in the direction of regional specialization.

Although no two economic regions or groups of countries which regard themselves as a region capable of economic integration are alike, we might begin by setting forth certain characteristics, some if not all of which under-developed regions tend to have in common. These characteristics are frankly based on those of the countries

206

making up the Latin American Free Trade Area or Montevideo Treaty Association. I have chosen this group as a model because it constitutes the most important group of under-developed countries that have formulated, and are actually in initial stages of carrying out, a free trade area plan. The only other group where significant progress has been made is the Central American group, the countries of which are in a much less advanced stage of development and whose domestic markets are smaller. The Central American group also differs from the Montevideo group in that a much larger proportion of the total income of the Central American countries is derived from foreign trade, and for most of them, at least, balance of payments problems have been less acute.

The characteristics of our model group of countries contemplating the formation of a customs union or free trade area are as follows :

(1) Intra-regional trade in primary commodities is not likely to be affected immediately by the regional trading arrangements either because (*a*) the countries are complementary and do not have significant restrictions if they are not substantial producers of these commodities, or (*b*) they are competitive and sell the same commodities in world markets. In addition, the agriculture escape clauses in the agreement may take agricultural products out of the regional trading arrangement.

It should also be noted that in Latin America, bilateral agreements and multiple exchange rates which favour exports to convertible currency areas (to say nothing of the U.S. dumping of surplus agricultural commodities) have tended to put trade within the region at a disadvantage compared with trade with the outside. Hence we begin with a system of trade restrictions which discriminates against trade among countries forming the free trade area or customs union.

(2) Trade in industrial products is virtually non-existent. Production for the domestic market is being initiated in more and more commodities and industrialization is moving into intermediate products and investment goods, especially in the more advanced members of the regional group. The expansion of output takes over a larger share of the market from imports in commodity after commodity, mainly as a consequence of trade restrictions, although in many cases domestic costs may be competitive with imports. Frequently former suppliers of imports from abroad with well-established distribution channels will have undertaken domestic production either directly or under licensing arrangements, perhaps including the provision of management and technical services. The same foreign firms may be suppliers of imports or may be producing

207

locally in other members of the region. Domestically owned firms usually lack marketing outlets in other members of the region even if they were permitted to compete on a cost basis in the markets of their regional partners.

(3) Slowly growing, if not stagnating, export proceeds from primary commodities, together with rapidly expanding import requirements and debt service plus the necessity of finding employment for unemployed workers, have directed national policies towards the promotion of rapid industrialization with special emphasis on the production of substitutes for imported commodities. The policy of directing or influencing production on the basis of achieving direct savings in foreign exchange, rather than on the basis of relative efficiency, usually leads to substantial cost and price disparities for the same products produced within the region and also to overcapacity sometimes for the same commodities in more than one country in the region. For example, there is substantial overcapacity in Argentina, Brazil, Mexico and certain other Latin-American countries for the production of consumers' durable goods such as refrigerators.

(4) Members of the regional trade group include relatively advanced countries with well-developed industrial sectors, such as Argentina, Brazil and Mexico, and less advanced and little-industrialized countries such as Paraguay and Uruguay. This creates the problem of assuring a balanced distribution of the welfare gains from regional trading arrangements or at least of preventing certain countries from gaining at the expense of others.

On the basis of the foregoing characteristics of our model regional trade group encompassing several developing countries, we shall examine the relevance of certain generalizations formulated by recent contributions to the theory of customs unions.

## II. THE BALANCE BETWEEN TRADE-DIVERTING AND TRADE-CREATING EFFECTS

We may begin with the well-known argument of Professor Jacob Viner that trade diversion tends to be harmful to welfare while trade creation is beneficial, and the net effects of a customs union on welfare will depend upon the balance of these opposing forces.[1] By and large the traditional primary exports of developing countries to the rest of the world will not be significantly affected by the

[1] Jacob Viner, *The Customs Union Issue*, Carnegie Endowment for International Peace, New York, 1950, Chapter 4.

creation of a regional trading arrangement. Moreover, their total purchases from the rest of the world will continue to depend very largely upon the growth of their primary commodity exports. However, the improved competitive position of their manufactures and semi-manufactures as a consequence of the creation of the competitive regional market may very well enable them to increase their total exports to the rest of the world. Also, if the regional market creates trade in other primary products not previously sold abroad, members may be able to broaden their primary commodity export base with respect to both regional and extra-regional trade. Hence, the long-run impact of a regional trading arrangement is not to decrease trade with the rest of the world but rather to change its pattern and possibly to enlarge it. In this sense, therefore, there is no over-all trade diversion, only trade creation. Thus, the basic questions which we must examine are : (i) whether the new regional pattern of trade with the rest of the world will become more economical as a consequence of the trading arrangements than would have otherwise have been the case ; and (ii) whether the newly created trade is economical or increases economic welfare.

As regards the first question, imports from the rest of the world will be determined by the effects of the regional trading arrangement on the pattern of production and trade within the region. In the absence of intra-regional trade in manufactures, each member will seek to produce as many commodities as possible for sale in the domestic market and import the rest from abroad. In order to save exchange, many commodities that cannot be produced domestically (because of limitations on investment capital and foreign exchange or otherwise) will be subjected to heavy duties or restrictive quotas or prohibitions. The creation of a regional market, however, will enable individual countries to obtain many of these goods from regional markets and to expand their own output for sale to the region. This will not only change the pattern of investment but will increase the total volume of investment. The additional foreign exchange required for the larger imports of capital goods, raw materials and fuel not produced by the regional partners will become available as a consequence of the reduced demand for consumers' goods and other commodities (including some capital goods) from outside the region. Thus, the new pattern of imports from the rest of the world will contribute to the process of greater specialization within the region.

There seems little doubt that a pattern of industrialization based on greater specialization within the region will be more economical than one based on production by each country for its own domestic

H        209

market. To the extent that greater specialization is permitted in agricultural commodities, there will also be gains from the removal of intra-regional trade barriers. The welfare gains will arise from the availability of a greater variety of goods at lower average cost, but substantial changes in price relationships may occur. Recent discussions of customs union theory which have emphasized *inter-commodity substitution* as against *inter-country substitution* have special relevance for the case of developing countries.[1] While a customs union will not establish the optimum relations between internal prices of domestic and internationally traded goods for maximizing welfare, intra-regional trade and specialization will change relative prices and consumption patterns toward optimum conditions. The increased consumption of commodity $x$ in country $A$ resulting from a lowering of the tariff on commodity $x$ supplied from regional partner $B$ as a consequence of increased investment and production for exploiting the larger regional market, will change consumption patterns in country $A$ in the direction of increased welfare. In turn, investment in country $A$ can be diverted to expanding output of commodity $y$ rather than towards the production of more commodity $x$ in which it is relatively less efficient.

It might be objected that resources will be transferred from production for the world market to production for the regional market and that this would result, in effect, in a reduction in the terms of trade since imports will be acquired from a higher-cost source. However, for reasons noted earlier, this is not likely to take place as a consequence of the creation of the regional market *per se,* although the urge to industrialize as a means of finding employment for labour has undoubtedly shifted capital resources out of primary production for world markets.[2] Of course, as incomes within the regional market grow, Chile and Peru may sell somewhat more copper to Argentina, Brazil and Mexico, and Brazil may sell more coffee to her southern neighbours. These exports will not be displacing the exports of copper and coffee from other areas of the world, and depending upon long-run supply conditions, they may not even be at the expense of exports of Chilean copper or Brazilian coffee to the rest of the world. Certainly it could not be argued that greater production and income promoted by the existence of a regional market, which in turn expands the demand for primary

---

[1] See R. G. Lipsey, 'The Theory of Customs Unions: a General Survey" *The Economic Journal,* Sept. 1960, p. 504, and J. E. Meade, *The Theory of Customs Unions,* North Holland Publishing Co., Amsterdam, 1955, pp. 34–41.

[2] Whether or not this is desirable for primary-producing countries as a whole depends upon the demand elasticities for primary goods and effects on their terms of trade. However, one can easily cite examples, e.g. Argentina, where this has been disastrous for the particular economy.

products from relatively efficient producers for the world market — which happen to be members of the regional trading area — are harmful to economic welfare.

All that we have said on this point with reference to developing economies reinforces the view of Professor Meade and others that there is a gain in welfare if there is a net expansion of trade. This point seems to be particularly evident when we consider that the alternative to directing investment to the production of those commodities in which countries have a relative competitive advantage within the regional market, is a haphazard directing of investment into production for the domestic market of those commodities which can most readily displace imports from the rest of the world.

## III. COMPLEMENTARY VERSUS COMPETITIVE PARTNERS AND THE PATTERN OF EXISTING TRADE

Recent contributions to customs union theory have evolved certain hypotheses with respect to the potential welfare gains from discriminatory regional trade arrangements which relate to the existing patterns of production in the member countries and to the proportion which trade among regional partners bears to their total trade. We shall present these generalizations without necessarily taking a position as to their correctness within the context of the assumptions under which they are made, and then seek to determine their relevance for our typical developing regional group.

(a) A regional trading arrangement is more likely to increase economic welfare if the economies of the members are very competitive but potentially very complementary.[1]

(b) Welfare is likely to be the greater, the higher the proportion of trade among the partners relative to their total trade.

(c) Welfare is likely to be the greater, the lower the proportion of the foreign trade of each member to purchases of domestic commodities.[2]

In our typical case of a regional group encompassing developing countries, members at the same stage of industrialization tend to be

---

[1] See Meade, *op. cit.* p. 107.

[2] See Lipsey, *op. cit.* pp. 508-9. Professor Meade concludes that 'a customs union between two countries will be the more likely to raise economic welfare, if each is the principal supplier to the other of the products which it exports to the other and if each is the principal market for the other of the products which it imports from the other'. He also concludes that 'the formation of a customs union is more likely to raise economic welfare the greater is the proportion of the world's production, consumption, and trade which is covered by the members of the union'. See Meade, *op. cit.* pp. 108-9.

producing many of the same manufactured goods, but there is little or no trade between them. So far as primary commodities are concerned, they may in some cases be highly competitive in the sense that both are producing the same commodities for world markets. For example, both Brazil and Mexico, which are members of the Montevideo group, produce both coffee and cotton for world markets. On the other hand, they may be quite complementary with respect to some primary commodities which each sells in world markets. The creation of a regional market is not likely to have much effect on trade in either of these two groups of primary commodities. Where they are competitive in primary commodities which they do not export on world markets, but each maintains import restrictions in order to support domestic output and prices, the elimination of trade restrictions within the group is likely to bring about a much more efficient utilization of resources and result in some displacement of imports from outside the group; the reduced prices and production costs for total regional output are likely to outweigh any loss from trade diversion for the region as a whole.

As regards developing countries, it might be said that because of the emphasis on industrialization, all members are actually or potentially competitive and certainly all members of a regional group are potentially complementary. Thus it would not be correct to say that the outlook for achieving economic welfare gains through a customs union of Central American states is poor because the members are at such a low stage of industrialization that they are actually not competitive at the present time ; nor would it make much sense to argue that because they all produce coffee and bananas and hence are actually competitive, this augurs well for a net increase in welfare from the creation of a customs union. They are not going to trade in coffee and bananas anyway, except possibly for some border trade. As industrialization proceeds, they are going to be more competitive ; but what these countries should strive for is a pattern of investment which will introduce a substantial degree of complementarity for the future.

When we come to consider the generalization noted in paragraph (*b*) above regarding the proportion of intra-regional to total trade, it might be concluded that there is little prospect for increasing welfare through a customs union or free trade area for our typical regional group. This conclusion would be wrong, however, because the alternative to increased intra-regional trade is not reduced trade with the outside world but, rather, the production of a larger proportion of each country's requirements within its borders, thus inevitably leading to a less efficient utilization of resources as compared

**212**

with regional specialization. Nor can we accept the implication in paragraph (c) above, that because the proportion of foreign trade to domestic expenditures is quite high for many developing countries, their chances for achieving welfare gains from the formation of regional trading groups is severely limited. The reason is that if the countries are going to develop on the basis of a rather slow growth of export proceeds (and perhaps a large part of these going to pay for debt service), the ratio of foreign trade to domestic expenditures will decline rapidly in the future. Welfare gains will be achieved through the creation of regional markets because they will tend to retard the rate of decline in the ratio of foreign trade (including intra-regional trade) to domestic expenditures.

It is for these reasons that I seriously question the applicability of the generalizations of the theory of customs unions which relate to complementarity, competitiveness and trade patterns, to the potential gains from regional trading arrangements for developing countries. It is necessary to look beyond the existing patterns of production and trade to those which are likely to emerge in the absence of the formation of a customs union or a free trade area.

## IV. PARTIAL VERSUS COMPLETE REMOVAL OF RESTRICTIONS ON INTRA-REGIONAL TRADE

Contrary to the traditional approach to customs unions and that which is embodied in the General Agreement on Tariffs and Trade,[1] recent theories of customs unions have suggested that a *partial* reduction of duties on imports from regional trading partners is more likely to increase welfare than is a *complete* removal of restrictions on trade within the preference area.[2] The basis for this generalization is that each successive reduction of duties within the preference area will contribute less to the gains from the expansion of trade between the partners, but the loss from trade diversion will continue as the degree of discrimination within the preference trade area continues to increase. This generalization is usually made on the assumption of an all-round reduction of tariffs affecting all commodities.[3]

[1] See Article XXIV of the General Agreement on Tariffs and Trade.
[2] See Meade, *op. cit.* pp. 110-11 ; see also Lipsey, *op. cit.* pp. 506-7.
[3] Closely related to this generalization is the one which states that the formation of regional preference arrangements is the more likely to increase welfare, 'the higher are the initial rates of duty on imports into the partner countries'. See Meade, *op. cit.* p. 108.

From the standpoint of the long-run effects on investment decisions within the regional trading bloc, and again considering the fact that individual countries will, over time, seek a maximum displacement of imports with domestic production in the absence of a regional arrangement, I seriously doubt the validity of the above generalization. I do not believe that a preferential trading arrangement can possibly have the same impact on resource distribution as one which looks towards the removal of *all* barriers to trade within a given time-period. Again, we are not concerned simply with the readjustment of existing trade patterns, but rather with alternative principles for the direction of investment which will establish the trade and production patterns a decade or so hence.

In this connection mention might be made of another generalization of Professor Meade's to the effect that 'a customs union is less likely to have adverse secondary repercussions upon economic welfare in a world in which trade barriers take the form of fixed quantitative restrictions rather than of taxes on imports'.[1] The reasoning here, of course, is that the removal of quantitative restrictions on trade among the partners, while maintaining the same quantitative restrictions against imports from the outside world, is likely not to affect the imports within the quotas from the outside world and hence there would be no trade diversion. On the other hand, the use of quantitative restrictions as against tariffs and discriminatory tariff treatment favouring imports from regional partners removes the necessity of competing for markets on a price and quality basis. In short, I would favour the use of tariffs over quantitative restrictions as a means of providing a discriminatory advantage to intra-regional trade.

To a considerable degree import restrictions of developing countries have taken the form of quotas or outright prohibitions on imports. Moreover, as developing countries become relatively self-sufficient in additional commodities in the future, they will, in the absence of a regional trade arrangement, restrict or eliminate foreign competition one way or the other. It might also be said that there has been a tendency to maintain the most restrictive import measures on the very commodities which might have been imported from neighbouring countries, since it is in these commodities — at least in the industrial field — that developing countries tend to be most competitive with their neighbours. Again, so far as the future is concerned, it is not so much a matter of trade diversion as between the outside world and the regional group, but rather whether policies will be adopted which favour regional specialization as against those

[1] *See* Meade, *op. cit.* p. 110.

214

which favour the maximum degree of self-sufficiency at whatever cost on the basis of domestic markets alone.

## V. THE ECONOMIES OF SCALE

While the gains from economies of scale are usually mentioned as a significant argument for the formation of customs unions or free trade areas, there is considerable difference of opinion as to its importance and some are frankly sceptical regarding its significance for Western Europe.[1] First of all, the possibilities of realizing economies of scale differ greatly for different types of commodities for the same market, and for the same commodity for countries of varying market size. For countries like Brazil and Argentina, the domestic market may be large enough to permit realization of economies of scale for a wide range of consumers' goods and even intermediate goods ; this is certainly not true for the countries of Central America, most of which could not support an economically sized soap factory or fertilizer plant on the basis of the domestic market alone. On the other hand, the domestic market even in the largest and most industrially advanced of less-developed countries is not large enough to justify a plant of economical size for a large number of items, such as specialized machinery, transport equipment, certain chemicals, and electronics.

In a recent study on 'Patterns of Industrial Growth',[2] Professor Chenery has shown that as *per capita* income rises from $100 to $600, the percentage of production of investment goods to total manufacturing output approximately triples according to the normal pattern based on a sample of some fifty countries, including both industrialized and under-developed. When allowance is made for variations in the size of the country, deviations from the normal pattern are 'smallest for services, agriculture, and most manufactured consumer goods', while the greatest variation from the normal is found in 'industries producing machinery, transport equipment, and intermediate goods, where economies of scale are important'.[3] Chenery points out that in modern developing countries, the leading sectors of the economies — or those which provide the impetus to growth — are 'likely to be the industries in which import substitution becomes profitable as markets expand and capital and skills

---

[1] See, for example, H. G. Johnson, 'The Criteria of Economic Advantage', *Bulletin of the Oxford University Institute of Statistics*, Feb. 1957, p. 35 ; see also 'The Economic Gains from Free Trade with Europe', *Three Banks Review*, Sept. 1958.

[2] See Hollis B. Chenery, 'Patterns of Industrial Growth', *American Economic Review*, Sept. 1960, pp. 624-54.　　　　[3] *Ibid.* pp. 650-1.

215

are acquired '.[1] Hence he concludes that limitations on market size are an important factor in preventing normal growth of developing countries by their being unable to move into the production of investment and intermediate goods where economies of scale are especially important. Thus he lays special emphasis on the creation of regional trading arrangements which will increase market size as a means of promoting development in accordance with the normal pattern of industrial growth.[2]

Any realistic discussion of the advantages to be derived from economies of scale must take into account the nature of the market in which producers are operating. Outside of Communist countries, few manufacturing industries are complete monopolies and in most cases the existence of several producers, some or all of which may already have excess capacity, prevents the realization of potential economies of scale on the basis of the domestic market alone. Imperfect competition, government restraints and private collusion of various kinds prevent individual producers from establishing new low-cost plants which would force competitors out of business by taking over a larger share of the market, or prevent new foreign enterprises from doing so. Hence the impetus for the establishment of new low-cost plants may need to come from the opening up of an external market where conditions of competition, either with firms in the export market or from third countries, may be such as to require lower-cost production achieved through economies of scale. Moreover, if production is in the hands of a foreign firm which has established distribution facilities throughout the regional market, the foreign firm may be able to supply its entire regional market by expanding the output of one country, or the foreign firm may be able to lower costs by producing certain components in individual countries while continuing to assemble the finished product in the plants of several individual countries within the common market. Finally, this same firm may as a consequence of reduced costs of the finished product or, more likely, of components, be able to supply markets outside the regional group, thereby broadening the export base of the regional group.

The contribution of Professor Tibor Scitovsky to the economies of scale which might be realized from economic integration is especially relevant for developing countries.[3] The high degrees of market imperfection, the factor of risk and uncertainty arising from political instability, and the tendency on the part of domestically

[1] See Hollis B. Chenery, 'Patterns of Industrial Growth', *American Economic Review*, Sept. 1960, pp. 624-54.    [2] *Ibid.*
[3] See Tibor Scitovsky, *Economic Theory and Western European Integration*, Stanford University Press, 1958, Chapter 3.

216

owned firms to favour high margins and low output, greatly limit the willingness of firms to build plants which will be optimal for the level of domestic demand, say, five or ten years hence. Therefore, accretions in demand tend to be supplied by the addition of sub-optimal equipment. On the other hand, an expansion of the market area to other countries in the region may lead some firms to establish plants with equipment permitting substantial economies of scale. Unfortunately, several of the countries in the Montevideo group are already highly competitive in a number of industries, such as durable consumers' goods and steel, in which significant economies of scale could probably be realized. Therefore, reductions in trade barriers affecting these commodities are likely to proceed very slowly if indeed much progress is made at all in the next few years. Greater progress will be made in the reduction of barriers to trade in goods, the domestic production of which is not yet substantial and imports still supply the vast bulk of the region's requirements. However, in the absence of a regional trade arrangement, relatively high-cost plants for the production of these commodities will eventually be established in some of the countries as the process of substitution continues. On the other hand, more economically sized plants might be established if production for a regional market could be assured. The reduction of intra-regional barriers on new goods, such as specialized capital equipment, will be determined on the basis of intra-regional bargaining, since the first interest of each country will be to preserve the potential domestic market for its own production of a given commodity, an interest that it will compromise only if each country is assured the opportunity of exploiting the regional market for other commodities. Tariff negotiations among the members of the Montevideo Free Trade Area are therefore likely to lead to agreements or understandings regarding the establishment of plants, and this will not always mean production in the most efficient country. It can be argued, of course, that this will increase the extent of trade diversion, but it should be kept in mind that trade diversion will take place in any case and that *total* trade with the outside world is not likely to be greatly affected.

## VI. THE EFFECTS OF COMPETITION

As we have already noted, any discussion of the gains from the economies of scale cannot be separated from the nature of the markets and the degree of competition within countries and competition between members of the regional group. Most discussions

H 2         217

of the benefits of customs unions have tended to emphasize the gains from competition that are likely to result from the more impersonal competitive forces arising from the creation of the regional market, impinging upon the imperfect or oligopolistic structure of domestic markets. This argument undoubtedly has significance for Western Europe and perhaps should apply with even greater force with respect to developing countries which are characterized by monopolistic elements of all kinds.[1] However, competition is by no means a popular principle in developing countries, and in Latin America regional trading arrangements are viewed more as mechanisms for development planning on a regional basis than as providing the basis for intra-regional competition. In fact, the term competition is not found in the text of the Montevideo Treaty and has been virtually absent from discussions of the gains from regional integration. The emphasis is on the principles of 'reciprocity' and of 'planned complementarity'.[2] Hence, we may find, initially at least, that the intra-regional trade liberalization measures within the Montevideo group will emphasize reductions in barriers on industrial commodities in which members are not currently competitive, while avoiding those in which they are competitive. In other words, the arrangements would seem to favour trade diversion over net trade expansion. However, as we have already indicated, we must look beyond the shorter-run impacts on the existing pattern of trade to the effects on future patterns of production and trade as determined by alternative policies affecting investment and resource allocation. If the Montevideo Treaty programme moves toward its long-term goal of complete free trade — at least in industrial commodities — members must begin undertaking reductions in barriers which will affect the commodities in which they are competitive. This will be easier to do if the reductions take place gradually so that, given the general accretion of demand, serious damage will not be done to existing firms, but, rather, there will be a gradual increase in the proportion of the market represented by intra-regional trade. Such a development cannot help but have an impact on breaking down internal market rigidities. In fact, the major gains in this respect may occur in the countries which become important exporters of certain commodities, since in order to compete abroad, these firms

---

[1] See Tibor Scitovsky, 'International Trade and Economic Integration as a Means of Overcoming the Disadvantages of a Small Nation', *Economic Consequences of the Size of Nations* (Proceedings of a Conference held by the International Economic Association), Macmillan, London, 1960, pp. 282-90.
[2] See Raymond F. Mikesell, 'The Movement Toward Regional Trading Groups in Latin America', *Latin America Issues : Essays and Comments* (edited by Albert O. Hirschman), Twentieth Century Fund, New York, 1961, pp. 125-51.

218

will inevitably undertake cost-reducing measures and introduce optimal equipment which will result in lower prices and perhaps the forcing out of marginal firms in the exporting country. The political repercussions of such developments will be less severe than in cases where marginal firms are forced out as a consequence of import competition.

Even the threat of eventual competition from abroad or a stepped-up pace of competitive activity for exporting abroad will shake a number of Latin American industries out of their lethargy and stagnation. They will see that sooner or later they must adopt new methods in both production and distribution. Moreover, the creation of a regional trading area will lead to greater contacts among business men with the consequent increase in the exchange of ideas. Finally, as a spur to a competitive activity within the region, competing foreign firms will enter on the expectation of being able to exploit a larger regional market, either directly with their own subsidiaries or through joint ventures. Such firms are accustomed to competing with one another in markets throughout the world and will spread the arena of their competitive activities to Latin America.

As has already been mentioned, Latin American policy-makers are hoping to work out complementary agreements in certain industries as a basis for trade, rather than simply lowering the barriers and letting competition take its course. In Central America this principle has been formalized in the General Treaty on Central American Integration, which, among other things, provides for joint planning and certification of manufacturing firms in particular industries which would be given free access to the Central American market under conditions which would avoid over-capacity and unrestricted competition. Although formal provisions for the certification of industries are not included under the Montevideo Treaty, the idea of special complementarity agreements is well established.

In a recent article Professor Jan Tinbergen [1] argues for regional planning as opposed to competition in the heavy industry field on grounds that free entry and competition in these industries are not likely to produce optimal development. Tinbergen favours planned production with the aid of economic models by means of which the optimum pattern of heavy industry development, including plant size and location, could be determined for the region as a whole.

[1] See Jan Tinbergen, 'Heavy Industry in the Latin American Common Market', *Economic Bulletin for Latin America*, United Nations Economic Commission for Latin America, Santiago, Chile, March 1960, pp. 1-5.

He gives several reasons why free entry and competition would not achieve these conditions, including : (1) the long construction period required for the individual projects, which would reduce the accuracy of decisions arising out of the market mechanism ; (2) the large amount of capital required to establish plants of optimum size which would not be forthcoming except under conditions of planning and assured demand ; and (3) the failure of free enterprise to establish heavy industry in optimum locations.[1]

It is undoubtedly true that regional planning of heavy industry based on Professor Tinbergen's economic models would provide a closer approximation to the optimal size and location of heavy industry than that which would be achieved under the operation of completely unfettered competition. However, freely competitive conditions in this field may be ruled out as unrealistic in any case because of the oligopolistic nature of heavy industry and of the role played by governments as providers of credits or of direct participants in the enterprises. Moreover, I do see a danger in leaving the planning of heavy industry to government negotiators. Agreements in this field are likely to be negotiated by political representatives with a view to achieving a kind of 'balance of industrialization' among the countries within the regional group, rather than on the basis of optimal size and location of plants in accordance with a rational programme. Also, I can think of few processes more stifling to growth than to leave the development of heavy industry to the almost interminable deliberations of government negotiators. This problem might be dealt with by the creation of an independent or supra-governmental authority such as the European Coal and Steel Community, which would have the power to control investments in heavy industry, but such an institution, so far as I am aware, has not been contemplated for the Latin American Free Trade Area. The price of achieving or attempting to achieve optimal solutions may be a considerably slower rate of investment and a prevention of the full operation of the inducement mechanisms. For countries in a hurry to develop, this is far too great a price to pay. However, the formulation of long-range economic programmes prepared by ECLA or other regional groups, which would serve as a guide to domestic and foreign private investors, to governments, and to external lending institutions, together with arrangements for regional consultations on plans for major investment expenditures in the heavy industry field, would be of immense value.

---

[1] See Jan Tinbergen, 'Heavy Industry in the Latin American Common Market', *Economic Bulletin for Latin America*, United Nations Economic Commission for Latin America, Santiago, Chile, March 1960, pp. 2-4.

## VII. UNEQUAL WELFARE EFFECTS AND ECONOMIC DUALISM

Customs union theory has recognized that some members of a customs union or free trade area may gain in terms of economic welfare while others may lose. For example, if one of the members does not appreciably increase its exports but simply shifts its imports from lower- to higher-cost sources, membership in a preferential trading area may mean little more than a deterioration in its terms of trade. An extreme case would be one in which a country, as a consequence of joining a customs union, would have to raise its import duties on commodities from the outside world in order to provide a market for higher-cost imports from its partners, while at the same time there was no offsetting export gain. Even if it did not raise its duties, and consumers were able to import at somewhat lower prices or at least no higher prices, from partner countries, the government would lose the tariff revenue on imports diverted from external sources and presumably would have to make up the revenue by taxing its citizens in some other way.

Looked at from the standpoint of the longer-run impact on developing countries, a customs union or other regional preference arrangement might have an even more adverse impact upon certain members. Capital, skills and entrepreneurs, both from within the preference region and from abroad, might be drawn to the major industrial centres of the more advanced partners in order to take advantage of external economies in these areas, and to locate their plants closer to the major markets. In his study of the *Strategy of Economic Development* Professor Hirschman warns against what he calls the *polarization* effects which operate in developing countries to create a situation in which progress in certain areas, mainly the rapidly industrializing regions, is accompanied by, and even contributes to, stagnation in other regions.[1] The creation of free trade areas or customs unions which include countries encompassing less advanced regions and partners representing the more industrially advanced regions may well reinforce these polarization effects. While there are some offsetting forces resulting from the increase in demand for primary commodities from the less advanced regions, these may not be strong enough to offset the polarization effects.

The problem of dualism can be handled by a single country or

---

[1] Albert O. Hirschman, *The Strategy of Economic Development*, Yale University Press, New Haven, 1958, Chapter 10.

221

an economic union with a strong central government and centralized fiscal system. Special encouragements can be given to the location of industries outside the metropolitan centres by means of tax inducements, and by heavy expenditures for transportation, power and other overhead facilities in advance of immediate industrial needs. Loan capital can be distributed in a way which favours the development of the hinterland. However, if economic integration programmes do not include some mechanism by means of which the less advanced countries are given somewhat more favourable treatment in the distribution of capital expenditures for economic overhead projects or possibly special measures for attracting direct private investment, the net results of combining less advanced and more advanced countries in a regional trade arrangement may very well be to increase the degree of dualism with its attendant political and social frictions and frustrations. This problem has been recognized in the EEC by the creation of special financial institutions such as the European Investment Bank and the European Social Fund, but it is likely to be much more serious among the developing countries forming common markets and free trade areas.

Although the Central American integration plan provides for the establishment of a Central American Integration Bank, the Montevideo Treaty has thus far not established financial facilities which would help to balance the advantages as between the less advanced and the more advanced partners. Provision is made in the Montevideo Treaty for the less advanced members to proceed more slowly with import liberalization than the more advanced members, but what is needed is something more positive which will help the less advanced members to broaden their export markets within the region, particularly in industrial commodities, rather than simply retard the impact of regional competition on their own markets. It would seem highly desirable, therefore, that either there be established a special long-term financing institution to operate as a part of the Montevideo Treaty Organization, or the Inter-American Development Bank play a special role in dealing with this problem in close co-operation with the Montevideo Organization.

There is little doubt that the less advanced members of the Latin American Free Trade Area have an actual or potential cost advantage in a number of industrial and agricultural products with respect to other members of the regional group. But for these advantages to be exploited there must be enterprise, capital, better transportation facilities, and perhaps distribution facilities in other members in addition to reduced trade barriers. Disadvantages arising from high transportation costs and location relative to the major markets may

**222**

well outweigh the cost advantages of producing in the less advanced members.

The fact that the Montevideo Treaty takes the form of a free trade area rather than a customs union undoubtedly reduces the extent of welfare loss on the part of the less advanced members, while at the same time it reduces the possibilities for trade expansion within the area. Countries like Uruguay and Paraguay, which cannot expect to produce their own tractors or capital equipment for a long time to come, tend to have low rates of duty on these commodities so that they can provide little margin of preference for imports of these goods from Brazil and Argentina even though they abolish all of their restrictions on industrial imports from partner countries. On the other hand, such industrial goods as they are likely to be able to export successfully in competition with producers in other members of the regional group are likely to have a high margin of preference. The tendency for countries to have low tariffs or few restrictions (except for balance of payments or revenue purposes) on commodities which they do not produce or expect to produce in the near future themselves, but a high degree of restriction on commodities they are producing for the domestic market, has led many Latin American economists to the position that the Montevideo Treaty Organization must be converted into a customs union if it is to be successful in expanding intra-regional trade. In other words, there is a fear that countries will not be willing to afford to partner members a discriminatory wall of protection on the commodities they are willing to import from them ; while they are not willing to make concessions to their partners on the commodities on which they are maintaining a high level of protection as a means of securing the market for their domestic producers.

This position seems to arise from a view that a Latin American preference area should be mainly trade-diverting and that intra-regional trade is possible only if a high and fairly uniform tariff wall around the entire region is maintained. I think this is a rather static and short-sighted view of the potential benefits from a regional free market. I suspect that the principal incentive to the expansion of investment in one country in order to market a portion of its output in neighbouring countries is the assurance or expectation that it will have free access to those markets with low or non-existent restrictions, rather than the expectation that it will have a substantial margin of preference over exports from third countries. Moreover, I do not believe that it is possible to create a successful regional trading area on a basis of raising prices to consumers in one country in order to provide a market for the goods of partner countries.

Finally, there is reason to believe that given equilibrium rates of exchange and adequate transportation facilities which will permit the realization of locational advantages, costs of production in developing countries will not be significantly higher (in fact, they might well be lower) for the goods which they are exporting within the region than those from industrially advanced countries outside of the preference region.

While admitting that a free trade area can and will provide a considerable degree of regional preference over external goods, the deliberate creation of a common high wall of protection against outside competition for all goods sold within the region does not appear to be either feasible or desirable from the standpoint of the long-run development of the region. In fact, the long-run aim of the region should be a gradual reduction of barriers on imports from outside as well as within the region itself.

## VIII. THE PAYMENTS PROBLEM

So much has been written regarding the payments problem in relation to common markets and free trade areas, and the subject has so many ramifications, that an adequate discussion of this problem for developing countries would require a separate paper in itself. Students of customs unions have evolved sharply conflicting positions with respect to the payments problem and the means of dealing with it. Dr. Thomas Balogh and Dr. Raul Prebisch (and his colleagues in the United Nations Economic Commission for Latin America) tend to favour a multilateral compensation system for financing trade among members of a regional preference area which would avoid, or largely avoid, the necessity for settlements in convertible exchange or gold.[1] At the other extreme are those who believe that a successful customs union or free trade area is not possible except under conditions of financing with freely convertible currencies and the maintenance of over-all balance of payments equilibrium by individual members without restrictions. According to this view, the attempt to achieve freedom of payments internally, while at the same time permitting individual countries to maintain balance of payments restrictions on trade and payments with the

[1] See T. Balogh, 'The Dollar Crisis Revisited', *Oxford Review Economic Papers*, Sept. 1954, and 'The Dollar Shortage Once More, a Reply', *Scottish Journal of Political Economy*, June 1955 ; for a discussion of the position of Dr. Prébisch and of the ECLA Secretariat, see *The Latin American Common Market*, United Nations Economic Commission for Latin America, Mexico, July 1959, pp. 17-22 ; see also Victor L. Urquidi, *Trayectoria del Mercado Comun Latino-americano*, Centro de Estudios Monetarios Latinoamericanos, Mexico City, 1960.

224

outside world, is unworkable since such an arrangement would lead to large imbalances within the preferential system. Even a common policy with regard to trade and payments relations with third countries is difficult to maintain in the absence of a full economic union.[1] Professor Meade favours a system of fluctuating exchange rates as perhaps the best means of maintaining balance of payments equilibrium and freedom from restrictions, while others, including myself, believe that the uncertainties resulting from frequent changes in exchange rates would greatly reduce the benefits from the formation of customs unions.

As regards developing countries which meet the conditions for our model, those who favour a multilateral compensation system involving no settlements in external currencies are usually identified with the position that it is necessary to establish a highly discriminatory system in which each member country's trade is balanced over time with the group, and no country is permitted to earn convertible exchange by increasing its exports to the group. This view is based in turn upon the conviction that manufactured exports from one partner to another simply cannot be competitive with external goods or, more generally, that there exists within the group a shortage of convertible currency which would lead members to avoid using their convertible exchange for purchases within the area. There is really little basis for this approach, which reflects adherence to a 'dollar shortage' philosophy long after there is any justification for it. There is no general shortage of convertible means of payment in the world today. Nearly all Latin American countries have adopted realistic exchange rates for the bulk of their trade with the outside world and they urgently need to remove existing price and cost disparities among themselves artificially created by subsidies, bilateral trade agreements and import controls of various kinds. If the problem is one of providing additional liquidity for financing an expanded volume of trade among themselves by means of convertible currencies rather than bilateral agreements, this can and should be handled by special assistance from external sources such as the International Monetary Fund.

Apart from this, the attempt to create a multilateral compensation scheme among countries whose existing intra-regional trade is relatively small is fraught with difficulties, since anything approaching a regional balance would be little short of a miracle. If the desire for a balanced expansion of trade is an important aspect of the

---

[1] For a discussion of the balance of payments problems of customs unions, see Scitovsky, *Economic Theory and Western European Integration, op. cit.* pp. 95-100 ; and Meade, *op. cit.* pp. 14-28 and pp. 116-19.

regional trading scheme, it should be achieved by means other than through a payments scheme. In fact, the results of the operation of a payments scheme is only a reflection of the operation of the fundamental trade liberalization programme, and if the payments positions cannot be compensated multilaterally over time, it will be necessary to change the basis of the trade liberalization programme. Hence all that is really needed is some means of keeping track of the intra-regional balances on current account, and an elaborate payments mechanism is unnecessary for this purpose.

A more fundamental question is whether the *trade* arrangements should be such as to achieve an approximate balance of each member with the group. Personally, I do not believe that this should be a fundamental aim since the existence of a surplus or a deficit of an individual member with the group does not measure the welfare gain or loss from membership. As the economies of partners progressively grow, the pattern of their production and trade will change and the determination of the long-run benefits of individual partners from membership in the preference area will require far more subtle means of measurement. For some countries, for example, there may be no loss whatsoever involved in increasing the share of imports from regional sources in their total imports, while at the same time they are enjoying the gains from a broader market for their export. Indeed, they may find their terms of trade improving even though they develop with their regional partners which they must finance with convertible currencies. Likewise, the achievement of a surplus on intra-regional account by one partner country is not a necessary measure of its relative benefits from membership in the regional preference area. Conceivably it could be paying too much in terms of a deterioration in its terms of trade from the intra-regional surplus that it is achieving.

All of this is not to say that there is not a problem in making sure that the benefits and losses from the creation of a regional preference area are equitably shared, but this must be done by means much more fundamental than setting up a multilateral compensation scheme.

## IX. THE INDUCEMENT EFFECTS ON INVESTMENT FROM BROADENING THE MARKET AREA

Development literature is full of examples of the impact of market growth upon investment and productivity and on the revitalization of stagnating industries serving a local market. As a

226

rule this has come about through the development of transportation within a country, through improved marketing methods and the expansion of incomes. For modern developing countries the expansion of external demand has not provided the basis for the growth of investment in manufacturing, and for many countries exports of primary commodities have been growing slowly or stagnating and have provided little inducement for increased investment. Also, the surplus of labour in the agricultural regions has provided little inducement for increasing productivity. Inducements to investment based solely on internal developments have certain limitations. First, for a large number of industries, income elasticities may be rather low and production of new goods as substitutes for imports depends upon whether or not the internal market has grown to the point where plants of an economical size can be established. Such substitution can be forced by high or prohibitive import restrictions, but this may mean an uneconomical use of capital through the creation of excess capacity or of very high-cost productive facilities. Once these industries are established, they tend to stagnate for lack of dynamic growth of demand.

A second difficulty with internally induced investment is that it does not provide any additional foreign exchange to meet import costs of investment goods, intermediate goods, raw materials and fuel, unless, of course, the industries are established by foreign capital. But even here, the actual foreign exchange contribution of foreign manufacturing enterprises is likely to be small since they depend for their growth upon reinvested profits and perhaps for a part of the capital for their initial establishment on domestic sources. There is, of course, an offset against these additional foreign exchange expenditures from increased supplies of the import-competing goods, but this does not occur fast enough for the country to maintain a high level of investment with slowly rising or stagnating export receipts. Hence, domestically induced investment is hampered by exchange shortages which result in the imposition of import restrictions and/or exchange depreciation. The import restrictions or steadily depreciating exchange rates in the face of growing domestic demand creates a condition of chronic inflation which brings about a misdirection of investment and a tendency for savings to flow into less productive uses or to find their way to foreign capital markets.

Investment induced by an expansion of external demand for export goods avoids these disadvantages. Export demand does not depend upon slowly growing domestic income and provides the foreign exchange for the increase in investment in productive facilities. The opening-up of markets in neighbouring countries

**227**

adds a new dimension to market growth. Demand and supply elasticities will increase substantially, particularly because they tend to be rather low in countries of limited industrial development. This will increase economic flexibility and open up new opportunities for investment, both foreign and domestic.

The literature dealing with the importance of broadening the export base for regional economic growth is of particular relevance for the creation of regional trading arrangements among developing countries.[1] Successful regional growth cannot be a 'bootstrap operation', but depends upon the creation of an export base which permits specialization in the production of those goods and services for which the region's resources are best suited and the creation of external economies. Of course, successful regional centres of economic growth soon develop local industries producing mainly for local consumption, but it is the exports to other regions that provide the external impetus which then has a multiplier effect. By analogy, the same reasoning can be applied to nations: for maximum growth they need the stimulus of an external demand for their products and the possibility of broadening their export base to include new export industries as the old ones lose their earlier vitality.

## X. CONCLUSION

In concluding this paper I would say that the principal way in which customs union theory needs to be modified for application to the problems of developing countries is by taking into account the likely long-term changes in the pattern of production. This is especially important for those countries where the export industries do not constitute the leading sector, but rather as a consequence of the slow growth of export proceeds there is a strong drive towards substitution. Growth is inhibited by the fact that (a) substitution cannot take place fast enough to keep import requirements within the limits of exchange availabilities; and (b) efforts to create new industries on the basis of supplying the domestic market alone result in high-cost production and misdirected resources. Because countries cannot or do not specialize in their industrial production, they either produce with sub-optimal equipment or create over-capacity, or both. In addition, they cannot take advantage of the opportunities to specialize in the production of commodities in which they have peculiar advantages resulting from access to raw

[1] See Douglass C. North, 'Location Theory and Regional Economic Growth', *Journal of Political Economy*, June 1955, pp. 243-58.

materials, location, etc. Moreover, the problem is not so much that countries may be producing the wrong things, since in time, and given broad enough markets and access to skills, techniques and know-how from abroad, they might become reasonably productive in any one of a very large range of industrial commodities. Rather, the problem stems from the fact that they are unable to specialize, and in trying to grow on the basis of limited exchange resources, they are seeking to produce too many things, including finished commodities, intermediate goods and, to an increasing extent, capital goods as well.

The creation of regional trading arrangements provides an opportunity for specialization and increased trade, thereby broadening the export base of individual countries and increasing the productivity of the trading region as a whole. It might, of course, be argued that the gains would be greater if each country broadened its export base by expanding its export of both primary commodities and of manufactures to the rest of the world. This is a good doctrine to preach, but it has not happened and it is not likely to happen until developing countries learn to trade and compete with one another on a regional basis. At a later stage, just as the countries of Western Europe soon began to compete actively in a wide range of commodities with the United States once they had learned to compete with one another, so also, I believe, trade and competition in industrial products among developing countries will provide the experience and discipline for them to sell their industrial products on world markets, thereby broadening the export base of the entire region. This process, of course, will be assisted by the operations of international corporations with distribution facilities throughout the world. Moreover, increased investment by foreign enterprise in developing countries will be greatly encouraged if they can produce finished products or components for sale throughout the preference region.

There is, of course, another pattern by which development can and will take place, at least for some countries : that is, for the export sectors to be the leading sectors and industrialization to develop, first for supplying the local market, and later for sale to world markets. Such countries may be able to develop along the same lines as the United States, Canada and Japan and certain Western European countries developed during the nineteenth century. However, I fear that this pattern of development may be the exception to the rule in the twentieth century.

# Part V

## International Cartels, Commodity Agreements, and the Oil Problem

Caves
Behrman
Lichtblau

# International Cartels and Monopolies in International Trade

RICHARD E. CAVES

The formal niceties of pure competition make it the stock-in-trade market structure in theoretical models of the international economy. Yet imperfect competition calls out for attention as a matter of international economic policy and, therefore, poses issues for theoretical and empirical research that cannot in good conscience be ignored. This chapter addresses two major ways in which imperfect competition impinges on the making of international economic policy. First, governments engage in or promote the formation of international cartels in order to maximize national monopoly gains, or they seek to evade exploitation by such cartels. Second, governments maintain competition in their national markets by means of policies that should recognize the presence or absence of international competition. The two main sections of this paper summarize the theory and empirical evidence relevant to governments' policy choices as exploiters of monopoly power and as enforcers of competition.

## I. CREATION AND MAINTENANCE OF INTERNATIONAL CARTELS

By an international cartel, I shall mean an agreement among producers of a given good or service located in different countries and covering the bulk of the market decision variables that must be manipulated or constrained in order to achieve significant joint monopoly profits. Cartels have often been organized among private-sector producers. For this analysis, however, I shall assume that they are either the direct work of national governments or that governments enforce them in the pursuit of national economic objectives. This approach is responsive to the recent role of governments in forming or seeking to form primary-product cartels, and it focuses our attention on the behavioral conditions necessary for forming and maintaining cartels. That is, we can take the key structural conditions

Reprinted from "International Economic Policy: Theory and Evidence" (R. Dornbusch and J.A. Frenkel, eds.), pp. 39–75, with permission of The Johns Hopkins University Press, Baltimore, Copyright 1979.

as given—a certain number of countries possessing bauxite deposits, for example—and concentrate on the game-theoretic depiction of their behavior that is the heart of traditional cartel theory. In the second part of this chapter, where we consider the problems of governments trying to maintain secure, effective, industrial performance, we shall be back in structuralist territory. The exposition of this section uses the microeconomic terminology that is standard for cartel theory. Some remarks about the process of translating the conclusions into the realm of national economies and general equilibrium come at the end.

Under what conditions do potential participants join a cartel? Once the cartel is under way, what are the incentives to defect, and what causes the parties to resist the temptation to cheat? It is useful to begin our answer by drawing upon the Cournot–Nash duopoly theory, in which Cournot's solution to the classic duopoly problem supplies a reference-point outcome for the case in which the parties achieve no cooperation. This solution of course is not in the core, but while we address the possibilities of a collusive bargain between sellers it is hardly unreasonable to put aside the possibility of a subsequent deal with the exploited buyers to supply them at marginal cost.[1]

The familiar "prisoner's dilemma" game explains why the Cournot solution emerges in single-period games. It would also emerge from a model involving trade in $n$ successive periods, if trade in all periods is arranged by means of binding contracts reached on the initial day. More surprising, the Cournot solution also emerges in a market operating for a finite number of trading days, if sales are determined afresh each period and there is an enforcement mechanism that penalizes each period's noncolluders in the next time period. As Shubik (1959) and Telser (1972) have shown, this proposition is easily proved by backward induction. In the last ($n$th) time period, no punishment is possible for violators of a monopolistic consensus, and so the Cournot solution prevails in the $n$th period. But given that expected outcome, nothing can deter a party from cheating on that consensus in period $n - 1$, on the assumption that enforcement consists of reverting to the Cournot solution, which will obtain in period $n$ in any case. The proof continues back to the initial period. The empirically interesting point that emerges from this analysis is that avoidance of the prisoner's-dilemma solution requires that the participants not know when the game is going to end. Telser develops this property by assuming that each duopolist supposes in each period that the game will continue for one (or more) additional periods with some exogenously

---

[1] Telser (1972, pp. 138–39) shows that the Cournot solution lies in the core of a reduced market—one in which the buyers extramarginal in the actual Cournot solution are assumed absent from the market.

given probability $a$. The temptation to cheat on a joint-monopoly consensus depends on the severity of punishment (modeled by Telser as retrogression for $k$ periods to the Cournot outcome), but given that severity it is greater the smaller is $a$. Telser also points out that, if cheating is expected to be profitable, it will in this model be undertaken immediately; a joint monopoly once attained should be stable indefinitely if $k$ and $a$ values are stationary and high enough to deter cheating.[2]

*The creation of cartels*

The Shubik–Telser analysis, reported so far, is useful for introducing the central concerns of cartel theory—the essential terms of agreement and the conditions under which an agreement once reached will continue to be honored. Yet there is something peculiar about distinguishing the conditions under which sellers can reach an agreement from those under which they can sustain it. Why not suppose that potential cartel members correctly anticipate the enforcement loopholes of an agreement when it is first drafted? With that rationality assumed, we could predict a cartel would be undertaken if and only if the present value of its joint monopoly profit exceeds the present value of its expected costs of operation and enforcement. Outside interference apart, a potentially profitable cartel then is blocked only by its members' lack of inventiveness with devices that bind their commitment to follow the terms of the agreement. As Schelling (1960, especially chap. 5) points out, the strategies for achieving binding commitment may be rather rich. Nonetheless, the received body of cartel theory puts most of its emphasis on the problems of punishment and defection from incomplete cartel agreements without explaining why—aside from observed fact—this contractual incompleteness should pertain.

It is useful at this point to remind ourselves what is required for a contractually complete cartel. It is not enough to divide markets or agree on a common price. Joint profits from the cartel's activity cannot simply pass to each member as the net revenues resulting from his apportioned sales. Without pooling profits and dividing them according to some pre-agreed formula, the members will find that the allocation of production among themselves must meet an inconsistent set of objectives: it must minimize the aggregate cost of producing the joint-profit-maximizing output, and it must also generate whatever division of profits is consistent with the bargaining power of the parties. Without profit pooling, the feasible locus of efficient profit outcomes will contain many points that fall short of maximum joint profit. Only with profit-pooling or side-payments can the

---

[2] See Telser (1972, pp. 142–45). Recent developments in game theory seem to identify devices capable of sustaining collusive outcomes in noncooperative games that may have some empirical counterparts. They include the use of randomized strategies and the presence of differing subjective priors. Unfortunately, it is beyond the scope of this paper to survey these developments.

parties attain the maximum joint profit and divide the spoils in a way consistent with their bargaining power.[3] Securing a cost-minimizing distribution of outputs that sum to the group's profit-maximizing joint output level poses its own problems for the optimizing cartel. If the participants have different cost curves, the efficient distribution of output requires that all active producing units operate at a common marginal cost and that the output level of the least efficient unit in operation be such that its average cost is at a minimum.[4] Cartel members might obtain this efficient allocation by creating rights to produce given quantities of the profit-maximizing joint output, issuing these to the charter members as part of the basic agreement, and allowing them to be traded among sellers. Fully efficient trade would produce the optimal allocation of output just described.[5] Many decision variables besides the division of output and profits may affect the attainment of maximum joint profits, when output is heterogeneous and the transaction offered to the buyer can vary in a number of dimensions besides price.

When we consider these requirements for agreement, it is evident that contractual costs and uncertainties are an important potential limit on the ability of sellers to reach accord. The heterogeneity of sellers' preferences and of their perceived opportunity sets contributes to these costs. Oligopoly theory in the Fellner–Chamberlin tradition deals with these influences as limitations on how closely the market bargain can approach full joint profit maximization, and the field of industrial organization contains a large literature on the structural conditions and sellers' strategies capable of sustaining a noncompetitive market bargain. However, these valuable insights into the structural conditions for a cartel agreement do not generally take the form of deterministic cartel models, and so I reluctantly put them aside. Consider instead two formal propositions about the terms of cartel agreements.

First, a cartel agreement can potentially be made enforceable without an elaborate mechanism to defuse what would otherwise be a strong temptation to cheat. Osborne (1976) and Spence (forthcoming) show that the adoption by duopolists of reaction functions that commit them to maintaining constant shares of the value of total sales can lead them to a point on the contract curve with respect to their profits. That is, an agreement on market shares has a certain superiority over other forms of agreement on the key market variables.[6] The cartel's charter members can make other choices of terms that ease the enforcement burden. One frequently observed empirically is to divide markets (classes of customers, geographic areas,

[3] Telser (1972, chap. 5, esp. pp. 192–94).
[4] Patinkin (1947); Telser (1972, chap. 5).
[5] See Stigler (1952, chap. 14).
[6] Of course, the problem remains of finding a mutually agreeable set of market shares. See Cross (1969, pp. 207–14).

etc.) among the various members. Enforcement costs are then reduced to those of assuring that no one sells to a forbidden customer. There has been some analysis of optimally imperfect terms of agreement for cartels that cannot achieve complete joint maximization. For instance, Comanor and Schankerman (1976) point out that, for industries selling on the basis of bids on individual transactions, identical bids are less costly to enforce than a scheme of rotated bids that requires explicit agreement on market shares, so that we expect (and find) schemes involving the rotation of bids typically to encompass smaller numbers of sellers.

Second, a cartel may have a positive value to its members even if it achieves no long-run departure from a competitive market outcome. Consider the simple story often told in the institutional literature on cartels operated between World Wars I and II: A cartel is formed without the accession of all actual or potential producers of the good in question. The price is raised. The outsiders find price comfortably in excess of their marginal costs and expand output. Newcomers observe the elevated price to exceed their minimum attainable average costs and enter the industry. The cartel members start to lose market share, and the cartel-managed price gives way. The usual account then concludes that the cartel failed, and a soporific moral is drawn about the ultimate triumph of pure competition. The trouble is that the cartel members did expropriate consumers' surplus and cause deadweight losses of welfare while the cartel was in operation. And it could be rational to enter a cartel expected to be temporary (whether due to entry by outsiders or the defection of nominal signatories), even if the profits of the cartel members in the competitive period after the cartel's collapse are expected to be less than the competitive profits they earned before its formation.

### The maintenance of cartels

Conventional cartel theory becomes more loquacious on the maintenance of a cartel agreement—the incentive to cheat, the chances that cheating will be detected, and the mechanisms of enforcement against the cheater. As the preceding discussion has suggested, this focus is unsatisfying because the enforcement problem results from the incompleteness of the cartel agreement—itself unexplained theoretically. Therefore, models of cartel enforcement float in an undefined structural context, and they are specific to an arbitrary set of initial conditions. I shall stress this problem of structural context in surveying these models.

The fundamental problem of maintaining cartel arrangements is that in order to obtain monopoly profits they must elevate price above producing members' marginal costs. In the absence of a common sales agency or binding profit-sharing arrangement, each member can potentially increase his own profits in the short run by any maneuver that lets him sell an extra unit for net revenue less than the official cartel price though greater than marginal cost. The factors affecting the net revenue gain expected by the potential cheater can be classified into three groups. First, consider the behavior of short-run marginal costs as the cartel members collectively restrict output and elevate the market price. The more steeply sloped their marginal cost curves in the neighborhood of the precartel output are, the greater becomes the gap between price and marginal cost. A familiar generalization is that a high level of fixed costs as a percentage of total costs should be associated with a steeply sloped marginal-cost curve.[7] Strictly speaking, the observation that fixed costs make up a large proportion of total costs tells us only about the magnitude of *total* variable costs— the integral under the marginal-cost curve up to the precartel output. Nonetheless, there is a probabilistic relation between high fixed costs and a steeply sloped marginal-cost curve, in that the average slope of the marginal-cost curve over its whole range must be steeper.

Another factor affecting the temptation to cheat is the elasticity of the cheater's demand curve, i.e., the responsiveness of the quantity he sells to whatever terms he offers to buyers. That responsiveness depends first on whether his product is a perfect substitute for those supplied by other members of the cartel. If it is differentiated in any way, or if transportation or other transaction costs differ for each buyer, depending on the seller chosen, the elasticity of the demand curve faced by the cheater will be reduced. The elasticity also depends on the method used by the price-cutter to lure extra business—an across-the-board announced price cut or clandestine price reductions to selected buyers. Between these extremes lie price cuts offered to all inquiring buyers but not publically announced and disguised price cuts in the form of quality or service improvements; these disguised improvements in the terms of the transaction can also be offered either selectively or to all buyers. The potential cheater presumably selects the method of defecting from the cartel agreement that has the highest expected present value. This is not necessarily the one corresponding to the most elastic demand curve for the cheater, because the behavior of costs and the likelihood of detection and enforcement also affect the calculation. The average sizes and size distribution of buyers are prominent among the factors determining the response of the quantity that the price-cutter sells to the terms he offers. A seller offering secret price cuts will certainly favor large buyers, if the likelihood that his cheating will become known depends on the number of buyers to whom cuts are offered and not their size. Conversely, the large buyer's threat to take his business elsewhere is more effective in forcing a selective price cut under any circumstances

[7] See Scherer (1970, pp. 192–98).

wherein the seller incurs some fixed contact cost per customer in securing new business.[8]

The third influence on the incentive to cheat is the probability of detection and the costliness of the enforcement that follows. One component is the expected lag between the offering of a reduced price and its detection by cartel members. The model developed by Orr and MacAvoy (1965) assumes that price information is transmitted only with a lag, so that the seller cutting price enjoys some increased profits before discovery, although reduced profits afterward. They show that if enforcement takes the form of matching the cheater's price cut, the potential cheater can calculate the optimal price cut (if any) to offer; and if the lag before detection is long enough, the present value of the profits expected from cheating will exceed those of remaining loyal to the cartel's terms. Besides the lag, the price-cutter's expected return will also depend on the chances of detection (within a given time) and the form of the punishment he expects. These features of cartel behavior demand treatment on their own.

*The detection of cheating*

The existence of a stochastic lag before cartel members detect cheating and a probability that they will discover it within a given period of time are alternative ways of formulating the same thing. The potential cheater who considers ways of offering price cuts in some sense seeks the one most effective in reaching buyers relative to its speed in reaching cartel-member competitors. Detection depends on the forms in which information passes through the market. If all prices are openly quoted, the same price must presumably be offered to all buyers (unless the structural conditions for price discrimination are present) and will become known to all buyers and sellers at the same time. If price quotations are made to individual buyers, the situation becomes more complicated. A favored buyer has an incentive to conceal his boon from other buyers, because their propensity to demand equally favorable treatment is probably hostile to the preservation of his own favored status. On the other hand, he may have an incentive to tattle about a below-market offer to other sellers in hope of getting a better price still. The seller who expects favored buyers to report the bonanza to other sellers will tend to refrain from cheating; the seller who expects them to switch to his custom and keep quiet will cheat more freely.

---

[8] Costs of switching to different trading partners, whether incurred by buyer, seller, or both, affect market equilibria in numerous ways that cannot be explored here. For example, the existence of switching costs for the seller are sufficient to guarantee large buyers a lower price, even if there are no scale economies in the transaction itself. If the buyer has no incentive to enter into a long-term supply contract with a particular seller, he can never be deprived of a credible threat to switch suppliers, and thus the seller in each market period rationally offers a discount that depends on the size of these contact costs and the difference between the sizes of the large and the average buyers.

In the face of these diverse possibilities, one model that has attracted much attention is Stigler's (1964) "A Theory of Oligopoly," which is really a theory of the detection of cheating under rather specific conditions. Stigler assumes that each seller recognizes his regular customers and can perceive whether or not he is losing an abnormally large percentage of them in a given time period. Stigler's sellers do not, however, know the prices that other sellers are charging, although these price offers are disseminated by any given seller both to the established customers of other sellers and to buyers new to the market. The Stiglerian seller may hear indirectly about price cutting, with a probability that increases with the number of buyers contacted—hence cheating grows more likely, the fewer buyers per seller. Cartel members' main defense against cheating, however, is to stand guard with probability tables in hand, inferring cheating from any movements of customers in the market that are sufficiently improbable if all sellers are maintaining the agreed price. These movements include the following: the loyal seller can lose too many old customers; the price-cutter can be observed to retain too large a fraction of his own old customers; or the price-cutter can be found attracting too many of the customers new to the market. The latter two tests of violation generally require that the loyal sellers pool their information, and there is always some gain to the loyalists from pooling. It emerges from this model that cheating is more likely to occur the more numerous the sellers, the fewer customers are present per seller, and the more random shifting of buyers among suppliers normally taken place (i.e., shifts not motivated by price cuts). The likelihood of cheating is reduced where large loyal firms are present, because the gains to equal-size firms from pooling information about customer movements are equivalent to the advantages in statistical confidence enjoyed by the large-firm observer of the market.

*The process of enforcement*

The process of detecting cheating in cartels holds no importance unless there is some mechanism to punish cheaters. And, as Yamey (1973) pointed out, even with a mechanism identified, we still require an explanation why it is in the interest of some cartel members to apply the indicated punishment.

In formal duopoly models it is usually assumed that the punishment for cheating on the cartel consensus is a reversion to the Cournot solution, either permanently or for some predetermined number of time periods. If there are lags in detecting a member's price cut, even a threat of permanent reversion to the Cournot solution will not deter cheating in all cases. To end the cartel, however, is an analytically uninteresting form of punishment, and there are other possibilities to consider:

1. The optimal price response of loyal cartel members is not necessarily to revert to the Cournot solution. In the model of Orr and MacAvoy (1965), where information lags permit price differentials between sellers to persist, matching the cutter's price is neither a sufficient deterrent nor optimal for the cartel. Rather, a reaction-function equilibrium can emerge between the price-cutter and the cartel members, and Orr and MacAvoy show that it is stable under certain conditions. This adaptive behavior between price-cutter and cartel members is reminiscent of a pattern observed empirically in industries with an oligopolistic core or dominant firm but also a fringe of price-taking small firms. The fringe cannot be kept from undercutting the core group's price by some amount. However, some constraints restrict the rate at which fringe firms can expand their joint market share; usually unspecified, these constraints might result either from information lags or adjustment costs in adding capacity for the fringe. The core holds to its collusive price (or perhaps chooses an optimal differential over the fringe) and loses market shares over time.[9] Finally, Salant (1976) develops the properties of an optimal reaction by a natural-resource cartel to the existence of noncooperative fringe producers. The essence of monopolistic exploitation of a natural resource is to charge a higher price initially than would competitive exploiters, but to raise it less rapidly over time and to make the resource last longer. Where a competitive fringe takes the cartel's behavior as parametric, Salant shows that the cartel maximizes its present value by producing nothing, or very little, until the competitive extractor's resources are exhausted, then producing on a schedule that maximizes the present value of the cartel's resource stock. An analysis somewhat similar to Salant's is provided by Hnyilicza and Pindyck (1976), who assume different discount rates for the fringe and core of the resource-extracting producers and thereby provide the basis for a Pareto-optimal extraction schedule that maximizes a weighted average of the welfare functions for the two groups.

2. A punishment more onerous than a return to a noncooperative equilibrium is a threat to ruin a price-cutter, and a formal analysis of games of survival can be employed to identify possible patterns.[10] Telser (1966) develops this analysis in the context of the capital structures of business firms. The firm in an uncertain environment, facing a rising marginal supply price of funds, must keep a portion of its capital in liquid assets for unforeseen contingencies. These liquid funds defend the value of the firm's fixed capital (to be exact, its value in use over its salvage value) by permitting it to operate for a period with its variable but not its fixed costs covered. If this firm is caught cutting prices, the cartel's enforcer may be able to put

[9] See Worcester (1957); Stigler (1968, chap. 9).
[10] See Shubik (1959, chap. 10).

prices lower still and force fatal losses on the price-cutter. The cost of predation to the potential monopolistic survivor is at least the liquid assets of the potential victim plus the present value of the profits lost by the enforcer during the period of price warfare. The feasibility of such enforcement is greater, the shorter the victim's liquid reserves and credit lines, and the higher the victim's minimum average variable costs relative to the long-run average costs of the enforcer. If the enforcer's fully allocated costs are actually less than the victim's variable costs, the victim's enterprise in any case has no economic value. Certain structural conditions in the market increase the feasibility of price warfare for eliminating a price-cutter. For instance, the price-cutter may operate in a more limited market segment that the enforcer; or the enforcer may be able to identify the price-cutter's "regular" customers and direct his retaliatory price cuts only at them. These conditions make predation a more feasible enforcement strategy, but they do not necessarily make it optimal; the factors determining the minimum cost of the attack to the enforcer are still the victim's liquid reserves plus foregone profits on some quantity of sales.

3. An enforcement strategy potentially preferable to predatory attacks on a price cutter is to buy him out, or to offer a bribe to induce him to leave the market. There is room for such a deal, as an alternative to a predatory attack, because the price-cutter is better off taking a price (or bribe) slightly greater than his liquid reserves, whereas the potential monopolist is better off paying anything up to these same liquid reserves plus the present value of the monopoly profits he expects to lose during a predatory attack. Telser (1966) shows, however, that the lower limit to the merger price or bribe just stated may be too small. This is because the cost to the potential monopolist of extending an actual price war for another time period always exceeds the cost to the price-cutter (because the former equals the latter plus the present value of the monopolist's foregone profits). Therefore, the price-cutter, as a blockade to monopoly profits, is a valuable market asset, and the price-cutter should be able to secure outside funds to enable him to hold out for at least the value of his total capital, not just the value of his liquid funds. The proposition remains that it is generally more profitable for a cartel to dispose of a single price-cutter through a bribe than through predatory action. However, a vital qualification is that the effects of the two strategies on the profits expected by potential entrants are quite opposite. The supply of potential entrants must be limited for the bribe strategy to be superior.

Although cartel theory identifies these possible enforcement strategies, it is reticent about what structural conditions might mark one seller a price-cutter, another an enforcer of a joint-maximization agreement. Among the few clear factors is relative size. The larger the firm's share of sales under the joint-maximization agreement, the smaller are its potential gains from diverting additional sales by price cutting (because there is less to divert), and the larger are its losses from foregoing monopoly profits.[11] Another discriminant sometimes listed is the relative efficiency of firms; the reasoning holds that for two otherwise identical cartel members, the one with the lower marginal cost curve will experience the greater gap between market price and marginal cost and thus the greater temptation to snatch at extra sales. The trouble with this inference is that the assumed difference in marginal costs at the outputs stipulated in the cartel arrangement indicates that the cartel members are not minimizing their total costs of production, and it also implies that they would not have identical preferences for the common cartel price. The cartel agreement thus is incomplete, to begin with. A few other factors might discriminate between price-cutters and loyalists. Short time horizons or high discount rates dispose participants toward price-cutting; the price-cutter trades short-run profits against long-run sacrifices, whereas the trade for the enforcer is of foregone short-run profits against the preservation of long-run monopoly returns.

In conclusion, these mechanisms for detecting and punishing cheaters are specific to the terms of the cartel agreement within which the behavior takes place. The mechanisms may be interesting if cartel agreements in fact do contain incomplete provisions for policing their terms. However, the reasons why the terms should be incomplete are left unexplained. If punishment *A* is an insufficient deterrent to cheating, why does the cartel not employ heavier punishment *B*? The loyalists can afford the punishment costs, because the seller who cheats on an optimal cartel destroys more surplus for the loyalists than he annexes for himself.

### Strategic responses to cartels

What of the buyers in a cartelized market? Traditional cartel theory offers them little but the option to bribe the cartel to sell at marginal cost. The possibility of a mutually beneficial arrangement rests on the familiar conclusion that the gain to the cartel (its monopoly profits) is less than the loss to the buyers (monopoly profits plus deadweight loss), except in the case of perfect price discrimination. Recently Nichols and Zeckhauser (1977) have revived Abba Lerner's (1944) proposal of government counterspeculation as an antidote to monopoly. Lerner pointed out that public authority could hold a stock of a monopoly's output, offering to sell unlimited quantities at long-run marginal cost, and thereby force the monopolist to accept marginal-cost pricing. Nichols and Zeckhauser deal with the problem that the government must first acquire a stock of the cartel's output before this strategy can be employed. Working with simulations based on two-period and multiperiod models, they analyze a strategy that basically reduces to the following. With the full knowledge of the cartel, the consuming-country government (or governments) purchases on

[11] This analysis assumes that the cartel members do not pool profits, so that individual members' profits are related to their sales.

the market in the first period and sells its holdings in the second period. This strategy can be mutually beneficial, despite the existence of storage and holding costs for the consumer-nation government, because the cartel gets its profits moved forward in time (from the second period to the first) and the consuming nation averts enough deadweight loss in the second period to offset the holding and storage costs and any increase in the cartel's profit-maximizing price that occurs during the first period. (It is not even necessarily optimal for the cartel to raise the first-period price when it knows stockpiling is taking place.) If several importing countries independently undertake counterspeculation, it will be underprovided, because part of the gain from *A*'s stockpile sales is in surplus for *B*'s consumers.

Because the cartel nations must benefit from the occurrence of stockpiling,[12] the cartel is actually worse off when stockpiling is underprovided and will lower its initial period price somewhat as a partial offset to this underprovision. To that extent, the "weakness" of divided (noncollusive) consuming nations becomes a virtue.

*Cartels in the international economy*

The preceding analysis has mainly dealt with firms as actors and proceeded in a microeconomic context. However, translating the conclusions to the circumstances of countries cartelizing their exports is generally a straightforward process. The process of monopolizing barter trade for a single country is familiar from the theory of optimal tariffs, and joint monopolization by several countries is merely a matter of calculating the optimal export tax from the joint foreign reciprocal demand curve. The analysis of countries combining to exploit joint monopoly power is closely related to customs-union theory, in which the welfare-maximizing country seeking a customs-union partner is simply a general-equilibrium discriminating monopolist looking for the submarket with the elastic demand curve, and the rest of the world becomes the loser from price discrimination.[13]

A good deal of current writing on international primary-product cartels implicitly employs the analysis of cartel agreements developed above. Without attempting a full survey, some points of contact can be noted.

1. The structural requisites for reaching initial agreement have been discussed in terms reminiscent of Fellner's (1949) analysis of oligopolistic bargains. Because the objectives of national governments are multiple and vaguely defined, a shared set of values (e.g., a common religion) may be more helpful than purely economic facilitating factors, such as a small number of participants. Sharing profits—or even defining them exactly—is

[12] They cannot be excluded from benefits because the cartel can always frustrate the stockpiling strategy by charging the no-stockpiling monopoly price each period.
[13] The relation between customs-union theory and price discriminating monopoly is developed by Caves (1974a).

not an easy matter when governments are traders, so the quest for cartels is constrained to agreements in which profits can be acceptably shared as the outcome of the division of output. Potential participants seem conscious that cartel efforts are unproductive where the demand curve and/or the supply of potential entrant producers is elastic.[14]

2. Because of the incompleteness of observed and practicable cartels, policing and enforcement become major problems. When the agreement is simply to raise price, total demand declines, while individual participants are motivated to expand their outputs; the cartel is visible only if dominant members possess extra-economic threat capabilities or are willing themselves to make the necessary output cutbacks. The latter course has apparently been followed by the largest oil-exporting countries. The international aluminum industry employs a consortium arrangement to buy up the otherwise unsold supplies of fringe producers.[15]

3. Another important implication of incomplete agreements is that the distribution of output cannot be rationalized and marginal costs equalized among the producers. This incompleteness either prevents agreement on a joint-maximizing price at the start, or brings about divisive disputes when the price is subsequently adjusted. Producers whose output has been expanded to its long-run equilibrium level, or the extraction of whose stock resource is far advanced, experience a high perceived marginal cost and therefore prefer high prices, even if all sellers are in agreement about the elasticity of demand. One systematic (and rational) divergence does occur, however, in estimates of the demand elasticity facing individual producers. For selling countries in which buyers have already invested in fixed plant to extract the resource, the relevant demand elasticity for price increases is reduced by the sunk character of this plant. For producer countries whose deposits are not yet developed, the operative elasticity of derived demand reflects the buyer's ability to vary the combination (and location) of complementary inputs.[16]

4. International governmental cartels were often formed between World Wars I and II at times when market price had dropped to a very low level, due to declining demand or the intrusion of substitutes. Although this source of cartelization efforts is often attributed to a psychological propensity to "do something" when and only when the existing situation has become unsatisfactory, a more economic rationale can be found in cartel theory. At such times the proportional gap (and probably the absolute

---

[14] See Rowe (1965, part IV), Bergsten (1974), Krasner (1974), and Mikdashi (1974).

[15] See Litvak and Maule (1975), on aluminum, and Blair (1967), on quinine.

[16] See Greene (1977), on the International Bauxite Association. Greene also shows that the presence of concentrated buyers complicates the formation of a sellers' cartel, because a series of arm's-length prices and price differentials between buyers' locations does not exist. The cartel must grope with limited information not only for a price level but also for a price structure.

gap) between Cournot profits and joint-maximum profits increases. It becomes rational for the potential cartel member to incur higher policing costs, higher risks of retaliation, or whatever costs might probabilistically be incurred in pursuit of elevated profits.

5. The debate over importing-country policies toward the Organization of Petroleum Exporting Countries touches upon the policing of cartel arrangements and the detection of cheating. Adelman (1972) has argued that the OPEC countries' use of a "tax" as a nonnegotiable base for pricing oil and the intermediary roles of the international oil companies have reduced the opportunity for the importing countries to exert their bargaining power as large buyers. There is ground for dispute, however, whether that bargaining power is better exercised by conversion of the crude-oil market into one of bilateral monopoly or by a number of large but independent buyers who can exploit the intrinsic difficulties of policing the cartel.[17]

6. Hexner (1945, chap. 4) demonstrates how the risks of failure in a cartel create distrust that further reduces the completeness of the cartel agreement that can be achieved. For instance, optimal allocation of output to a changing group of customers generally requires a collective mechanism to assign customers to suppliers. However, individual suppliers are apt to prefer keeping their old customers in order to build loyalty against the day of the cartel's collapse.

## II. POLICY TOWARD COMMERCIAL MONOPOLY

In the balance of this chapter I consider the policy-making country not as a participant in an international cartel or an adversary of other nations forming cartels, but rather as a welfare maximizer dealing with commercial monopoly in national product markets. Market monopoly may pose three issues for the national policy-maker: (1) it may affect the nation's gains from trade and its efficient participation in international trade; (2) trade policy provides an instrument of competition policy by limiting or extinguishing monopoly in national markets; (3) patterns of collusion in noncompetitive markets may affect the short-run dynamics of the economy's response to changes in the international economy.

### Monopoly and the gains from trade

A small country participating in international trade by definition faces an externally determined world price ratio. If one of its domestic industries is monopolized, it therefore follows that exposure to trade will eliminate the monopolist's perceived distinction between price and marginal revenue and force him to act as a pure competitor on the world market. This effect, which amplifies the conventional gains from trade, will be

[17] Compare Adelman (1972) and Roberts (1974).

considered in the next section. What must be recognized first, however, is that the standard propositions about gains from trade acquire some important qualifications if monopoly extends beyond a single industry in a small country. Melvin and Warne (1973) analyze the conventional two-by-two general-equilibrium model in which one or both of the two sectors is monopolized in each country. (It turns out not to matter whether monopolistic control affects both industries in each of the two countries, or only one industry—the same one—in each.) Their model employs explicit utility functions which assure that at any relative price ratio demand elasticities for a given good are the same in both countries.[18]

In this model, one trading country must wind up worse off than with purely competitive free trade, and both can wind up worse off. More alarmingly, one country can experience a lower level of welfare with trade carried on by monopolies than it would with monopolized production and no trade at all. The potential loser can be identified—the country having a comparative advantage in the good with the higher elasticity of demand at equilibrium world prices. If only one sector (the same one) is monopolized in each country, the nation with a comparative advantage in the competitively produced good may suffer from the introduction of trade. These conclusions from a two-by-two model do not translate easily into policy conclusions for an *n*-sector trading nation, but they provide a suitable cautionary note about the effect of monopolies that persist in the presence of international trade.

Some significant findings about monopoly and the gains from trade have been developed in a partial-equilibrium context. Corden (1967) considered the problem of a decreasing-cost industry—a natural monopoly—facing import competition. When should such an industry be established, given that the country can instead acquire its product through international trade? Corden shows that there is a critical import price that makes its establishment just socially desirable; that price lies below the one that would allow the monopoly to cover its average costs and above the one that would prevail in an equilibrium with price equal to its marginal costs. When the world price lies below the monopoly's average-cost price, however, the gain from establishing the natural monopoly is contingent on a lump-sum transfer arrangement that permits (forces) it to price at marginal cost. Corden demonstrates that a tariff can play no useful role in this situation. Raising the landed price of imports by less than the amount needed to establish the industry merely costs consumers' surplus. Raising it to or above the price that permits the monopoly to cover its costs permits it to price at or above average cost and impose a welfare loss.

[18] In this model the introduction of trade does not change the degree of monopoly; the two national producers of any given good behave as if they were jointly monopolizing the world market, with their market shares determined by the general-equilibrium adjustment of trade to tastes and factor endowments.

Corden assumed that his decreasing-cost firm was somehow probihited from expanding to minimum-cost scale and charging forth into export markets. Basevi (1970) and Pursell and Snape (1973) point out that an export market affects the decision to establish a decreasing-cost industry, even if the net world price received by the exporter is below his minimum attainable long-run average cost with all scale economies exhausted. Their case is essentially that of the decreasing-cost producer who can cover his costs only through price discrimination. Domestic demand by itself is not substantial enough to permit the industry to cover its costs at any level of output. And the world-market price, as mentioned, lies below the minimum attainable average cost. Nonetheless, if the activity is carried on by a discriminating monopolist, output may be sufficiently expansible through foreign sales that a higher price in the domestic market will yield enough profits to cover total costs. Thus, there can be cases in which welfare is increased by permitting a decreasing-cost monopolist to discriminate between home and foreign markets, as the price of bringing him into existence. As in Corden's case, a tariff is no help for reaching the optimum, and indeed an import subsidy may be desirable to curb excess monopoly price in the domestic market. Frenkel (1971) points out that discrimination against the domestic market is not always necessary; the cost reduction attainable through selling on price-elastic export markets may make a natural monopoly viable, charging a single price at home and abroad. White (1974) indicates some consequences of monopolized production of exportables for the volume of goods exported. The discriminating monopolist who sells at a competitive world price will supply more exports than would a competitive industry experiencing the same costs, because he elevates the domestic price, reducing demand at home and freeing more goods for the world market. If the monopolist of exportable goods cannot price-discriminate, however, the quantity he exports will be at most equal to what a competitive industry would supply, and he may opt for exporting nothing at all.[19]

In this section we have taken as a given either a monopolized market or decreasing costs as a structural basis for natural monopoly. Empirical research in industrial organization treats markets not as dichotomously monopolistic or competitive, but as capable of showing degrees of "market power," i.e., departing by varying amounts from conditions of pure competition. There is, in fact, a feedback loop running from the structural conditions that create market power to the pattern of a country's foreign trade, because certain fundamental characteristics of technology and demand underlie both comparative advantage and market power. The specific

---

[19] White also points out that a risk-neutral import-competing monopolist will produce less than a risk-neutral competitive industry facing the same distribution of expected import supply prices.

relation between trade patterns and market structure must, of course, vary from country to country because not all countries can export the same thing, and so the evidence will not be summarized here.[20] Instead we turn to international trade as a limit on market power.

*International trade and competition policy*

Eliminating a market distortion due to monopoly is an extra dividend that may be associated with the gains from trade. Put the other way around is the old American saying, "the tariff is mother of the Trusts." The two-sector general equilibrium model provides some simple comparative-statics findings about the effect of opening an economy to competitive trade when one industry is initially monopolized. If the monopolized sector is the export industry at externally given world prices, that sector's output will definitely expand, but the pretrade domestic price could either rise or fall. If the monopolized sector turns out to be import-competing, the domestic price will definitely fall, but the pretrade output could either expand or contract. In any of these cases, the world's gains from trade are greater than if the economy had been initially competitive.[21] In each instance, unrestricted international trade is a sufficient remedy for monopolistic distortions of domestic markets, in the sense that the marginal conditions for a competitive welfare optimum will hold after the introduction of trade. Feenstra (1977) points out, however, that international trade is not a complete remedy to monopolistic distortions in this general-equilibrium case, if the monopolist also exercises his monopsony power in national factor markets.[22]

The last few years have brought a burst of statistical research on the effect of international trade on monopoly or—more generally—the extent of market power exercised by oligopolistic industries. These studies examine rates of profit on equity capital or price-cost margins as indicators of the fruits of exercised monopoly power. These dependent variables are related by means of multivariate regression analysis to assorted measures of trade exposure for a sample of manufacturing industries. There is a problem of how to evaluate the extent of import competition facing an industry—

[20] Relevant studies are Pagoulatos and Sorenson (1976b), Caves and Khalilzadeh-Shirazi (1977), and Caves et al. (1977).

[21] See Caves (1974c); also Melvin and Warne (1973).

[22] Feenstra (1977) demonstrates that monopsony in the two-by-two general equilibrium case could be exercised in two ways. The conventional monopsonist recognizes his influence on individual factor markets, depressing the relative price of the factor used intensively in the monopolized industry and forcing the economy onto an inferior transformation curve. The "multiplant monopolist" buys factors of production competitively (say, because branch plant managers fail to recognize their collective influence on the price of each individual factor); however, the monopolist recognizes (as numerous pure competitors in the same industry would not) that expansion of his output along a convex transformation surface drives up his marginal cost. The multiplant monopolist's decisions leave the economy on its competitive transformation curve, though not at the optimal output, either with or without free trade.

specifically, how to proxy the position and elasticity of the excess-supply curve of competing goods importable into the national market. Reliable econometric estimates of import-supply elasticities are not found in droves. Most researchers have settled for using the share that imports comprise of domestic production or domestic disappearance as a proxy for the missing parameters. The proxy could obviously be faulty; e.g., the import-supply curve might be perfectly elastic, and yet domestic supply might fall just slightly short of domestic demand at the world price, making the import share very small. Still, given that the "industries" identified in published statistics and used in these statistical analyses only roughly approximate homogeneous markets, the import-share proxy is probably not a bad indicator of what proportion of the finely defined goods marketed by a group of sellers actually face close competition from foreign suppliers.

In any case, this variable has been found to have a statistically significant negative influence on measures of monopolistic distortion in several studies. Effects of import competition for U.S. manufacturing industries were reported by Esposito and Esposito (1971) and Pagoulatos and Sorenson (1976a). Both studies cover relatively large samples of manufacturing industries and secure statistical results that seem robust to minor changes in statistical specifications.[23]

With such strong results found for the large and relatively closed U.S. economy, one would expect even clearer findings for the smaller industrial nations. The pattern has been a bit murky, however. Khalilzadeh-Shirazi (1974) undertook a similar analysis of U.K. manufacturing industries and secured a regression coefficient for the import-share variable that was correctly signed, but only marginally significant. Hart and Morgan (1977) found no significant relation in their analysis of U.K. data for 1968. A very likely explanation for their negative result is the 1967 devaluation of the pound sterling, which raised the landed price of competing imports and should have given import-competing industries a temporary windfall. Pagoulatos and Sorenson (1976c) studied the determinants of price-cost margins for France, Italy, Germany, the Netherlands, and Belgium–Luxembourg in 1965, reporting a significant negative influence of import competition for all but Italy. Adams's (1976) coefficient for a transnational sample of large companies is negative, but not always significant. For Canada, Schwartzman (1959) employed a different research design that involved comparing the performance of more concentrated Canadian manufacturing industries to their less concentrated U.S. counterparts. He recognized that higher concentration in Canada should lead to higher price-cost margins only when the Canadian industry's trade exposure is

---

[23] The Espositos report separate regressions for consumer and producer-good industries. When a correction has been made for heteroscedasticity, the import variable appears significant for the consumer goods but not the producer goods.

attenuated. However, the concentration-profit relation that he found for the trade-sheltered industries became even more significant when the import-competing industries were added to the sample. Jones, Laudadio, and Percy (1973) chose to represent import competition by dummy variables designating high (imports over 30 percent of domestic shipments) and medium (imports between 15 and 30 percent) import competition with Canadian manufacturing industries. Their medium-imports dummy proved insignificant, their high-imports dummy *positive* and significant.[24]

Two modifications of these analyses may help to explain their incomplete support for the hypothesis that import competition limits the exploitation of monopoly positions. Most important of these is Bloch's (1974) development of the proposition that import competition should affect industries' profitability only if their seller concentration is indeed high enough that excess profits would be taken in the absence of foreign rivals. Bloch's own statistical work, discussed below, deals with the effect of tariffs rather than import competition. In research in progress on Canadian industrial organization,[25] we find that monopoly profits are significantly related to a variable that is a measure of seller concentration divided by imports as a percentage of domestic-industry shipments. There is not a significant relation to either concentration or the import share when they are included separately in a regression equation. Because the import share is a small fraction with a skewed distribution, the interaction variable tends to "turn on" concentration as an influence on profits only when import competition is low.[26] Another modification recognizes the weakness of the import-share variable as a proxy for the parameters of the excess-supply curve of imports. Turner (1976, chap. 4) utilized the 1967 devaluation of sterling, which provided a substantial disturbance to the import competition faced by U.K. manufacturing industries, as an opportunity to improve this specification. Given that the devaluation should have elevated world prices relative to U.K. domestic prices and costs by about the same proportion in all industries, variations among industries in the change in imports' share of the market in the years immediately following devaluation years should be correlated with the elasticity of the unobserved excess-supply function.

[24] For this they offer an unsatisfactory explanation that might apply to a short-run time-series analysis, but not to a cross-sectional analysis in which (one normally assumes) the entities are observed only randomly displaced from their long-run equilibria.

[25] This study, undertaken jointly with M. E. Porter, M. Spence, and J. T. Scott, draws upon a data base constructed with the support of the Royal Commission on Corporate Concentration. Caves et al. (1977) provides a preliminary report on this project that does not include the result mentioned in the text.

[26] Pagoulatos and Sorenson (1976c) also employ an interaction between concentration and import share, but they inappropriately formulate it as a product rather than as a quotient. Naturally, it is not significant.

Turner seeks to explain variations in price-cost margins in 1973 both by imports as a percentage of domestic disappearance in 1973 and by the recent change in a somewhat similar variable, the proportional change in import share. He finds that the change in imports has much more explanatory power.[27] Pagoulatos and Sorenson (1976a) employ a somewhat similar variable, the proportional change in the level of imports 1963–67 as an alternative to the 1973 share of imports for explaining price-cost margins in 1967; the rate of import growth is marginally significant, but appears to have considerably less explanatory power than the level of the import share.

Some evidence from surveys and case studies supports and extends these statistical findings about imports and monopoly in the United States. Sichel (1975) queried large manufacturing firms in the United States as to the identity of their three principal competitors—foreign or domestic. At least one foreign company was listed among the principal rivals of 23 percent of his respondents, although only 5 percent listed a foreign firm as the leading rival. When the respondents were classified to their principal industries, at least one response in 42 of 69 industries designated a foreign company among the chief competitors. Frederiksen (1975) studied several highly concentrated U.S. industries that had experienced increased foreign competition since World War II, finding that foreign rivals had increased price competition in two industries selling undifferentiated products, but that foreign rivalry had been less effective in two differentiated-product industries for which a significant proportion of imports are "captive" purchases by the leading companies in the U.S. industry.

The evidence on import competition, taken together, suggests that imports are a substantial limit on monopolistic distortions. A principal implication of this finding is that tariffs facilitate the collusive behavior of domestic sellers in concentrated industries and can thereby cause welfare losses in such markets that are greater than the familiar deadweight losses expected when a purely competitive industry receives protection. The proposition that tariffs increase the incidence of monopolistic distortion has been tested directly in some statistical studies. Pagoulatos and Sorenson (1976a) report a significant and correctly signed regression coefficient for a variable indicating the proportion of competing imports that are subject to nontariff barriers, but nominal tariff rates are not significant. For Canada, McFetridge (1973) found no influence of effective rates of protection on price-cost margins. Bloch (1974) likewise found that the gross profits of heavily protected Canadian industries, relative to the gross profits of their U.S. counterparts, were no higher than for industries with low tariffs. He did find, however, that selling prices of the heavily protected industries

---

[27] It makes no important difference whether the change is calculated for 1968–73, 1970–73, or 1971–73.

were higher, suggesting that the effect of tariff protection may be on efficiency rather than profitability.[28] We return to this question below.

This analysis provides a strong case against tariffs because they can amplify the scope for monopolistic distortion. But the policy implications go beyond a preference for free trade over tariffs. Vicas and Deutsch (1964) point out that the government could force even a monopoly not facing import competition to price at marginal cost by offering a subsidy to imports equal to the difference between the higher world price and the monopolist's marginal cost at the output that equates his marginal cost to price. If his average costs are covered, the monopolist would produce the "competitive" output, and no imports would actually enter.

The theory of the effect of trade exposure on monopoly indicates that the consequences for a small country's domestic monopoly should be the same whether the industry emerges as competing with imports or making net exports. Either way, it faces a parametric price on the world market. In a sample of actual industries, however, two factors could upset the implied prediction of a negative relation between the proportion of an industry's output exported and its rate of profit. For a concentrated and collusive industry, dumping may be possible, so that the presence of international trade is associated with discriminating monopoly. This would imply (if anything) a positive relation between exporting and profitability. Second, under some microeconomic assumptions efficiency rents could accrue to exporting firms, even if they lack shared monopoly power. For instance, if the industry's product is differentiated, the presence of foreign markets should shift the demand curve facing the average seller outward and lift the profit rate above the normal rate of return implied by the Chamberlinean tangency solution. Hence we have no determinate empirical prediction about the effect of export-market participation on profitability.

Actual statistical results have turned out correspondingly diverse. For the United States, Pagoulatos and Sorenson (1976a) report no consistent relation (even as to sign) between profit rates and exports as a percentage of the industry's value of shipments. In their study of five European countries, Pagoulatos and Sorenson (1976c) get negative signs for all five countries, but the coefficients are significant only for France and Italy. Consistent with this, Jenny and Weber (1976) secure a significant negative relation for France, and the result is robust when several alternative measures of profit are empolyed as the dependent variable. Adams's (1976)

---

[28] There are reasons, having to do with sample properties, why the theoretically certain effect of tariffs on an industry's potential market power fails to show up in cross-section studies. As Bloch's (1974) analysis suggests, protecting an industry results in excess profits only if free entry of domestic sellers does not compete away the resulting rents. Nonetheless, governments may choose to ignore this fact and award high tariffs to many sectors that are purely competitive or employ large quantities of low-wage labor. Some of the highest tariffs therefore may generate no monopoly profits.

transnational sample gives an insignificant negative result. The outlier is the United Kingdom, for which Khalilzadeh-Shirazi (1974) found a significant positive relation between exports and price-cost margins. Additional exploration of this result by Caves and Khalilzadeh-Shirazi (1977) suggests the following interpretation: (1) for the United Kingdom, like other countries,[29] there is a strong relation between export participation and the sizes of companies and manufacturing establishments; (2) because U.K. exports run heavily to differentiated goods, efficiency rents associated with larger scale might be captured by the companies rather than dissipated through competitive entry; (3) the positive statistical relation between exports and profits, rationalized in this way, is weaker in the more concentrated industries, consistent with export-market participation having some dampening effect on the exercise of monopoly power. Taken together, the evidence on the exports and profitability of manufacturing industries suggests that extensive participation in foreign markets is (*ceteris paribus*) hostile to effective achievement of monopoly power in the domestic economy, and that it may also bring dividends in the achievement of more efficient scales of production.

A brief account is needed of the relation of the multinational company to monopoly and international trade.[30] Foreign direct investment tends to occur in industries where the average firm is large and sellers are concentrated. Although there are ways in which high concentration can promote foreign investment, and the presence of multinationl companies can increase the degree of monopoly, probably the most important fact is that concentration and foreign investment share a number of common fundamental causes. We can develop, however, some more positive propositions about the behavior of the multinational company that are relevant to international economic policy. It is seldom recognized that the multinational company is a favored entrant to industries with high barriers to entry. Monopoly is a long-run problem only where entry barriers deter the elimination of monopoly profits through entry, and analysis of the differing incidence of multinational companies from industry to industry establishes that the firm-specific assets that induce them to invest abroad tend to be what is needed for scaling the principal sources of barriers to entry (ample supplies of funds, established ability to differentiate their products, and perhaps other forms of technical and marketing skill). Thus the multinational company is a likely potential entrant into national industries that might otherwise be cloistered by even higher entry barriers.[31] When a foreign auto company begins assembly in the U.S. market, or a steel company develops

[29] Sue, for example, Scherer et al. (1975, p. 396).
[30] Brief because I have dealt with the subject at length elsewhere; see Caves (1974b) and Caves (1974c, pp. 17–28).
[31] Gorecki (1976a) demonstrates that multinational companies are not halted by the same entry barriers that affect other companies.

an iron ore deposit in a difficult piece of terrain, it may be providing an additional market participant that could come from few other sources. Besides its ability to enter a market, there is also some possibility that the multinational may be an entrant particularly disruptive of an oligopolistic consensus, especially in the early period of its presence. Its alien status may make it initially less sensitive to signals about an oligopolistic consensus emanating from established native firms. And its superior access to information about alternative returns to resources placed elsewhere in the world may make it less risk-averse than firms dependent on a single market.

The analysis so far supports a general policy of openness to market entry by multinational companies. The case is not completely clear-cut, however, for some considerations run the opposite direction. If the multinational company is good at scaling existing industrial barriers to the entry of new firms, it is also good at building up such barriers. The resources required to contrive such barriers (maintaining excess capacity, integrating forward to control distributive outlets, advertising heavily, accelerating the frequency of "model changes," etc.) are often found in the portfolios of multinational companies. It is also true that the multinational possesses the "long purse" that might drive out single-market rivals (see preceding section).[32] Of course, the most direct approach of public policy to such offensive forms of market behavior is to regulate or prohibit the behavior directly rather than blocking the very presence of international ownership links. Conduct designed to reduce the competitiveness of an industry is socially undesirable whatever sort of firm undertakes it.

One policy instrument that appears to have strong leverage on the activity of multinational companies is the tariff, in the case of companies whose foreign subsidiaries normally produce the same line of goods as the parent. A tariff elevates a company's cost of landing its goods within a national market relative to the cost of producing them there through a subsidiary, and hence tends to increase the flow of foreign investment. Many historical accounts affirm this effect of increased tariff rates. The statistical evidence, except for Horst (1972), is less consistent, but a cross-section statistical analysis is ill-suited for testing the hypothesis.[33] If we nonetheless accept the hypothesis, it has interesting implications about monopoly in national markets. Multinational-company entrants are likely to be a significant competitive force in industries that are protected from import competition by tariffs. The same holds for industries subject to

[32] Statistical research on the profitability of U.S. manufacturing industries suggests that profit rates are higher, the larger is the extent of foreign investment by member firms in the industry, after we control for the extent of monopoly or market power. This profit increment could represent the return to intangible assets garnered by working them in foreign markets, or it could measure the effect of additional monopolistic distortion due to international entry-barrier building or similar practices.

[33] Because some industries lack the structural requisites for direct investment to occur, and so no tariff is high enough to induce significant foreign investment.

product differentiation of the Chamberlinean stripe, in which multinational companies flourish on the basis of their success in establishing intangible good-will assets. Conversely, international trade is a more effective curb on monopoly where tariffs are low and where products are homogeneous, so that elaborate marketing organizations are not necessary to sell substantial quantities in a foreign market.

In considering international trade as a restraint on monopoly, economic analysis habitually concentrates on costs of allocative inefficiency—the deadweight loss due to monopoly. However, where competition is imperfect, trade—and trade restrictions—can have important effects on the degree to which costs are minimized. It is a commonplace that the welfare gains from policy changes that reduce unit costs can easily exceed those due to the recapture of deadweight-loss triangles. I shall concentrate on the relation between trade and cost minimization through attaining efficient scales of production.[34]

Trade changes the effective size of the market in which the firm sells. The connections can be illustrated by the firm in the position of the Chamberlinean monopolistic competitor, facing its individual, downward-sloping demand curve though not necessarily taking part in oligopoly. The total size of the market affects the typical firm's scale of production by changing the slope or position of the demand curve that it faces. Access to export markets shifts the curve outward and may also render it more elastic, if the firm's output faces closer substitutes in the international market than in the domestic market. Both changes tend to move the monopolistic competitor's profit-maximizing scale toward that which minimizes long-run average cost. The presence of competing imports has an ambiguous effect, shifting the import-competing seller's demand curve to the left, but also making it flatter, with an indeterminate effect on its profit-maximizing scale of production.[35]

This analysis can also be applied to the organization of Chamberlinean industries marked by either of the following conditions: (1) each "variety" of the product is subject to the same production function, but buyers' preferences are distributed unevenly among the varieties (some are popular, some are not); (2) each of $n$ varieties is preferred by $1/n$ buyers when all varieties sell at the same price, but production functions differ so that some varieties are subject to greater scale economies than others (i.e., must

---

[34] This is only one channel through which competition and technical efficiency may be interrelated. Others are the outright inefficiency of the enterprise that employs more inputs than necessary to produce a given output and—a special case of this—chronic excess capacity.

[35] In referring to a Chamberlinean industry, I am assuming that entry by new firms propels existing ones *toward* the tangency solution but not invariably *to* it. Oligopoly (mutual dependence recognized) is assumed absent, but the average firm can still command a rent. Enlargement of market size can enlarge the rents of those firms not pressed to the tangency solution.

attain larger volumes to achieve minimum average cost). With free trade, the international distribution of production will be influenced from the cost side by the classic forces of comparative advantage. Given those forces, however, production in the small national market will tend to be confined to popular varieties and those subject to minimal scale economies (in the sense just defined). Large countries will tend to specialize in the unpopular varieties and those subject to extensive scale economies.[36] The effect of tariff protection for any national market—though especially a small one— is to make viable the production of less popular varieties demanded domestically in smaller quantities. Also, domestic production may become viable for varieties subject to more extensive economies of scale. Taking these effects together, it is possible that surrounding a small national market with tariff protection actually reduces the average size of enterprises. And it necessarily follows that tariffs imposed by any country tend to reduce the average scale of production for the world industry as a whole.[37]

A good deal of statistical evidence has accumulated that provides at least indirect support for these propositions about trade, scale, and efficiency. Scherer et al. (1975, pp. 117–20), analyzing the branches of twelve manufacturing industries located in six industrial countries, found that the extent to which plants attain minimum efficient scale in manufacturing industries depends sensitively on the proportion of total shipments that is exported. Eastman and Stykolt (1976, chap. 3) and Gorecki (1976b, chap. 5) report similar results for Canadian manufacturing industries. Owen (1976) found an association for pairs of European countries between the relative sizes of manufacturing plants and the balance of trade. Relatively larger plants (though not larger firms) are associated with larger net exports, although Owen's analysis does not clearly identify the direction of causation between the variables. Other studies have noted a simple correlation between the proportion of output exported and the average sizes of plants and firms,[38] and Pryor (1973, chaps. 5, 6) finds for a large sample of countries that the average sizes of both plants and companies (measured in various ways) are associated with the share of output exported after controlling for the size of the national market. On the import side, we saw that the effect of trade (and its impediment by tariffs) on efficient scale is ambiguous. Scherer's six-country study correspondingly finds that the extent to which an industry's plants achieve efficient scale is negatively but in-

---

[36] The analysis assumes that transportation and transactions costs are greater between than within national economies.

[37] This relation between trade policy and efficient scale depends on the individual firm facing a downward-sloping demand curve, and economists enchanted with the siren song of pure competition may suppose that this is an uncommon market condition. Therefore, I must stress the abundance of evidence supporting the principal corollary of the preceding analysis, namely, that the sizes of plants and firms will be related to the size of the national market in which they are embedded; for example, Pryor (1973, chaps. 5, 6).

[38] Caves and Khalilzadeh-Shirazi (1977); Caves et al. (1977, chap. 7).

significantly related to the extent of import competition. For Canada, however, several studies lean toward the conclusion that tariffs reduce the average plant scale of production or the degree to which efficient scale is attained.[39] Broadly speaking, the evidence confirms that trade restrictions reduce the average scale of production in an industry worldwide, but they may increase it within the markets of some tariff-imposing countries.

*Monopoly and short-run adjustments*

A final consequence of monopoly in open economies is that it may change the path of short-run adjustment to disturbances from what would prevail in the presence of pure competition. The difference arises not from the behavior of the theoretical pure monopolist but rather mainly from the pricing practices employed in oligopolies with incomplete collusive arrangements. Such industries' prices may be relatively sticky in the short run, because each change in list prices incurs the risk of a breakdown in the oligopolistic consensus. Such pricing behavior can affect both the imports and exports of the country in question.

If an industry's comparative advantage is deteriorating, or if the nation's money price and cost structure is getting too high relative to its fixed exchange rate, the short-run inflexibility of a domestic price can have the effect of inflating the volume of imports. This consequence has been documented for U.S. steel imports by Krause (1962), who found the prices of domestically produced steel products insensitive to changes in import prices and market shares. Rowley (1971, pp. 220–21) describes a period in which the U.K. steel industry showed similar behavior.

Other possibilities pertain to the concentrated exporting industry. Suppose that prices in the domestic market are maintained at a sticky collusive level, while the producers also sell as pure competitors on the world market. First of all, this behavior inflates the average volume of exports above what a competitive industry would sell abroad, as White (1974) pointed out. It can also influence the variability of exports. If the domestic price is inflexible in the face of shifts in the demand curve for the product, the induced fluctuations in the volume of exports will have a greater amplitude than if the domestic price adjusted competitively. Such competitive adjustments in the domestic price would reduce the variability of the quantity sold in the domestic market by a sticky-price oligopoly. Given the position of the sellers' marginal-cost curves, larger fluctuations in the volume of exports are implied by greater fluctuations in the oligopoly's domestic sales.

The presence of product differentiation also affects the adjustment of trade flows. Differentiated goods by definition lack perfect substitutes in

---

[39] Gorecki (1976b, chap. 5) reports an insignificant negative influence of tariffs. Eastman and Stykolt (1967) and English (1964) stressed the tendency of the tariff to hold an umbrella over firms that were inefficient for whatever reason, and the proclivity of multinational companies to locate inefficiently small-scale production facilities behind a tariff wall.

foreign markets. What adjustments are made to their export prices following a devaluation therefore can depend on pricing practices and market conditions in the domestic market. Turner (1976, chap. 3) found that after the 1967 devaluation U.K. producers of differentiated goods raised their domestic-currency prices of exports significantly less than producers of homogeneous goods.[40] The lower increase by itself does not prove a non-competitive response, of course, but it does establish a potential influence for collusive practices in the domestic market.

The normative significance of monopoly for short-run adjustments in international trade is not clear and probably varies from one situation to the next. The point is simply that paths of adjustment can be different where elements of monopoly are present.

*National and international policy toward monopoly*

An "optimum tariff" serves the interest of the single country imposing it, but it imposes a net cost on the rest of the trading world. A similar problem of national versus international welfare arises in the making of national policy toward competition in an open economy. Consider first an industry that is monopolized and sells all of its output abroad. Suppose that some application of antitrust policy by the nation's government can potentially force the industry to sell competitively at a price equal to marginal cost. If, for simplicity, we assume that marginal cost is constant, application of the policy will have two effects. It relieves foreign consumers of a dead-weight loss, and it transfers the monopoly profits formerly earned by the monopoly to the consumers' surplus of foreign buyers. The nation enforcing competition is necessarily a net loser, although the world as a whole is better off. If the monopoly has been selling some of its output at home (without import competition) and some abroad, the same antitrust policy now eliminates some deadweight loss at home. Given elasticities of demand in the home and foreign markets, one could evidently identify a share of output exported just small enough that the country would become a net gainer by enforcing competition for its exportable-goods monopolist.[41] The optimal national policy would, of course, be to establish competition in the industry, while applying an optimal export tax.[42] Without that

[40] Also see Hague, Oakeshott, and Strain (1974).

[41] This critical export share could be larger if the country attaches some utility to the redistribution of income from domestic profit recipients to domestic final buyers.

[42] That governments are not indifferent to the joys of monopoly profits taken from foreigners hardly needs to be argued in the days of OPEC. Less familiar, however, is the common practice of permitting domestic companies to collude on export sales, even when such behavior is illegal in the domestic market. In the United States the Webb–Pomerene Act is the vehicle for the exercise of joint monopoly in international trade. Part of the Act's rationale was to assist small companies to meet the heavy transaction costs of exporting through joint associations—a sensible policy by itself. However, Larson (1970) has shown that the Act has primarily benefitted already concentrated industries, and that the Webb–Pomerene associations have assisted in monopolizing the domestic as well as foreign markets.

option, however, the government must choose between the potential national gains from cartel participation outlined in the first part of this chapter and the advantages of domestic-market competition attainable partly through appropriate international economic policies.

This problem can be analyzed formally in terms of optimal competition policy for the exporting industry, on the assumption that policy instruments can secure any outcome from pure competition to pure monopoly, but must accept the same degree of monopoly in both domestic and export markets (Auquier and Caves, 1978). The larger the export market relative to the domestic market, the less competition should be enforced to maximize national welfare. The optimal degree of competition is greater, the more elastic is domestic demand and the less elastic is demand in the export market.[43] These partial-equilibrium conclusions can be transplanted to a general-equilibrium setting, using the model of Melvin and Warne (1973). For instance, if the same degree of competition characterizes a given industry in each country (the degrees may differ between industries, though), a nation can lose by entering into international trade either because its exportable good faces an elastic demand (Melvin–Warne's conclusion) or because a high level of competition in its export industry allows that industry to claim little of its potential monopoly profit. If the home country's industries are monopolized, while those abroad are competitive, the home country can lose from entering into international trade.

The general clash between national and international interests considered in this formal analysis has many simple implications for competition policy. A familiar form of international cartel agreement is for sellers in countries $A$ and $B$ to divide up world markets, with $C$'s market assigned to $A$, and $D$'s market assigned to $B$. Countries $A$ and $B$ may well be net beneficiaries of the cartel. $C$ and $D$ could potentially bribe them to terminate it, but handing over voluntary tribute to a foreign exploiter is not a policy proposal that commonly wins elections. The divergence of national interests bears not just on policy toward outright monopolies and collusive arrangements but also toward potentially monopolistic practices engaged in by multinational companies. It was pointed out above that the practices generating good-will assets that permit companies profitably to invest abroad also can create or augment contrived barriers to entry into their industry—whether in the multinational's home market or in the market where its subsidiary operates. Once again, prohibiting the practices might not be in the interest of the home-country government.[44]

A different problem of divergent national interests can arise when a multinational company in country $A$ acquires or merges with a national firm in a country $B$ engaged in producing the same line of goods. Suppose

[43] A related analysis demonstrates the use of an optimal export tax or subsidy on the assumption that competition policy is inoperable and the degree of monopoly must be taken as given.

[44] Other examples are discussed in Caves (1975).

that $A$'s producer has not previously been exporting to country $B$, and that each country's competition authorities regulate mergers only on the basis of the effect on seller concentration in the national market. Concentration conventionally measured is unaffected by the merger in either of the national economies, yet there has been an increase for the two markets taken together. If concentration should be high enough in the international market, this merger could impose real costs on the trading world as a whole. Even if the producer in $A$ has previously been exporting to $B$, the same misperception could occur if $B$'s competition authorities follow the common practice of watching concentration ratios calculated only over domestic production and not over all sellers present in the domestic market.

Actual antitrust policy in the United States has been fairly sensitive to such spillovers from anticompetitive actions—even in cases where the United States is the beneficiary from monopoly rents. United States companies have been stopped from acquiring foreign companies that are their actual or potential competitors in the U.S. market (and elsewhere). United States and British companies in one instance were prosecuted for agreeing jointly to monopolize the Canadian market.[45] The internationalized competition policy of the European Economic Community has dealt with a number of cases that probably had divergent effects on the national economic welfare of Common Market members (see de Jong, 1975). Despite these favorable patterns, divergent national and international interests in competition create a general problem similar to the problem of tariff reduction addressed by the General Agreement on Tariffs and Trade. A possible (though not necessary) interpretation of that agreement is that countries agree to multinational tariff reduction on the conjecture that the losses they incur from reducing their own tariffs below "optimal" levels will be more than offset by their gains in consumers' surplus (including reduced deadweight losses) from foreign countries' reductions of monopolistic tariffs. One can imagine a similar declaration of faith in the averaging out of gross losses that could occur if all countries agreed that each would apply its competition-policy instruments to whatever monopolistic structures or practices lay within its reach, wherever the resulting social benefits might be felt.

The provisions of competition policy dealing specifically with international trade pose a somewhat different set of issues. All industrial countries allow their domestic sellers greater freedom to collude in the export market than at home, and most of them restrict monopolization of their export trades little or not at all.[46] This policy represents a consistent pursuit of national welfare maximization, in the sense that the prices charged abroad

---

[45] For details see Brewster (1958).

[46] See Organization for Economic Cooperation and Development (1974) and Gribbin (1976). Jacquemin (1974) suggests that the EEC takes a symmetrically tough line against foreign firms with market power in member-country markets.

by monopolistic exporters could in principle be identical to the "optimum tax" on exports. The only qualm about this policy for a government maximizing its own national welfare lies in the possibility that collusion among exporters may unavoidably spill over and increase distortions in the domestic market. In that case, the country faces the same tradeoff identified above between capturing consumers' surplus abroad and suffering deadweight losses at home. There has been little or no public recognition of the divergence between national and global welfare resulting from these policies, and the policies open to the individual country to combat foreign monopolization of its imports are limited (see the first section of this chapter).

The familiar applications of competition policy to imports are generally concerned not with getting them more cheaply but with restricting the sale of imported goods whose prices are affected by dumping or export subsidies. There is generally little or no foundation for such policies in the maximization of national welfare. The restriction of dumping from a nationalistic point of view makes sense only in the special case of short-run predatory dumping.[47] From an international viewpoint the case is more complicated, because dumping as a form of monopolistic price discrimination is efficient only if total costs can be covered in that way and no other. The case of an export subsidy is somewhat different, assuming that the exporting industry itself is competitive and the subsidy creates rather than removes a market distortion. The importing country maximizes its national welfare by doing nothing and accepting the improvement in its terms of trade. From an international viewpoint, however, all importing countries should impose a counter-vailing duty to offset the distortion induced by the subsidy, and this duty should not be contingent on the occurrence of injury to import-competing domestic producers.

## III. CONCLUSIONS

This chapter has surveyed policy toward monopoly in the open economy. The national government may find itself dealing with two classes of issues. First, the nation may possess monopoly power over its export goods, if it colludes successfully with other producing nations. The literature of cartel theory, which concentrates on the conditions for sustainable collusion, supplies a number of useful predictions about the circumstances that render such alliances stable and supply countries with a self-interest in joining them. Some helpful hints are also available for nations whose terms of trade have been worsened by cartels; a version of counterspeculation may retrieve a portion of their losses.

Countries may also find themselves using international economic policy

[47] See Barcelo (1972).

to deal with imperfections in their national markets for goods and services. Where natural monopoly exists due to extensive scale economies, tariffs are no help; however, dumping may sometimes be desirable, and the analysis indicates conditions under which it is desirable to offer lump-sum subsidy to an import-competing monopolist. The power of unrestricted international trade to eliminate market distortions due to monopoly has recently been subject to extensive empirical testing. The general conclusion is that import competition (definitely) and export opportunities (probably) reduce the ability of concentrated industries to exercise their joint monopoly power. Conversely, tariffs augment or preserve this power and may also induce organizational patterns in industries that are inconsistent with cost minimization. The multinational company offers the advantage of being a well-equipped entrant into national product markets surrounded by high entry barriers, but it may also contribute in some ways to long-run anticompetitive conditions.

These two strands of national policy-making can be brought together around the situation of a national authority contemplating its policy toward a monopolized export industry. If the industry sells only in foreign markets, the national (though not global) welfare is served by allowing it to extract the rents available to it. On the other hand, if it sells partly in the domestic market, there is an offsetting deadweight loss to domestic consumers that might be greater than the profits corresponding to surplus captured from foreign buyers. The best nationalistic policy is to make the industry competitive, but levy an optimal tax on exports. If this cannot be done (e.g., because the monopoly power is holistic, due to patents or trademarks, and cannot be preserved one place but not another), a choice must be made by weighing the foreign loss against the domestic gain from enforcing competition. National competition policies implicitly recognize this dilemma by permitting collusion more freely in foreign than domestic markets. United States antitrust policy shows some tendency to recognize international competition and treat it in order to maximize global rather than national welfare.

## REFERENCES

Adams, W. J. 1976. "International Differences in Corporate Profitability." *Economica* 43: 367–79.

Adelman, M. A. 1972. *The World Petroleum Market*. Baltimore: The Johns Hopkins University Press for Resources for the Future.

Auquier, A., and Caves, R. E. 1978. "Monopolistic Export Industries and Optimal Competition Policy." Harvard Institute of Economic Research, Discussion Paper No. 607.

Barcelo, J. J. 1970. "Antidumping Laws as Barriers to Trade—the United States and the International Antidumping Code." *Cornell Law Review* 57: 491–560.

Basevi, G. 1970. "Domestic Demand and Ability to Export." *Journal of Political Economy* 78: 330–37.

Bergsten, C. F. 1974. "The Threat Is Real." *Foreign Policy*, no. 14, pp. 84–90.

Blair, J. M. 1967. "Statement," in U.S. Senate, Committee on the Judiciary, Subcommittee on Antitrust and Monopoly, *Prices of Quinine and Quinidine*, Part 2, Hearings pursuant to S. Res. 26, 90th Cong., 1st sess., pp. 180–223. Washington, D.C.: Government Printing Office.

Bloch, H. 1974. "Prices, Costs, and Profits in Canadian Manufacturing: The Influence of Tariffs and Concentration." *Canadian Journal of Economics* 7: 594–610.

Brewster, K. 1958. *Antitrust and American Business Abroad*. New York: McGraw–Hill.

Caves, R. E. 1974a. "The Economics of Reciprocity: Theory and Evidence on Bilateral Trading Arrangements." In *International Trade and Finance: Essays in Honour of Jan Tinbergen*, edited by W. Sellekaerts, pp. 17–54. London: Macmillan & Co.

––––––. 1974b. "Industrial Organization." In *The Multinational Enterprise and Economic Analysis*, edited by J. H. Dunning, pp. 115–46. London: George Allen & Unwin.

––––––. 1974c. *International Trade, International Investment, and Imperfect Markets*, Special Papers in International Economics, no. 10. Princeton, N.J.: International Finance Section, Princeton University.

––––––. 1975. "International Enterprise and National Competition Policy: An Economic Analysis." In *International Conference on International Economy and Competition Policy*, edited by M. Ariga, pp. 183–90. Tokyo.

Caves, R. E., and Khalilzadeh-Shirazi, J. 1977. "International Trade and Industrial Organization: Some Statistical Evidence." In *Welfare Aspects of Industrial Markets: Scale Economies, Competition and Policies of Control*, edited by A. P. Jacquemin and H. W. de Jong, pp. 111–27. Leiden: Martinus Nijhoff.

Caves, R. E., et al. 1977. *Studies in Canadian Industrial Organization*. Ottawa: Information Canada.

Comanor, W. S., and Schankerman, M. A. 1976. "Identical Bids and Cartel Behavior." *Bell Journal of Economics* 7: 281–86.

Corden, W. M. 1967. "Monopoly, Tariffs and Subsidies." *Economica* 34: 50–58.

Cross, J. G. 1969. *The Economics of Bargaining*. New York: Basic Books.

de Jong, H. W. 1975. "EEC Competition Policy towards Restrictive Practices." In *Competition Policy in the UK and EEC*, edited by K. D. George and C. Joll, chap. 2. Cambridge: Cambridge University Press.

Eastman, H. C., and Stykolt, S. 1967. *The Tariff and Competition in Canada*. Toronto: Macmillan & Co.

English, H. E. 1964. *Industrial Structure in Canada's International Competitive Position*. Montreal: Canadian Trade Committee.

Esposito, L., and Esposito, F. F. 1971. "Foreign Competition and Domestic Industry Profitability." *Review of Economics and Statistics* 53: 343–53.

Feenstra, R. C. 1977. "Trade, Competition, and Efficiency: A General Equilibrium Analysis." Senior honors thesis, University of British Columbia.

Fellner, W. 1949. *Competition Among the Few*. New York: Knopf.

Frederiksen, P. G. 1975. "Prospects of Competition from Abroad in Major Manufacturing Oligopolies." *Antitrust Bulletin* 20: 339–76.

Frenkel, J. A. 1971. "On Domestic Demand and Ability to Export." *Journal of Political Economy* 79: 668–72.

Gorecki, P. K. 1976*a*. "The Determinants of Entry by Domestic and Foreign Enterprises in Canadian Manufacturing Industries: Some Comments and Empirical Results." *Review of Economics and Statistics* 58: 485–88.

————. 1976*b*. *Economies of Scale and Efficient Plant Size in Canadian Manufacturing Industries*, Research Monograph No. 1, Bureau of Competition Policy. Ottawa: Department of Consumer and Corporate Affairs.

Greene, R. S. 1977. "Cartel Action and Forward Integration by the Bauxite Producing Nations." Senior honors thesis, Harvard College.

Gribbin, J. D. 1976. "Review of OECD, *Export Cartels*." *Antitrust Bulletin* 21: 341–50.

Hague, D. C., Oakeshott, A., and Strain, A. 1974. *Devaluation and Pricing Decisions*. London: Allen & Unwin.

Hart, P., and Morgan, E. 1977. "Market Structure and Economic Performance in the United Kingdom." *Journal of Industrial Economics* 25: 177–93.

Hexner, E. 1945. *International Cartels*. Chapel Hill, N.C.: University of North Carolina Press.

Hnyilicza, E., and Pindyck, R. S. 1976. "Pricing Policies for a Two-Part Exhaustible Resource Cartel: The Case of OPEC." *European Economic Review* 8: 139–54.

Horst, T. 1972. "The Industrial Composition of U.S. Exports and Subsidiary Sales in the Canadian Market." *American Economic Review* 62: 37–45.

Hu, S. C. 1975. "Uncertainty, Domestic Demand, and Exports." *Canadian Journal of Economics* 8, 258–68.

Jacquemin, A. P. 1974. "Application to Foreign Firms of European Rules on Competition." *Antitrust Bulletin* 19: 157–79.

Jenny, F., and Weber, A. P. 1976. "Profit Rates and Structural Variables in French Manufacturing Industries." *European Economic Review* 7: 187–206.

Jones, J. C. H., Laudadio, L., and Percy, M. 1973. "Market Structure and Profitability in Canadian Manufacturing Industry: Some Cross-Section Results." *Canadian Journal of Economics* 6: 356–68.

Khalilzadeh-Shirazi, J. 1974. "Market Structure and Price-Cost Margins in United Kingdom Manufacturing Industries." *Review of Economics and Statistics* 56: 67–76.

Knickerbocker, F. T. 1973. *Oligopolistic Reaction and Multinational Enterprise*. Boston: Division of Research, Harvard Business School.

Krasner, S. D. 1974. "Oil Is the Exception." *Foreign Policy*, no. 14, pp. 68–83.

Krause, L. B. 1962. "Import Discipline: the Case of the United States Steel Industry," *Journal of Industrial Economics* 11: 33–47.

Larson, D. A. 1970. "An Economic Analysis of the Webb–Pomerene Act." *Journal of Law and Economics* 13: 461–500.

Lerner, A. P. 1944. *The Economics of Control*. New York: Macmillan & Co.

Litvak, L. A., and Maule, C. J. 1975. "Cartel Strategies in the International Aluminum Industry." *Antitrust Bulletin* 20: 641–63.

McFetridge, D. G. 1973. "Market Structure and Price-Cost Margins: An

Analysis of the Canadian Manufacturing Sector." *Canadian Journal of Economics* 6: 344–55.

Melvin, J. R., and Warne, R. D. 1973. "Monopoly and the Theory of International Trade." *Journal of International Economics* 3: 117–34.

Mikdashi, Z. 1974. "Collusion Could Work." *Foreign Policy*, no. 14, pp. 57–67.

Nichols, A. L., and Zeckhauser, R. J. 1977. "Stockpiling Strategies and Cartel Prices." *Bell Journal of Economics* 8: 66–96.

Organization for Economic Cooperation and Development. 1974. *Export Cartels—Report of the Committee of Experts on Restrictive Business Practices.* Paris: OECD.

Orr, D., and MacAvoy, P. W. 1965. "Price Strategies to Promote Cartel Stability." *Economica* 32: 186–97.

Osborne, D. K. 1976. "Cartel Problems." *American Economic Review* 66: 835–44.

Owen, N. 1976. "Scale Economies in the EEC: An Approach Based on Intra-EEC Trade." *European Economic Review* 7: 143–63.

Pagoulatos, E., and Sorenson, R. 1976*a*. "International Trade, International Investment and Industrial Profitability of U.S. Manufacturing." *Southern Economic Journal* 42: 425–34.

———. 1976*b*. "Domestic Market Structure and International Trade: An Empirical Analysis." *Quarterly Review of Economics and Business* 16: 45–60.

———. 1976*c*. "Foreign Trade, Concentration and Profitability in Open Economies." *European Economic Review* 8: 255–67.

Patinkin, D. 1947. "Multi-Plant Firms, Cartels, and Imperfect Competition." *Quarterly Journal of Economics* 61: 173–205.

Pryor, F. L. 1973. *Property and Industrial Organization in Communist and Capitalist Nations.* Bloomington, Ind.: Indiana University Press.

Pursell, G., and Snape, R. H. 1973. "Economies of Scale, Price Discrimination and Exporting." *Journal of International Economics* 3: 85–91.

Roberts, M. J. 1974. "Review of Adelman, *The World Petroleum Market.*" *Journal of Economic Literature* 12: 1363–68.

Rowe, J. W. F. 1965. *Primary Commodities in International Trade.* Cambridge and London: Cambridge University Press.

Rowley, C. K. 1971. *Steel and Public Policy.* London: McGraw–Hill.

Salant, S. W. 1976. "Exhaustible Resources and Industrial Structure: A Nash–Cournot Approach to the World Oil Market." *Journal of Political Economy* 84: 1079–93.

Schelling, T. C. 1960. *The Strategy of Conflict.* Cambridge, Mass.: Harvard University Press.

Scherer, F. M. 1970. *Industrial Market Structure and Economic Performance.* Chicago: Rand McNally.

Scherer, F. M., et al. 1975. *The Economics of Multi-Plant Operation: An International Comparisons Study.* Cambridge: Harvard University Press.

Schwartzman, D. 1959. "The Effect of Monopoly on Price." *Journal of Political Economy* 67: 352–67.

Shubik, M. 1959. *Strategy and Market Structure: Competition, Oligopoly, and the Theory of Games.* New York: John Wiley.

Sichel, W. 1975. "The Foreign Competition Omission in Census Concentration Ratios: An Empirical Evaluation." *Antitrust Bulletin* 20: 89–105.

Spence, M. (forthcoming). "Tacit Coordination and Imperfect Information." *Canadian Journal of Economics.*

Stigler, G. J. 1952. *The Theory of Price,* rev. ed. New York: Macmillan & Co.

———. 1964. "A Theory of Oligopoly." *Journal of Political Economy* 72: 44–61.

———. 1968. *The Organization of Industry.* Homewood, Ill.: Richard D. Irwin.

Telser, L. G. 1966. "Cutthroat Competition and the Long Purse." *Journal of Law and Economics* 9: 259–77.

———. 1972. *Competition, Collusion, and Game Theory.* Chicago: Aldine-Atherton.

Turner, P. P. 1976. "Some Effects of Devaluation: A Study Based on the U.K.'s Trade in Mañufactured Goods." Ph.D. dissertation, Harvard University.

Vicas, A. G., and Deutsch, A. 1964. "The Paradox of Employment Creation through Import Subsidies." *Economic Journal* 74: 228–30.

White, L. J. 1974. "Industrial Organization and International Trade: Some Theoretical Considerations." *American Economic Review* 64: 1013–20.

Worcester, D. A., Jr. 1957. "Why 'Dominant Firms' Decline." *Journal of Political Economy* 65: 338–46.

Yamey, B. S. 1973. "Notes on Secret Price-Cutting in Oligopoly." In *Studies in Economics and Economic History: Essays in Honour of Professor H. M. Robertson,* edited by M. Kooy, pp. 280–300. London: Macmillan & Co.

# Comment

ALAN V. DEARDORFF

Professor Caves has provided us with a useful survey of the literature on the role of imperfect competition in the international economy. From his remarks there appears to be a difference between the two topics that he treats, in terms of the extent to which distinctly international considerations are important for understanding and dealing with the market imperfections discussed. In the second part of his paper, he cites considerable evidence of how international trade can influence domestic market power and of how such market power, in turn, can affect the gains from trade. But in the first part of the paper he analyzes international cartels with the tools that are used for the study of purely domestic cartel arrangements, and leaves the impression that there is little that is distinctly international in the character of international cartels. While I do not dispute his claim that the same analytical framework is appropriate for both domestic and international cartels, I would like to amplify and extend his short section on "Cartels in the international economy." Specifically I shall suggest three issues that have occurred to me as posing difficulties for the maintenance of international cartels, difficulties that would not arise in a purely domestic context.

1. Caves mentions briefly the difficulty of defining the objectives of an international cartel when, as is likely to be the case, the participants either are themselves, or are abetted by, national governments. A "shared set of values" may, as he says, be helpful, but these values may well contradict the basic nature of a cartel, which is to restrict output so as to raise price. A concern for employment is the obvious example. Governments are often notoriously willing to stimulate inefficient production in order to sustain employment and may therefore be reluctant to enter or abide by a cartel agreement which requires that output be reduced. This suggests that international cartels are most likely to succeed among governments that feel no such responsibility to their people, and in industries where labor requirements are minimal.

2. Members of an international cartel face another difficulty that is reminiscent of one long faced by economists in determining the efficient pattern of trade: they must decide among various measures of cost in determining the output to be provided by each member. This problem does not arise in a domestic cartel where all members share a common currency. But the distinction between money cost and opportunity cost can be troublesome for cartel members that operate out of different countries. Suppose, for example, that money wages rise in the country of one of the members. A member firm, looking at money costs, will react to the change by reducing output, but from the national point of view opportunity cost is unchanged and output should not be curtailed. This suggests the need for coordination of macroeconomic policies among cartel member governments if the cartel is to function smoothly.

3. Finally, Caves points out that "joint monopolization by several countries is merely a matter of calculating the optimal export tax from the joint foreign reciprocal demand curve." This particularly simple means of setting up a cartel is, of course, not available to domestic cartels, and therefore suggests that cartels are more easily formed in the international context. It is not even necessary that the number of producing firms be small—only that the number of governments that are party to an agreement be limited. But if an export tax can be used to form a cartel, this also suggests a devious means of cheating on the cartel that would also not be available in the domestic context. For we know from Lerner's symmetry theorem that anything that can be done by an export tax can be undone by a reduction in tariffs.[1] For symmetry to be perfect, of course, the export tax must apply to all exports, but cartel member countries are often highly specialized in their exports, and even if they were not a tariff reduction

---

[1] See A. P. Lerner (August 1936) "The Symmetry between Import and Export Taxes," *Economica* 3: 306–13.

could be expected to partially offset the effect of an industry-specific export tax. Thus the enforcers of an international cartel must not only see that export taxes are effectively maintained in the member countries but also that tariff reductions (and perhaps other policy changes as well) are not used to undermine the agreement. One could object that member governments are unlikely to have read Lerner and to be aware that this means of cheating is at their disposal. But if the cartel succeeds in increasing export revenue, tariff reductions may be undertaken without the knowledge that they interfere with the cartel.

These remarks are only intended to be suggestive of the special problems that may face an international cartel. I would hope that these and other issues that are unique to the international cartel will be given further scrutiny by those economists, such as Caves, whose expertise covers both industrial organization and international trade.

# Simple Theoretical Analysis of International Commodity Agreements

Jere Behrman

Simple economic theory provides useful guidelines for considering some important issues about which there is some confusion in much of the speculation about commodity market agreements. Therefore it is useful to discuss briefly the following four theoretical questions: What are the implications of price stabilization attempts for producers' revenues? Who benefits from stabilization? What are the normative implications of market solutions to economic problems? Under what conditions is it probable that collusive action by producers alone can raise market prices? Each of the first four sections in this chapter explores one of these questions. The last section gives conclusions.

## 3.1 IMPLICATIONS OF PRICE STABILIZATION ATTEMPTS FOR VARIABILITY AND LEVEL OF PRODUCERS' REVENUES

Advocates of international commodity agreements recognize that stabilization of export revenues probably is of much more interest to the developing countries than is stabilization of prices. In principle, of course, a buffer stock authority might buy and sell with the intent of stabilizing revenues.[1] Such an operation would be much more difficult than price stabilization,

---

1 It could attempt to act so that the total market demand curve facing producers approached a unit elastic curve with constant revenue implications. Elasticity is defined below as the percentage change in quantity in response to a given percentage change in price. For a unit elastic demand curve this value is unitary or one. Therefore as one moves along the demand curve every change of x percent in price causes a change of x percent in the opposite direction of the quantity demand, so revenue (the product of the two) remains constant.

Reprinted from "Development, the Internationl Economic Order, and Commodity Agreements," pp. 29–46, with permission of Addison-Wesley, Reading, Massachusetts, Copyright 1978.

however, for several reasons. Day-to-day operations would be harder because of the greater lags in the availability of quantity than price data. If such an arrangement were successful, strong inducements would exist for supply reductions because the same revenues could be earned with lower sales, which would release factors of production for other uses.[2] The concurrence of importing nations with a revenue-stabilizing scheme, finally, seems unlikely.

For such reasons, advocates of international commodity agreements argue for price stabilization instead of revenue stabilization. But this strategy raises the question: What are the implications of price stabilization attempts for revenues?

A preeminent international economist, Harry G. Johnson [1], states that "elementary economic analysis" suggests that international commodity agreements are dubious on these grounds. His argument is illustrated in Fig. 3.1. The basic average supply and demand curves for a purely competitive international commodity market are given by solid straight lines. The average supply curve ($SS$) gives the average quantity supplied for each possible price.[3] The average demand curve ($DD$) gives the average quantity demanded for each possible price. The assumption of pure competition implies that the total market supply is the sum of the quantities supplied at various prices by a large number of individual producers, each of whose production is so small that it cannot perceptibly change the market price by altering its quantity supplied. There is a parallel assumption on the demand side. The solid supply curve is an average curve in the sense that it is halfway between the two equally likely dashed-line actual supply curves, where the different locations reflect differences in some nonprice supply determinant (say, good and bad weather). A parallel situation holds for demand due to two different and equally likely values of some determinant of demand other than price (say, high and low income). $P_0$ is the average equilibrium price, at which level average quantity demanded just equals average quantity supplied (and both equal $Q_0$). $P_0$ also is the price at which the buffer stock is assumed to stabilize prices by purchasing the commodity if otherwise the price would fall, and selling it if otherwise the price would rise.

---

2 Such an outcome probably would not occur if there were a large number of relatively small producers each operating independently, but only if supply were organized in decision units of large enough size so that their individual impact on market prices was noticeable.

3 In Fig. 3.1a, for example, only the demand curve shifts so that the supply curve is traced out. When the demand curve is high the equilibrium price is $P_2$ and the quantity supplied is $Q_2$. When the demand curve is low the equilibrium price is $P_1$ and the quantity supplied is $Q_1$. The loci of quantities supplied at different prices (such as $Q_2$ at $P_2$ and $Q_1$ at $P_1$) is the supply curve.

Similar comments apply for the demand curve in Fig. 3.1b.

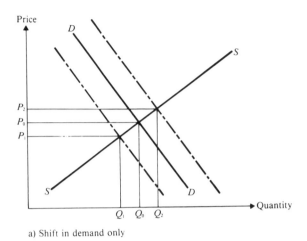

a) Shift in demand only

**Fig. 3.1** Impact on revenues of shifts in demand and supply curves with and without price stabilization at $P_0$ by a buffer stock

Consider first the case of instability due to demand shifts alone (Fig. 3.1a). Without price stabilization, producers' revenues are $P_2 * Q_2$ when the demand curve is shifted up and $P_1 * Q_1$ when the demand curve is shifted down.[4] The average is $(P_1 * Q_1 + P_2 * Q_2)/2$. With a buffer stock stabilization scheme, the buffer stock sells $Q_2 - Q_0$ units when the demand curve shifts down in order to keep the price at $P_0$. Whether the demand curve is shifted up or down, producers receive $P_0 * Q_0$ when the buffer stock operates, so this also is the value of their average revenues. Therefore price stabilization clearly implies producers' revenue stabilization in this case. But it also causes a reduction in the average value of producers' revenues since $P_0 * Q_0$ is smaller than $(P_1 * Q_1 + P_2 * Q_2)/2$, as can be seen by comparing in Fig. 3.1a the size of two rectangles each of which is $P_0 * Q_0$ with the sum of the areas in the rectangles that are $P_1 * Q_1$ and $P_2 * Q_2$. In the case of instability due to demands shifts alone, therefore, price stabilization causes producers' revenue stabilization, but at the cost of a reduction in those revenues.

Consider next the case of instability due to supply shifts alone. Here we must distinguish between various subcases that differ depending upon the supply and demand responsiveness to price changes. To summarize this price responsiveness it is useful to define the concept of elasticity. The price *elasticity of a curve* indicates by what percentage the quantity changes along a curve when the price changes by 1 percent. If the quantity changes by a

---

4 Here and below the standard notation of an "*" to mean multiplication is used. $P_1 * Q_1$, for example, should be read as $P_1$ multiplied by $Q_1$.

larger percentage than does the price, the absolute value[5] of the price elasticity for that curve is greater than one and the curve is price elastic for that range of price changes (for example, the demand curve and the supply curve in Fig. 3.1b). If the quantity changes by a smaller percentage than does the price, the absolute value of the price elasticity for that curve is less than one and the curve is price inelastic for that range of price changes (for example, the demand curve and the supply curve in Fig. 3.1c). If the quantity does not change at all when the price changes, the price elasticity is zero and the curve is completely price inelastic (for example, the supply curve in Fig. 3.1c).

Now let us consider the case of instability due to supply shifts alone. Johnson considers the most normal subcase to be one with price-elastic supply and demand curves (Fig. 3.1b). Following reasoning parallel to the case of demand shifts alone, we can find the average producers' revenues by considering what they are for both equally likely positions of the supply curve. Without price stabilization, producers' average revenues are $(P_2 * Q_2 + P_3 * Q_3)/2$. With stabilization they are $P_0 * (Q_1 + Q_4)/2$. In this subcase price stabilization increases producers' revenues, as can be seen by comparing the sizes of the relevant rectangles once again.

What about the stability of revenues under price stabilization when supply curves alone shift? In the subcase of price-elastic supply and demand curves, price stabilization increases the instability of revenues.

Of course one also can consider mixed cases in which both demand and supply curves shift. The net result depends on the size of the two shifts and the size of the price elasticities. The tradeoff between level and instability of revenues nevertheless seems to persist in a number of theoretical cases.

Therefore Johnson concludes that price stabilization generally leads to a tradeoff between revenue stabilization and the level of revenues and that advocates of international commodity agreements lump together two different economic problems (instability of demand and instability of supply) that require quite different solutions. He is quite critical of the UNCTAD proposal and of the analysis underlying it.

Is Johnson right? This depends on exactly what are the objectives and what is the empirical reality regarding the relative importance of shifts in supply and demand and regarding the shapes of these curves. We can identify several subcases in which Johnson, not the advocates of international commodity agreements, apparently is wrong:

1. Suppose that, while the developing economies desire higher revenues, *ceteris paribus*,[6] they are *very* risk averse in wanting very much to avoid

---

5 We refer to the absolute value because normally along a demand curve the quantity changes in the opposite direction from the price, while along a supply curve both change in the same direction.

6 *Ceteris paribus* means "everything else being equal" or "everything else held constant."

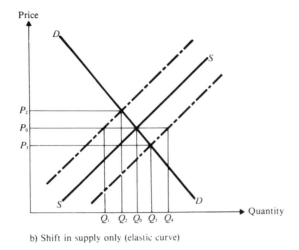

b) Shift in supply only (elastic curve)

**Figure 3.1 [continued].**

fluctuations in producers' revenues because of the perceived great disruptive effects of such fluctuations on their own economies. Then the proposal makes sense if either demand shifts dominate (Fig. 3.1a) or the curves are sufficiently inelastic (Fig. 3.1c).

2. Suppose that the objectives are weighted in the reverse order: While the developing countries would like revenue stabilization, they really care *much* more about increasing revenues. Then the proposal makes sense if shifts in supply curves are dominant (Figs. 3.1b-c). If the underlying curves are sufficiently price inelastic (Fig. 3.1c), moreover, producers' revenues may be increased at the same time fluctuations in those revenues are reduced.

3. Suppose that the assumptions of linear curves and/or parallel shifts are not valid. Then some of the conclusions of Johnson's "elementary economic analysis" may be changed. For example, consider the case in which the demand curve is very price inelastic above $P_0$ but very price elastic below this price and the completely price inelastic supply curve shifts (Fig. 3.2). Price stabilization may reduce revenues but increase their stability. This is the opposite outcome from what Johnson considers to be the normal result based on a shift in the supply curve with linear and price elastic curves (Fig. 3.1b).

4. Yet another possibility is that destabilizing speculation causes large price fluctuations that lower the long-run demand curve by inducing substitution of synthetics and other goods for the commodities of concern by risk-

---

7 Note that in the subcase of a supply shift and sufficiently price-inelastic curves, price stabilization leads to revenue stabilization and increased revenues (Fig. 3.1c).

averse manufacturers. Commodity producers, therefore, might rationally prefer price stabilization in order to limit the downward long-run shift in the market demand curve even if the short-term result may be lower immediate revenues or greater instability in revenues.

These possibilities all emphasize that in important respects the manner is an empirical question. Johnson's "elementary economic analysis" is not enough. Without empirical knowledge concerning preferences, long-run movements, the shapes of the curves, risk aversion, the elasticities, and the causes of shifts, whether they are additive or multiplicative, and so on, we cannot state with assurance what is the impact of price stabilization on producers' revenues.

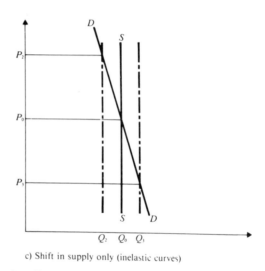

c) Shift in supply only (inelastic curves)

**Figure 3.1 [continued].**

At this point, it is useful to refer to some available empirical evidence: (1) For many of the relevant commodities, existing estimates indicate that short-run nonprice shifts in the supply curves tend to be larger than those in the demand curves, suggesting that Figs. 3.1b-c generally are more relevant than is Fig. 3.1a. (2) The estimated supply and demand price elasticities indicate for most of the relevant commodities quite low short-run price responsiveness. Therefore, the subcase of low price elasticities with price stabilization leading to larger revenues and less fluctuations in them (Fig. 3.1c) may be "normal," rather than the high price elasticities subcase with a tradeoff between the levels and instability of revenue (Fig. 3.1b) that Johnson emphasizes. (3) For foodgrains, Sarris, Abbott, and Taylor [2]

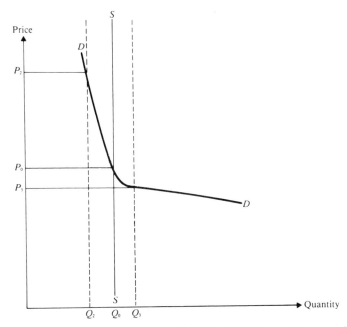

**Fig. 3.2** Impact on producers' revenues of shift in inelastic supply curve with nonlinear demand curve, with and without price stabilization at $P_0$ by a buffer stock

maintain that the nonlinear demand curve in Fig. 3.2 better represents reality than does the linear case.

And where does this empirical evidence lead us? For many of the UNCTAD core commodities it suggests that price stabilization may lead to revenue increases *and* greater revenue stability because of the dominance of supply shifts and low price elasticities. This may result in substantial benefits to the developing countries that, as a group, are net exporters of these commodities, independently of whether higher revenues or lower fluctuations in revenues are valued more highly. The possibility of forestalling long-run substitution for their exports by risk-averse users, also mentioned above, may increase the benefits of price stabilization to the developing countries.

For foodgrains, producers' revenues may be lowered, together with fluctuations therein, by price stabilization programs if Sarris, Abbott, and Taylor are right about the shape of the demand curve. However, Third World countries taken as a whole[8] still might benefit because they are net

---

8  A few developing countries (eg. Thailand, Burma, Argentina) are net exporters of foodgrains and thus, would not benefit under these assumptions.

importers of foodgrains. The lower level of producers' revenues in this case means lower consumer expenditures and import bills for them.

Thus, contrary to the assertions made by Johnson, simple economic theory in conjunction with this empirical evidence suggests that the developing countries as a whole well might benefit from effective price stabilization programs for many of the UNCTAD core commodities and for foodgrains. This is but a tentative conclusion, however, because we have not yet incorporated the dynamic adjustments of the interaction between supply and demand into our analysis. We return to such questions in Chapter 5, where we simulate the impact of price stabilization programs with models that incorporate empirical estimates of the relevant elasticities and of the dynamic adjustment paths.

## 3.2  WHO GAINS FROM PRICE STABILIZATION?

This question is related to the subject of the previous section, but the impact on consumers also needs to be incorporated. We explore it here under simplifying assumptions that ignore risk aversion, the question of distributional effects among consumers or among producers, storage and transaction costs for the buffer stock, and general-equilibrium aspects outside of the market of interest. We measure the benefits (losses) to producers by the additional (lessened) revenues they receive. We measure the benefits (losses) to consumers by the additional (lessened) consumer surplus they receive.

Consumer surplus is measured by the sum, for all units of a commodity, of the difference between what consumers would be willing to pay for each unit and what they have to pay. To illustrate, consider the downward sloping demand curve in Fig. 3.3. To purchase the first unit consumers are willing to pay a price $P_3$. To purchase the next unit they are willing to pay a price slightly less than $P_3$. To purchase the $Q_2$th unit they are willing to pay $P_2$. If the market price is $P_2$, then $P_2$ must be paid for each of the $Q_2$ units demanded. To measure the consumer surplus given a market price of $P_2$, we subtract $P_2$ from what consumers would be willing to pay for each of the $Q_2$ units actually purchased. But that is just the difference between the demand curve and the horizontal price line at $P_2$, or the area indicated by the triangle labeled $J$ in Fig. 3.3. Likewise, with a market price of $P_0$, the consumer surplus would be $J + F + G$.

Now let us return to the question of who gains and who loses from price stabilization. Let us consider the case in Fig. 3.3 in which the completely price inelastic supply curve is equally likely to be at $Q_1$ or $Q_2$, so on average it is at $Q_0$. Assume that the demand curve is fixed, so the only source of instability is the shifting supply curve. $P_0$ is the average price and the one at which the buffer stock stabilizes the price when it is in operation.

When the supply curve shifts out to $Q_1$, the buffer stock purchases $Q_1 - Q_0$ units. The change in consumer surplus due to paying $P_0$ instead of

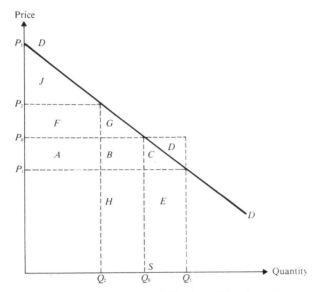

**Fig. 3.3** Gains and losses from price stabilization (shifts in inelastic supply curve only)

the price $P_1$ that would have prevailed without a buffer stock is negative, $-A - B - C$. The producers' revenue gain due to the higher prices is positive, $A + B + C + D$. The cost to the buffer stock of purchasing $Q_1 - Q_0$ units is $-C - D - E$. The total benefit (summing these three components) is $-C - E$.

When the supply curve shifts to $Q_2$, $Q_0 - Q_2$ units are sold by the buffer stock at price $P_0$. This precludes the price from rising to $P_2$, as it otherwise would. The benefit to the consumers is $F + G$ due to the lower price and the larger quantity. The benefit to the producers is $-F$ since they receive a lower price for their $Q_2$ units than they would without the buffer stock. The financial inflow to the buffer stock is $B + H$. The total benefit is the sum of these three components, $B + G + H$.

If the sequencing over time of the supply shifts is ignored,[9] the total benefits to each of the three groups is the sum of those obtained from buffer stock operation with supply at $Q_1$ and at $Q_2$. For consumers the sum is $F + G - A - B - C$. For producers the sum is $A + B + C + D - F$. For the buffer stock the sum is $B + H - C - D - E$. For the total benefit the sum is $B + G + H - C - E$. Under these assumptions the sum for the buffer

---

9 In Section 5.1 we discuss how events in the distant future might be discounted to make them comparable to current events. In the terminology of that discussion, we here are assuming a zero discount rate.

stock is zero and the overall sum is positive.[10] However, whether or not consumers or producers, respectively, benefit depends on the exact shape of the curves. The issue basically is an empirical one.

The thoughtful reader will realize that Fig. 3.3 represents only one of the alternative cases and subcases considered in the previous section. We could examine each of these and others in which the curves are not linear or the shifts are not parallel. To do so would only reinforce the conclusion that either the producers or the consumers might gain, but the issue is basically empirical. Instead of examining each theoretical possibility, therefore, in Chapters 5 and 6 we focus on the gains that are implied by the empirical estimates of actual supply and demand curves for the commodities of interest.

## 3.3 NORMATIVE IMPLICATIONS OF MARKET SOLUTIONS[11]

Pure competition is defined to be the situation in which no single participant has the capacity to affect market prices more than infinitesimaly (in other words, no single participant has market power). From the point of view of individual entities in the market place, prices seem to be given (not necessarily fixed over time) parameters independent of their own behavior. Pure competition generally is considered an interesting paradigm for reasons summarized below, but not of very general applicability in the real world. However, most of the initial producers of agricultural commodities that enter into international commodity markets (and agricultural products account for about 85 percent of total nonpetroleum commodity exports from developing countries) and most of the ultimate consumers (generally in processed form of both the agricultural and nonagricultural internationally traded commodities sell and purchase these goods, respectively, under conditions approximating pure competition. At both ends of the marketchain for the relevant international commodities (but not in the middle, where marketing boards, other government agencies, and large companies dominate), therefore, the purely competitive model has substantial applicability.

What are the advantages of pure competition? Some answers are in the area of political economy and thus derivative of particular value systems. Under pure competition the basic economic problems (for example, what is produced, how is it produced, and for whom is it produced) are solved in an

---

10 To see this, note in Fig. 3.3 that $B + H = C + D + E$ (so the sum for the buffer stock is zero) and $B + G + H$ is greater than $C + E$ (so the total benefit is positive).

11 The distinction often is made between positive and normative economics. Positive economic analysis reveals what happens under certain conditions (for example, if the demand curve is downward sloping, the quantity demanded declines as the price increases, everything else held constant). Normative economics pertains to what should occur, and, therefore, incorporates value judgments. Appendix A provides a more detailed analysis of the issues in this section for the interested reader.

impersonal manner, independently of personal ties or characteristics such as race or national origin.[12] The atomistic structure of buyers and sellers required for competition also decentralizes and disperses power. Moreover, if the conditions necessary for pure competition to exist do in fact prevail, freedom of entry into various industries and individual mobility will both be high.

In response to the question about the advantages of pure competition, most economists focus on answers related to economic efficiency. In a world with the correct initial distribution of input ownership for a given social welfare function, with easy entry (for instance, due to a lack of legal restrictions and limited increasing returns to scale relative to the size of industries), with no externalities, with no uncertainty, and with pure competition everywhere else, pure competition in international commodity markets results in maximization of the social welfare function.[13]

This is a strong result. But what does it really mean? Appendix A explores this questions in some depth, and interested readers are encouraged to study it for more details. Here it suffices to expand on the notion of efficiency by distinguishing among efficiency in production, efficiency in exchange, and overall efficiency. Under the assumptions of the previous paragraph, pure competition leads to all three kinds of efficiency, as well as to maximization of social welfare.

*Efficiency in production* occurs if production of one good cannot be increased without lessening the production of some other good. That is, the economy is on the production frontier of Fig. 1.1 so that no more manufactured products can be made without reducing the output of agricultural products (or vice versa). Under pure competition each firm chooses to sell the number of units of product at which its marginal cost (the cost of producing the last unit) just equals the market price of the product in order to maximize its profits. Also, each firm selects its inputs so that the marginal products (i.e. the additional products obtained from using the last units of inputs) for the last dollars spent on all inputs are identical in order to minimize the costs of producing the profit-maximizing level of output. This means that an individual firm satisfying this condition could not gain by substituting one input for another. It also means that society as a whole could not increase the output of one good without reducing the output of some other good since every firm minimizes its cost by equalizing across inputs the marginal product for the last dollar spent on each input and every firm faces the same input prices. The last unit of each input is everywhere valued the same, so no overall output increase can result from merely shifting inputs around. Such behavior assures efficiency in production.

---

12 Given certain sets of values, this impersonality is a negative dehumanizing feature.

13 The assumptions made here are defined and discussed below.

*Efficiency in exchange* means that for a given level of production of all goods, no one individual can be made better off merely by exchanging goods with someone else without making at least one other person worse off. Under pure competition all individuals maximize their satisfactions (or utilities) by choosing a combination of goods so that the last bit of satisfaction (that is, marginal utility) obtained from the last dollar spent on the good is the same for all goods. Since every person faces the same product prices, every person values the last unit of one good that they purchase relative to the last unit of another in the same relative way. Therefore everyone could not be made better off merely by switching given levels of goods among individuals. Such behavior assures efficiency in exchange.

*Overall efficiency* exists when the rate at which the last unit of product of one good can be transformed into another by moving along the production possibility frontier is the same as the rate of which individuals substitute the last unit of one good for the other. The rate at which the last unit of one commodity can be transformed into the other is given by the ratio of the marginal costs of the two commodities, or the slope of the production possibility frontier. Given that purely competitive firms choose an output at which the marginal cost is equal to the product price, this rate of transformation between two goods is equal to the ratio of the product prices. But satisfaction-maximizing consumers also choose combinations of goods so that their ratios of marginal satisfactions (or marginal utilities) are equal to these same ratios of product prices. Marginal consumption decisions among goods are made on the basis of true relative marginal costs of production for society. Therefore, pure competition assures overall efficiency. In such a situation no one can be made better off—by changing inputs among firms, by changing the composition of output among commodities, or by changing the distribution of output among consumers—without making someone else worse off.

*Social welfare maximization* occurs if there is a social welfare function that depends on the levels of satisfactions of all individuals and if the maximum value of this function is obtained for a given supply of inputs, technology, and preferences of individuals. Given any particular social welfare function, pure competition leads to its maximization if there is overall efficiency *and* if the initial ownership of inputs is exactly right so that the market solutions lead to just the right incomes for that particular social welfare function.[14]

---

14 For example, if the social welfare function weighted everyone's preferences equally, a very unequal distribution of ownership of inputs probably would not lead to maximization of the function. On the other hand, if the social welfare function put very high weight on the satisfaction of one individual, it probably would not be maximized by an equal distribution of ownership of inputs.

## Reservations

The result that pure competition leads to social welfare maximization is a strong result. But the necessary conditions are very strong too, and obviously not even approximately satisfied in the real world. Let us consider them one by one.

First, maximization of a social welfare function depends upon having exactly the right distribution of income and therefore of ownership of inputs. Within a static framework, much of the conflict between the developing and developed nations may arise at this point. Even if all the other conditions given above are satisfied so that economic efficiency is attained, the initial distribution of assets is seen by many to be so inequitable that the world is far away from a welfare maximization. Efficiency concerns may be unimportant in light of this maldistribution.

Second, leaving aside the question of welfare maximization, pure competition leads to overall efficiency if all of the other conditions are satisfied. No shift in resources and so on exists that would improve the welfare of any one individual without reducing the welfare of at least one other individual. Attainment of this state seems desirable, everything else being equal, and its virtues are emphasized (perhaps overemphasized) by many economists. But the existence of pure competition alone is not enough to guarantee even efficiency. To obtain efficiency in the above discussion we had to assume that there are no externalities. That is, we had to assume that the production of one product depends only on the market inputs used directly in the production process for it and that the utility of an individual depends only on the market goods he or she consumes. In the real world, however, externalities abound. Individuals' satisfactions depend not only on their consumption of purchased items, but also on such factors as the consumption of others (for example, "keeping up with the Jones") or non-market products like pollution. Likewise, production of one good may depend not only on the inputs purchased for use in its production process, but also on other nonmarket factors such as pollution. The existence of these nonmarket interdependencies or externalities precludes overall efficiency even if the conditions exist to permit pure competition.

Third, even if all of the conditions for the existence of pure competition are satisfied and there are no externalities, the behavioral assumptions assumed above may not be satisfied. For example, it is assumed that firms maximize profits. But firms may have other objectives or considerations in mind. Possibilities include avoiding risk in an uncertain world or providing perquisites (such as nice offices and company cars) for the managers. If such other considerations are important, behavior of these firms may not be approximated well by profit maximization—and efficient outcomes do not result.

Fourth, the conditions for pure competition to exist often are not satisfied even if profit maximization and the lack of externalities are both assumed. Entry into an industry is not easy in many cases due to legal, natural, and technological monopolies or due to increasing returns to scale that are large relative to the industry. If there are relatively few firms in an industry, they perceive correctly that they have market power in that they can affect their product price by changing their output. Fig. 3.4 illustrates the profit-maximizing behavior of a monopolist in the $M$ industry. Since the demand curve is sloping downward, to sell more the monopolist must lower the price. As a result, the marginal revenue (or additional revenue generated from selling one more unit) is below the demand curve since the price must be lowered on all previous units to sell one more unit. The profit-maximizing level of output is $M_1$, at which level the marginal revenue equals the marginal cost.[15] At lower outputs the marginal revenue exceeds the marginal cost, so profits can be increased by expanding output and sales (and vice versa at higher levels of output). To sell $M_1$ units of output, the monopolist must charge the price $PM_1$, as is indicated by the demand curve. But this profit-maximizing condition for the monopolist implies a price greater than marginal cost. Therefore, if industry $A$ is purely competitive and industry $M$ is a monopoly, the economy no longer satisfies the overall efficiency condition. Instead, people are substituting between $A$ and $M$ at the margin at a different rate than the relative marginal cost of $A$ to $M$.

Let us pursue one implication of this example further. Ignore the possibilities of externalities for the moment and presume that the only nonpure competition in the system is the existence of monopoly in the production and sale of $M$. To obtain overall efficiency, the "first best" solution would be to make $M$ act as a pure competitor, perhaps by breaking it up into enough small units so that none has perceptible influence on the market price. But suppose that such an option is not available? What is best to do? Under the assumptions we have made, the "second-best" solution is for $A$ to charge the same ratio of price to marginal cost as does $M$. Then the overall efficiency condition is satisfied.

This is an illustration of the "theory of second best": If all of the conditions for an optimum cannot be satisfied, efficient outcomes can be ensured by introducing particular new distortions. The problem is that the desirable new distortions depend on the exact nature of the particular situation. The second-best solution indicated in the previous paragraph, for example, works under the particular assumptions indicated, but would not be efficient if one or both industries produced inputs for the other or if total vari-

---

15  The next section discusses the case of limit pricing to preclude entry by firms with market power, in which case the optimal strategy may not be to maximize short-run profits.

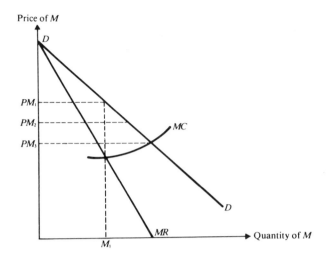

**Fig. 3.4** Market behavior of a monopoly

able input supplies were not fixed. Thus, to devise second-best solutions in the real world requires much more knowledge than policy-makers normally have.

What does this discussion of the possibility of nonpurely competitive markets mean in the consideration of international commodity markets? Clearly, substantial market power exists in markets other than the commodity markets in the real world. In principle, if such market power in the rest of the world cannot be eliminated, in the interests of efficiency the theory of the second-best suggests that it *might* be desirable to introduce it into international markets (if it does not already exist), not resist interfering in these markets because of supposed advantages of pure competition!

Realistically, substantial market power already exists in most international commodity markets. In the majority of cases, a relatively few buyers (governments or large firms) account for most of the purchases. In many cases sales are also quite concentrated. If there is no new attempt to regulate these markets, therefore, the alternatives are not an efficient, purely competitive system as many of the critics of the international commodity agreements seem to assume, but systems with considerable existing market imperfections.

Fifth, the whole analysis above is at a point of time without uncertainty. Incorporation of risk aversion and dynamic considerations well might vitiate the claim that pure competition leads to efficient outcomes over time. Schumpeter, for example, placed great emphasis on the importance of market power to lead to new technological developments that are much more important in a dynamic context than is static efficiency (see Behrman [3]). Although such claims are not uncontested, the possibility that they are right raises further doubts about advocating unregulated markets on short-run efficiency grounds above.

## Conclusion

Where does this discussion leave us? Unhindered markets may be relatively efficient devices for processing a great deal of information to signal shortages or surpluses through price or inventory changes. However, one has to be quite careful in regard to their normative implications. They lead to maximization of a given social welfare function only with the correct distribution of assets and the satisfaction of all of the other conditions discussed above. They lead to economic efficiency only under strong and unrealistic assumptions. The "theory of second best" at worst implies that, in the real world, policies directed at economic efficiency should be abandoned. At best it suggests the advocation of "third-best," very general, policies that have a reasonable probability of leading to greater economic efficiency but that do not guarantee a step in that direction when applied in a specific case. Dynamic considerations may weaken further the argument for unhindered markets leading to overall efficiency.

Economic theory leads us to this highly qualified view of the normative properties of unhindered market solutions. It is a much weaker view than that of many who oppose international commodity agreements on the basis that they would lessen the gains from free market operations. One can understand why economists from developing countries might wonder if the position of the strongest advocates of unhindered free market operations is based on a lack of understanding of underlying economic theory, or a disguised defense of vested interests in the status quo.

## 3.4 CONDITIONS UNDER WHICH COLLUSIVE ACTION BY PRODUCERS ALONE CAN RAISE MARKET PRICES

The basic motivation behind the advocacy of international commodity agreements may have little to do with stabilization per se. Instead, the major concern may be to raise the real resources of the developing countries who export the affected commodities. Under certain conditions discussed in Section 3.1 above (for example, dominant supply shifts), stabilization itself may lead to increased revenues for the exporters. The content of the UNCTAD documents, however, suggests that the concern goes further than this to a desire to raise market prices (or prevent real market prices from falling) to levels above that which otherwise would prevail. If market demand curves are price inelastic, successful price raising is rewarded by greater revenues since the quantities demanded do not decline by as large a percentage as prices rise.

This leads us into a much less rigorous area of economic theory: oligopoly formation and behavior. An oligopoly is an industry than has a small number of producers. Each member of an oligopoly can perceive that not only does its own output decision affect the market price, but also the output decisions of all its fellow oligopolists affect the market price. This interdependency creates a rivalry (although not necessarily price competition) among the oligopolists in regard to the division of the existing market. If the oligopolists are able to agree how to divide the market shares, they can maximize their joint profits by acting as a monopolist and selecting that output for which industry marginal cost is equal to marginal revenue ($M_1$ in Fig. 3.4).

However, this is a very static view. If the oligopolists collude to raise prices, new firms may be induced to enter into the industry or new substitutes may be developed to compete with the product of the industry. This brings us to the possibility of limit pricing to discourage entry. In Fig. 3.4 the collusive short-run static profit-maximizing price is $PM_1$. Suppose, however, that at any price above $PM_2$, new entrants are induced. To maximize long-run profits it may be desirable from the point of view of the colluding oligopolists to set the price below the short-run profit maximizing level in order to limit the inducement for new firms to enter. The effective long-run demand curve for the current colluding oligopolists has a horizontal segment at $PM_2$ until it hits the downward-sloping demand curve $DD$. The colluders have to decide how to balance off higher short-run and long-run prices and profits. The easier is entry or the possibility of others substituting for the product, the less are the colluding oligopolists able both to gain large current profits and to maintain longer-run market power.

The theory of limit pricing and a number of theories[16] of how oligopolists divide a given market share lead to a checklist of conditions that seem to facilitate oligopolistic coordination of pricing and output decisions: (1) the perception that joint action will lead to greater returns for producers, (2) common output preferences due to similar cost structures and market shares, (3) cheap and rapid communication, (4) high concentration of production in relatively few firms, (5) a small (or no) competitive fringe, (6) repetitive small transactions, (7) homogeneity and simplicity of products, (8) the willingness to utilize inventory and order backlogs as buffers instead of making overly sensitive price adjustments, (9) limited or no substitution for the product, and (10) high barriers to entry (for example, restricted technological knowledge, restricted control over exhaustible resources, legal re-

---

16 See any standard intermediate economic theory or industrial organization textbook. None of these theories is completely persuasive.

strictions on new firms, returns to scale at a high level of production relative to market size).

How do the UNCTAD core commodities stack up against such a checklist? In some respects they do rather well: product homogeneity, frequent transactions, and a common perceived interest—at least currently among the developing country producers. But in other respects they generally fare poorly: low barriers to entry, limited returns to scale, an active competitive fringe, substantial current and potential substitution for the products. And for the commodity that probably is most promising by these criteria—copper—the developing countries account for about only 40 percent of world production. Thus the developing countries indeed may be advocating international commodity agreements that include both producers and consumers because they perceive little likelihood of developing successful producer cartels for the ten core commodities on their own. In an important sense, therefore, the oil cartel of OPEC may seem *not* to be a model that can be imitated by producers of the UNCTAD core commodities.

## 3.5  CONCLUSIONS

The theoretical considerations of this chapter give important insights into the arguments for and against international commodity agreements. They suggest that the developing-country producers of the ten UNCTAD core commodities may be advocating such agreements because the chances of success of producers' cartels for these commodities are not high. They also indicate that normative arguments against international commodity agreements on the grounds of efficiency or social welfare maximization are *not* well based. Finally, they imply that whether producers or consumers gain from international commodity agreements and whether or not there is a tradeoff between the level and instability of producers' revenues cannot be established on the basis of economic theory. Instead, empirical analysis is required. The rest of this book attempts to provide such analysis.

## REFERENCES

1. H. G. Johnson. "Commodities: Less Developed Countries' Demands and Developed Countries Response." Paper presented at MIT Workshop on Specific Proposals and Desirable DC Response to LDC Demands Regarding the New International Economic Order, 17-20 May 1976.

2. A. H. Sarris, P. Abbott, and L. Taylor. *World Grain Reserves*. Washington, D. C.: Overseas Development Council, 1977.

3. J. R. Behrman. "Development Economics." *Modern Economic Thought*. Ed. by S. Weintraub. Philadelphia: University of Pennsylvania Press, 1977.

# WHAT FACTORS INFLUENCE OIL PRICES?

JOHN H. LICHTBLAU

To get a measure of the hazardousness of long-term oil price forecasts, one only needs to look at the 10-year predictions made in 1970. None of the experts came even remotely close to the actual world price of last year. Nor did they do much better on supply-and-demand projections. By 1972 at least the direction of some of the structural changes in the market that underlay the subsequent price revolutions began to emerge.

Perhaps we are in a similar situation regarding the remainder of the 1980s. We can see some structural changes in the marketplace that should make the 1980s very different from the 1970s. But to guess at the specific price in 1990 is still quite hazardous. An analysis of the world oil price in the 1980s, therefore, must be confined to the principal trends that will be influential.

The 750% increase in the real world oil price between 1972 and 1981 was made possible by the interaction of three principal factors: market trends, extraneous political events, and cartel policies. Let us look at how each of these factors operated in the 1970s and then speculate on their likely roles in the current decade.

*Market trends* in the 1960s and early 1970s were such that a substantial increase in the real price of oil would have been inevitable even without the Organization of Petroleum Exporting Countries (OPEC) from the mid-1970s on. The sustained growth rate in world oil demand at the pre-1973 price level was simply too fast relative to physically available supply and thus had to be substantially curtailed by means of higher prices to restore market equilibrium. By 1978, after the price increases of 1973–74 had eroded somewhat in real terms, oil demand began to rise again at a rate which, if continued, would have led to supply tightness within 5–6 years. Thus, by the early 1980s a further real price increase to balance supply and demand would have been required even in the absence of the Iranian revolution. The revolution advanced the need for the increase because it caused a long-term, perhaps permanent, reduction in OPEC's producing capacity on the order of about 10%. Hence, directionally in both 1973–1974 and 1979–1980 the underlying market price trend was clearly upward.

In 1981–1982, the market trend was, of course, in the opposite direction. World oil demand (outside the Communist Bloc and China) declined in both 1980 and 1981 by 5%. These are the biggest annual drops, by far, in the entire post-war period. There is a debate over what part of the decline is cyclical and what part is structural. The cyclical part is ephemeral and will be reversed once the industrial

Reprinted with permission from *Oil and Gas Journal* **79**(45), 198–204, Copyright 1981.

economies of the world recover from their present slump. This is just what happened after the 1974–1975 oil demand drop that followed the first world oil price shock. The structural changes underlying the demand reduction are, of course, more enduring because they reflect economically motivated trends to conserve oil, use it more efficiently, and substitute other fuels for it. In the short run these trends are not readily reversible nor even arrestable.

Clearly, both cyclical and structural factors have been at work in reducing world oil demand by almost 5 million barrels/day between 1979 and 1981. OPEC oil exports have been under additional downward pressure from the current large-scale extraordinary world inventory reduction, which is by definition self-limited. Thus, the demand drop of the 1980–1981 period will be reversed when the inventory reduction is completed and economic recovery is underway. But the increase will be strongly tempered by a continuation of the structural downward trend in oil consumption. Automobile fuel efficiency in the United States will keep rising throughout the 1980s because of existing legislation; atomic power plants currently under construction throughout the world will progressively reduce fuel oil requirements for power generation; most new commercial airplanes are more fuel efficient; most new homes and buildings are designed with much more attention to fuel conservation than those built before 1975; and coal will maintain a substantial price advantage over fuel oil under any realistic price scenario and, hence, will continue to displace oil in certain markets. One could easily find other examples of structural reductions in oil demand that will be carried forward for a number of years by their historic momentum even if there should be some decline in the real price of oil.

Taking account of both the cyclical and the structural factors and assuming for the moment that real world oil prices will remain approximately unchanged, world oil demand will start recovering in 1982 but will climb at a very slow rate and is unlikely to exceed the 1979 level of about 52 million barrels/day in this decade. Meanwhile, the existing production constraints resulting from the Iranian-Iraqi war can be expected to end, potentially freeing substantial additional OPEC supplies for export. Also, non-OPEC supplies will continue to increase, perhaps at an annual growth rate of about 500,000 barrels/day to 1985 and somewhat less thereafter.

These factors indicate that in the absence of major extraneous supply interruptions (discussed in the following) readily available world oil supplies will generally be comfortably in excess of demand until at least the late-1980s at unchanged real prices. Foreign and domestic requirements for OPEC crude oil should remain below the organization's current collective allowable crude-producing capacity of about 28 million barrels/day until at least the late-1980s and will not come near its current maximum sustainable crude-producing capacity of around 33 million barrels/day (not including the existing military restrictions on production in Iran and Iraq) even by 1990.

Thus, market conditions do not require an increase in the real price of crude oil for the next 7–8 years to balance supply and demand, since throughout this period there will be excess producing capacity. In fact, for the next 4 years the magnitude of the excess will be such that, on the basis of market conditions, the real world oil price should decline from its present level. This is in sharp contrast to the 1970s when market conditions, as we have seen, put upward pressure on prices.

The second factor affecting price is the *extraneous political events* that could cause major oil *supply* disruptions. The principal reason for the price explosions in the 1970s was, of course, the physical disruptions that created actual or perceived temporary oil shortages.

Since supply disruptions are by their very nature unpredictable, speculations about their probability are not particularly useful. Neither are projections based on historic occurrences. There is now a tendency to downgrade the likely impact of future supply disruptions because the most recent one, the Iranian–Iraqi war, has had only a minimal effect on prices even though the war caused a substantial reduction in world production. It is certainly true that the Iranian–Iraqi disruption has laid bare the previous axiomatic assumption that any major oil supply disruption would inevitably cause a price explosion. But this revisionist view must not be carried too far. The latest disruption occurred when world oil stocks were exceptionally high and world oil demand was rapidly falling. Had it occurred at a different phase of the inventory or business cycle, it might have had a very different impact on prices.

Nevertheless, supply disruptions are not as likely in the 1980s as in the 1970s. The principal factors are higher OPEC excess producing capacity and higher levels of government oil inventories for security purposes. As to the first, OPEC's effective, sustainable excess producing capacity during the 1973–1979 period was 10% or less of total producing capacity (except in 1975 when it was much larger). Currently it is around 30%, and if the military restrictions on Iranian and Iraqi production end, it may range between 15% and 20% throughout the 1980s.

The second factor cushioning price from supply disruptions is government oil inventories. There were none in the United States at the time of the Arab oil embargo and virtually none when the Iranian revolution broke out. Currently, the Strategic Petroleum Reserve is slightly above 200 million barrels. By the beginning of 1984 it could be as high as 400 million barrels, enough to offset a 2-million barrels/day shortfall in the United States imports for nearly 7 months. Japan, too, had no strategic government reserve in the two previous oil disruptions but now has over 50 million barrels, which is separate from the industry's inventory.

It appears, therefore, that if any oil-exporting country, with the exception of Saudi Arabia, were to cease all of its exports for the better part of a year, this would not have a major effect on world oil supplies because the production loss could be offset by increased output from other OPEC countries, particularly Saudi Arabia. This was clearly not the case in the 1970s.

The situation is, of course, very different for Saudi Arabia. A Saudi output reduction of, for example, 50% caused by nonmarket factors, would bring about a world oil shortage with major price implications if maintained for several months. Whether and under what conditions such a reduction might occur is beyond the scope of this chapter. It would present a global strategic problem and would have to be dealt with on that basis.

Now we come to the *cartel policies*. It has been suggested that OPEC has lost the ability and willingness to influence prices in the 1980s. However, this is not so. With a disproportionately large contribution from Saudi Arabia, the organization has been able to maintain a floor price for its oil at $32 per barrel for Saudi Light quality crude. This is substantially higher than what the current price would be under free market conditions. Since no OPEC member is attempting to sell below the floor price, the organization remains effective, although its upward price mobility is severely curtailed by market forces.

OPEC's future as an effective price-setting organization depends on its ability to maintain enough cohesiveness among its members to continue to set and enforce a floor price that is a multiple of production cost. This, in turn, depends on its members' excess producing capacity. As we have seen, under conditions of approximately unchanged real prices, OPEC's excess producing capacity may amount to 15–20% throughout most of the 1980s.

If the real price should rise by an annual average rate of, for instance, 3.5% over the next 10 years, which would be about in line with OPEC's long-term pricing strategy announced in early 1980, export demand for OPEC oil could well drop by 3.5–4.0 million barrels/day given the fact that the worldwide reduction in oil demand resulting from the price increase would be concentrated on imported oil and within the import sector on OPEC oil. This means that total OPEC oil demand could fall toward 20 million barrels/day. OPEC would then operate at about 60% of its technical capacity. This would make it very difficult for the organization to continue to maintain its price cohesiveness. With 13 million barrels/day of readily producible OPEC oil overhanging the market, some members may succumb to the temptation to sell more oil by offering hidden or open discounts to their customers. Once this process spreads, it would rapidly undermine OPEC's floor-price defense and cause prices to tumble.

Probably OPEC's long-term survival will depend on its ability to maintain its crude output between 22 million and 28 million barrels/day. If production drops below the lower level of this range for more than a year, the organization's price cohesion, which is its raison d'être, would come under extraordinary pressure. This would be particularly true after the cessation of Iranian–Iraqi hostilities and the consequent significant increase in OPEC's exportable supply. At the upper end of the range—28 million barrels/day of crude—OPEC's remaining effective spare capacity would be quite small. Hence, if demand exceeds this level, market forces would enable OPEC to raise its real prices once again at a rapid rate, possible even in the face of Saudi opposition. Initially, this would benefit OPEC but over time it would accelerate existing measures to conserve and replace imported oil.

As of now, it seems that an OPEC production level within this range is more likely during the 1980s than production below or above it. If this is so and if Saudi Arabia maintains its current oil policy, which is by no means certain, real world oil prices should neither rise nor fall substantially in the 1980s, though they may fall somewhat between now and 1985 and rise somewhat thereafter. Compared to the 1970s, this would be a very tolerable development.

# Part VI

## Multinationals and International Investment

Caves
Vernon
Hymer

# FOREIGN DIRECT INVESTMENT
# AND MARKET PERFORMANCE

Richard E. Caves

The other international force affecting the performance of national markets is the multinational corporation. In this section I consider the sorts of markets in which we can expect the multinational firm to appear, and then outline the probable market behavior of these firms in the context of both national and international markets. This analysis leads to predictions of the effects of the multinational firm on market performance.

## Causes of Foreign Direct Investment

In recent years, economists' analysis of the causes of foreign direct investment has moved away from macroeconomic explanations (e.g., national gluts or shortages of entrepreneurship) to sector-specific explanations. Rather than losing generality, this shift in focus has allowed us to explain many phenomena—the large interindustry differences in the importance of direct investment and the significant gross exports of equity capital from many industrial countries—that had previously resisted understanding. We can now explain the occurrence of the multinational firm, starting from a coherent model of profit-maximizing behavior and moving to empirical predictions about its incidence, behavior, and welfare significance.[31]

Briefly, the analysis starts from the proposition that the entrepreneurial unit has a natural national identity. Economically, this means that it automatically comes by a large stock of knowledge about the language, laws, and customs of its native land—intangible capital that is productive in guiding the firm toward profit-maximizing decisions. A firm that invests in a foreign market is at an intrinsic disadvantage, because it must consciously recruit this information (or run the risk associated with action under relative ignorance). On this view, the dice are loaded to some degree against the multinational firm, and its emergence thus demands an explanation.

The explanation for much foreign investment—certainly that in manufacturing industries[32]—lies in the fact that the successful firm

---

[31] For a synthesis with references to earlier contributions, see Caves (1971).

[32] The following discussion concentrates exclusively on what I call "horizontal" direct investment—the firm produces abroad the same general line of goods as at home. An important volume of foreign investment instead involves backward vertical integration, to provide the parent with components or raw materials at minimum cost or risk. The explanation of this sort of direct investment is rather

Reprinted from "International Trade, International Investment, and Imperfect Markets—Special Papers in International Economics No. 11, Nov. 1974," pp. 17–29, with permission of Princeton University Press, Princeton, Copyright 1974.

also gains intangible capital in the form of patents, trademarks, or general knowledge about how to produce and distribute its products. Being intangible, these assets can in some measure be moved from one national market to another, gaining rents in new locations without impairing the stock left in service at the home base. And the advantage to the firm investing abroad can offset the disadvantage noted above—the cost of gathering intangible capital for the foreign subsidiary.[33]

A firm that has acquired such intangible capital chooses among several methods of exploiting it in foreign markets. One is simply to export goods that embody the design, formula, trademark, or reputation that the firm has established. Another is to license producers abroad to employ the firm's technology or replicate its product. The third is to establish a producing subsidiary to exploit these assets directly in the foreign market. The choice among these alternatives will depend on many factors. An explanation of foreign direct investment thus must answer not only the question "Why invest abroad at all?" but also "Why investment, rather than exports or licensing?" The answers to the second question will depend on characteristics specific to the firm and its industry—the realm of industrial organization—and on the nation's factor endowment—the realm of international trade.

Take first the market-specific characteristics. Intangible capital is heterogeneous: knowledge about production processes, knowledge about adapting the firm's basic product to local demand conditions, etc. Certain components of intangible capital are much more suited than others to employment via direct investment—notably, knowledge about how to serve a market. When the product must be adapted to local tastes and conditions and when the existence of nearby production facilities is complementary to servicing the product after sale, or even just to forging a reputation for quality, direct investment tends to be the preferred alternative. In terms of the standard concepts of

---

different from that for horizontal investment. In the case of large natural-resource investments in the industrial countries, it turns on the role of capital costs and the avoidance of uncertainty that would otherwise surround bilateral oligopoly bargaining among firms with high fixed costs and long-lived investments. Vertical integration, including that via direct investment, can have its disfunctional consequences in such situations, but market failures of one kind or another are hard to avoid. See Caves (1971, pp. 10-11, 27) and Caves (1974c). The latter paper develops the industrial-organization framework of direct investment in more detail than the present essay.

[33] The analysis of the role of intangible capital leans heavily on Harry G. Johnson, "The Efficiency and Welfare Implications of the International Corporation," in Kindleberger, ed. (1970, Chap. 2).

market structure, a strong affinity exists between direct investment and *product differentiation*.[34]

Conversely, if the knowledge takes the form of specific production techniques that can be written down and transmitted objectively, licensing may be a prime vehicle. Exporting stands as a contender when the intangible advantage can be embodied in the firm's product only at its primary locus of production, or at least when no special need arises for local adaptation.

Another leading market-specific determinant that favors direct investment is size of the parent firm. Direct investment clearly entails a larger and riskier fixed cost than the alternatives, exporting or licensing, because of the substantial and relatively fixed information and search costs that must precede any actual investment abroad. Licensing entails much lower costs of search and real investment, but it is also a more rough-edged method of extracting quasi-rents. Given the presence of lender's risk or outright imperfections of capital markets, direct investment becomes the province of the large firm with substantial internally generated funds to finance the initial fixed charges. On a probabilistic basis, this requirement of large size for the investing firm implies that foreign investment will occur principally in industries where sellers are few in number.[35] Putting all this together, we expect to find direct investment in manufacturing industries marked by differentiation and fewness of sellers, or *differentiated oligopoly*.

The other characteristics determining direct investment lie in the realm of international trade. One characteristic is evidently the position of the parent's industry in the nation's scale of comparative advan-

[34] In connection with Porter's (1973) research, I noted evidence of an important subdivision of differentiated consumer goods according to the relative monopsony power of the distributive channels and scale economies in nationwide sales promotion. It was suggested that trade is apt to be the more effective international market constraint in the case of specialty and shopping goods. Conversely, the multinational firm should be relatively more important where convenience goods are involved. The economies of scale in nationwide sales promotion require a large-scale market entry if the multinational firm is to make any effective use of its intangible assets, and the uncertainties of international trade probably impel the seller toward local production facilities if high-density distribution activities are to be carried out successfully. Statistical research (Caves, 1974b) does not directly confirm this prediction, but it does show that foreign investment in the two sectors responds to various determinants in ways consistent with Porter's model.

[35] The importance of size of firm as a predictor of direct investment is shown by Horst (1972b). Other connections may exist between concentration and foreign investment. A firm with market power must diversify if it wishes to grow faster than its "base" market; otherwise, its growth entails a struggle for market share. The diversification might take the form of expansion into a foreign market, rather than into other domestic markets.

tage. A favorable position encourages exporting, against the alternatives of direct investment and licensing. Because two-way trade can clearly occur in direct investment as well as in a differentiated industry's flow of merchandise, we can say that an industry that is a net importer on trade account is apt (*ceteris paribus*) to be a net exporter on the balace of international indebtedness. An important corollary is that direct investment, unlike other forms of international capital flow, should be sensitive to the exchange rate (as the link between nations' production-cost levels, and apart from any expectations concerning future exchange rates); a country that devalues can hope for improvement in its balance of payments on direct investment as well as on goods and services.

Another trade-related variable that should influence direct investment is tariffs and transport costs. A finding from many of the surveys of foreign subsidiaries is that the initial investment was often made after the parent firm had established an export trade that was threatened by a higher tariff; with a substantial goodwill asset already created in the market, the parent chose to establish local production facilities rather than abandon its goodwill entirely. Transport costs should affect the choice in similar ways. A firm that makes a product that is costly to ship per unit value and that requires only ubiquitous raw materials will be disposed toward establishing multiple plants close to its customers. In this and other respects, the multinational firm may be viewed simply as a multiplant enterprise that happens to sprawl across national boundaries.[36]

One conclusion evident from this analysis of the interindustry distribution of direct investment is that it goes where trade does not. Not only are exporting and direct investment alternative strategies for the individual firm, but direct investment tends also to occur in differentiated products where international trade may be a relatively ineffective constraint on poor market performance. We return to this proposition after examining the probable effects of the multinational firm on market behavior and performance.

## Multinational Firms and Market Behavior

Is the multinational firm a constraint on market distortions, in the same sense that a perfectly elastic world supply of cabbage at 10 cents a pound constrains what the domestic cabbage monopoly can charge?

[36] This and the preceding predictions, save for the influence of tariffs, find strong statistical support in Caves (1974b). On the influence of tariffs, see Horst (1972a). The multiplant hypothesis is developed in Eastman and Stykolt (1967, Chap. 4) and McManus (1972).

It may be, for two reasons. First, multinational firms tend to develop in just those industries where barriers to the entry of new firms tend to be high. The formation of a new subsidiary, at least on a "green field" basis without takeover of a going enterprise or establishment, amounts to entry by an established firm in another (geographical) industry; counting as a specially well-endowed potential entrant, the multinational renders the supply of potential entrants larger and in effect makes the barriers to entry lower than they otherwise would be.[37]

Second, multinational firms actually operating in a given national industry may behave differently from domestic firms holding equivalent shares of the market. These possible differences in conduct are vital to the multinational's effects on market performance, and I shall thus consider them in some detail. A national branch of a multinational firm might behave differently from an equal-sized independent company for three reasons:

1. *Motivation.* I accept the conventional view that, as a first approximation, the maximization of profit can safely be assumed to be the prevailing motivation of the firm, multinational or not. Indeed, the available evidence supports the view that the multinational maximizes profits from its activities as a whole, rather than, say, telling each subsidiary to maximize independently and ignoring the profit interdependences among them (Stevens, 1969). But overall maximization by the multinational can lead its subsidiary to behave differently from an independent firm. The rate of earnings retention is a possible example. A subsidiary might pass up an otherwise profitable local use of funds if the expected yield would be higher elsewhere in the global corporation, whereas a local firm would make the local commitment. Another difference arises because the multinational firm almost automatically spreads its risks, and could therefore behave quite differently in an uncertain situation from an independent having the same risk/return preference function (Shearer, 1964; Dunning and Steuer, 1969; Schwartzman, 1970, p. 205). It is hard to generalize, though, about the consequences of such differences in opportunity costs and allocative choices.

2. *Cognition and information.* Its corporate family relations give the multinational unit access to more information about markets located in other countries—or (what is equivalent in effect) information to which it can attach a higher degree of certainty. This information need not be unavailable or even more costly to the national firm. The point is that at any given time the multinational has this stock in hand; its

[37] The reasons why the multinational firm has a potential advantage against each of the conventional barriers to entry are set forth in Caves (1974c).

national rival may or may not. The national firm therefore is probably more dependent on its home base in the national market, and this difference could color the multinational's view of actions that might increase its market share (and its own profits) at the expense of its rivals' shares (and total profits). With better information about extranational alternative uses of its resources and less dependence on the local scene for organizational survival, it has less to lose from rivalrous market actions. It could be more disposed than a national enterprise toward strategies yielding a larger profit but with a larger variance. The multinational unit could also collaborate less closely with its local rivals, especially in its early years, for the simple reason than its ear is less attuned to the "focal points" of tacit collusion among its unfamiliar new neighbors. The young subsidiary, even if formed by the acquisition of a going national firm, stands outside the network of tacit understandings and rules of the game developed by the previous market occupants, and is more apt to rock the boat. As the subsidiary ages and loses its parvenu status, however, it tends to play by the rules of the local game; also, it may tread softly at any time because of its political vulnerability.[38]

3. *Opportunity set.* Each player in a complex oligopoly game is apt to hold a somewhat different set of assets and to seek to slant the game along lines that will make his own asset bundle most productive of profits. The asset bundle of the multinational unit can differ from its national rival's in various ways. Its skill in differentiating its product, arguably a precondition for direct investment, inclines it to prefer nonprice forms of rivalry. Holding other traits of market structure constant, the presence of subsidiaries thus disposes an industry toward venting its competitive animal spirits through nonprice rather than price competition. Another asset that the subsidiary holds is the option to call on the financial assets of its corporate siblings—the "long purse" that makes it relatively secure from the predatory conduct of its rivals and a possible predator itself (Telser, 1966).

What do these behavioral differences mean for the performance of markets populated by multinational firms? The conclusions evidently will be ambiguous. Let us see where they fall. Because of the multinational's advantages over new firms, it provides a clear increase in the supply of potential entrants to the industries in which it operates and should thus constrain the departure of industry profits from normality —lowering the "limit price," to use the concepts of industrial organiza-

---

[38] The range of behavior patterns suggested in this paragraph seems to match those reported in surveys of subsidiaries' behavior (see Brash, 1966, pp. 182-192; Stonehill, 1965, pp. 98-99).

tion.[39] Furthermore, the cognitive and information resources of the multinational may dispose it to exhibit less collusive and restrictive conduct in the national market than would a similar domestic firm. There is thus some chance that multinational firms reduce allocative distortions in a certain range of industries. This effect is subject to some offsets, however. The multinational's predisposition toward product rivalry and advertising may cause it to devote excessive resources to these activities. Differentiated varieties of a good originating abroad do offer users genuine welfare-increasing expansions of the choices available to them. But the commitment of resources to sales promotion may count at least in part as a minus, especially when we consider that such nonprice competition feeds back to augment barriers to the entry of new firms and thus raises the long-run potential for market distortions.[40]

The presence of multinational firms may also change an industry's probable quality of performance in two other dimensions—technical efficiency and progressiveness. The multinational probably tends to be a technically efficient firm itself—if only on the assumption that the market tends to deny inefficient firms the chance to go multinational. Furthermore, it enjoys an option for avoiding diseconomies of small scale that may not be open to its domestic rivals: producing components subject to extensive economies of scale at a single world location. However, operating in industries subject to product differentiation, the profit-maximizing multinational need not always build efficient-scale facilities. Where a nation's market for manufactures is relatively small and heavily protected by tariffs (e.g., Canada), multinationals may crowd in with inefficiently small facilities; each firm profits from a small group of loyal customers and none is induced to lower its price-cost margin and expand its scale of operations (English, 1964).

Any favorable rating of the multinational on technical progressiveness probably turns on its role as a conduit for transferring new productive knowledge from one country to another. Does it raise the speed (or lower the cost) of technology transfers, considering the al-

---

[39] A weak negative relation has been found, among Canadian manufacturing industries, between the profit rate of domestic firms and the share of sales accounted for by foreign subsidiaries. The effect is partly explained, though, by variations in the relative size of the domestic firms; i.e., where their profits are relatively low, it is also because their size is relatively small (see Caves, 1974a).

[40] Another adverse structural feedback could result from the multinational's "long purse." The size of American multinationals serves as reason—or excuse—for horizontal mergers, often government-encouraged, among relatively large European firms. Whether or not the multinational is by nature a predatory species, the fact of this reaction is itself important.

ternative channels through which they can take place? Both the analytical issues and the empirical evidence are complex. My tentative impression from both survey and statistical evidence is that the multinational firm, in some countries and industries, probably does speed the transfer of technology (see Brash, 1966, Chap. 8).[41]

## Multinational Firms and International Industries

In considering the effects of international trade on competition, I argued that its salutary influence is limited by any oligopolistic interdependence that spreads across national boundaries. The national boundary, however, was found to be a fairly effective insulator against international collusion. But the multinational firm may promote international collusion. It extends the tendrils of ownership from one national market to another. Clearly, corporate siblings are not likely to compete with each other. Furthermore, the multinational could serve as a vehicle for extending oligopoly behavior across national boundaries.

Consider an international industry populated by a number of national and multinational firms, the multinationals based in diverse parent countries. What patterns of oligopolistic interdependence might arise within and among the various national markets? There are two limiting cases:

1. Multinational status (parent or subsidiary) makes no difference in the patterns of conduct adopted by firms in a national market. In this case, the member units of a multinational firm serve as independent profit centers with full autonomy over national price and product decisions. Any cross-national links in market conduct would be due to factors other than ownership status.

2. Multinational enterprises recognize their interdependence comprehensively wherever concentration is high enough and their perceptions sharp enough to permit it. That is to say, firm $A$ sets its actions in market $X$ taking account of $B$'s expected reactions not only in $X$ but also in any other market $Y$ where they both operate. $A$ expects $B$ to react wherever $B$'s interests are best served by so doing.

Between these extremes of cross-national independence and full cross-national interdependence, a variety of patterns could permit dependence across some boundaries but also maintain cordons along others. National origin might tell; firms domiciled in country $X$ might recognize their interdependence in the $X$ market and in host countries $Y, Z, \ldots$, but firms domiciled in $X$ and $Y$ might not perceive their in-

[41] For statistical evidence of the effects of subsidiaries on productivity in competing home-owned firms in Australia, see Caves (1974a).

terdependence comprehensively. Interdependence might be recognized among units (parents or subsidiaries) producing in the largest single national market (as a "home base") and also with parallel operations in other national markets. Interdependence might run outward from a national market where law and custom smile most kindly on overt collusion.

If we try coupling these patterns of possible interdependence with the possible behavior patterns of multinational units outlined above, the taxonomy quickly overtaxes patience. Let us concentrate on one facet of behavior, the decision to establish a subsidiary in a national market. Whatever the international interdependence among firms, the formation of a new subsidiary is clearly a rivalrous or independent move. Even if the parent was previously exporting to the market, local production facilities make it a more effective rival and a greater threat to other sellers. International collusive arrangements of the "sphere of influence" sort should entail nonaggression pacts between firms to keep subsidiaries out of each other's territory. On the other hand, an obvious form of retaliation when $A$ founds a subsidiary in $B$'s home base is for $B$ to invest in $A$'s. Firms domiciled in the same national market might well tend to follow the leader in starting subsidiaries. Assume that $A$ and $B$ both are domiciled in $X$ and have been exporting to $Y$. When $A$ starts a subsidiary in $Y$, not only are $B$'s exports to that market threatened, but also it is possible that $A$'s experience with the subsidiary will yield feedback that makes $A$ a more formidable competitor back in $X$.[42]

Indeed, the empirical evidence does document a good deal of this parallel and reactive behavior in founding subsidiaries. The entry of foreign firms into the U.S. market has sometimes followed on the establishment of U.S. subsidiaries abroad—and the foreign parents' discovery that they could compete with the U.S. giants (Daniels, 1971, p. 47). American industries have shown a strong tendency to parallel behavior in starting subsidiaries in foreign countries. Knickerbocker (1973), studying the dates when subsidiaries were established by U.S. manufacturing companies during the years 1948-1967, found that they were bunched in individual countries more than one would expect on a chance basis (note that scale economies should cut against the simultaneous start-up of new facilities). Furthermore, the tendency to tight parallel action was stronger for firms not highly diversified in the U.S. market and thus exposed to greater risks if rivals should successfully steal a march via foreign investment.

[42] Evidence suggests that this feedback in fact occurs in a majority of cases (see Reddaway et al., 1968, pp. 322-324).

The industry studies are even more reticent on the role of foreign investment than they are on the role of international trade. There is some indication that, while multinational firms have been effective in promoting product rivalry and innovation, they have also bestirred defensive mergers among national firms.[43] Whether these mergers weigh more heavily as a step toward increased technical efficiency or as a sinister move toward a higher level of oligopolistic collusion is, alas, unknown.

## IV. SUMMARY: INTERNATIONAL FORCES AND MARKET PERFORMANCE

I have suggested that competitive forces in the international economy complement one another in limiting the distortions that can occur in national markets. Whatever an industry's structural traits, we can pick out some international force as the most likely potential constraint on departures from a reasonable competitive outcome. (One of Panglossier disposition than mine might say that some international force will always ensure competitive performance.) This knowledge of the most probable source of market discipline is valuable for purposes of both research and policy.

Consider how the pieces fit together. Under certain assumptions, the effects of foreign trade via import competition and export opportunities are symmetrical in limiting departures from competitive outcomes. If they are, an industry will tend to face one constraint or another—depending on its comparative-advantage position. This proposition is sharply limited, however, because the disciplining force of export opportunities can easily evaporate when tariffs are present and dumping possible.

The disciplining force of trade flows is probably less when product differentiation is present, but in just those circumstances the multinational firm becomes a more prominent actor. Furthermore, because the firm itself makes a choice between direct investment and export, we conclude that the industry with a comparative disadvantage will face less threat from the entry of multinational firms but more from imports, and vice versa for the industry with a comparative advantage. Both natural and artificial trade impediments blunt the disciplining force of trade, but they may encourage that of foreign investment. Industries producing nontraded goods and services, sheltered from direct

[43] Information on the automobile industry is at least suggestive (see Silberston, 1958, p. 33; Ensor, 1971; Sundelson, "U.S. Automotive Investments Abroad," in Kindleberger, ed., 1970, Chap. 10).

foreign competition, thus are also potential prey to foreign subsidiaries.

An important qualification to this universal harmony is that the multinational firm is a mixed blessing as a market force. It is well equipped for scaling barriers to entry and may be less disposed toward oligopoly consensus in a national market than a domestic firm; also, it may speed the transfer of technology and attain (and encourage in its rivals) higher levels of technical efficiency. But it can slant market behavior to an undesirable degree toward advertising and product competition, and it may promote increased concentration and collusion running across national boundaries.

Let me close with some suggestions for economic research and policy. Empirical research on international market forces has at last become active in the field of industrial organization, as the statistical inputs have become available—and not just for the United States. We have some relatively strong conclusions about the role of trade. But even with the copious survey evidence on the multinational firm, it has not been examined very closely as a market force. This is despite the experiments that Nature and statesmen have obligingly performed in recent years—greatly increasing the multilateral penetration of foreign investment, removing tariff barriers within the European Economic Community, etc. The opportunities for research seem great, and the appropriate direction clear. In the international-trade camp, the research performance has been far less satisfactory. Lulled by the mathematical convenience of purely competitive conditions, theoretical research has paid little attention to the causes or consequences of imperfect competition, save for the obligatory bow to optimal tariffs and taxes on capital. And empirical research—lavished on a few safe topics such as the Leontief Paradox and financial capital flows—has elsewhere either been nonexistent or followed its nose with but slender guidance from economic theory. One hopes for both a redirection of research and a more fruitful interchange between theory and empirical investigation.

It is clear that many issues of antitrust and commercial policy turn closely on the results of testing the hypotheses discussed above. Paradoxically, the relation between these branches of policy, once a staple of American political economy ("the tariff is the mother of trusts"), has nearly disappeared from sight. One finds, instead, such spectacles as the U.S. government, bent on restricting certain imports at minimum annoyance to foreign nations, encouraging the cartelization of foreign exporters to reduce competition in U.S. markets! If we can avert our gaze from such squalor, more subtle issues remain to be dealt with. International market links—trade and the multinational firm—logically preclude our dealing with issues of competition on the basis of one-national-market-at-a-time. Policies toward competition, trade, and foreign investment in one country spill over and affect market performance in its trading partners. Issues of policy assignment and interdependence, familiar in international macroeconomics, are clearly present at a microeconomic level as well. One hopes that the further development of theory and empirical research in this area will lead to their due recognition.

# REFERENCES

Adams, William James, "Corporate Power and Profitability in the North Atlantic Community," unpublished Ph.D. dissertation, Harvard University, 1973.

Adler, Michael, "Specialization in the European Coal and Steel Community," *Journal of Common Market Studies*, 8 (March 1970), pp. 175-191.

Bain, Joe S., *Industrial Organization*, 2d ed., New York, Wiley, 1968.

Balassa, Bela "Tariff Reduction and Trade in Manufactures among the Industrial Countries," *American Economic Review*, 56 (June 1966), pp. 466-473.

Barker, H. P., "Home and Export Trade," *Economic Journal*, 61 (June 1951), pp. 276-278.

Basevi, Giorgio, "Domestic Demand and Ability to Export," *Journal of Political Economy*, 78 (March/April 1970), pp. 330-337.

Brash, Donald T., *American Investment in Australian Industry*, Cambridge, Mass., Harvard University Press, 1966.

Bright, A. A., Jr., *The Electric-Lamp Industry: Technological Change and Economic Development from 1800 to 1947*, New York, Macmillan, 1949.

Burenstam Linder, S., *An Essay on Trade and Transformation*, Stockholm, Almqvist & Wiksell, 1961.

Caves, Richard E., "International Corporations: The Industrial Economics of Foreign Investment," *Economica*, 38 (February 1971), pp. 1-27.

———, "Multinational Firms, Competition, and Productivity in Host-Country Industries," *Economica*, 41 (May 1974a), pp. 176-193.

———, "Causes of Direct Investment: Foreign Firms' Shares in Canadian and United Kingdom Manufacturing Industries," *Review of Economics and Statistics*, 56 (August 1974b), pp. 279-293.

———, "Industrial Organization," in John H. Dunning, ed., *The Multinational Enterprise and Economic Analysis*, London, George Allen & Unwin, 1974c, pp. 115-146.

Caves, Richard E., and Ronald W. Jones, *World Trade and Payments*, Boston, Little, Brown, 1973.

# INTERNATIONAL INVESTMENT AND INTERNATIONAL TRADE IN THE PRODUCT CYCLE *

## RAYMOND VERNON

Anyone who has sought to understand the shifts in international trade and international investment over the past twenty years has chafed from time to time under an acute sense of the inadequacy of the available analytical tools. While the comparative cost concept and other basic concepts have rarely failed to provide some help, they have usually carried the analyst only a very little way toward adequate understanding. For the most part, it has been necessary to formulate new concepts in order to explore issues such as the strengths and limitations of import substitution in the development process, the implications of common market arrangements for trade and investment, the underlying reasons for the Leontief paradox, and other critical issues of the day.

As theorists have groped for some more efficient tools, there has been a flowering in international trade and capital theory. But the very proliferation of theory has increased the urgency of the search for unifying concepts. It is doubtful that we shall find many propositions that can match the simplicity, power, and universality of application of the theory of comparative advantage and the international equilibrating mechanism; but unless the search for better tools goes on, the usefulness of economic theory for the solution of problems in international trade and capital movements will probably decline.

The present paper deals with one promising line of generalization and synthesis which seems to me to have been somewhat neglected by the main stream of trade theory. It puts less emphasis upon comparative cost doctrine and more upon the timing of innovation, the effects of scale economies, and the roles of ignorance and uncertainty in influencing trade patterns. It is an approach

* The preparation of this article was financed in part by a grant from the Ford Foundation to the Harvard Business School to support a study of the implications of United States foreign direct investment. This paper is a by-product of the hypothesis-building stage of the study.

with respectable sponsorship, deriving bits and pieces of its inspiration from the writings of such persons as Williams, Kindleberger, MacDougall, Hoffmeyer, and Burenstam-Linder.[1]

Emphases of this sort seem first to have appeared when economists were searching for an explanation of what looked like a persistent, structural shortage of dollars in the world. When the shortage proved ephemeral in the late 1950's, many of the ideas which the shortage had stimulated were tossed overboard as prima facie wrong.[2] Nevertheless, one cannot be exposed to the main currents of international trade for very long without feeling that any theory which neglected the roles of innovation, scale, ignorance and uncertainty would be incomplete.

## LOCATION OF NEW PRODUCTS

We begin with the assumption that the enterprises in any one of the advanced countries of the world are not distinguishably different from those in any other advanced country, in terms of their access to scientific knowledge and their capacity to comprehend scientific principles.[3] All of them, we may safely assume, can secure access to the knowledge that exists in the physical, chemical and biological sciences. These sciences at times may be difficult, but they are rarely occult.

It is a mistake to assume, however, that equal access to scientific principles in all the advanced countries means equal probability of the application of these principles in the generation of new products. There is ordinarily a large gap between the knowledge of a scientific principle and the embodiment of the principle in

1. J. H. Williams, "The Theory of International Trade Reconsidered," reprinted as Chap. 2 in his *Postwar Monetary Plans and Other Essays* (Oxford: Basil Blackwell, 1947); C. P. Kindleberger, *The Dollar Shortage* (New York: Wiley, 1950); Erik Hoffmeyer, *Dollar Shortage* (Amsterdam: North-Holland, 1958); Sir Donald MacDougall, *The World Dollar Problem* (London: Macmillan, 1957); Staffan Burenstam-Linder, *An Essay on Trade and Transformation* (Uppsala: Almqvist & Wicksells, 1961).

2. The best summary of the state of trade theory that has come to my attention in recent years is J. Bhagwati, "The Pure Theory of International Trade," *Economic Journal*, LXXIV (Mar. 1964), 1–84. Bhagwati refers obliquely to some of the theories which concern us here; but they receive much less attention than I think they deserve.

3. Some of the account that follows will be found in greatly truncated form in my "The Trade Expansion Act in Perspective," in *Emerging Concepts in Marketing*, Proceedings of the American Marketing Association, December 1962, pp. 384–89. The elaboration here owes a good deal to the perceptive work of Se'ev Hirsch, summarized in his unpublished doctoral thesis, "Location of Industry and International Competitiveness," Harvard Business School, 1965.

a marketable product. An entrepreneur usually has to intervene to accept the risks involved in testing whether the gap can be bridged.

If all entrepreneurs, wherever located, could be presumed to be equally conscious of and equally responsive to all entrepreneurial opportunities, wherever they arose, the classical view of the dominant role of price in resource allocation might be highly relevant. There is good reason to believe, however, that the entrepreneur's consciousness of and responsiveness to opportunity are a function of ease of communication; and further, that ease of communication is a function of geographical proximity.[4] Accordingly, we abandon the powerful simplifying notion that knowledge is a universal free good, and introduce it as an independent variable in the decision to trade or to invest.

The fact that the search for knowledge is an inseparable part of the decision-making process and that relative ease of access to knowledge can profoundly affect the outcome are now reasonably well established through empirical research.[5] One implication of that fact is that producers in any market are more likely to be aware of the possibility of introducing new products in that market than producers located elsewhere would be.

The United States market offers certain unique kinds of opportunities to those who are in a position to be aware of them.

First, the United States market consists of consumers with an average income which is higher (except for a few anomalies like Kuwait) than that in any other national market — twice as high as that of Western Europe, for instance. Wherever there was a chance to offer a new product responsive to wants at high levels of income, this chance would presumably first be apparent to someone in a position to observe the United States market.

Second, the United States market is characterized by high unit labor costs and relatively unrationed capital compared with practically all other markets. This is a fact which conditions the demand for both consumer goods and industrial products. In the case of consumer goods, for instance, the high cost of laundresses contributes to the origins of the drip-dry shirt and the home washing machine. In the case of industrial goods, high labor cost leads to the early

---

4. Note C. P. Kindleberger's reference to the "horizon" of the decision-maker, and the view that he can only be rational within that horizon; see his *Foreign Trade and The National Economy* (New Haven: Yale University Press, 1962), p. 15 *passim*.

5. See, for instance, Richard M. Cyert and James G. March, *A Behavioral Theory of the Firm* (Englewood Cliffs, N.J.: Prentice-Hall, 1963), esp. Chap. 6; and Yair Aharoni, *The Foreign Investment Decision Process*, to be published by the Division of Research of the Harvard Business School, 1966.

development and use of the conveyor belt, the fork-lift truck and the automatic control system. It seems to follow that wherever there was a chance successfully to sell a new product responsive to the need to conserve labor, this chance would be apparent first to those in a position to observe the United States market.

Assume, then, that entrepreneurs in the United States are first aware of opportunities to satisfy new wants associated with high income levels or high unit labor costs. Assume further that the evidence of an unfilled need and the hope of some kind of monopoly windfall for the early starter both are sufficiently strong to justify the initial investment that is usually involved in converting an abstract idea into a marketable product. Here we have a reason for expecting a consistently higher rate of expenditure on product development to be undertaken by United States producers than by producers in other countries, at least in lines which promise to substitute capital for labor or which promise to satisfy high-income wants. Therefore, if United States firms spend more than their foreign counterparts on new product development (often misleadingly labeled "research"), this may be due not to some obscure sociological drive for innovation but to more effective communication between the potential market and the potential supplier of the market. This sort of explanation is consistent with the pioneer appearance in the United States (conflicting claims of the Soviet Union notwithstanding) of the sewing machine, the typewriter, the tractor, etc.

At this point in the exposition, it is important once more to emphasize that the discussion so far relates only to innovation in certain kinds of products, namely to those associated with high income and those which substitute capital for labor. Our hypothesis says nothing about industrial innovation in general; this is a larger subject than we have tackled here. There are very few countries that have failed to introduce at least a few products; and there are some, such as Germany and Japan, which have been responsible for a considerable number of such introductions. Germany's outstanding successes in the development and use of plastics may have been due, for instance, to a traditional concern with her lack of a raw materials base, and a recognition that a market might exist in Germany for synthetic substitutes.[6]

6. See two excellent studies: C. Freeman, "The Plastics Industry: A Comparative Study of Research and Innovation," in *National Institute Economic Review*, No. 26 (Nov. 1963), p. 22 *et seq.*; G. C. Hufbauer, *Synthetic Materials and the Theory of International Trade* (London: Gerald Duckworth, 1965). A number of links in the Hufbauer arguments are remarkably similar to

Our hypothesis asserts that United States producers are likely to be the first to spy an opportunity for high-income or labor-saving new products.[7] But it goes on to assert that the first producing facilities for such products will be located in the United States. This is not a self-evident proposition. Under the calculus of least cost, production need not automatically take place at a location close to the market, unless the product can be produced and delivered from that location at lowest cost. Besides, now that most major United States companies control facilities situated in one or more locations outside of the United States, the possibility of considering a non-United States location is even more plausible than it might once have been.

Of course, if prospective producers were to make their locational choices on the basis of least-cost considerations, the United States would not always be ruled out. The costs of international transport and United States import duties, for instance, might be so high as to argue for such a location. My guess is, however, that the early producers of a new product intended for the United States market are attracted to a United States location by forces which are far stronger than relative factor-cost and transport considerations. For the reasoning on this point, one has to take a long detour away from comparative cost analysis into areas which fall under the rubrics of communication and external economies.

By now, a considerable amount of empirical work has been done on the factors affecting the location of industry.[8] Many of these studies try to explain observed locational patterns in conventional cost-minimizing terms, by implicit or explicit reference to labor cost and transportation cost. But some explicitly introduce problems of communication and external economies as powerful locational forces. These factors were given special emphasis in the analyses which were a part of the New York Metropolitan Region Study of the 1950's. At the risk of oversimplifying, I shall try to summarize what these studies suggested.[9]

some in this paper; but he was not aware of my writings nor I of his until after both had been completed.

7. There is a kind of first-cousin relationship between this simple notion and the "entrained want" concept defined by H. G. Barnett in *Innovation: The Basis of Cultural Change* (New York: McGraw-Hill, 1953) p. 148. Albert O. Hirschman, *The Strategy of Economic Development* (New Haven: Yale University Press, 1958), p. 68, also finds the concept helpful in his effort to explain certain aspects of economic development.

8. For a summary of such work, together with a useful bibliography, see John Meyer, "Regional Economics: A Survey," in the *American Economic Review*, LIII (Mar. 1963), 19–54.

9. The points that follow are dealt with at length in the following publications: Raymond Vernon, *Metropolis, 1985* (Cambridge: Harvard Uni-

In the early stages of introduction of a new product, producers were usually confronted with a number of critical, albeit transitory, conditions. For one thing, the product itself may be quite unstandardized for a time; its inputs, its processing, and its final specifications may cover a wide range. Contrast the great variety of automobiles produced and marketed before 1910 with the thoroughly standardized product of the 1930's, or the variegated radio designs of the 1920's with the uniform models of the 1930's. The unstandardized nature of the design at this early stage carries with it a number of locational implications.

First, producers at this stage are particularly concerned with the degree of freedom they have in changing their inputs. Of course, the cost of the inputs is also relevant. But as long as the nature of these inputs cannot be fixed in advance with assurance, the calculation of cost must take into account the general need for flexibility in any locational choice.[1]

Second, the price elasticity of demand for the output of individual firms is comparatively low. This follows from the high degree of production differentiation, or the existence of monopoly in the early stages.[2] One result is, of course, that small cost differences count less in the calculations of the entrepreneur than they are likely to count later on.

Third, the need for swift and effective communication on the part of the producer with customers, suppliers, and even competitors is especially high at this stage. This is a corollary of the fact that a considerable amount of uncertainty remains regarding the ultimate dimensions of the market, the efforts of rivals to preempt that market, the specifications of the inputs needed for production, and the specifications of the products likely to be most successful in the effort.

All of these considerations tend to argue for a location in which communication between the market and the executives directly concerned with the new product is swift and easy, and in which a wide

---

versity Press, 1960), pp. 38–85; Max Hall (ed.), *Made in New York* (Cambridge: Harvard University Press, 1959), pp. 3–18, 19 *passim*; Robert M. Lichtenberg, *One-Tenth of a Nation* (Cambridge: Harvard University Press, 1960), pp. 31–70.

1. This is, of course, a familiar point elaborated in George F. Stigler, "Production and Distribution in the Short Run," *Journal of Political Economy*, XLVII (June 1939), 305, *et seq*.

2. Hufbauer, *op. cit.*, suggests that the low price elasticity of demand in the first stage may be due simply to the fact that the first market may be a "captive market" unresponsive to price changes; but that later, in order to expand the use of the new product, other markets may be brought in which are more price responsive.

variety of potential types of input that might be needed by the production unit are easily come by. In brief, the producer who sees a market for some new product in the United States may be led to select a United States location for production on the basis of national locational considerations which extend well beyond simple factor cost analysis plus transport considerations.

## THE MATURING PRODUCT [3]

As the demand for a product expands, a certain degree of standardization usually takes place. This is not to say that efforts at product differentiation come to an end. On the contrary; such efforts may even intensify, as competitors try to avoid the full brunt of price competition. Moreover, variety may appear as a result of specialization. Radios, for instance, ultimately acquired such specialized forms as clock radios, automobile radios, portable radios, and so on. Nevertheless, though the subcategories may multiply and the efforts at product differentiation increase, a growing acceptance of certain general standards seems to be typical.

Once again, the change has locational implications. First of all, the need for flexibility declines. A commitment to some set of product standards opens up technical possibilities for achieving economies of scale through mass output, and encourages long-term commitments to some given process and some fixed set of facilities. Second, concern about production cost begins to take the place of concern about product characteristics. Even if increased price competition is not yet present, the reduction of the uncertainties surrounding the operation enhances the usefulness of cost projections and increases the attention devoted to cost.

The empirical studies to which I referred earlier suggest that, at this stage in an industry's development, there is likely to be considerable shift in the location of production facilities at least as far as internal United States locations are concerned. The empirical materials on international locational shifts simply have not yet been analyzed sufficiently to tell us very much. A little speculation, however, indicates some hypotheses worth testing.

Picture an industry engaged in the manufacture of the high-income or labor-saving products that are the focus of our discussion. Assume that the industry has begun to settle down in the United States to some degree of large-scale production. Although the first

3. Both Hirsch, *op. cit.*, and Freeman, *op. cit.*, make use of a three-stage product classification of the sort used here.

mass market may be located in the United States, some demand for the product begins almost at once to appear elsewhere. For instance, although heavy fork-lift trucks in general may have a comparatively small market in Spain because of the relative cheapness of unskilled labor in that country, some limited demand for the product will appear there almost as soon as the existence of the product is known.

If the product has a high income elasticity of demand or if it is a satisfactory substitute for high-cost labor, the demand in time will begin to grow quite rapidly in relatively advanced countries such as those of Western Europe. Once the market expands in such an advanced country, entrepreneurs will begin to ask themselves whether the time has come to take the risk of setting up a local producing facility.[4]

How long does it take to reach this stage? An adequate answer must surely be a complex one. Producers located in the United States, weighing the wisdom of setting up a new production facility in the importing country, will feel obliged to balance a number of complex considerations. As long as the marginal production cost plus the transport cost of the goods exported from the United States is lower than the average cost of prospective production in the market of import, United States producers will presumably prefer to avoid an investment. But that calculation depends on the producer's ability to project the cost of production in a market in which factor costs and the appropriate technology differ from those at home.

Now and again, the locational force which determined some particular overseas investment is so simple and so powerful that one has little difficulty in identifying it. Otis Elevator's early proliferation of production facilities abroad was quite patently a function of the high cost of shipping assembled elevator cabins to distant locations and the limited scale advantages involved in manufacturing elevator cabins at a single location.[5] Singer's decision to invest in Scotland as early as 1867 was also based on considerations of a sort sympathetic with our hypothesis.[6] It is not unlikely that the

4. M. V. Posner, "International Trade and Technical Change," *Oxford Economic Papers*, Vol. 13 (Oct. 1961), p. 323, *et seq.* presents a stimulating model purporting to explain such familiar trade phenomena as the exchange of machine tools between the United Kingdom and Germany. In the process he offers some particularly helpful notions concerning the size of the "imitation lag" in the responses of competing nations.

5. Dudley M. Phelps, *Migration of Industry to South America* (New York: McGraw-Hill, 1963), p. 4.

6. John H. Dunning, *American Investment in British Manufacturing Industry* (London: George Allen & Unwin, 1958), p. 18. The Dunning book

overseas demand for its highly standardized product was already sufficiently large at that time to exhaust the obvious scale advantages of manufacturing in a single location, especially if that location was one of high labor cost.

In an area as complex and "imperfect" as international trade and investment, however, one ought not anticipate that any hypothesis will have more than a limited explanatory power. United States airplane manufacturers surely respond to many "noneconomic" locational forces, such as the desire to play safe in problems of military security. Producers in the United States who have a protected patent position overseas presumably take that fact into account in deciding whether or when to produce abroad. And other producers often are motivated by considerations too complex to reconstruct readily, such as the fortuitous timing of a threat of new competition in the country of import, the level of tariff protection anticipated for the future, the political situation in the country of prospective investment and so on.

We arrive, then, at the stage at which United States producers have come around to the establishment of production units in the advanced countries. Now a new group of forces are set in train. In an idealized form, Figure I suggests what may be anticipated next.

As far as individual United States producers are concerned, the local markets thenceforth will be filled from local production units set up abroad. Once these facilities are in operation, however, more ambitious possibilities for their use may be suggested. When comparing a United States producing facility and a facility in another advanced country, the obvious production-cost differences between the rival producing areas are usually differences due to scale and differences due to labor costs. If the producer is an international firm with producing locations in several countries, its costs of financing capital at the different locations may not be sufficiently different to matter very much. If economies of scale are being fully exploited, the principal differences between any two locations are likely to be labor costs.[7] Accordingly, it may prove wise for the international firm to begin servicing third-country markets from the new location. And if labor cost differences are large enough to offset transport

is filled with observations that lend casual support to the main hypotheses of this paper.

7. Note the interesting finding of Mordecai Kreinin in his "The Leontief Scarce-Factor Paradox," *The American Economic Review*, LV (Mar. 1965), 131–39. Kreinin finds that the higher cost of labor in the United States is not explained by a higher rate of labor productivity in this country.

# UNITED STATES

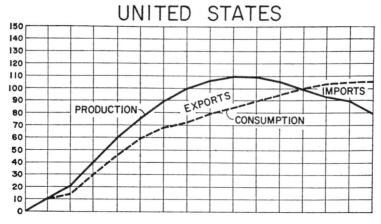

# OTHER ADVANCED COUNTRIES

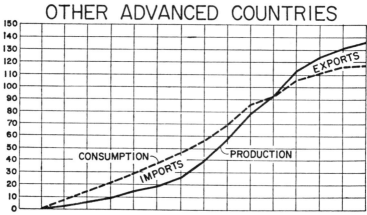

# LESS DEVELOPED COUNTRIES

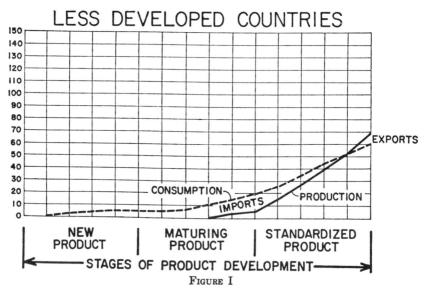

FIGURE I

costs, then exports back to the United States may become a possibility as well.

Any hypotheses based on the assumption that the United States entrepreneur will react rationally when offered the possibility of a lower-cost location abroad is, of course, somewhat suspect. The decision-making sequence that is used in connection with international investments, according to various empirical studies, is not a model of the rational process.[8] But there is one theme that emerges again and again in such studies. Any threat to the established position of an enterprise is a powerful galvanizing force to action; in fact, if I interpret the empirical work correctly, threat in general is a more reliable stimulus to action than opportunity is likely to be.

In the international investment field, threats appear in various forms once a large-scale export business in manufactured products has developed. Local entrepreneurs located in the countries which are the targets of these exports grow restive at the opportunities they are missing. Local governments concerned with generating employment or promoting growth or balancing their trade accounts begin thinking of ways and means to replace the imports. An international investment by the exporter, therefore, becomes a prudent means of forestalling the loss of a market. In this case, the yield on the investment is seen largely as the avoidance of a loss of income to the system.

The notion that a threat to the status quo is a powerful galvanizing force for international investment also seems to explain what happens after the initial investment. Once such an investment is made by a United States producer, other major producers in the United States sometimes see it as a threat to the status quo. They see themselves as losing position relative to the investing company, with vague intimations of further losses to come. Their "share of the market" is imperiled, viewing "share of the market" in global terms. At the same time, their ability to estimate the production-cost structure of their competitors, operating far away in an unfamiliar foreign area, is impaired; this is a particularly unsettling state because it conjures up the possibility of a return flow of products to the United States and a new source of price competition, based on cost differences of unknown magnitude. The uncertainty can be reduced by emulating the pathfinding investor and by investing in the same area; this may not be an optimizing investment

8. Aharoni, *op. cit.*, provides an excellent summary and exhaustive bibliography of the evidence on this point.

pattern and it may be costly, but it is least disturbing to the status quo.

Pieces of this hypothetical pattern are subject to empirical tests of a sort. So far, at any rate, the empirical tests have been reassuring. The office machinery industry, for instance, has seen repeatedly the phenomenon of the introduction of a new product in the United States, followed by United States exports,[9] followed still later by United States imports. (We have still to test whether the timing of the commencement of overseas production by United States subsidiaries fits into the expected pattern.) In the electrical and electronic products industry, those elements in the pattern which can be measured show up nicely.[1] A broader effort is now under way to test the United States trade patterns of a group of products with high income elasticities; and, here too, the preliminary results are encouraging.[2] On a much more general basis, it is reassuring for our hypotheses to observe that the foreign manufacturing subsidiaries of United States firms have been increasing their exports to third countries.

It will have occurred to the reader by now that the pattern envisaged here also may shed some light on the Leontief paradox.[3] Leontief, it will be recalled, seemed to confound comparative cost theory by establishing the fact that the ratio of capital to labor in United States exports was lower, not higher, than the like ratio in the United States production which had been displaced by competitive imports. The hypothesis suggested in this paper would have the United States exporting high-income and labor-saving products in the early stages of their existence, and importing them later on.[4] In the early stages, the value-added contribution of industries engaged in producing these items probably contains an

9. Reported in U.S. Senate, Interstate and Foreign Commerce Committee, *Hearings on Foreign Commerce*, 1960, pp. 130–39.

1. See Hirsch, *op. cit.*

2. These are to appear in a forthcoming doctoral thesis at the Harvard Business School by Louis T. Wells, tentatively entitled" International Trade and Business Policy."

3. See Wassily Leontief, "Domestic Production and Foreign Trade: The American Capital Position Re-examined," *Proceedings of the American Philosophical Society*, Vol. 97 (Sept. 1953), and "Factor Proportions and the Structure of American Trade: Further Theoretical and Empirical Analysis," *Review of Economics and Statistics*, XXXVIII (Nov. 1956).

4. Of course, if there were some systematic trend in the inputs of new products — for example, if the new products which appeared in the 1960's were more capital-intensive than the new products which appeared in the 1950's — then the tendencies suggested by our hypotheses might be swamped by such a trend. As long as we do not posit offsetting systematic patterns of this sort, however, the Leontief findings and the hypotheses offered here seem consistent.

unusually high proportion of labor cost. This is not so much because the labor is particularly skilled, as is so often suggested. More likely, it is due to a quite different phenomenon. At this stage, the standardization of the manufacturing process has not gotten very far; that is to come later, when the volume of output is high enough and the degree of uncertainty low enough to justify investment in relatively inflexible, capital-intensive facilities. As a result, the production process relies relatively heavily on labor inputs at a time when the United States commands an export position; and the process relies more heavily on capital at a time when imports become important.

This, of course, is an hypothesis which has not yet been subjected to any really rigorous test. But it does open up a line of inquiry into the structure of United States trade which is well worth pursuing.

### The Standardized Product

Figure I, the reader will have observed, carries a panel which suggests that, at an advanced stage in the standardization of some products, the less-developed countries may offer competitive advantages as a production location.

This is a bold projection, which seems on first blush to be wholly at variance with the Heckscher-Ohlin theorem. According to that theorem, one presumably ought to anticipate that the exports of the less-developed countries would tend to be relatively labor-intensive products.

One of the difficulties with the theorem, however, is that it leaves marketing considerations out of account. One reason for the omission is evident. As long as knowledge is regarded as a free good, instantaneously available, and as long as individual producers are regarded as atomistic contributors to the total supply, marketing problems cannot be expected to find much of a place in economic theory. In projecting the patterns of export from less-developed areas, however, we cannot afford to disregard the fact that information comes at a cost; and that entrepreneurs are not readily disposed to pay the price of investigating overseas markets of unknown dimensions and unknown promise. Neither are they eager to venture into situations which they know will demand a constant flow of reliable marketing information from remote sources.

If we can assume that highly standardized products tend to have a well-articulated, easily accessible international market and to sell largely on the basis of price (an assumption inherent in the definition), then it follows that such products will not pose the problem of market information quite so acutely for the less-developed countries. This establishes a necessary if not a sufficient condition for investment in such industries.

Of course, foreign investors seeking an optimum location for a captive facility may not have to concern themselves too much with questions of market information; presumably, they are thoroughly familiar with the marketing end of the business and are looking for a low-cost captive source of supply. In that case, the low cost of labor may be the initial attraction drawing the investor to less-developed areas. But other limitations in such areas, according to our hypothesis, will bias such captive operations toward the production of standardized items. The reasons in this case turn on the part played in the production process by external economies. Manufacturing processes which receive significant inputs from the local economy, such as skilled labor, repairmen, reliable power, spare parts, industrial materials processed according to exacting specification, and so on, are less appropriate to the less-developed areas than those that do not have such requirements. Unhappily, most industrial processes require one or another ingredient of this difficult sort. My guess is, however, that the industries which produce a standardized product are in the best position to avoid the problem, by producing on a vertically-integrated self-sustaining basis.

In speculating about future industrial exports from the less-developed areas, therefore, we are led to think of products with a fairly clear-cut set of economic characteristics.[5] Their production function is such as to require significant inputs of labor; otherwise there is no reason to expect a lower production cost in less-developed countries. At the same time, they are products with a high price elasticity of demand for the output of individual firms; otherwise, there is no strong incentive to take the risks of pioneering with production in a new area. In addition, products whose production process did not rely heavily upon external economies would be more obvious candidates than those which required a more elaborate industrial environment. The implications of remoteness also would be critical; products which could be precisely described by standardized specifications and which could be produced for inventory without fear of obsolescence would be more relevant than those

5. The concepts sketched out here are presented in more detail in my "Problems and Prospects in the Export of Manufactured Products from the Less-developed Countries," U.N. Conference on Trade and Development, Dec. 16, 1963 (mimeo.).

which had less precise specifications and which could not easily be ordered from remote locations. Moreover, high-value items capable of absorbing significant freight costs would be more likely to appear than bulky items low in value by weight. Standardized textile products are, of course, the illustration par excellence of the sort of product that meets the criteria. But other products come to mind such as crude steel, simple fertilizers, newsprint, and so on.

Speculation of this sort draws some support from various interregional experiences in industrial location. In the United States, for example, the "export" industries which moved to the low-wage south in search of lower costs tended to be industries which had no great need for a sophisticated industrial environment and which produced fairly standardized products. In the textile industry, it was the grey goods, cotton sheetings and men's shirt plants that went south; producers of high-style dresses or other unstandardized items were far more reluctant to move. In the electronics industry, it was the mass producers of tubes, resistors and other standardized high-volume components that showed the greatest disposition to move south; custom-built and research-oriented production remained closer to markets and to the main industrial complexes. A similar pattern could be discerned in printing and in chemicals production.[6]

In other countries, a like pattern is suggested by the impressionistic evidence. The underdeveloped south of Italy and the laggard north of Britain and Ireland both seem to be attracting industry with standardized output and self-sufficient process.[7]

Once we begin to look for relevant evidence of such investment patterns in the less-developed countries proper, however, only the barest shreds of corroboratory information can be found. One would have difficulty in thinking of many cases in which manufacturers of standardized products in the more advanced countries had made significant investments in the less-developed countries with a view of exporting such products from those countries. To be sure, other

6. This conclusion derives largely from the industry studies conducted in connection with the New York Metropolitan Region study. There have been some excellent more general analyses of shifts in industrial location among the regions of the United States. See e.g., Victor R. Fuchs, *Changes in the Location of Manufacturing in the United States Since 1929* (New Haven: Yale University Press, 1962). Unfortunately, however, none has been designed, so far as I know, to test hypotheses relating locational shifts to product characteristics such as price elasticity of demand and degree of standardization.

7. This statement, too, is based on only impressionistic materials. Among the more suggestive, illustrative of the best of the available evidence, see J. N. Toothill, *Inquiry into the Scottish Economy* (Edinburgh: Scottish Council, 1962).

types of foreign investment are not uncommon in the less-developed countries, such as investments in import-replacing industries which were made in the face of a threat of import restriction. But there are only a few export-oriented cases similar to that of Taiwan's foreign-owned electronics plants and Argentina's new producing facility, set up to manufacture and export standard sorting equipment for computers.

If we look to foreign trade patterns, rather than foreign investment patterns, to learn something about the competitive advantage of the less-developed countries, the possibility that they are an attractive locus for the output of standardized products gains slightly more support. The Taiwanese and Japanese trade performances are perhaps the most telling ones in support of the projected pattern; both countries have managed to develop significant overseas markets for standardized manufactured products. According to one major study of the subject (a study stimulated by the Leontief paradox), Japanese exports are more capital-intensive than is the Japanese production which is displaced by imports; [8] this is what one might expect if the hypothetical patterns suggested by Figure I were operational. Apart from these cases, however, all that one sees are a few provocative successes such as some sporadic sales of newsprint from Pakistan, the successful export of sewing machines from India, and so on. Even in these cases, one cannot be sure that they are consistent with the hypothesis unless he has done a good deal more empirical investigation.

The reason why so few revelant cases come to mind may be that the process has not yet advanced far enough. Or it may be that such factors as extensive export constraints and overvalued exchange rates are combining to prevent the investment and exports that otherwise would occur.

If there is one respect in which this discussion may deviate from classical expectations, it is in the view that the overall scarcity of capital in the less-developed countries will not prevent investment in facilities for the production of standardized products.

There are two reasons why capital costs may not prove a barrier to such investment.

First, according to our hypotheses, the investment will occur in industries which require some significant labor inputs in the production process; but they will be concentrated in that subsector of the

8. M. Tatemoto and S. Ichimura, "Factor Proportions and Foreign Trade: the Case of Japan," *Review of Economics and Statistics*, XLI (Nov. 1959), 442–46.

industry which produces highly standardized products capable of self-contained production establishments. The net of these specifications is indeterminate so far as capital-intensiveness is concerned. A standardized textile item may be more or less capital-intensive than a plant for unstandardized petro-chemicals.

Besides, even if the capital requirements for a particular plant are heavy, the cost of the capital need not prove a bar. The assumption that capital costs come high in the less-developed countries requires a number of fundamental qualifications. The reality, to the extent that it is known, is more complex.

One reason for this complexity is the role played by the international investor. Producers of chemical fertilizers, when considering whether to invest in a given country, may be less concerned with the going rate for capital in that country than with their opportunity costs as they see such costs. For such investors the alternatives to be weighed are not the full range of possibilities calling for capital but only a very restricted range of alternatives, such as the possibilities offered by chemical fertilizer investment elsewhere. The relevant capital cost for a chemical fertilizer plant, therefore, may be fairly low if the investor is an international entrepreneur.

Moreover, the assumption that finance capital is scarce and that interest rates are high in a less-developed country may prove inapplicable to the class of investors who concern us here.[9] The capital markets of the less-developed countries typically consist of a series of water-tight, insulated, submarkets in which wholly different rates prevail and between which arbitrage opportunities are limited. In some countries, the going figures may vary from 5 to 40 per cent, on grounds which seem to have little relation to issuer risk or term of loan. (In some economies, where inflation is endemic, interest rates which in effect represent a negative real cost are not uncommon.)

These internal differences in interest rates may be due to a number of factors: the fact that funds generated inside the firm usually are exposed to a different yield test than external borrowings; the fact that government loans are often floated by mandatory levies on banks and other intermediaries; and the fact that funds borrowed by governments from international sources are often re-

9. See George Rosen, *Industrial Change in India* (Glencoe, Ill.: Free Press, 1958). Rosen finds that in the period studied from 1937 to 1953, "there was no serious shortage of capital for the largest firms in India." Gustav F. Papanek makes a similar finding for Pakistan for the period from 1950 to 1964 in a book about to be published.

loaned in domestic markets at rates which are linked closely to the international borrowing rate, however irrelevant that may be. Moreover, one has to reckon with the fact that public international lenders tend to lend at near-uniform rates, irrespective of the identity of the borrower and the going interest rate in his country. Access to capital on the part of underdeveloped countries, therefore, becomes a direct function of the country's capacity to propose plausible projects to public international lenders. If a project can plausibly be shown to "pay its own way" in balance-of-payment and output terms at "reasonable" interest rates, the largest single obstacle to obtaining capital at such rates has usually been overcome.

Accordingly, one may say that from the entrepreneur's viewpoint certain systematic and predictable "imperfections" of the capital markets may reduce or eliminate the capital-shortage handicap which is characteristic of the less-developed countries; and, further, that as a result of the reduction or elimination such countries may find themselves in a position to compete effectively in the export of certain standardized capital-intensive goods. This is not the statement of another paradox; it is not the same as to say that the capital-poor countries will develop capital-intensive economies. All we are concerned with here is a modest fraction of the industry of such countries, which in turn is a minor fraction of their total economic activity. It may be that the anomalies such industries represent are systematic enough to be included in our normal expectations regarding conditions in the less-developed countries.

\* \* \* \* \*

Like the other observations which have preceded, these views about the likely patterns of exports by the less-developed countries are attempts to relax some of the constraints imposed by purer and simpler models. Here and there, the hypotheses take on plausibility because they jibe with the record of past events. But, for the most part, they are still speculative in nature, having been subjected to tests of a very low order of rigorousness. What is needed, obviously, is continued probing to determine whether the "imperfections" stressed so strongly in these pages deserve to be elevated out of the footnotes into the main text of economic theory.

HARVARD GRADUATE SCHOOL OF BUSINESS ADMINISTRATION

# THE MULTINATIONAL CORPORATION AND
# THE LAW OF UNEVEN DEVELOPMENT

## by Stephen Hymer

"The settlers' town is a strongly-built town, all made of stone and steel. It is a brightly-lit town; the streets are covered with asphalt, and the garbage-cans swallow all the leavings, unseen, unknown and hardly thought about. The settler's feet are never visible, except perhaps in the sea; but there you're never close enough to see them. His feet are protected by strong shoes although the streets of his town are clean and even, with no holes or stones. The settler's town is a well-fed town, an easy-going town, its belly is always full of good things. The settler's town is a town of white people, of foreigners.

The town belonging to the colonized people, or at least the native town, the Negro village, the medina, the reservation, is a place of ill fame peopled by men of evil repute. They are born there, it matters little where or how; they die there, it matters not where nor how. It is a world without spaciousness: men live there on top of each other, and their huts are built one on top of the other. The native town is a hungry town, starved of bread, of meat, of shoes, of coal, of light. The native town is a crouching village, a town on its knees, a town wallowing in the mire. It is a town of niggers and dirty Arabs. The look that the native turns on the settler's town is a look of lust, a look of envy . . ." Fanon, *The Wretched of the Earth.*

We have been asked to look into the future towards the year 2000. This essay attempts to do so in terms of two laws of economic development: the Law of Increasing Firm Size and the Law of Uneven Development.[1]

Since the beginning of the Industrial Revolution, there has been a tendency for the representative firm to increase in size from the *workshop* to the *factory* to the *national corporation* to the *multi-divisional corporation* and now to the *multinational corporation*. This growth has been qualitative as well as quantitative. With each step, business enter-

Reprinted from "Economics and the World Order: From the 1970s to the 1990s" (J.N. Bhagwati, ed.), pp. 113–140, with permission of The Free Press, New York, Copyright 1972.

prises acquired a more complex administrative structure to coordinate its activities and a larger brain to plan for its survival and growth. The first part of this essay traces the evolution of the corporation stressing the development of a hierarchical system of authority and control.

The remainder of the essay is concerned with extrapolating the trends in business enterprise (the microcosm) and relating them to the evolution of the international economy (the macrocosm). Until recently, most multinational corporations have come from the United States, where private business enterprise has reached its largest size and most highly developed forms. Now European corporations, as a by-product of increased size, and as a reaction to the American invasion of Europe, are also shifting attention from national to global production and beginning to "see the world as their oyster."[2] *If* present trends continue, multinationalization is likely to increase greatly in the next decade as giants from both sides of the Atlantic (though still mainly from the U.S.) strive to penetrate each other's markets and to establish bases in underdeveloped countries, where there are few indigenous concentrations of capital sufficiently large to operate on a world scale. This rivalry may be intense at first but will probably abate through time and turn into collusion as firms approach some kind of oligopolistic equilibrium. A new structure of international industrial organization and a new international division of labor will have been born.[3]

What will be the effect of this latest stage in the evolution of business enterprise on the Law of Uneven Development, *i.e.,* the tendency of the system to produce poverty as well as wealth, underdevelopment as well as development? The second part of this essay suggests that a regime of North Atlantic Multinational Corporations would tend to produce a hierarchical division of labor between geographical regions corresponding to the vertical division of labor within the firm. It would tend to centralize high-level decision-making occupations in a few key cities in the advanced countries, surrounded by a number of regional sub-capitals, and confine the rest of the world to lower levels of activity and income, *i.e.,* to the status of towns and villages in a new Imperial system. Income, status, authority, and consumption patterns would radiate out from these centers along a declining curve, and the existing pattern of inequality and dependency would be perpetuated. The pattern would be complex, just as the structure of the corporation is complex, but the basic relationship between different countries would be one of superior and subordinate, head office and branch plant.

How far will this tendency of corporations to create a world in their own image proceed? The situation is a dynamic one, moving dialectically. Right now, we seem to be in the midst of a major revolution in international relationships as modern science establishes the technological basis for a major advance in the conquest of the material world and the beginnings of truly cosmopolitan production.[4] Multinational corporations are in the vanguard of this revolution, because of their great financial and administrative strength and their close contact with the new technology. Governments (outside the military) are far behind, because of their narrower horizons and perspectives, as are labor organizations and most non-business institutions and associations. (As John Powers, President of Charles Pfizer Corporation, has put it, "Practise is ahead of theory and policy.") Therefore, in the first round, multinational corporations are likely to have a certain degree of success in organizing markets, decision making, and the spread of information in their own interest. However, their very success will create tensions and conflicts which will lead to further development. Part III discusses some of the contradictions that are likely to emerge as the multinational corporate system overextends itself. These contradictions provide certain openings for action. Whether or not they can or will be used in the next round to move towards superior forms of international organization requires an analysis of a wide range of political factors outside the scope of this essay.

## Part I. THE EVOLUTION OF THE MULTINATIONAL CORPORATION

### The Marshallian Firm and the Market Economy

What is the nature of the "beast?" It is called many names: Direct Investment, International Business, the International Firm, the International Corporate Group, the Multinational Firm, the Multinational Enterprise, the Multinational Corporation, the Multinational Family Group, World Wide Enterprise, La Grande Entreprise Plurinationale, La Grande Unité Interterritoriale, La Grande Entreprise Multinationale, La Grande Unité Pluriterritoriale; or, as the French Foreign Minister called them, "The U.S. corporate monsters." (Michel Debré quoted in *Fortune*, August 1965, p. 126.)

Giant organizations are nothing new in international trade. They were a characteristic form of the mercantilist period when large joint-stock companies, *e.g.*, The Hudson's Bay Co., The Royal African Co., The East India Co., to name the major English merchant firms, organized long-distance trade with America, Africa and Asia. But neither these firms, nor the large mining and plantation enterprises in the production sector, were the forerunners of the multinational corporation. They were like dinosaurs, large in bulk, but small in brain, feeding on the lush vegetation of the new worlds (the planters and miners in America were literally *Tyrannosaurus rex*).

The activities of these international merchants, planters and miners laid the groundwork for the Industrial Revolution by concentrating capital in the metropolitan centre, but the driving force came from the small-scale capitalist enterprises in manufacturing, operating at first in the interstices of the feudalist economic structure, but gradually emerging into the open and finally gaining predominance. It is in the small workshops, órganized by the newly emerging capitalist class, that the forerunners of the modern corporation are to be found.

The strength of this new form of business enterprise lay in its power and ability to reap the benefits of cooperation and division of labor. Without the capitalist, economic activity was individualistic, small-scale, scattered and unproductive. But a man with capital, *i.e.*, with sufficient funds to buy raw materials and advance wages, could gather a number of people into a single shop and obtain as his reward the increased productivity that resulted from social production. The reinvestment of these profits led to a steady increase in the size of capitals, making further division of labor possible and creating an opportunity for using machinery in production. A phenomenal increase in productivity and production resulted from this process, and entirely new dimensions of human existence were opened. The growth of capital revolutionized the entire world and, figuratively speaking, even battered down the Great Wall of China.

The hallmarks of the new system were *the market* and *the factory,* representing the two different methods of coordinating the division of labor. In the factory entrepreneurs consciously plan and organize cooperation, and the relationships are hierarchical and authoritarian; in the market coordination is achieved through a decentralized, unconscious, competitive process.[5]

To understand the significance of this distinction, the new system should be compared to the structure it replaced. In the pre-capitalist system of production, the division of labor was hierarchically structured at the *macro* level, *i.e.* for society as a whole, but unconsciously structured at the *micro* level *i.e.,* the actual process of production. Society as a whole was partitioned into various castes, classes, and guilds, on a rigid and authoritarian basis so that political and social stability could be maintained and adequate numbers assured for each industry and occupation. Within each sphere of production, however, individuals by and large were independent and their activities only loosely coordinated, if at all. In essence, a guild was composed of a large number of similar individuals, each performing the same task in roughly the same way with little cooperation or division of labor. This type of organization could produce high standards of quality and workmanship but was limited quantitatively to low levels of output per head.

The capitalist system of production turned this structure on its head. The macro system became unconsciously structured, while the micro system became hierarchically structured. The market emerged as a self-regulating coordinator of business units as restrictions on capital markets and labor mobility were removed. (Of course the State remained above the market as a conscious coordinator to maintain the system and ensure the growth of capital.) At the micro level, that is the level of production, labor was gathered under the authority of the entrepreneur capitalist.

Marshall, like Marx, stressed that the internal division of labor within the factory, between those who planned and those who worked (between "undertakers" and laborers), was the "chief fact in the form of modern civilization, the 'kernel' of the modern economic problem."[6] Marx, however, stressed the authoritarian and unequal nature of this relationship based on the coercive power of property and its anti-social characteristics. He focused on the irony that concentration of wealth in the hands of a few and its ruthless use were necessary historically to demonstrate the value of cooperation and the social nature of production.[7]

Marshall, in trying to answer Marx, argued for the voluntary cooperative nature of the relationship between capital and labor. In his view, the *market* reconciled individual freedom and collective production. He argued that those on top achieved their position because of their superior organizational ability, and that their relation to the workers below them was essentially harmonious and not exploitative. "Undertakers" were not captains of industry because they had capital; they could obtain capital because they had the ability to be captains of industry. They retained their authority by merit, not by coercion; for according to Marshall, natural selection, operating through the market, constantly destroyed inferior organizers and gave everyone who had the ability—including workers—a chance to rise to managerial positions. Capitalists earned more than workers because they contributed more, while the system as a whole provided all its members, and especially the workers, with improved standards of living and an ever-expanding field of choice of consumption.[8]

## The Corporate Economy

The evolution of business enterprise from the small workshop (Adam Smith's pin factory) to the Marshallian family firm represented only the first step in the development of business organization. As total capital accumulated, the size of the individual concentrations composing it increased continuously, and the vertical division of labor grew accordingly.

It is best to study the evolution of the corporate form in the United States environment, where it has reached its highest stage.[9] In the 1870s, the United States industrial structure consisted largely of Marshallian type, single-function firms, scattered over the country. Business firms were typically tightly controlled by a single entrepreneur or small family group who, as it were, saw everything, knew everything and decided everything. By the early twentieth century, the rapid growth of the economy and the great merger movement had consolidated many small enterprises into large national corporations engaged in many functions over many regions. To meet this new strategy of continent-wide, vertically integrated production and marketing, a new administrative structure evolved. The family firm, tightly controlled by a few men in close touch with all its aspects, gave way to the administrative pyramid of the corporation. Capital acquired new powers and new horizons. The domain of conscious coordination widened and that of market-directed division of labor contracted.

According to Chandler the railroad, which played so important a role in creating the national market, also offered a model for new forms of business organization. The need to administer geographically dispersed operations led railway companies to create an administrative structure which distinguished field offices from head offices. The field offices managed local operations; the head office supervised the field offices. According to Chandler and Redlich, this distinction is important because "it implies that the executive responsible for a firm's affairs had, for the first time, to supervise the work of other executives."[10]

This first step towards increased vertical division of labor within the management function was quickly copied by the recently-formed national corporations which faced the same problems of coordinating widely scattered plants. Business developed an organ system of administration, and the modern corporation was born. The functions of business administration were sub-divided into *departments* (organs)—finance, personnel, purchasing, engineering, and sales—to deal with capital, labor, purchasing, manufacturing, etc. This horizontal division of labor opened up new possibilities for rationalizing production and for incorporating the advances of physical and social sciences into economic activity on a systematic basis. At the same time a "brain and nervous" system, *i.e.*, a vertical system of control, had to be devised to connect and coordinate departments. This was a major advance in decision-making capabilities. It meant that a special group, the Head Office, was created whose particular function was to coordinate, appraise, and plan for the survival and growth of the organism as a whole. The organization became conscious of itself as organization and gained a certain measure of control over its own evolution and development.

The corporation soon underwent further evolution. To understand this next step we must briefly discuss the development of the United States market. At the risk of great oversimplification, we might say that by the first decade of the twentieth century, the problem of production had essentially been solved. By the end of the nineteenth century, scientists and engineers had developed most of the inventions needed for mass producing at a low cost nearly all the main items of basic consumption. In the language of systems analysis, the problem became one of putting together the available components in an organized fashion. The national corporation provided *one* organizational solution, and by the 1920s it had demonstrated its great power to increase material production.

The question was which direction growth would take. One possibility was to expand mass production systems very widely and to make basic consumer goods available on a broad basis throughout the world. The other possibility was to concentrate on continuous innovation for a small number of people and on the introduction of new consumption goods even before the old ones had been fully spread. The latter course was in fact chosen, and we now have the paradox that 500 million people can receive a live TV broadcast from the moon while there is still a shortage of telephones in many advanced countries, to say nothing of the fact that so many people suffer from inadequate food and lack of simple medical help.

This path was associated with a choice of capital-deepening instead of capital-widening in the productive sector of the economy. As capital accumulated, business had to choose the degree to which it would expand labor proportionately to the growth of capital or, conversely, the degree to which they would substitute capital for labor. At one extreme business could have kept the capital-labor ratio constant and accumulated labor at the same rate they accumulated capital. This horizontal accumulation would soon have exhausted the labor force of any particular country and then either capital would have had to migrate to foreign countries or labor would have had to move into the industrial centers. Under this system, earnings per employed worker would have remained steady and the composition of output would have tended to remain constant as similar basic goods were produced on a wider and wider basis.

However, this path was not chosen, and instead capital per worker was raised, the rate of expansion of the industrial labor force was slowed down, and a dualism was created between a small, high wage, high productivity sector in advanced countries, and a large, low wage, low productivity sector in the less advanced.[11]

The uneven growth of per capita income implied unbalanced growth and the need on the part of business to adapt to a constantly changing composition of output. Firms in the producers' goods sectors had continuously to innovate labor-saving machinery because the capital output ratio was increasing steadily. In the consumption goods sector, firms had continuously to introduce new products since, according to Engel's Law, people do not generally consume proportionately more of the same things as they get richer, but rather reallocate their consumption away from old goods and towards new goods. This non-proportional growth of demand implied that goods would tend to go through a life-cycle, growing rapidly when they were first introduced and more slowly later. If a particular firm were tied to only one product, its growth rate would follow this same life-cycle pattern and would eventually slow down and perhaps even come to a halt. If the corporation was to grow steadily at a rapid rate, it had continuously to introduce new products.

Thus, product development and marketing replaced production as a dominant problem of business enterprise. To meet the challenge of a constantly changing market, business enterprise evolved the multidivisional structure. The new form was originated by General Motors and DuPont shortly after World War I, followed by a few others during the 1920s and 1930s, and was widely adopted by most of the giant U.S. corporations in the great boom following World War II. As with the previous stages, evolution involved a process of both differentiation and integration. Corporations were decentralized into several *divisions,* each concerned with one product line and organized with its own head office. At a higher level, a *general office* was created to coordinate the division and to plan for the enterprise as a whole.

The new corporate form has great flexibility. Because of its decentralized structure, a multidivisional corporation can enter a new market by adding a new division, while leaving the old divisions undisturbed. (And to a lesser extent it can leave the market by dropping a division without disturbing the rest of its structure.) It can also create competing product-lines in the same industry, thus increasing its market share while maintaining the illusion of competition. Most important of all, because it has a cortex specializing in strategy, it can plan on a much wider scale than before and allocate capital with more precision.

The modern corporation is a far cry from the small workshop or even from the Marshallian firm. The Marshallian capitalist ruled his factory from an office on the second floor. At the turn of the century, the president of a large national corporation was lodged in a higher building, perhaps on the seventh floor, with greater perspective and power. In today's giant corporation, managers rule from the top of skyscrapers; on a clear day, they can almost see the world.

U.S. corporations began to move to foreign countries almost as soon as they had completed their continent-wide integration. For one thing, their new administrative structure and great financial strength gave them the power to go abroad. In becoming national firms, U.S. corporations learned how to become international. Also, their large size and oligopolistic position gave them an incentive. Direct investment became a new weapon in their arsenal of oligopolistic rivalry. Instead of joining a cartel (prohibited under U.S. law), they invested in foreign customers, suppliers, and competitors. For example, some firms found they were oligopolistic buyers of raw materials produced in foreign countries and feared a monopolization of the sources of supply. By investing directly in foreign producing enterprises, they could gain the security implicit in control over their raw material requirements. Other firms invested abroad to control marketing outlets and thus maximize quasi-rents on their technological discoveries and differentiated products. Some went abroad simply to forestall competition.[12]

The first wave of U.S. direct foreign capital investment occurred around the turn of the century followed by a second wave during the 1920s. The outward migration slowed down during the depression but resumed after World War II and soon accelerated rapidly. Between 1950 and 1969, direct foreign investment by U.S. firms expanded at a rate of about 10 percent per annum. At this rate it would double in less than ten years, and even at a much slower rate of growth, foreign operations will reach enormous proportions over the next 30 years.[13]

Several important factors account for this rush of foreign investment in the 1950s and the 1960s. First, the large size of the U.S. corporations and their new multidivisional structure gave them wider horizons and a global outlook. Secondly, technological developments in communications created a new awareness of the global challenge and threatened established institutions by opening up new sources of competition. For reasons noted above, business enterprises were among the first to recognize the potentialities and dangers of the new environment and to take active steps to cope with it.

A third factor in the outward migration of U.S. capital was the rapid growth of Europe and Japan. This, combined with the slow growth of the United States economy in the 1950s, altered world market shares as firms confined to the U.S. market found themselves falling behind in the competitive race and losing ground to European and Japanese firms, which were growing rapidly because of the expansion of their markets. Thus, in the late 1950s, United States corporations faced a serious "non-American" challenge. Their answer was an outward thrust to establish sales production and bases in foreign territories. This strategy was possible in Europe, since government there provided an open door for United States investment, but was blocked in Japan, where the government adopted a highly restrictive policy. To a large extent, United States business was thus able to redress the imbalances caused by the Common Market, but Japan remained a source of tension to oligopoly equilibrium.

What about the future? The present trend indicates further multinationalization of all giant firms, European as well as American. In the first place, European firms, partly as a reaction to the United States penetration of their markets, and partly as a natural result of their own growth, have begun to invest abroad on an expanded scale and will probably continue to do so in the future, and even enter into the United States market. This process is already well underway and may be expected to accelerate as time goes on. The reaction of United States business will most likely be to meet foreign investment at home with more foreign investment abroad. They, too, will scramble for market positions in underdeveloped countries and attempt to get an even larger share of the European market, as a reaction to European investment in the United States. Since they are large and powerful, they will on balance succeed in maintaining their relative standing in the world as a whole—as their losses in some markets are offset by gains in others.

A period of rivalry will prevail until a new equilibrium between giant U.S. firms and giant European and Japanese firms is reached, based on a strategy of multinational operations and cross-penetration.[14] We turn now to the implications of this pattern of industrial organization for international trade and the law of uneven development.

## Part II. UNEVEN DEVELOPMENT

Suppose giant multinational corporations (say 300 from the U.S. and 200 from Europe and Japan) succeed in establishing themselves as the dominant form of international enterprise and come to control a significant share of industry (especially modern industry) in each country. The world economy will resemble more and more the United States economy, where each of the large corporations tends to spread over the entire continent and to penetrate almost every nook and cranny. What would be the effect of a world industrial organization of this type on international specialization, exchange and income distribution? The purpose of this section is to analyze the spatial dimension of the corporate hierarchy.

A useful starting point is Chandler and Redlich's[15] scheme for analyzing the evolution of corporate structure. They distinguish "three levels of business administration, three horizons, three levels of task, and three levels of decision making . . . and three levels of policies." Level III, the lowest level, is concerned with managing the day-to-day operations of the enterprise, that is with keeping it going within the established framework. Level II, which first made its appearance with the separation of head office from field office, is responsible for coordinating the managers at Level III. The functions of Level I—top management— are goal-determination and planning. This level sets the framework in which the lower levels operate. In the Marshallian firm, all three levels are embodied in the single entrepreneur or undertaker. In the national corporation a partial differentiation is made in which the top two levels are separated from the bottom one. In the multidivisional corporation, the differentiation is far more complete. Level I is completely split off from Level II and concentrated in a general office whose specific function is to plan strategy rather than tactics.

The development of business enterprise can therefore be viewed as a process of centralizing and perfecting the process of capital accumulation. The Marshallian entrepreneur was a jack-of-all-trades. In the modern multidivisional corporation, a powerful general office consciously plans and organizes the growth of corporate capital. It is here that the key men who actually allocate the corporation's available resources (rather than act within the means allocated to them, as is true for the managers at lower levels) are located. Their power comes from their ultimate control over *men* and *money* and although one should not overestimate the ability to control a far-flung empire, neither should one underestimate it.

> The senior men could take action because they controlled the selection of executive personnel and because, through budgeting, they allocated the funds to the operating divisions. In the way they allocated their resources—capital and personnel—and in the promotion, transferral and retirement of operating executives, they determined the framework in which the operating units worked and thus put into effect their concept of the long term goals and objectives of the enterprise . . . Ultimate authority in business enterprise, as we see it, rests with those who hold the purse strings, and in modern large-scale enterprises, those persons hold the purse strings who perform the functions of goal setting and planning.[16]

What is the relationship between the structure of the microcosm and the structure of the macrocosm? The application of location theory to the Chandler-Redlich scheme suggests a *correspondence principle* relating centralization of control within the corporation to centralization of control within the international economy.

Location theory suggests that Level III activities would spread themselves over the globe according to the pull of manpower, markets, and raw materials. The multinational corporation, because of its power to command capital and technology and its ability to rationalize their use on a global scale, will probably spread production more evenly over the world's surface than is now the case. Thus, in the first instance, it may well be a force for diffusing industrialization to the less developed countries and creating new centers of production. (We postpone for a moment a discussion of the fact that location depends upon transportation, which in turn depends upon the government, which in turn is influenced by the structure of business enterprise.)

Level II activities, because of their need for white-collar workers, communications systems, and information, tend to concentrate in large cities. Since their demands are similar, corporations from different industries tend to place their coordinating offices in the same city, and Level II activities are consequently far more geographically concentrated than Level III activities.

Level I activities, the general offices, tend to be even more concentrated than Level II activities, for they must be located close to the capital market, the media, and the government. Nearly every major corporation in the United States, for example, must have its general office (or a large proportion of its high-level personnel) in or near the city of New York because of the need for face-to-face contact at higher levels of decision making.

Applying this scheme to the world economy, one would expect to find the highest offices of the multinational corporations concentrated in the world's major cities—New York, London, Paris, Bonn, Tokyo. These, along with Moscow and perhaps Peking, will be the major centers of high-level strategic planning. Lesser cities throughout the world will deal with the day-to-day operations of specific local problems. These in turn will be arranged in a hierarchical fashion: the larger and more important ones will contain regional corporate headquarters, while the smaller ones will be confined to lower level activities. Since business is usually the core of the city, geographical specialization will come to reflect the hierarchy of corporate decision making, and the occupational distribution of labor in a city or region will depend upon its function in the international economic system. The "best" and most highly paid administrators, doctors, lawyers, scientists, educators, government officials, actors, servants and hairdressers, will tend to concentrate in or near the major centers.

The structure of income and consumption will tend to parallel the structure of status and authority. The citizens of capital cities will have the best jobs—allocating men and money at the highest level and planning growth and development—and will receive the highest rates of re-muneration. (Executives' salaries tend to be a function of the wage bill of people under them. The larger empire of the multinational corpora-tion, the greater the earnings of top executives, to a large extent inde-pendent of their performance.[17] Thus, growth in the hinterland sub-sidiaries implies growth in the income of capital cities, but not *vice versa.*)

The citizens of capital cities will also be the first to innovate new products in the cycle which is known in the marketing literature as trickle-down or two-stage marketing. A new product is usually first intro-duced to a select group of people who have "discretionary" income and are willing to experiment in their consumption patterns.[18] Once it is accepted by this group, it spreads, or trickles down to other groups via the demonstration effect. In this process, the rich and the powerful get more votes than everyone else; first, because they have more money to spend, second, because they have more ability to experiment, and third, because they have high status and are likely to be copied. This special group may have something approaching a choice in consumption pat-terns; the rest have only the choice between conforming or being iso-lated.

The trickle-down system also has the advantage—from the center's point of view—of reinforcing patterns of authority and control. Accord-ing to Fallers,[19] it helps keep workers on the treadmill by creating an illusion of upward mobility even though relative status remains un-changed. In each period subordinates achieve (in part) the consumption standards of their superiors in a previous period and are thus torn in two directions: if they look backward and compare their standards of living through time, things seem to be getting better; if they look up-ward they see that their relative position has not changed. They receive a consolation prize, as it were, which may serve to keep them going by softening the reality that in a competitive system, few succeed and many fail. It is little wonder, then, that those at the top stress growth rather than equality as the welfare criterion for human relations.

In the international economy trickle-down marketing takes the form of an international demonstration effect spreading outward from the metropolis to the hinterland.[20] Multinational corporations help speed up this process, often the key motive for direct investment, through their control of marketing channels and communications media.

The development of a new product is a fixed cost; once the expendi-ture needed for invention or innovation has been made, it is forever a by-gone. The actual cost of production is thus typically well below selling price and the limit on output is not rising costs but falling demand due to saturated markets. The marginal profit on new foreign markets is thus high, and corporations have a strong interest in maintaining a system which spreads their products widely. Thus, the interest of multinational corporations in underdeveloped countries is larger than the size of the market would suggest.

It must be stressed that the dependency relationship between major and minor cities should not be attributed to technology. The new technology, because it increases interaction, implies greater interdependence but not necessarily a hierarchical structure. Communications linkages could be arranged in the form of a grid in which each point was directly connected to many other points, permitting lateral as well as vertical communication. This system would be polycentric since messages from one point to another would go directly rather than through the center; each point would become a center on its own; and the distinction between center and periphery would disappear.

Such a grid is made *more* feasible by aeronautical and electronic revolutions which greatly reduce costs of communications. It is not technology which creates inequality; rather, it is *organization* that imposes a ritual judicial asymmetry on the use of intrinsically symmetrical means of communications and arbitrarily creates unequal capacities to initiate and terminate exchange, to store and retrieve information, and to determine the extent of the exchange and terms of the discussion. Just as colonial powers in the past linked each point in the hinterland to the metropolis and inhibited lateral communications, preventing the growth of independent centers of decision making and creativity, multinational corporations (backed by state powers) centralize control by imposing a hierarchical system.

This suggests the possibility of an alternative system of organization in the form of national planning. Multinational corporations are private institutions which organize one or a few industries across many countries. Their polar opposite (the antimultinational corporation, perhaps) is a public institution which organizes many industries across one region. This would permit the centralization of capital, *i.e.*, the coordination of many enterprises by one decision-making center, but would substitute regionalization for internationalization. The span of control would be confined to the boundaries of a single polity and society and not spread over many countries. The advantage of the multinational corporation is its global perspective. The advantage of national planning is its ability to remove the wastes of oligopolistic anarchy, *i.e.*, meaningless product differentiation and an imbalance between different industries within a geographical area. It concentrates *all* levels of decision-making in one locale and thus provides each region with a full complement of skills and occupations. This opens up new horizons for local development by making possible the social and political control of economic decision-making. Multinational corporations, in contrast, weaken political control because they span many countries and can escape national regulation.

A few examples might help to illustrate how multinational corporations reduce options for development. Consider an underdeveloped country wishing to invest heavily in education in order to increase its stock of human capital and raise standards of living. In a market system it would be able to find gainful employment for its citizens within its *national boundaries* by specializing in education-intensive activities and selling its surplus production to foreigners. In the multinational corporate system, however, the demand for high-level education in low-ranking areas is limited, and a country does not become a world center simply by having a better educational system. An outward shift in the supply of educated people in a country, therefore, will not create its own demand but will create an excess supply and lead to emigration. Even then, the employment opportunities for citizens of low-ranking countries are restricted by discriminatory practices in the center. It is well-known that ethnic homogeneity increases as one goes up the corporate hierarchy; the lower levels contain a wide variety of nationalities, the higher levels become successively purer and purer. In part this stems from the skill differences of different nationalities, but more important is the fact that the higher up one goes in the decision-making process, the more important mutual understanding and ease of communications become; a common background becomes all-important.

A similar type of specialization by nationality can be expected within the multinational corporation hierarchy. Multinational corporations are torn in two directions. On the one hand, they must adapt to local circumstances in each country. This calls for decentralized decision making. On the other hand, they must coordinate their activities in various parts of the world and stimulate the flow of ideas from one part of their empire to another. This calls for centralized control. They must, therefore, develop an organizational structure to balance the need for coordination with the need for adaptation to a patch-work quilt of languages, laws and customs. One solution to this problem is a division of labor based on nationality. Day-to-day management in each country is left to the nationals of that country who, because they are intimately familiar with local conditions and practices, are able to deal with local problems and local government. These nationals remain rooted in one spot, while above them is a layer of people who move around from country to country, as bees among flowers, transmitting information from one subsidiary to another and from the lower levels to the general office at the apex of the corporate structure. In the nature of things, these people (reticulators) for the most part will be citizens of the country of the parent corporation (and will be drawn from a small, culturally homogeneous group within the advanced world), since they will need to have the confidence of their superiors and be able to move easily in the higher management circles. Latin Americans, Asians and Africans will at best be able to aspire to a management position in the intermediate coordinating centers at the continental level. Very few will be able to get much higher than this, for the closer one gets to the top, the more important is "a common cultural heritage."

Another way in which the multinational corporations inhibit economic development in the hinterland is through their effect on tax capacity. An important government instrument for promoting growth is expenditure on infrastructure and support services. By providing transportation and communications, education and health, a government can create a productive labor force and increase the growth potential of its economy. The extent to which it can afford to finance these intermediate outlays depends upon its tax revenue.

However, a government's ability to tax multinational corporations is limited by the ability of these corporations to manipulate transfer prices and to move their productive facilities to another country. This means that they will only be attracted to countries where superior infrastructure offsets higher taxes. The government of an underdeveloped country will find it difficult to extract a surplus (revenue from the multinational corporations, less cost of services provided to them) from multinational corporations to use for long-run development programs and for stimulating growth in other industries. In contrast, governments of the advanced countries, where the home office and financial center of the multinational corporation are located, can tax the profits of the corporation as a whole, as well as the high incomes of its management. Government in the metropolis can, therefore, capture some of the surplus generated by the multinational corporations and use it to further improve their infrastructure and growth.

In other words, the relationship between multinational corporations and underdeveloped countries will be somewhat like the relationship between the national corporations in the United States and state and municipal governments. These lower-level governments tend always to be short of funds compared to the federal government which can tax a corporation as a whole. Their competition to attract corporate investment eats up their surplus, and they find it difficult to finance extensive investments in human and physical capital even where such investment would be productive. This has a crucial effect on the pattern of government expenditure. For example, suppose taxes were first paid to state government and then passed on to the federal government. What chance is there that these lower level legislatures would approve the phenomenal expenditures on space research that now go on? A similar discrepancy can be expected in the international economy with overspending and waste by metropolitan governments and a shortage of public funds in the less advanced countries.

The tendency of the multinational corporations to erode the power of the nation state works in a variety of ways, in addition to its effect on taxation powers. In general, most governmental policy instruments (monetary policy, fiscal policy, wage policy, etc.) diminish in effectiveness the more open the economy and the greater the extent of foreign investments. This tendency applies to political instruments as well as economic, for the multinational corporation is a medium by which laws, politics, foreign policy and culture of one country intrude into another. This acts to reduce the sovereignty of all nation states, but again the relationship is asymmetrical, for the flow tends to be from the parent to the subsidiary, not *vice versa*. The United States can apply its antitrust laws to foreign subsidiaries or stop them from "trading with the enemy" even though such trade is not against the laws of the country in which the branch plant is located. However, it would be illegal for an underdeveloped country which disagreed with American foreign policy to hold a U.S. firm hostage for acts of the parent. This is because legal rights are defined in terms of property-ownership, and the various subsidiaries of a multinational corporation are not "partners in a multinational endeavor" but the property of the general office.

In conclusion, it seems that a regime of multinational corporations would offer underdeveloped countries neither national independence nor equality. It would tend instead to inhibit the attainment of these goals. It would turn the underdeveloped countries into branch-plant countries, not only with reference to their economic functions but throughout the whole gamut of social, political and cultural roles. The subsidiaries of multinational corporations are typically amongst the largest corporations in the country of operations, and their top executives play an influential role in the political, social and cultural life of the host country. Yet these people, whatever their title, occupy at best a medium position in the corporate structure and are restricted in authority and horizons to a lower level of decision making. The governments with whom they deal tend to take on the same middle management outlook, since this is the only range of information and ideas to which they are exposed.[21] In this sense, one can hardly expect such a country to bring forth the creative imagination needed to apply science and technology to the problems of degrading poverty. Even so great a champion of liberalism as Marshall recognized the crucial relationship between occupation and development.

For the business by which a person earns his livelihood generally fills his thoughts during the far greater part of those hours in which his mind is at its best; during them his character is being formed by the way in which he uses his facilities in his work, by the thoughts and feelings which it suggests, and by his relationship to his associates in work, his employers to his employees.[22]

## Part III. THE POLITICAL ECONOMY OF THE MULTINATIONAL CORPORATION

The viability of the multinational corporate system depends upon the degree to which people will tolerate the unevenness it creates. It is well to remember that the "New Imperialism" which began after 1870 in a spirit of Capitalism Triumphant, soon became seriously troubled and after 1914 was characterized by war, depression, breakdown of the international economic system, and war again, rather than Free Trade, Pax Britannica and Material Improvement.

A major, if not the major, reason was Great Britain's inability to cope with the byproducts of its own rapid accumulation of capital; *i.e.*, a class conscious labor force at home; a middle class in the hinterland; and rival centers of capital on the Continent and in America. Britain's policy tended to be atavistic and defensive rather than progressive, more concerned with warding off new threats than creating new areas of expansion. Ironically, Edwardian England revived the paraphernalia of the landed aristocracy it had just destroyed. Instead of embarking on a "big push" to develop the vast hinterland of the Empire, colonial administrators often adopted policies to slow down rates of growth and arrest the development of either a native capitalist class or a native proletariat which could overthrow them.

As time went on, the center had to devote an increasing share of government activity to military and other unproductive expenditures; they had to rely on alliances with an inefficient class of landlords, officials and soldiers in the hinterland to maintain stability at the cost of development. A great part of the surplus extracted from the population was thus wasted locally.

The new Mercantilism (as the Multinational Corporate System of special alliances and privileges, aid and tariff concessions is sometimes called) faces similar problems of internal and external division. The center is troubled: excluded groups revolt and even some of the affluent are dissatisfied with their roles. (The much talked about "generation gap" may indicate the failure of the system to reproduce itself.) Nationalistic rivalry between major capitalist countries (especially the challenge of Japan and Germany) remains an important divisive factor, while the economic challenge from the socialist bloc may prove to be of the utmost significance in the next thirty years. Russia has its own form of large-scale economic organizations, also in command of modern technology, and its own conception of how the world should develop. So does China to an increasing degree.[23] Finally, there is the threat presented by the middle classes and the excluded groups of the underdeveloped countries.

The national middle classes in the underdeveloped countries came to power when the center weakened but could not, through their policy of import substitution manufacturing, establish a viable basis for sustained growth. They now face a foreign exchange crisis and an unemployment (or population) crisis—the first indicating their inability to function in the international economy, and the second indicating their alienation from the people they are supposed to lead. In the immediate future, these national middle classes will gain a new lease on life as they take advantage of the spaces created by the rivalry between American and non-American oligopolists striving to establish global market positions. The native capitalists will again become the champions of national independence as they bargain with multinational corporations. But the conflict at this level is more apparent than real, for in the end the fervent nationalism of the middle class asks only for promotion within the corporate structure and not for a break with that structure. In the last analysis their power derives from the metropolis and they cannot easily afford to challenge the international system. They do not command the loyalty of their own population and cannot really compete with the large, powerful, aggregate capitals from the center. They are prisoners of the taste patterns and consumption standards set at the center, and depend on outsiders for technical advice, capital, and when necessary, for military support of their position.

The main threat comes from the excluded groups. It is not unusual in underdeveloped countries for the top 5 percent to obtain between 30 and 40 percent of the total national income, and for the top one-third to obtain anywhere from 60 to 70 percent.[24] At most, one-third of the population can be said to benefit in some sense from the dualistic growth that characterizes development in the hinterland. The remaining two-thirds, who together get only one-third of the income, are outsiders, not because they do not contribute to the economy, but because they do not share in the benefits. They provide a source of cheap labor which helps keep exports to the developed world at a low price and which has financed the urban-biased growth of recent years. Because their wages are low, they spend a moderate amount of time in menial services and are sometimes referred to as underemployed as if to imply they were not needed. In fact, it is difficult to see how the system in most under-developed countries could survive without cheap labor, since removing it (*e.g.,* diverting it to public works projects as is done in socialist countries) would raise consumption costs to capitalists and professional elites. Economic development under the Multinational Corporation does not offer much promise for this large segment of society and their antagonism continuously threatens the system.

The survival of the multinational corporate system depends on how fast it can grow and how much trickles down. Plans now being formu-lated in government offices, corporate headquarters and international organizations, sometimes suggest that a growth rate of about 6 percent per year in national income (3 percent per capita) is needed. (Such a target is, of course, far below what would be possible if a serious effort were made to solve basic problems of health, education and clothing.) To what extent is it possible?

The multinational corporation must solve four critical problems for the underdeveloped countries, if it is to foster the continued growth and survival of a "modern" sector. First, it must break the foreign-exchange constraint and provide the underdeveloped countries with imported goods for capital formation and modernization. Second, it must finance an expanded program of government expenditure to train labor and provide support services for urbanization and industrialization. Third, it must solve the urban food problem created by growth. Finally, it must keep the excluded two-thirds of the population under control.

The solution now being suggested for the first is to restructure the world economy allowing the periphery to export certain manufactured goods to the center. Part of this program involves regional common markets to rationalize the existing structure of industry. These plans typically do not involve the rationalization and restructuring of the en-tire economy of the underdeveloped countries but mainly serve the small manufacturing sector which caters to higher income groups and which, therefore, faces a very limited market in any particular country. The solution suggested for the second problem is an expanded aid pro-gram and a reformed government bureaucracy (perhaps along the lines of the Alliance for Progress). The solution for the third is agri-business and the green revolution, a program with only limited benefits to the rural poor. Finally, the solution offered for the fourth problem is popu-lation control, either through family planning or counterinsurgency.

It is doubtful whether the center has sufficient political stability to finance and organize the program outlined above. It is not clear, for example, that the West has the technology to rationalize manufacturing abroad or modernize agriculture, or the willingness to open up market-ing channels for the underdeveloped world. Nor is it evident that the center has the political power to embark on a large aid program or to readjust its own structure of production and allow for the importation of manufactured goods from the periphery. It is difficult to imagine labor accepting such a re-allocation (a new repeal of the Corn Laws as it were[25]), and it is equally hard to see how the advanced countries could create a system of planning to make these extra hardships unnecessary.

The present crisis may well be more profound than most of us imagine, and the West may find it impossible to restructure the international economy on a workable basis. One could easily argue that the age of the Multinational Corporation is at its end rather than at its beginning. For all we know, books on the global partnership may be the epitaph of the American attempt to take over the old international economy, and not the herald of a new era of international cooperation.

## CONCLUSION:

The multinational corporation, because of its great power to plan economic activity, represents an important step forward over previous methods of organizing international exchange. It demonstrates the social nature of production on a global scale. As it eliminates the anarchy of international markets and brings about a more extensive and productive international division of labor, it releases great sources of latent energy.

However, as it crosses international boundaries, it pulls and tears at the social and political fabric and erodes the cohesiveness of national states.[26] Whether one likes this or not, it is probably a tendency that cannot be stopped.

Through its propensity to nestle everywhere, settle everywhere, and establish connections everywhere, the multinational corporation destroys the possibility of national seclusion and self-sufficiency and creates a universal interdependence. But the multinational corporation is still a private institution with a partial outlook and represents only an imperfect solution to the problem of international cooperation. It creates hierarchy rather than equality, and it spreads its benefits unequally.

In proportion to its success, it creates tensions and difficulties. It will lead other institutions, particularly labor organizations and government, to take an international outlook and thus unwittingly create an environment less favorable to its own survival. It will demonstrate the possibilities of material progress at a faster rate than it can realize them, and will create a worldwide demand for change that it cannot satisfy.

The next round may be marked by great crises due to the conflict between national planning by governments and international planning by corporations. For example, if each country loses its power over fiscal and monetary policy due to the growth of multinational corporations (as some observers believe Canada has), how will aggregate demand be stabilized? Will it be possible to construct super-states? Or does multinationalism do away with Keynesian problems? Similarly, will it be possible to fulfill a host of other government functions at the supranational level in the near future? During the past twenty five years many political problems were put aside as the West recovered from the depression and the war. By the late sixties the bloom of this long upswing had begun to fade. In the seventies, power conflicts are likely to come to the fore.

Whether underdeveloped countries will use the opportunities arising from this crisis to build viable local decision-making institutions is difficult to predict. The national middle class failed when it had the opportunity and instead merely reproduced internally the economic dualism of the international economy as it squeezed agriculture to finance urban industry. What is needed is a complete change of direction. The starting point must be the needs of the bottom two-thirds, and not the demands of the top third. The primary goal of such a strategy would be to provide minimum standards of health, education, food and clothing to the entire population, removing the more obvious forms of human suffering. This requires a system which can mobilize the entire population and which can search the local environment for information, resources and needs. It must be able to absorb modern technology, but it cannot be mesmerized by the form it takes in the advanced countries; it must go to the roots. This is not the path the upper one-third chooses when it has control.

The wealth of a nation, wrote Adam Smith two hundred years ago, is determined by "first, the skill, dexterity and judgement with which labor is generally applied; and, secondly by the proportion between the number of those who are employed in useful labor, and that of those who are not so employed."[27] Capitalist enterprise has come a long way from his day, but it has never been able to bring more than a small fraction of the world's population into useful or highly productive employment. The latest stage reveals once more the power of social cooperation and division of labor which so fascinated Adam Smith in his description of pin manufacturing. It also shows the shortcomings of concentrating this power in private hands.

## EPILOGUE

Many readers of this essay in draft form have asked: Is there an alternative? Can anything be done? The problem simply stated is to go beyond the multinational corporation. Scholarship can perhaps make the task easier by showing how the forms of international social production devised by capital as it expanded to global proportions can be used to build a better society benefiting all men. I have tried to open up one avenue for explanation by suggesting a system of regional planning as a positive negation of the multinational corporation. Much more work is needed to construct alternative methods of organizing the international economy. Fortunately businessmen in attacking the problem of applying technology on a world level have developed many of the tools and conditions needed for a socialist solution, if we can but stand them on their head. But one must keep in mind that the problem is not one of ideas alone.

A major question is how far those in power will allow the necessary metamorphosis to happen, and how far they will try to resist it by violent means. I do not believe the present structure of uneven development can long be maintained in the light of the increased potential for world development demonstrated by corporate capital itself. But power at the center is great, and the choice of weapons belongs in the first instance to those who have them.

Theodor Mommsen summed up his history of the Roman Republic with patient sadness.

> It was indeed an old world, and even the richly gifted patriotism of Caesar could not make it young again. The dawn does not return till after the night has run its course.[28]

I myself do not view the present with such pessimism. History moves more quickly now, the forces for positive change are much stronger, and the center seems to be losing its will and self confidence. It is becoming increasingly evident to all that in contrast to corporate capitalism we must be somewhat less "efficient" within the microcosm of the enterprise and far more "efficient" in the macrocosm of world society. The dysutopia of the multinational corporate system shows us both what is to be avoided and what is possible.

## NOTES

1. See Marx, *Capital*, Vol. 1, Chapter XXV, "On the General Law of Capitalist Accumulation," Chapter XII, "Co-operation" and Chapter XIV, part 4, "Division of Labour in Manufacturing and Division of Labour in Society," and Vol. 3, Chapter XXIII.

2. Phrase used by Anthony M. Salomon in *International Aspects of Antitrust*, Part I. Hearings before the Sub-Committee on Antitrust and Monopoly of the Senate Committee on the Judiciary. April 1966, p. 49.

3. These trends are discussed in Stephen Hymer and Robert Rowthorn, "Multinational Corporations and International Oligopoly: the Non-American Challenge" in C. P. Kindleberger, ed., *The International Corporation* (Cambridge, M.I.T. Press, 1970).

4. Substituting the word *multinational corporation* for *bourgeois* in the following quote from *The Communist Manifesto* provides a more dynamic picture of the multinational corporation than any of its present day supporters have dared to put forth.

The need of a constantly expanding market for its products chases the multinational corporation over the whole surface of the globe. It must nestle everywhere, settle everywhere, establish connections everywhere. The bourgeoisie has through its exploitation of the world-market given a cosmopolitan character to production and consumption in every country. To the great chagrin of Reactionists, it has drawn from under the feet of industry the national ground on which it stood. All old-established national industries have been destroyed or are daily being destroyed. They are dislodged by new industries, whose introduction becomes a life and death question for all civilized nations, by industries that no longer work up indigenous raw material, but raw material drawn from the remotest zones; industries whose products are consumed, not only at home, but in every quarter of the globe. In place of the old wants, satisfied by the production of the country, we find new wants, requiring for their satisfaction the products of distant lands and climes. In place of the old local and national seclusion and self-sufficiency, we have intercourse in every direction, universal interdependence of nations. And as in material, so also in intellectual production. The intellectual creations of individual nations become common property. National one-sidedness and narrow-mindedness become more and more impossible, and from the numerous national and local literatures there arises a world literature.

The multinational corporation, by the rapid improvement of all instruments of production, by the immensely facilitated means of communication, draws all, even the most barbarian, nations into civilization. The cheap prices of its commodities are the heavy artillery with which it batters down all Chinese walls, with which it forces the barbarians' intensely obstinate hatred of foreigners to capitulate. It compels all nations, on pain of extinction, to adopt the bourgeois mode of production, it compels them to introduce what it calls civilization into their midst, i.e., to become bourgeois themselves. In a word, it creates a world after its own image.

The multinational corporation has subjected the country to the rule of the towns. It has created enormous cities, has greatly increased the urban population as compared with the rural, and has thus rescued a considerable part of the population from the idiocy of rural life. Just as it has made the country dependent on the towns, so it has made barbarian and semi-barbarian countries dependent on the civilized ones, nations of peasants on nations of bourgeois, the East on the West.

The multinational corporation keeps more and more doing away with the scattered state of the population, of the means of production, and of property. It has agglomerated population, centralized means of production, and has concentrated property in a few hands. The necessary consequence of this was political centralization. Independent, or but loosely connected provinces, with separate interests, laws, systems of taxation, and governments, became lumped together in one nation, with one government, one code of laws, one national class-interest, one frontier, and one customs tariff.

5. See R. H. Coase for an analysis of the boundary between the firm and the market: "outside the firm, price movements direct production which is coordinated through a series of exchange transactions on the market. Within the firm these market transactions are eliminated and in place of the complicated market structure with exchange transactions, is substituted the entrepreneur co-ordinator who directs production." R. H. Coase, "The Nature of the Firm," reprinted in G. J. Stigler and K. E. Boulding *Readings in Price Theory* (Homewood, Richard D. Irwin, Inc., 1952).

6. "Even in the very backward countries we find highly specialized trades; but we do not find the work within each trade so divided up that the planning and arrangement of the business, its management and its risks, are borne by one set of people, while the manual work required for it is done by higher labour. This form of division of labour is at once characteristic of the modern world generally and of the English race in particular. It may be swept away by the further growth of that free enterprise which has called it into existence. But for the present it expands out for good and for evil as the chief fact in the form of modern civilization, the 'kernel' of the modern economic problem." Marshall, *Principles of Economics*, 8th edition, pp. 74-75. Note that Marshall preferred to call businessmen Undertakers rather than Capitalists (p. 74).

7. "Division of labour within the workshop implies the undisputed authority of the capitalist over men that are but parts of a mechanism that belongs to him . . . The same bourgeois mind which praises division of labour in the workshop, lifelong annexation of the labourer to a partial operation, and his complete subjection to capital, as being an organisation of labour that increases its productiveness —that same bourgeois mind denounces with equal vigour every conscious attempt to socially control and regulate the process of production, as an inroad upon such sacred things as the rights of property, freedom and unrestricted play for the bent of the individual capitalist. It is very characteristic that the enthusiastic apologists of the factory system have nothing more damning to urge against a general organization of the labour of society, than that it would turn all society into one immense factory." K. Marx, *Capital*, Volume I (Moscow, Foreign Language Publishing House, 1961), p. 356.

8. The following analysis by E. S. Mason of current attempts to justify hierarchy and inequality by emphasizing the skill and knowledge of managers and the technostructure is interesting and of great significance on this connection:

> "As everyone now recognizes, classical economics provided not only a system of analysis, or analytical 'model,' intended to be useful to the ex-

planation of economic behaviour but also a defense—and a carefully reasoned defense—of the proposition that the economic behaviour promoted and constrained by the institutions of a free-enterprise system is, in the main, in the public interest.

It cannot be too strongly emphasized that the growth of the nineteenth-century capitalism depended largely on the general acceptance of a reasoned justification of the system on moral as well as on political and economic grounds.

It seems doubtful whether, to date, the managerial literature has provided an equally satisfying apologetic for big business.

The attack on the capitalist apologetic of the nineteenth century has been successful, but a satisfactory contemporary apologetic is still to be created. I suspect that, when and if an effective new ideology is devised, economics will be found to have little to contribute. Economists are still so mesmerized with the fact of choice and so little with its explanations, and the concept of the market is still so central to their thought, that they would appear to be professionally debarred from their important task. I suspect that to the formulation of an up-to-date twentieth-century apologetic the psychologists, and possibly, the political scientists will be the main contributors. It is high time they were called to their job."

Edward S. Mason, "The Apologetics of Managerialism," *The Journal of Business of the University of Chicago,* January 1958, Vol. XXXI, No. 1, pp. 1-11.

9. This analysis of the modern corporation is almost entirely based on the work of Alfred D. Chandler, *Strategy and Structure* (New York, Doubleday & Co., Inc., 1961) and Chester Barnard, *The Functions of Executives* (Cambridge, Harvard University Press, 1938).

10. Alfred D. Chandler and Fritz Redlich, "Recent Developments in American Business Administration and Their Conceptualization," *Business History Review,* Spring 1961, pp. 103-128.

11. Neoclassical models suggest that this choice was due to the exogenously determined nature of technological change. A Marxist economic model would argue that it was due in part to the increased tensions in the labor market accompanying the accumulation of capital and the growth of large firms. This is discussed further in S. Hymer and S. Resnick, "International Trade and Uneven Development," in J. N. Bhagwati, R. W. Jones, R. A. Mundell, Jaroslave Vanek, eds., *Kindleberger Festschrift* (Cambridge, M.I.T. Press, 1970).

12. The reasons for foreign investment discussed here are examined in more detail in S. Hymer, "La Grande Corporation Multinationale," *Revue Economique,* Vol. XIX, No. 6, Novembre 1968, pp. 949-973, and in Hymer and Rowthorn, *op. cit.*

13. At present, U.S. corporations have about 60 billion dollars invested in foreign branch plants and subsidiaries. The total assets of these foreign operations are much larger than the capital invested and probably equal 100 billion dollars at book value. (American corporations, on the average, were able to borrow 40 percent of their subsidiaries' capital requirements locally in the country of operation.) The total assets of 500 large U.S. firms are about 300-350 billion dollars, while the total assets of the 200 largest non-U.S. firms are slightly less than 200 billion dollars. See U.S. Department of Commerce, *Survey of Current Business,* September 1969 and *Fortune* list of the 500 largest U.S. corporations and 200 largest non-American.

14. At present unequal growth of different parts of the world economy upsets the oligopolistic equilibrium because the leading firms have different geographical distributions of production and sales. Thus, if Europe grows faster than the United States, European firms tend to grow faster than American firms, unless American firms engage in heavy foreign investment. Similarly, if the United States grows faster than Europe, U.S. firms will grow faster than European firms because Europeans have a lesser stake in the American market. When firms are distributed evenly in all markets, they share equally in the good and bad fortunes of the various submarkets, and oligopolistic equilibrium is not upset by the unequal growth of different countries.

15. Chandler and Redlich, *op. cit.*

16. Chandler and Redlich, *op. cit.*, p. 120.

17. See H. A. Simon, "The Compensation of Executives," *Sociometry*, March 1957.

18. Sean Gervasi, "Publicité et Croissance Economique," *Economie et Humanisme*, (Novembre/Decembre, 1964).

19. Lloyd A. Fallers, "A Note on the Trickle Effect," in Perry Bliss, ed., *Marketing and the Behavioural Sciences*, (Boston, Allyn and Bacon, 1963), pp. 208-216.

20. See Raymond Vernon, "International Investment and International Trade in the Product Cycle," *Quarterly Journal of Economics*, LXXX, May 1966.

21. An interesting illustration of the asymmetry in horizons and prospectives of the big company and the small country is found in these quotations from *Fortune*. Which countries of the world are making a comparable analysis of the Multinational Corporation?

A Ford economist regularly scans the international financial statistics to determine which countries have the highest rates of inflation; these are obviously prime candidates for devaluation. He then examines patterns of trade. If a country is running more of an inflation than its chief trading partners and competitors and its reserves are limited, it is more than a candidate; it is a shoo-in. His most difficult problem is to determine exactly when the devaluation will take place. Economics determines whether and how much, but politicians control the timing. So the analyst maintains a complete library of information on leading national officials. He tries to get "into the skin of the man" who is going to make the decision. The economist's forecasts have been correct in sixty-nine of the last seventy-five crisis situations.

DuPont is one company that is making a stab in the direction of formally measuring environmental incertainty, basically as a tool for capital budgeting decisions. The project is still in the research stage, but essentially the idea is to try to derive estimates of the potential of a foreign market, which is, of course, affected by economic conditions. The state of the economy in turn is partly a function of the fiscal and monetary policies the foreign government adopts. Policy decisions depend on real economic forces, on the attitudes of various interest groups in the country, and on the degree to which the government listens to these groups.

In the fiscal and monetary part of their broad economic model, the DuPont researchers have identified fifteen to twenty interest groups per country, from small land-owners to private bankers. Each interest group has a "latent influence," which depends on its size and educational level and the group's power to make its feelings felt. This influence, subjectively measured, is multiplied by an estimate of "group cohesiveness": i.e., how likely the group is to mobilize its full resources on any particular issue. The product is a measure of "potential influence." This in turn must be multiplied by a factor representing the government's receptivity to each influence group.

> Sanford Rose, "The Rewarding Strategies of Multinationalism," *Fortune*, September 15, 1968, p. 105.

22. This quote is taken from the first page of Marshall's *Principles of Economics.* In the rest of the book, he attempted to show that the economic system of laissez-faire capitalism had an overall positive effect in forming character. As we noted above, his argument rested upon the existence of competitive markets (and the absence of coercion). Because multinational corporations substitute for the international market they call into question the liberal ideology which rationalized it. (See footnote 9 above, quoting E. S. Mason).

23. A. A. Berle, Jr., has put the problem most succinctly:

> The Industrial Revolution, as it spread over twentieth-century life, required collective organization of men and things . . . As the twentieth century moves into the afternoon, two systems—and (thus far) two only—have emerged as vehicles of modern industrial economics. One is the socialist commissariat; its highest organization at present is in the Soviet Union, the other is the modern corporation, most highly developed in the United States.
>
> Foreword to *The Corporation in Modern Society,* E. S. Mason, ed. (New York, Atheneum, 1967), p. IX.

24. S. Kuznets, *Modern Economic Growth* (New Haven, Yale University Press, 1966), pp. 423-24.

25. See K. Polanyi, *The Great Transformation* (New York, Farrar and Rinehart, Inc. 1944), on the consequences after 1870 of the repeal of the Corn Laws in England.

26. See Kari Levitt, *Silent Surrender: The Multinational Corporation in Canada* (Toronto, Macmillan Company of Canada, 1970) and Norman Girvan and Owen Jefferson "Corporate vs. Caribbean Integration" *New World Quarterly*, Vol. IV, No. 2.

27. See A. Smith, *The Wealth of Nations* (New York, The Modern Library, 1937), p. 1 vii.

28. See Theodor Mommsen, *The History of Rome* (New York, Meridian Books, Inc., 1958), p. 587.

# Part VII

## Commercial Policies

### Developing Market Economies

Balassa
Myint

### Center Countries, Periphery, and Centrally Planned Economies

Prebisch
Portes

### Developed Market Economies, GATT, and Foreign Aid

The GATT Secretariat
Chenery and Strout

# The Choice of a Development Strategy: Lessons and Prospects

Bela Belassa

## Inward- vs. Outward-Oriented Development Strategies

The evidence is quite conclusive: countries applying outward-oriented development strategies performed better in terms of exports, economic growth, and employment than countries with continued inward orientation, which encountered increasing economic difficulties. At the same time, policy reforms aimed at greater outward orientation brought considerable improvement to the economic performance of countries that had earlier applied inward-oriented policies.

It has been suggested, however, that import substitution was a necessary precondition for the development of manufactured exports in present-day developing countries. In attempting to provide an answer to this question, a distinction needs to be made between first-stage and second-stage import substitution.

I have noted that, except in Britain and Hong Kong, the exportation of nondurable consumer goods and their inputs was preceded by an import-substitution phase. At the same time, there were differences among the countries concerned as regards the length of this phase and the level of protection applied. First-stage import substitution was of relatively short duration in the present-day industrial countries and in the three Far Eastern developing countries that subsequently adopted an outward-oriented strategy; it was longer in most other developing countries, and these countries also generally had higher levels of protection.

Nor did all nondurable consumer goods and their inputs go through an import-substitution phase before the Far Eastern countries began to export them. Synthetic textiles in Korea, plastic shoes in Taiwan, and fashion clothing in Singapore all began to be produced largely for export markets. Plywood and wigs, which were Korea's leading exports in the late sixties and early seventies, did not go through an import-substitution phase either.

Wigs provide a particularly interesting example, because they reflect the responses of entrepreneurs to incentives. Korea originally exported human hair to the industrial countries, especially the United States. Recognizing that human hair was made into wigs by a labor-intensive process, entrepreneurs began to exploit what appeared to be a profitable opportunity to export wigs, given the favorable treatment of exports in Korea and the limitations imposed by the United States on wigs originating from Hong Kong. The supply of human

Reprinted from "The Process of Industrial Development and Alternative Development Strategies—Essays in International Finance No. 141, December 1980," pp. 18–27, with permission of Princeton University Press, Princeton, Copyright 1981.

hair soon proved to be insufficient, however, and firms turned to exporting wigs made of synthetic hair. Wigs made with synthetic hair were for a time Korea's second-largest single export commodity, after plywood.

The example indicates that entrepreneurs will export the commodities that correspond to the country's comparative advantage if the system of incentives does not discriminate against exports. It also points to the need to leave the choice of exports to private initiative. It is highly unlikely that government planners would have chosen wigs as a potential major export or that they would have effected a switch from human to synthetic hair in making them. Even if a product group such as toys were identified by government planners, the choice of which toys to produce would have to be made by the entrepreneur, who has to take the risks and reap the rewards of his actions. At the same time, providing similar incentives to all export commodities other than those facing market limitations abroad and avoiding a bias against exports will ensure that private profitability corresponds to social profitability. This was, by and large, the case in countries pursuing an outward strategy.

These considerations may explain why Singapore and Taiwan did not need a planning or targeting system for exports. Export targets were in effect in Korea, but the fulfillment of these targets was not a precondition of the application of the free-trade regime to exports or of the provision of export incentives. While successful exporters were said to enjoy advantageous treatment in tax cases and export targets may have exerted pressure on some firms, these factors merely served to enhance the effects of export incentives without introducing discrimination among export products. At any rate, most firms continually exceeded their targets. A case in point is the increase in Korean exports by two-thirds between the second quarter of 1975 and the second quarter of 1976, exceeding the targets by a very large margin.

The reliance on private initiative in countries that adopted an outward-oriented development strategy can be explained by the need of exporters for flexibility to respond to changing world market conditions. Furthermore, government cannot take responsibility for successes and failures in exporting that will affect the profitability of firms. For these reasons Hungary, among socialist countries, gave firms the freedom to determine the product composition of their exports after the 1968 economic reform and especially after 1977.

In the Latin-American countries that reformed their incentive systems in the period preceding the 1973 oil crisis, the expansion of manufactured exports was not based on export targets either. The question remains, however, whether the development of exports in these countries was helped by the fact that they had undertaken second-stage import substitution.

This question can be answered in the negative as far as nondurable consumer goods and their inputs are concerned. Had appropriate incentives been provided, these commodities could have been exported as soon as first-stage import substitution was completed, as was the case in the Far Eastern countries. In fact, to the extent that the products in question had to use some domestic inputs produced at higher than world market costs, exporters were at a disadvantage in foreign markets. It can also be assumed that the inability to exploit fully economies of scale and the lack of sufficient specialization in the production of parts, components, and accessories in the confines of the protected domestic markets retarded the development of exports of intermediate products and producer and consumer durables.

More generally, as a Hungarian economist has pointed out, there is the danger that second-stage import substitution will lead to the establishment of an industrial structure that is "prematurely old," in the sense that it is based on small-scale production with inadequate specialization and outdated machinery. Should this be the case, any subsequent move toward outward orientation will encounter difficulties. Such difficulties were apparent in the case of Hungary and may also explain why, although exports grew rapidly from a low base, their share in manufacturing output remained small in the Latin-American countries that moved toward outward orientation from the second stage of import substitution.

In contrast, in the period following the oil crisis the Far Eastern countries increasingly upgraded their exports of nondurable consumer goods and began exporting machinery, electronics, and transport equipment. For several of these products, including shipbuilding in Korea, photographic equipment in Singapore, and other electronic products in Taiwan, exporting was not preceded by an import-substitution phase. There are even examples, such as color television sets in Korea, where the entire production was destined for foreign markets.

Intermediate goods, machinery, and automobiles require special attention, given the importance of economies of scale on the plant level for the first; the need for product (horizontal) specialization for the second; and the desirability of vertical specialization in the form of the production of parts, components, and accessories on an efficient scale for the third. In all these cases, production in protected domestic markets will involve high costs in most developing countries, and the establishment of small-scale and insufficiently specialized firms will make the transition to exportation difficult. This contrasts with the case of nondurable consumer goods and their inputs, where efficient production does not require large plants or horizontal and vertical specialization.

It follows that, rather than enter into second-stage import substitution as a prelude to subsequent exports, it is preferable to undertake the manufacture of intermediate goods and producer and consumer durables for domestic and foreign markets simultaneously. This will permit the exploitation of economies of scale and ensure efficient import substitution in some products, while others continue to be imported. At the same time, it will require the provision of equal incentives to exports and to import substitution instead of import protection that discriminates against exports.

## Vulnerability and Policy Responses to External Shocks

Outward orientation involves increasing the share of exports in GNP, and the high share of exports in the national economies of countries undertaking such a strategy has been said to increase their vulnerability to foreign events. The experience of the post-1973 period casts some light on the validity of this claim.

Available evidence indicates that the Far Eastern countries applying an outward-oriented strategy weathered the effects of the quadrupling of oil prices in 1973-74 and the world recession of 1974-75 better than countries with continued inward orientation. This may be explained by differences in the "compressibility" of imports and in the flexibility of the national economies of countries applying different strategies. Outward orientation is associated with high export *and* import shares that permit reductions in nonessential imports without serious adverse effects on the functioning of the economy. By contrast, continued inward orientation involves limiting imports to an unavoidable minimum, so that any further reduction will impose a considerable cost in terms of growth. Furthermore, the greater flexibility of the national economies of countries pursuing an outward-oriented strategy, under which firms learn to live with foreign competition, makes it possible to change the product composition of exports in response to changes in world market conditions, whereas inward orientation entails establishing a more rigid economic structure.

I come next to policy responses to external shocks. In the Far Eastern countries there were pressures for a shift toward inward orientation in the immediate aftermath of the oil crisis and the world recession. These countries nevertheless continued their outward-oriented development strategy, which made it possible for them to maintain high rates of growth of exports and GNP. Taking the 1973-79 period as a whole, per capita GNP rose at average annual rates of 8.3 per cent in Korea, 6.1 per cent in Singapore, and 5.5 per cent in Taiwan. Growth rates declined, however, after 1978 in Korea as its currency became increasingly overvalued and some large capital-intensive investments were undertaken.

Brazil attempted to maintain past rates of economic growth by relying on foreign borrowing and increased import protection. The high capital intensity of import-substitution projects, however, raised capital-output ratios and led to a decline in the rate of economic growth. Per capita incomes rose 5.2 per cent a year in 1966-73, 4.5 per cent in 1973-76, and 2.4 per cent in 1976-79. At the same time, the servicing of foreign loans imposed an increasing burden on Brazil's balance of payments.

Policy changes in the opposite direction were made in Chile and Uruguay, which had applied an inward-oriented strategy until the 1973 oil crisis. These countries responded to the deterioration of their terms of trade and the slowdown in the growth of foreign demand for their export products by reforming the system of incentives. The reforms involved eliminating quantitative restrictions, reducing the bias against exports, liberalizing financial markets, and adopting positive real interest rates.

In Uruguay, which had had a stagnant economy in the previous decade, the reforms led to rapid increases in exports and GNP; per capita GNP rose 3.1 per cent a year between 1973 and 1976 and 4.3 per cent a year between 1976 and 1979. The growth of exports and GNP accelerated in Chile after a period of dislocation caused by the application of a severe deflationary policy that was aggravated by rapid reductions in tariffs.

Argentina and Colombia rely on domestically produced oil and hence were unaffected by the quadrupling of petroleum prices. Colombia also enjoyed higher coffee prices, which more than offset the shortfall in exports due to the slowdown in the growth of foreign demand. But it reduced incentives to nontraditional exports, with attendant losses in export market shares, and it was not able to translate increases in foreign-exchange earnings from traditional exports into higher GNP growth rates. The distortions caused by rapid inflation were largely responsible for low GNP growth rates in Argentina.

Mexico lost export market shares in both traditional and non-traditional exports following the adoption of domestic expansionary policies, financed in large part by the inflow of foreign capital. And while the discovery of large oil deposits benefited Mexico's balance of payments, it increased the overvaluation of the currency, thus discriminating against agricultural and manufacturing activities. Finally, no substantive policy changes occurred in India, which continued to lose export market shares.

*Policy Prescriptions and Prospects for the Future*

The experience of developing countries in the postwar period leads to certain policy prescriptions. First, while infant-industry considerations call for the preferential treatment of manufacturing activities, such treatment should be applied on a moderate scale, both to avoid the establishment and maintenance of inefficient industries and to ensure the continued expansion of primary production for domestic and foreign markets.

Second, equal treatment should be given to exports and to import substitution in the manufacturing sector, in order to ensure resource allocation according to comparative advantage and the exploitation of economies of scale. This is of particular importance in the case of intermediate goods and producer and consumer durables, where the advantages of large plant size and horizontal and vertical specialization are considerable and where import substitution in the framework of small domestic markets makes the subsequent development of exports difficult. The provision of equal incentives will contribute to efficient exportation and import substitution through specialization in particular products and in their parts, components, and accessories.

Third, infant-industry considerations apart, variations in incentive rates within the manufacturing sector should be kept to a minimum. This amounts to the application of the "market principle" in allowing firms to decide on the activities to be undertaken. In particular, firms should be free to choose their export composition in response to changing world market conditions.

Fourth, in order to minimize uncertainty for the firm, the system of incentives should be stable and automatic. Uncertainty will also be reduced if the reform of the system of incentives necessary to apply the principles just described is carried out according to a timetable made public in advance.

It has been objected that the application of these principles—characteristic of an outward-oriented development strategy—would encounter market limitations, aggravated by protectionist policies, in the industrial countries. To address this issue, one needs to examine recent and prospective trends in trade in manufactured goods between the industrial and the developing countries.

Notwithstanding protectionist pressures in the industrial countries, their imports of manufactured goods from the developing countries rose at a rapid rate during the period following the oil crisis, averaging 10.2 per cent a year in volume terms between 1973 and 1978. Moreover, the "apparent" income elasticity of demand for these imports, calculated as the ratio of the growth rate of imports to that of gross domestic product, increased from 3.6 in 1963-73 to 4.1 in 1973-78.

Given the increased volume of manufactured imports from the developing countries, the apparent income elasticity of demand for manufactured goods originating in these countries can be expected to decline in the future. Assuming an elasticity of 3.2 and a GDP growth rate of 3.9 per cent in the industrial countries, I have projected their imports of manufactured goods from the developing countries to rise at an average annual rate of 12.5 per cent between 1978 and 1990. This projection assumes unchanged policies in the industrial countries, including the maintenance of the Multifiber Arrangement.

If this import growth rate were realized, the share of the developing countries in the consumption of manufactured goods in the industrial countries would rise from 1.5 per cent in 1978 to 4.0 per cent in 1990, with an incremental share of 8.9 per cent. The incremental share would be the highest in clothing, 28.1 per cent; it would be 7.2 per cent in textiles and 6.6 per cent in other consumer goods. Nonetheless, the production of textiles and clothing would rise at an average annual rate of 2 per cent in the industrial countries. And these countries would have a rising export surplus in trade in manufactured goods with the developing countries, which would contribute to the growth of their manufacturing sector.

At the same time, in accordance with the "stages" approach to comparative advantage, changes would occur in the product composition of the manufactured exports of the developing countries as they proceeded to higher stages of industrial development. This process is exemplified by Japan, which shifted from unskilled-labor-intensive to skill-intensive to physical-capital-intensive exports and is increasingly expanding its technology-intensive exports.

Shifts in export composition are now occurring in the newly industrializing developing countries, including the Far Eastern and Latin-American countries that carried out policy reforms after the mid-sixties. The Far Eastern countries that have a relatively high educational level may increasingly take the place of Japan in exporting skill-intensive products, while Latin-American countries may expand their exports of relatively capital-intensive products. Countries at lower stages of industrial development, in turn, may take the place of the newly industrializing countries in exporting products that require chiefly unskilled labor.

To the extent that the exports of newly industrializing countries replace Japanese exports, and their exports are in turn replaced by the exports of countries at lower stages of industrial development, the threat to the domestic manufacturing industries of the industrial countries is reduced. Nor does the upgrading and diversification of manufactured exports by the newly industrializing countries represent a serious threat, inasmuch as the exports of individual commodities would account for a relatively small proportion of the consumption and production of these commodities in the industrial countries. This conclusion also applies to the international division of the production process, exemplified by the development of Ford's "world car," which will entail manufacturing in nineteen countries.

It follows that it is in the interest of the newly industrializing developing countries to upgrade and diversify their exports in line with their changing comparative advantage. This is also in the interest of countries at lower stages of industrial development, as they can replace exports of unskilled-labor-intensive commodities from the newly industrializing countries to industrial-country markets.

There are also considerable opportunities for the expansion of trade in manufactured goods among the developing countries themselves. With increased oil earnings, the largely open markets of the OPEC countries will experience rapid growth. Furthermore, the newly industrializing countries can trade skill-intensive and physical-capital-intensive goods among themselves and exchange these commodities for the unskilled-labor-intensive products of countries at lower stages of industrial development.

The expansion of this trade requires the pursuit of outward-oriented strategies by the newly industrializing countries, so as to provide appropriate incentives to exports and to allow imports from other developing countries. The pursuit of such a strategy would also contribute to efficient import substitution by ensuring low-cost manufacture through international specialization and the international division of the production process. Similar conclusions apply to countries at lower stages of industrial development.

Finally, lowering protection in the industrial countries would lead to increases in their imports of manufactured goods from the developing countries over and above projected levels. This would also be in the interest of the industrial countries, properly conceived. They would benefit from shifts to high-technology products within the manufactured sector as higher export earnings permitted the developing countries to increase their imports of these products.

Trade liberalization in the industrial countries could proceed over a ten-year horizon without involving excessively large adjustment costs. One could accept, for example, a decline in the production of textiles and clothing over time that would involve not replacing the normal attrition of workers and depreciated equipment in branches that utilized largely unskilled labor. In turn, new entrants into the industrial labor force would increasingly enter technologically advanced industries where productivity levels are substantially higher.

Apart from expanding the volume of trade, then, the pursuit of appropriate policies by developed and developing countries would permit shifts in the pattern of international specialization in response to the changing structure of comparative advantage in countries at different levels of industrial development. As a result, the efficiency of resource allocation would improve and rates of economic growth would accelerate, with benefits to all concerned.

# THE "CLASSICAL THEORY" OF INTERNATIONAL TRADE AND THE UNDERDEVELOPED COUNTRIES [1]

## Hal Myint

THERE has recently been a considerable amount of controversy concerning the applicability of the "classical theory" of international trade to the underdeveloped countries.[2] The twists in this controversy may be set out as follows. The critics start with the intention of showing that the "nineteenth-century pattern" of international trade, whereby the underdeveloped countries export raw materials and import manufactured goods, has been unfavourable to the economic development of these countries. But instead of trying to show this directly, they concentrate their attacks on the "classical theory," which they believe to be responsible for the unfavourable pattern of trade. The orthodox economists then come to the defence of the classical theory by reiterating the principle of comparative costs which they claim to be applicable both to the developed and the underdeveloped countries. After this, the controversy shifts from the primary question whether or not the nineteenth-century pattern of international trade, as a historical reality, has been unfavourable to the underdeveloped countries to the different question whether or not the theoretical model assumed in the comparative-costs analysis is applicable to these countries. Both sides then tend to conduct their argument as though the two questions were the same and to identify the "classical theory" with the comparative-costs theory.

It will be argued in this paper that this has led to the neglect of those other elements in the classical theory of international trade which are much nearer to the realities and ideologies of the nineteenth-century expansion of international trade to the underdeveloped countries. In Sections I and II we shall outline these elements and show that they are traceable to Adam Smith and to some extent to J. S. Mill. In Section III we shall show how one of Adam Smith's lines of approach can be fruitfully developed to throw a more illuminating light on the past and present patterns of the international trade of the underdeveloped countries than the conventional theory. In Section IV we shall touch upon some policy implications of our analysis and show certain weaknesses in the position both of the orthodox economists

---

[1] This paper has benefited from comments by Sir Donald MacDougall, Professor H. G. Johnson, R. M. Sundrum and G. M. Meier.

[2] Of the very extensive literature on the subject, we may refer to two notable recent works, the first stating the orthodox position and the second the position of the critics: J. Viner, *International Trade and Economic Development*, and G. Myrdal, *An International Economy*.

in relation to the ideological than to the actual economic forces which characterised the nineteenth-century expansion of international trade to the underdeveloped countries. It is true, as we shall see later,[1] that both the total value and the physical output of the exports of these countries expanded rapidly. In many cases the rate of increase in export production was well above any possible rate of increase in population, resulting in a considerable rise in output per head. But it is still true to say that this was achieved not quite in the way envisaged by Smith, viz., a better division of labour and specialisation leading on to innovations and cumulative improvements in skills and productivity per man-hour. Rather, the increase in output per head seems to have been due: (i) to once-for-all increases in productivity accompanying the transfer of labour from the subsistence economy to the mines and plantations, and (ii) what is more important, as we shall see later, to an increase in working hours and in the proportion of gainfully employed labour relatively to the semi-idle labour of the subsistence economy.

The transfer of labour from the subsistence economy to the mines and plantations with their much higher capital–output ratio and skilled management undoubtedly resulted in a considerable increase in productivity. But this was mostly of a once-for-all character for a number of reasons. To begin with, the indigenous labour emerging from the subsistence economy was raw and technically backward. Moreover, it was subject to high rates of turnover, and therefore not amenable to attempts to raise productivity. Unfortunately, this initial experience gave rise to or hardened the convention of " cheap labour," which regarded indigenous labour merely as an undifferentiated mass of low-grade man-power to be used with a minimum of capital outlay.[2] Thus when the local labour supply was exhausted the typical reaction was not to try to economise labour by installing more machinery and by reorganising methods of production but to seek farther afield for additional supplies of cheap labour. This is why the nineteenth-century process of international trade in the underdeveloped countries was characterised by large-scale movements of cheap labour from India and China.[3] This tendency was reinforced by the way in which the world-market demand for raw materials expanded in a series of waves. During the booms output had to be expanded as quickly as possible along existing lines, and there was no time to introduce new techniques or reorganise production; during the slumps it was difficult to raise capital for such purposes.

[1] See footnotes on pp. 324 and 327 below. See also Sir Donald MacDougall's *The World Dollar Problem*, pp. 134–43. Sir Donald's argument that the productivity of labour in the underdeveloped countries has been rising faster than is generally assumed is mainly based on figures for productivity *per capita*. These figures are not inconsistent with our argument that on the whole the expansion of the export production has been achieved on more or less constant techniques and skills of indigenous labour, by increasing working hours and the proportion of gainfully employed labour rather than by a continuous rise in productivity per man-hour.

[2] Cf. S. H. Frankel, *Capital Investment in Africa*, pp. 142–6, and W. M. Macmillan, *Europe and West Africa*, pp. 48–50.

[3] Cf. Knowles, *op. cit.*, pp. viii and 182–201.

This failure to achieve Adam Smith's ideal of specialisation leading on to continuous improvements in skills can also be observed in the peasant export sectors.   Where the export crop happened to be a traditional crop (*e.g.*, rice in South-East Asia), the expansion in export production was achieved simply by bringing more land under cultivation with the same methods of cultivation used in the subsistence economy.   Even where new export crops were introduced, the essence of their success as peasant export crops was that they could be produced by fairly simple methods involving no radical departure from the traditional techniques of production employed in subsistence agriculture.[1]

Thus instead of a process of economic growth based on continuous improvements in skills, more productive recombinations of factors and increasing returns, the nineteenth-century expansion of international trade in the underdeveloped countries seems to approximate to a simpler process based on constant returns and fairly rigid combinations of factors.   Such a process of expansion could continue smoothly only if it could feed on *additional* supplies of factors in the required proportions.

## II

Let us now turn to Smith's " vent for surplus " theory of international trade.   It may be contrasted with the comparative-costs theory in two ways.

(*a*) The comparative-costs theory assumes that the resources of a country are given and fully employed before it enters into international trade.   The function of trade is then to reallocate its given resources more efficiently between domestic and export production in the light of the new set of relative prices now open to the country.   With given techniques and full employment, export production can be increased only at the cost of reducing the domestic production.   In contrast, the " vent for surplus " theory assumes that a previously isolated country about to enter into international trade possesses a surplus productive capacity [2] of some sort or another.   The function of trade here is not so much to reallocate the given resources as to provide the new effective demand for the output of the surplus resources which would have remained unused in the absence of trade.   It follows that export production can be increased without necessarily reducing domestic production.

---

[1] Thus A. McPhee wrote about the palm-oil and ground-nut exports of West Africa: " They made little demand on the energy and thought of the natives and they effected no revolution in the society of West Africa.   That was why they were so readily grafted on the old economy and grew as they did " (*The Economic Revolution in West Africa*, pp. 39–40).   Some writers argue that there was a studied neglect of technical improvements in the peasant sector to facilitate the supply of cheap labour to other sectors.   Cf., for example, W. A. Lewis, " Economic Development with Unlimited Supplies of Labour," *Manchester School*, May 1954, pp. 149–50.   For a description of imperfect specialisation in economic activity in West Africa see P. T. Bauer and B. S. Yamey, " Economic Progress and Occupational Distribution," ECONOMIC JOURNAL, December 1951, p. 743.

[2] A surplus over domestic requirements and *not* a surplus of exports over imports.

(b) The concept of a surplus productive capacity above the requirements of domestic consumption implies an inelastic domestic demand for the exportable commodity and/or a considerable degree of internal immobility and specificness of resources. In contrast, the comparative-costs theory assumes either a perfect or, at least, a much greater degree of internal mobility of factors and/or a greater degree of flexibility or elasticity both on the side of production and of consumption. Thus the resources not required for export production will not remain as a surplus productive capacity, but will be reabsorbed into domestic production, although this might take some time and entail a loss to the country.

These two points bring out clearly a peculiarity of the " vent-for-surplus " theory which may be used either as a free-trade argument or as an anti-trade argument, depending on the point of view adopted. (a) From the point of view of a previously isolated country, about to enter into trade, a surplus productive capacity suitable for the export market appears as a virtually " costless " means of acquiring imports and expanding domestic economic activity. This was how Adam Smith used it as a free-trade argument. (b) From the point of view of an established trading country faced with a fluctuating world market, a sizeable surplus productive capacity which cannot be easily switched from export to domestic production makes it " vulnerable " to external economic disturbances. This is in fact how the present-day writers on the underdeveloped countries use the same situation depicted by Smith's theory as a criticism of the nineteenth-century pattern of international trade. This concept of vulnerability may be distinguished from that which we have come across in discussing the " productivity " theory of trade. There, a country is considered " vulnerable " because it has adapted and reshaped its productive structure to meet the requirements of the export market through a genuine process of " specialisation." Here, the country is considered " vulnerable " simply because it happens to possess a sizeable surplus productive capacity which (even without any improvements and extensions) it cannot use for domestic production. This distinction may be blurred in border-line cases, particularly in underdeveloped countries with a large mining sector. But we hope to show that, on the whole, while the " vulnerability " of the advanced countries, such as those in Western Europe which have succeeded in building up large export trades to maintain their large populations, is of the first kind, the " vulnerability " of most of the underdeveloped countries is of the second kind.

Let us now consider the " vent-for-surplus " approach purely as a theoretical tool. There is a considerable amount of prejudice among economists against the " vent-for-surplus " theory, partly because of its technical crudeness and partly because of its mercantilist associations. This may be traced to J. S. Mill, who regarded Smith's " vent-for-surplus " doctrine as " a surviving relic of the Mercantile Theory " (*Principles*, p. 579).

The crux of the matter here is the question: why should a country isolated from international trade have a surplus productive capacity? The answer which suggests itself is that, given its random combination of natural resources, techniques of production, tastes and population,[1] such an isolated country is bound to suffer from a certain imbalance or disproportion between its productive and consumption capacities. Thus, take the case of a country which starts with a sparse population in relation to its natural resources. This was broadly true not only of Western countries during their mercantilist period but also of the underdeveloped countries of South-East Asia, Latin America and Africa when they were opened up to international trade in the nineteenth century. Given this situation, the conventional international-trade theory (in its Ohlin version) would say that this initial disproportion between land and labour would have been equilibrated away by appropriate price adjustments: i.e., rents would be low and relatively land-using commodities would have low prices, whereas wages would be high and relatively labour-using commodities would have high prices. In equilibrium there would be no surplus productive capacity (although there might be surplus land by itself) because the scarce factor, labour, would have been fully employed. Thus when this country enters into international trade it can produce the exports only by drawing labour away from domestic production. Now this result is obtained only by introducing a highly developed price mechanism and economic organisation into a country which is supposed to have had no previous economic contacts with the outside world. This procedure may be instructive while dealing with the isolated economy as a theoretical model. But it is misleading when we are dealing with genuinely isolated economies in their proper historical setting; it is misleading, in particular, when we are dealing with the underdeveloped countries, many of which were subsistence economies when they were opened to international trade. In fact, it was the growth of international trade itself which introduced or extended the money economy in these countries. Given the genuine historical setting of an isolated economy, might not its initial disproportion between its resources, techniques, tastes and population show itself in the form of surplus productive capacity?

Adam Smith himself thought that the pre-existence of a surplus productive capacity in an isolated economy was such a matter of common observation that he assumed it implicitly without elaborating upon it. But he did give some hints suggesting how the " narrowness of the home market," which causes the surplus capacity, is bound up with the underdeveloped economic organisation of an isolated country, particularly the lack of a good internal transport system and of suitable investment opportunities.[1] Further his concept of surplus productive capacity is not merely a matter of surplus land by itself but surplus land combined with surplus labour; and the

---

[1] Op. cit., Vol. I, pp. 21 and 383. This is similar to what Mrs. J. Robinson has described as " primitive stagnation." Cf. The Accumulation of Capital, pp. 256–8.

surplus labour is then linked up with his concept of "unproductive" labour. To avoid confusion, this latter should not be identified with the modern concept of "disguised unemployment" caused by an acute shortage of land in overpopulated countries. Although Smith described some cases of genuine "disguised unemployment" in the modern sense, particularly with reference to China, "unproductive" labour in his sense can arise even in thinly populated countries, provided their internal economic organisation is sufficiently underdeveloped. In fact, it is especially in relation to those underdeveloped countries which started off with sparse populations in relation to their natural resources that we shall find Smith's "vent-for-surplus" approach very illuminating.

### III

Let us now try to relate the "vent-for-surplus" theory to the nineteenth-century process of expansion of international trade to the underdeveloped countries. Even from the somewhat meagre historical information about these countries, two broad features stand out very clearly. First the underdeveloped countries of South-East Asia, Latin America and Africa, which were to develop into important export economies, started off with sparse populations relatively to their natural resources. If North America and Australia could then be described as "empty," these countries were at least "semi-empty." Secondly, once the opening-up process had got into its stride, the export production of these countries expanded very rapidly, along a typical growth curve,[1] rising very sharply to begin with and tapering off afterwards. By the Great Depression of the 1930s, the expansion process seems to have come to a stop in many countries; in others, which had a later start, the expansion process may still be continuing after the Second World War.

There are three reasons why the "vent-for-surplus" theory offers a more effective approach than the conventional theory to this type of expansion of international trade in the underdeveloped countries.

(i) The characteristically high rates of expansion which can be observed in the export production of many underdeveloped countries cannot really be explained in terms of the comparative-costs theory based on the assumption of given resources and given techniques. Nor can we attribute any significant part of the expansion to revolutionary changes in techniques and increases in productivity. As we have seen in Section I, peasant export

---

[1] For instance, the annual value of Burma's exports, taking years of high and low prices, increased at a constant proportional rate of 5% per annum on the average between 1870 and 1900. Similar rates of expansion can be observed for Siam and Indonesia (Cf. J. S. Furnivall, *Colonial Policy and Practice*, Appendix I; J. H. Boeke, *The Structure of Netherlands Indian Economy*, p. 184; and J. C. Ingram, *Economic Change in Thailand since 1850*, Appendix C). African export economies started their expansion phase after 1900, and the official trade returns for the Gold Coast, Nigeria and Uganda show similar rates of increase after that date, although the expansion process was arrested by the depression of the 1930s.

production expanded by extension of cultivation using traditional methods of production, while mining and plantation sectors expanded on the basis of increasing supplies of cheap labour with a minimum of capital outlay. Thus the contributions of Western enterprise to the expansion process are mainly to be found in two spheres: the improvements of transport and communications [1] and the discoveries of new mineral resources. Both are methods of increasing the total volume of resources rather than methods of making the given volume of resources more productive. All these factors suggest an expansion process which kept itself going by drawing an increasing volume of hitherto unused or surplus resources into export production.

(ii) International trade between the tropical underdeveloped countries and the advanced countries of the temperate zone has grown out of sharp differences in geography and climate resulting in absolute differences of costs. In this context, the older comparative-costs theory, which is usually formulated in terms of qualitative differences [2] in the resources of the trading countries, tends to stress the obvious geographical differences to the neglect of the more interesting quantitative differences in the factor endowments of countries possessing approximately the same type of climate and geography. Thus while it is true enough to say that Burma is an exporter of rice because of her climate and geography, the more interesting question is why Burma should develop into a major rice exporter while the neighbouring South India, with approximately the same type of climate and geography, should develop into a net importer of rice. Here the "vent-for-surplus" approach which directs our attention to population density as a major determinant of export capacity has an advantage over the conventional theory.[3]

(iii) Granted the importance of quantitative differences in factor endowments, there still remains the question why Smith's cruder "vent-for-surplus" approach should be preferable to the modern Ohlin variant of the comparative-costs theory. The main reason is that, according to the Ohlin theory, a country about to enter into international trade is supposed already to possess a highly developed and flexible economic system which can adjust its methods of production and factor combinations to cope with a wide range of possible variations in relative factor supplies (see Section II above). But in fact the economic framework of the underdeveloped countries is a

---

[1] This is what Professor L. C. A. Knowles described as the "Unlocking of the Tropics" (*op. cit.*, pp. 138–52).

[2] Cf. J. Viner, *International Trade and Economic Development*, pp. 14–16.

[3] Those who are used to handling the problem in terms of qualitative differences in factors and differential rent may ask: why not treat the surplus productive capacity as an extreme instance of "differential rent" where the transfer cost of the factors from the domestic to export production is zero? But this does not accurately portray the situation here. The transfer cost of the factors is zero, not because land which is used for the export crop is not at all usable for domestic subsistence production but because with the sparse population in the early phase there is no demand for the surplus food which could have been produced on the land used for the export crop. As we shall see, at a later stage when population pressure begins to grow, as in Java, land which has been used for export is encroached upon by subsistence production.

much cruder apparatus which can make only rough-and-ready adjustments. In particular, with their meagre technical and capital resources, the under-developed countries operate under conditions nearer to those of fixed technical coefficients than of variable technical coefficients. Nor can they make important adjustments through changes in the outputs of different commodities requiring different proportions of factors because of the inelastic demand both for their domestic production, mainly consisting of basic foodstuff, and for their exportable commodities, mainly consisting of industrial raw materials. Here again the cruder " vent-for-surplus " approach turns out to be more suitable.

Our argument that, in general, the " vent-for-surplus " theory provides a more effective approach than the comparative-costs theory to the inter-national trade of the underdeveloped countries does not mean that the " vent-for-surplus " theory will provide an exact fit to all the particular patterns of development in different types of export economies. No simple theoretical approach can be expected to do this. Thus if we interpret the concept of the surplus productive capacity strictly as pre-existing surplus productive capacity arising out of the original endowments of the factors, it needs to be qualified, especially in relation to the mining and plantation sectors of the underdeveloped countries. Here the surplus productive capacity which may have existed to some extent before the country was opened to inter-national trade is usually greatly increased by the discovery of new mineral resources and by a considerable inflow of foreign capital and immigrant labour. While immigrant labour is the surplus population of other under-developed countries, notably India and China, the term " surplus " in the strict sense cannot be applied to foreign capital. But, of course, the exist-ence of suitable surplus natural resources in an underdeveloped country is a pre-condition of attracting foreign investment into it. Two points may be noted here. First, the complication of foreign investment is not as damaging to the surplus-productive-capacity approach as it appears at first sight, because the inflow of foreign investment into the tropical and semi-tropical underdeveloped countries has been relatively small both in the nineteenth century and the inter-war period.[1] Second, the nineteenth-century pheno-menon of international mobility of capital and labour has been largely neglected by the comparative-costs theory, which is based on the assumption of perfect mobility of factors within a country and their imperfect mobility between different countries. The surplus-productive-capacity approach at least serves to remind us that the output of mining and plantation sectors can expand without necessarily contracting domestic subsistence output.

The use of the surplus-productive-capacity approach may prove in particular to be extremely treacherous in relation to certain parts of Africa,

[1] Cf. R. Nurkse, " International Investment To-day in the Light of Nineteenth Century Experience," ECONOMIC JOURNAL, December 1954, pp. 744–58, and the United Nations Report on *International Capital Movements during the Inter-war Period.*

where mines, plantations and other European enterprises have taken away from the tribal economies the so-called " surplus " land and labour, which, on a closer analysis, prove to be no surplus at all.    Here the extraction of these so-called " surplus " resources, by various forcible methods in which normal economic incentives play only a part, entails not merely a reduction in the subsistence output but also much heavier social costs in the form of the disruption of the tribal societies.[1]

When we turn to the peasant export sectors, however, the application of the " vent-for-surplus " theory is fairly straightforward.    Here, unlike the mining and plantation sectors, there has not been a significant inflow of foreign investment and immigrant labour.    The main function of the foreign export–import firms has been to act as middlemen between the world market and the peasants, and perhaps also to stimulate the peasants' wants for the new imported consumers' goods.    As we have seen, peasant export production expanded by using methods of production more or less on the same technical level as those employed in the traditional subsistence culture. Thus the main effect of the innovations, such as improvements in transport and communications [2] and the introduction of the new crops, was to bring a greater area of surplus land under cultivation rather than to raise the physical productivity per unit of land and labour.    Yet peasant export production usually managed to expand as rapidly as that of the other sectors while remaining self-sufficient with respect to basic food crops.    Here, then, we have a fairly close approximation to the concept of a pre-existing surplus productive capacity which can be tapped by the world-market demand with a minimum addition of external resources.

Even here, of course, there is room for differences in interpretation.    For instance, there is evidence to suggest that, in the early decades of expansion, the rates of increase in peasant export production in South-East Asian and West African countries were well above the possible rates of growth in their working population.[3]    Given the conditions of constant techniques, no significant inflow of immigrant foreign labour and continuing self-sufficiency with respect to the basic food crops, we are left with the question how these peasant economies managed to obtain the extra labour required to

[1] Cf. The United Nations Report on the *Enlargement of the Exchange Economy in Tropical Africa*, pp. 37 and 49–51.

[2] It may be noted that the expansion of some peasant export crops, notably rice in South-East Asia, depended to a much greater extent on pre-existing indigenous transport facilities, such as river boats and bullock carts, than is generally realised.

[3] For instance, cocoa output of the Gold Coast expanded over forty times during the twenty-five year period 1905–30.    Even higher rates of expansion in cocoa production can be observed in Nigeria combined with a considerable expansion in the output of other export crops.    Both have managed to remain self-sufficient with regard to basic food crops (cf. West African Institute of Economic Research, *Annual Conference*, Economic Section, Achimota, 1953, especially the chart between pp. 96 and 98; *The Native Economies of Nigeria*, ed. M. Perham, Vol. I, Part II).    In Lower Burma, for the thirty-year period 1870–1900, the area under rice cultivation increased by more than three times, while the population, including immigrants from Upper Burma, doubled. (Cf. also, Furnivall, *op. cit.*, pp. 84–5.)

expand their export production so rapidly. A part of this labour may have been released by the decline in cottage industries and by the introduction of modern labour-saving forms of transport in place of porterage, but the gap in the explanation cannot be satisfactorily filled until we postulate that even those peasant economies which started off with abundant land relatively to their population must have had initially a considerable amount of under-employed or surplus labour. This surplus labour existed, not because of a shortage of co-operating factors, but because in the subsistence economies, with poor transport and little specialisation in production, each self-sufficient economic unit could not find any market outlet to dispose of its potential surplus output, and had therefore no incentive to produce more than its own requirements. Here, then, we have the archetypal form of Smith's " unproductive " labour locked up in a semi-idle state in the underdeveloped economy of a country isolated from outside economic contacts. In most peasant economies this surplus labour was mobilised, however, not by the spread of the money-wage system of employment, but by peasant economic units with their complement of " family " labour moving *en bloc* into the money economy and export production.

The need to postulate a surplus productive capacity to explain the rapid expansion in peasant export production is further strengthened when we reflect on the implications of the fact that this expansion process is inextric-ably bound up with the introduction of the money economy into the sub-sistence sectors. To the peasant on the threshold of international trade, the question whether or not to take up export production was not merely a question of growing a different type of crop but a far-reaching decision to step into the new and unfamiliar ways of the money economy.

Thus let us consider a community of self-sufficient peasants who, with their existing techniques, have just sufficient land and labour to produce their minimum subsistence requirements, so that any export production can be achieved only by reducing the subsistence output below the minimum level. Now, according to the conventional economic theory, there is no reason why these peasants should not turn to export production if they have a differential advantage there, so that they could more than make up for their food deficit by purchases out of their cash income from the export crop. But, in practice, the peasants in this situation are unlikely to turn to export production so readily. Nor is this " conservatism " entirely irrational, for by taking up export production on such a slender margin of reserves, the peasants would be facing the risk of a possible food shortage for the sake of some gain in the form of imported consumers' goods which are " luxuries " to them. Moreover, this gain might be wiped off by unfavourable changes in the prices of both the export crop they would sell and the food-stuffs they would have to buy and by the market imperfections, which would be considerable at this early stage. Thus, where the margin of resources is very small above that required for the minimum subsistence output, we should expect the spread of export production to be inhibited or very slow, even if there were some genuine possibilities of gains on the comparative costs principle.[1]

In contrast, the transition from subsistence agriculture to export production is made much easier when we assume that our peasants start with some surplus resources which enable them to produce the export crop *in addition* to their subsistence production. Here the surplus resources perform two functions: first they enable the peasants to hedge their position completely and secure their subsistence minimum before entering into the risks of trading; and secondly, they enable them to look upon the imported goods they obtain from trade in the nature of a clear net gain obtainable merely for the effort of the extra labour in growing the export crop. Both of these considerations are important in giving the peasants just that extra push to facilitate their first plunge into the money economy.

Starting from this first group of peasants, we may picture the growth of export production and the money economy taking place in two ways. Firstly, the money economy may grow extensively, with improvements in transport and communications and law and order, bringing in more and more groups of peasants with their complements of family labour into export production on the same " part-time " basis as the first group of peasants. Secondly, the money economy may grow intensively by turning the first group of peasants from " part-time " into " whole-time " producers of the export crop.[2]    In the first case, surplus resources are necessary as a lubricant to push more peasants into export production at each round of the widening circle of the money economy.    Even in the second case, surplus resources are necessary if the whole-time export producers buy their food requirements locally from other peasants, who must then have surplus resources to produce the food crops above their own requirements.    Logically, there is no reason why the first group of peasants who are now whole-time producers of the

[1] Of course, this argument can be countered by assuming the differences in comparative costs to be very wide.  But, so long as export production requires withdrawing some resources from subsistence production, some risks are unavoidable.  Further, remembering that the middlemen also require high profit margins at this stage, the gains large enough to overcome the obstacles are likely to arise out of surplus resources rather than from the differential advantages of the given fully employed resources.  The risk of crop-failure is, of course, present both in subsistence and export production.

[2] In either case the expansion process may be looked upon as proceeding under conditions approximating to constant techniques and fixed combinations between land and labour once equilibrium is reached.  The distinctive feature of peasant export economies is their failure to develop new and larger-scale or extensive methods of farming.  It is true that in subsistence agriculture " fixed factors," such as a plough and a pair of bullocks, were frequently used below capacity, and one important effect of cash production was to increase the size of the holding to the full capacity of these " fixed factors."  But this may be properly looked upon as equilibrium adjustments to make full use of surplus capacity rather than as the adoption of new and more land-using methods of production.  Increasing the size of holding to make a more effective use of a pair of bullocks is different from the introduction of a tractor!  Our assumption of constant techniques does not preclude the development of large-scale ownership of land as distinct from large-scale farming.

export crop should buy their food requirements locally instead of importing them. But, as it happens, few peasant export economies have specialised in export production to such an extent as to import their basic food requirements.

The average economist's reaction to our picture of discrete blocks of surplus productive capacity being drawn into a widening circle of money economy and international trade is to say that while this " crude " analysis may be good enough for the transition phase, the conventional analysis in terms of differential advantages and continuous marginal productivity curves must come into its own once the transition phase is over. Here it is necessary to distinguish between the expansion phase and the transition phase. It is true that in most peasant export economies the expansion process is tapering off or has come to a stop, as most of the surplus land suitable for the export crop has been brought under cultivation. This, of course, brings back the problem of allocating a fixed amount of resources, as we shall see in the next section when we consider issues of economic policy. But even so, the surplus-productive-capacity approach is not entirely superseded so long as the transition from a subsistence to a fully developed money economy remains incomplete. In most underdeveloped countries of Asia and Africa [1] this transition seems not likely to be over until they cease to be underdeveloped.

The continuing relevance of the surplus-productive-capacity approach may be most clearly seen in the typical case of a peasant export economy which with its natural resources and methods of production has reached the limit of expansion in production while its population continues to grow rapidly. According to the surplus-productive-capacity approach, we should expect the export capacity of such a country to fall roughly in proportion as the domestic requirement of resources to feed a larger population increases. This common-sense result may, however, be contrasted with that obtainable from the conventional theory as formulated by Ohlin. First, it appears that the Ohlin theory puts to the forefront of the picture the *type* of export, *i.e.*, whether it is more labour-using or land-using as distinct from the total export capacity measured by the ratio of total exports to the total national output of the trading country. Secondly, in the Ohlin theory there is no reason why a thickly populated country should not also possess a high ratio of (labour-intensive) exports to its total output.

The ideal pattern of trade suggested by the Ohlin theory has a real counterpart in the thickly populated advanced countries of Europe, which for that very reason are obliged to build up a large export trade in manufactures or even in agriculture as in the case of Holland. But when we turn to the thickly populated underdeveloped countries, however, the ideal

[1] Cf. the United Nations Report cited above on the *Enlargement of the Exchange Economy.* Even in the most developed peasant export economies the money economy has not spread to the same extent in the market for factors of production as in the market for products.

and the actual patterns of international trade diverge widely from each other.   Indeed, we may say that these countries remain underdeveloped precisely because they have not succeeded in building up a labour-intensive export trade to cope with their growing population.   The ratio of their export to total production could, of course, be maintained at the same level and the pressure of population met in some other way.   But given the existing conditions, even this neutral pattern may not be possible in many underdeveloped countries.   Thus, in Indonesia there is some evidence to suggest that the volume of agricultural exports from the thickly populated Java and Madura is declining absolutely and also relatively to those of the Outer Islands, which are still sparsely populated.[1]   Of course, there are other causes of this decline, but population pressure reducing the surplus productive capacity of Java seems to be a fundamental economic factor; and the decline spreads from peasant to plantation exports as more of the plantation lands, which were under sugar and rubber, are encroached upon by the peasants for subsistence production.[2]   In general, given the social and economic conditions prevailing in many underdeveloped countries, it seems fair to conclude that the trend in their export trade is likely to be nearer to that suggested by the surplus-productive-capacity approach than to that suggested by the theory of comparative costs.[3]

## IV

This paper is mainly concerned with interpretation and analysis, but we may round off our argument by touching briefly upon some of its policy implications.

(i) We have seen that the effect of population pressure on many under-developed countries, given their existing social and economic organisation, is likely to reduce their export capacity by diverting natural resources from export to subsistence production.   If we assume that these natural resources have a genuine differential advantage in export production, then population pressure inflicts a double loss: first, through simple diminishing returns, and secondly, by diverting resources from more to less productive use.

---

[1] Cf. J. H. Boeke, *Ontwikkelingsgang en toekomst van bevolkings-en ondernemingslandbouw in Neder-landsch-Indie* (Leiden, 1948), p. 91.   I owe this reference to an unpublished thesis by Mr. M. Kidron.

[2] The same tendency to transfer land from plantation to subsistence agriculture may be observed in Fiji with the growing population pressure created by the Indian immigrant labour originally introduced to work in the sugar plantations.   The outline is blurred here by the decline in the sugar industry.   The reason why this tendency does not seem to operate in the West Indies is complex. But it may be partly attributable to the tourist industry, which helps to pay for the food imports of some of the islands.

[3] The surplus-productive-capacity approach also partly helps to explain why underdeveloped countries, such as India, which started off with a thick population tend to retain large and persistent pockets of subsistence sectors in spite of their longer contacts with the world economy, while the subsistence sectors in thinly populated countries, such as those in West Africa, tend to disappear at a faster rate in spite of their much later start in international trade.

Thus, if Java has a genuine differential advantage in growing rubber and sugar, she would obtain a greater amount of rice by maintaining her plantation estates instead of allowing them to be encroached upon by peasants for subsistence rice cultivation. The orthodox liberal economists, confronted with this situation, would, of course, strongly urge the removal of artificial obstacles to a more systematic development of the money economy and the price system. Now there are still many underdeveloped countries which are suffering acutely from the economic rigidities arising out of their traditional social structure and/or from discriminatory policies based on differences in race, religion and class. Here the removal of barriers, for instance, to the horizontal and vertical mobility of labour, freedom to own land and to enter any occupation, etc., may well prove to be a great liberating force.[1] But our analysis has suggested that it is much easier to promote the growth of the money economy in the early stage when a country is newly opened up to international trade and still has plenty of surplus land and labour rather than at a later stage, when there are no more surplus resources, particularly land, to feed the growth of the money economy. Thus in a country like Java there is a considerable amount of artificial restriction, customary or newly introduced, which the liberal economists can criticise, e.g., restriction on land ownership. But given the combination of population pressure, large pockets of subsistence economy and traditional methods of production which can no longer be made more labour-intensive, it seems very doubtful whether the mere removal of artificial restrictions can do much by itself without a more vigorous policy of state interference. The truth of the matter is that in the underdeveloped countries where, for various reasons described above, the exchange economy is still an extremely crude and imperfect apparatus which can make only rough-and-ready responses to economic differentials, it may require a considerable amount of state interference to move toward the comparative-costs equilibrium. Thus given that Java has genuine differential advantages in the production of rubber and sugar, a more optimal reallocation of her resources may require, for instance, the removal of her surplus population either to the thinly populated Outer Islands or to industries within Java and a vigorous export-drive policy supplemented by bulk purchase and subsidies on the imported rice. Here we come to a fundamental dilemma which is particularly acute for the orthodox liberal economists. On a closer examination it turns out that their free-trade argument, although ostensibly based on the comparative-costs principle, is buttressed by certain broad classical presumptions against protection and state interference:[2] e.g., the difficulty of selecting the right

---

[1] This is why the case for the "liberal" solution is strong in places such as East and Central Africa, where due both to the general backwardness of the indigenous population and the presence of a white settler population, both types of rigidity prevail (cf. *The Royal Commission Report on East Africa*).

[2] Cf. J. Viner, *International Trade and Economic Development*, pp. 41–2. See also Sidgwick, *Principles of Political Economy*, Book III, Chapter V.

industry to protect, the virtual impossibility of withdrawing protection once given, the tendency of controls to spread promiscuously throughout the economic system strangling growth, and so on. These presumptions gain an added strength from the well-known administrative inefficiency and sometimes corruption of the governments of some underdeveloped countries. Thus even if we believe in the " nineteenth-century pattern " of international trade based on natural advantages, how can we be sure that the state is competent enough to select the right commodities for its export-drive policy when it is considered incompetent to select the right industry for protection?

(ii) We have seen that the rapid expansion in the export production of the underdeveloped countries in the nineteenth century cannot be satisfactorily explained without postulating that these countries started off with a considerable amount of surplus productive capacity consisting both of unused natural resources and under-employed labour. This gives us a common-sense argument for free trade which is especially relevant for the underdeveloped countries in the nineteenth century: the surplus productive capacity provided these countries with a virtually " costless " means of acquiring imports which did not require a withdrawal of resources from domestic production but merely a fuller employment for their semi-idle labour. Of course, one may point to the real cost incurred by the indigenous peoples in the form of extra effort and sacrifice of the traditional leisurely life [1] and also to the various social costs not normally considered in the comparative-costs theory, such as being sometimes subject to the pressure of taxation and even compulsory labour and frequently of having to accommodate a considerable inflow of immigrant labour creating difficult social and political problems later on. One may also point to a different type of cost which arises with the wasteful exploitation of natural resources.[2] But for the most part it is still true to say that the indigenous peoples of the underdeveloped countries took to export production on a voluntary basis and enjoyed a clear gain by being able to satisfy their developing wants for the new imported commodities. Thus our special argument for free trade in this particular context still remains largely intact. The orthodox economists, by rigidly insisting on applying the comparative-costs theory to the underdeveloped countries in the nineteenth century, have therefore missed this simpler and more powerful argument.

---

[1] It may be formally possible to subsume the surplus-productive-capacity approach under the opportunity-cost theory, by treating leisure instead of foregone output as the main element of cost. But this would obscure the important fact that the underdeveloped countries have been able to expand their production very rapidly, not merely because the indigenous peoples were willing to sacrifice leisure but also because there were also surplus natural resources to work upon.

[2] The social cost of soil erosion can be very great, but this may be caused not merely by an expansion of export production but also by bad methods of cultivation and population pressure. The problem of adequately compensating the underdeveloped countries for the exploitation of their non-replaceable mineral resources belongs to the problem of the distribution of gains from trade. Here we are merely concerned with establishing that the indigenous peoples do obtain some gains from trade.

(iii) We have seen in Section I that the deep-rooted hostility of the critics towards the " classical theory " and the nineteenth-century pattern of international trade may be partly traced back to the time when Western colonial powers attempted to introduce export-drive policies in the tropical underdeveloped countries; and tried to justify these policies by invoking the " classical theory " of free trade and the Adam Smithian doctrine of international trade as a dynamic force generating a great upward surge in the general level of productivity of the trading countries. To the critics, this appears as a thinly disguised rationalisation of the advanced countries' desire for the markets for their manufactured products and for raw materials. Thus it has become a standard argument with the critics to say that the nineteenth-century process of international trade has introduced a large " export bias " into the economic structure of the underdeveloped countries which has increased their "vulnerability" to international economic fluctuations.

In Section II we have seen that once we leave the ideal world of the comparative costs theory in which the resources not required for the export market can be re-absorbed into domestic production, every country with a substantial export trade may be considered " vulnerable." Thus a country may be said to be vulnerable because it has built up a large ratio of export to its total production simply by making use of its pre-existing surplus productive capacity. A fortiori, it is vulnerable when it has genuinely improved upon its original surplus productive capacity. How does the idea of " export bias " fit into our picture?

The term " export bias " presumably means that the resources of the underdeveloped countries which could have been used for domestic production have been effectively diverted into export production by deliberate policy. The implication of our surplus-productive-capacity approach is to discount this notion of " export bias." In the peasant export sectors, at the early stage with sparse populations and plenty of surplus land, the real choice was not so much between using the resources for export production or for domestic production as between giving employment to the surplus resources in export production or leaving them idle. In the later stage, when the population pressure begins to increase as in the case of Java, we have seen that the bias is likely to develop against, rather than in favour of, the export sector. Even when we turn to the mining and plantation sectors, it is difficult to establish a significant " export bias " in the strict sense. Here the crucial question is: how far would it have been possible to divert the foreign capital and technical resources which have gone into these sectors into the domestic sector? The answer is clear. For a variety of reasons, notably the smallness of domestic markets, few governments of the under-developed countries, whether colonial or independent, have so far succeeded in attracting a significant amount of foreign investment away from the extractive export industries to the domestic industries. In criticising the colonial governments it should be remembered that the only choice open to them was whether to attract a greater or a smaller amount of foreign investment within the export sector and not whether to attract investment for the domestic or the export sector.

This is not to deny that the colonial governments had a strong motive for promoting export production. Apart from the interests of the mother country, the individual colonial governments themselves had a vested interest in the expansion of foreign trade because they derived the bulk of their revenues from it.[1] In their search for revenue they have pursued various policies designed to attract foreign investment to the mining and plantation sectors, such as granting favourable concessions and leases, favourable tariff rates for rail transport, taxation policy designed to facilitate the supply of labour, provision of various technical services, etc.[2] But on the whole it is still true to say that the most important contribution of the colonial governments towards the expansion of the colonial exports is to be found, not in these export-drive policies, but in their basic services, such as the establishment of law and order and the introduction of modern transport, which enabled the pre-existing surplus productive capacity of the colonies to be tapped by the world market demand. If we wish to criticise the export-drive policies of the colonial governments it would be more appropriate to do so, not on the ground of " export bias " but on the ground that they may have diverted too great a share of the gains from international trade and of the public services of the colonies to the foreign-owned mines and plantations at the expense of indigenous labour and peasant export producers.

It may be argued that we have given too strict an interpretation of the " export-bias " doctrine which is merely meant to convey the general proposition that, whatever the exact cause, the nineteenth-century process of international trade has landed many underdeveloped countries with a large ratio of raw materials exports to their total national products, making it desirable to reduce their " vulnerability " to international economic fluctuations. But the trouble is that the " export bias " doctrine tends to suggest that the raw-materials export production of the underdeveloped countries has been artificially over-expanded, not merely in relation to their domestic sector, but absolutely. Given the strong feelings of economic nationalism and anti-colonialism in the underdeveloped countries, this can be a very mischievous doctrine strengthening the widespread belief that to go on producing raw materials for the export market is tantamount to preserving the " colonial " pattern of trade. Thus already many underdeveloped countries are giving too little encouragement to their peasant

---

[1] This is true for the governments of most underdeveloped countries, whether colonial or independent, past or present.

[2] For a discussion of the question of the possible export bias through the operation of the 100% sterling exchange system of the colonies, see A. D. Hazlewood, " Economics of Colonial Monetary Arrangements," *Social and Economic Studies*, Jamaica, December 1954.

export sectors by diverting too much of their capital and technical resources
to industrial-development projects, and are also crippling their mining and
plantation export sectors by actual or threatened nationalisation and
various restrictions and regulations.  The effect is to reduce their foreign-
exchange earnings so urgently needed for their economic development.
Of course, no competent critic of the nineteenth-century pattern of international trade would ever suggest the drastic step of reducing exports
absolutely; some would even concede the need for vigorous export drive
policies.[1]  But having built up a pervasive feeling of hostility and suspicion
against the " nineteenth-century " or the " colonial " pattern of international
trade, they are not in a position to ram home the obvious truths:  (a) that,
even on an optimistic estimate of the possibilities of international aid, the
underdeveloped countries will have to pay for the larger part of the cost of
their economic plans aiming either at a greater national self-sufficiency or
at the export of manufactured goods;  (b) that the necessary foreign exchange
for these development plans can be earned by the underdeveloped countries
at the present moment only by the export of raw materials (though not
necessarily the same commodities for which they were supposed to have a
differential advantage in the nineteenth century); and (c) that therefore to
pursue their development plans successfully it is vitally important for them
to carry out the " export-drive " policies, which in their technical properties
may not be very different from those of the colonial governments in the
past.[2]  In trying to carry out their development plans on the foreign-
exchange earnings from raw-materials export they would, of course, still
be " vulnerable "; but this should be considered separately as a problem
in short-term economic stability [3] and not as a criticism of the nineteenth-
century pattern of international trade in relation to the long-term development of the underdeveloped countries.  From a long-term point of view,
even countries which have successfully industrialised themselves and are
therefore able to maintain their population at a higher standard of living
by building up a large export trade in manufactures, such as Japan or the

---

[1] Cf., for example, Gunnar Myrdal, *An International Economy*, p. 274.

[2] Colonial governments have frequently defended their export-drive policies as the means of
taxing foreign trade to finance services needed for internal development.  But because they were
colonial governments, their motives were suspect.  At first sight we might imagine that the new
independent governments of the underdeveloped countries would be free from this disability.
But unfortunately, given the atmosphere of intense nationalism and anti-colonialism, this is not
true.  In some cases the hands of the newly independent governments seem to be tied even more
tightly, and economic policies admitted to be desirable are turned down as " politically impossible."
Here those economists who regard themselves as the critics of the classical theory and the nineteenth-
century pattern of international trade have a special responsibility.  Instead of dealing tenderly
with the " understandable " emotional reactions which they have partly helped to create, they
ought to be emphatic in pointing out the conflicts between rational considerations and " understandable " mental attitudes.  The underdeveloped countries are too poor to enjoy the luxury of
harbouring their emotional resentments.

[3] Cf. the United Nations Report on *Measures for International Economic Stability* and Myrdal's
comments on it, *op. cit.*, pp. 238–53.

thickly populated countries of Western Europe, will continue to be " vulnerable." [1]

H. MYINT

*Oxford.*

[1] It is particularly in relation to the thickly populated advanced countries of Western Europe
which have specialised and adapted their economic structure to the requirements of the export
market that Professor J. H. Williams found Adam Smith's " vent-for-surplus " approach illuminating. We have, in this paper, interpreted the " surplus " more strictly in its pre-existing form without the improvements and augmentation in productive capacity due to genuine " specialisation."
(Cf. J. H. Williams, " International Trade Theory and Policy—Some Current Issues," *American
Economic Review, Papers and Proceedings,* 1951, pp. 426–7.)

# The Latin American periphery in the global system of capitalism*

## Raúl Prebisch**

In a series of articles, appearing above all in this *Review*, the author has gradually been giving form to his mature view of the economic, social and political structure and transformations of Latin America. In this process of further perfecting his ideas by giving them greater depth and coherence, the present article represents a major step, being a concise summary of the main lines of thought which he is developing in three closely interrelated spheres.

To begin with, he returns to his long-standing concern for the relationship between the centres and the periphery, which he analyses in the light of a number of salient features of the contemporary scene. In his opinion, the topic is of the utmost importance, in that the nature of those relations conditions, limits and orients the Latin American countries' forms and possibilities of development. Secondly, he broaches the question of the internal dynamics of peripheral capitalism in order to throw light upon its main components, contradictions and trends. Thus, he asserts that peripheral capitalism is driven by its internal contradictions towards structural crises which it can overcome only by turning to authoritarian political régimes. This thesis has a corollary which is the starting point for his third line of thought: a stable and democratic solution to those structural crises calls for a profound change in the bases of peripheral capitalism, and particularly of its predominant forms of appropriation and use of the surplus. As a contribution to thinking on this controversial topic, he outlines his theory of change, guided by the hope of finding a synthesis of liberal and socialist ideals.

*This article was especially prepared for the Seminar on Latin American Development Policies held between September 1980 and May 1981 by the Development Training Centre (CECADE) of the Ministry of Planning and the Budget of the Mexican Government.
**Director of the *CEPAL Review*.

# I

## The dynamics of the centres

Peripheral development is an integral part of the world system of capitalism, but the conditions in which it takes place are different from those in the centres, whence the specificity of peripheral capitalism.

Technology plays a fundamental role in this: its development in the centres is accompanied by continuous changes in their social structure, and this is also true of the peripheral countries when the same technology penetrates them much later. The relations between the two correspondingly alter.

In the course of these continuous changes, some highly important constants are to be found. We shall mention the main ones.

While exerting considerable influence on peripheral development, the dynamics of the centres is limited in scope, on account of the centripetal nature of capitalism. Thus it fosters peripheral development only to the extent that concerns the interests of the dominant groups in the centres.

The centripetal nature of capitalism is constantly manifested in the relations between the centres and the periphery. It is in the former that technical progress originates and that the benefits of the concomitant rise in productivity tend to be concentrated. Thanks to the higher demand which accompanies the rise in productivity, industrialization is likewise concentrated there, spurred on by ceaseless technological innovation which diversifies the production of goods and services to an ever greater extent.

Thus, in the spontaneous course of development the periphery tends to be left on the margin of this industrialization process in the historical evolution of capitalism.

Rather than deliberate, this exclusion is the consequence of the play of market laws at the international level.

At a later stage, when becoming industrialized as a result of international crises the periphery again tends to be shut off from the major trade flows in manufactures of the centres. The periphery has had to learn to export, and it is doing so primarily through its

own efforts, as the transnationals have contributed far more to the internationalization of
forms of consumption than to the internationalization of production through trade with the
centres.

This largely explains the inherent tendency towards external disequilibrium in past and
present peripheral development: an attempt
has been made to correct this tendency first
through import substitution and subsequently
through the export of manufactures.

The centres have by no means encouraged
this process through changes in their production structure; and by failing to open their doors
to manufacturing imports from the periphery,
they force the latter to continue with import
substitution. Substitution is not the result of
any doctrinaire preference, but rather something imposed by the centripetal nature of capitalism. However, it has been taking place within narrow national compartments, at the expense of economic efficiency and of vigorous
development.

The economic interest of the dominant
groups of the centres form a cluster with strategic, ideological and political interests in the
centres, giving rise to stubborn forms of dependence in centre-periphery relations.

In those relations, the economic interests
of the dominant groups of the centres are articulated with those of the peripheral countries, and in the play of these power relations
the technical and economic superiority of the
former weighs heavily. The structural changes
which accompany the development and spread
of technology are highly important. In the
periphery, besides their significance for its
development these changes eventually give
rise to disruptive pressures when the internal
conflictive tendencies characteristic of development spill over towards the centres, where
they arouse an adverse reaction from the power
cluster. This is a clear manifestation of the
above-mentioned dependence.

The economic interest of the dominant
groups continues to prevail in the centres as in
the periphery. Its efficiency in the market, at
the national and international level, cannot be
denied. But the market, despite its enormous
economic and political importance, neither is
nor can be the supreme regulator of the development of the periphery and of its relations
with the centres.

This is patently clear in the present crisis
of those relations. The market has not been able
to cope with the ambivalence of technology,
which has had an incalculable effect on material wellbeing, but has also brought irresponsi-
ble exploitation of non-renewable natural resources and a striking deterioration of the
biosphere, not to mention other serious consequences.

Nor have the laws of the market remedied
the major flaws in centre-periphery relations,
nor still less the exclusive and conflictive tendencies in peripheral development.

Individual decisions in the market-place
must be combined with collective decisions
outside it which override the interest of the
dominant groups. All this, however, calls for a
great vision, a vision of change, both in peripheral development and in relations with the
centres; a vision based on far-reaching projects
combining farsighted economic, social and political considerations.

# II

# The internal dynamics of peripheral capitalism

The dynamics of the centres does not tend to
penetrate deeply the social structure of the periphery; it is essentially limited.

In contrast, the centres propagate and
spread in the periphery their technology, forms
of consumption and lifestyles, institutions,
ideas and ideologies. Peripheral capitalism
increasingly draws its inspiration from the
centres and tends to develop in their image and
likeness.

This imitative development takes place
belatedly in a social structure which differs in

major respects from the developed structures of the centres.

The penetration of technology takes place through capital accumulation, in terms both of physical means and of the training of human beings. As the process develops, changes continuously take place in the social structure, which embraces a series of partial structures linked together by close relations of interdependence; the technical, production and employment structures, the power structure and the distribution structure. These changes must be analysed to throw light on the complex internal dynamics of peripheral capitalism.

## 1. *Structural changes, surplus and accumulation*

The penetration of technology gradually creates successive layers of rising productivity and efficiency which are superimposed upon less productive and efficient technical layers, while at the base of this technological structure precapitalist or semicapitalist layers usually persist. These changes in the technical structure are accompanied by changes in the employment structure, as labour is continuously shifting from the less to the more productive layers. However, the income structure does not develop in line with the changes in technology and occupation. Thus, the mass of the labour force does not increase its earnings correlatively with the growth of productivity in the play of market forces.

This is explained by the regressive competition of the new manpower in the technical layers of low productivity, or else unemployed, which is seeking to enter productive activity. Only a part of the fruits of technical progress are transferred to a limited fraction of the labour force which, above all through its social power, has been able to acquire the ever greater skills required by technology.

The part of the fruits of higher productivity which is not transferred constitutes the surplus, which is appropriated primarily by the upper social strata, where most physical capital as well as land ownership are concentrated.

The surplus does not tend to disappear through a fall in prices resulting from competition among enterprises —even if this were unrestricted— but rather is retained and circulates among them. This is a structural and dynamic phenomenon. The growth of the production of final goods, thanks to the continuous accumulation of capital, means that there must be a preceding growth of production in process which will later give rise to the final goods. For this purpose, enterprises pay higher incomes, giving rise to the greater demand which absorbs the final supply increased by the growth of productivity, without prices falling.

In fact, the incomes thus paid in the successive stages of the process (including the surplus) through the creation of money are much greater than would be necessary to prevent prices from falling. The reason for this is that only part of those incomes immediately becomes demand for final goods. Another part is diverted towards demand for services, in the market and the State spheres, where it circulates and gradually returns to demand for goods. In addition to the incomes paid to factors of production, enterprises purchase imported goods, and thus the exporting countries recover the incomes they paid in producing them as well as the corresponding surplus. The opposite occurs in the case of exports.

There is no strict correspondence between demand for goods and supply, but the necessary adjustments are made spontaneously or through the precautionary corrective intervention of the monetary authority when the capacity for sharing out the surplus has not yet developed.

The unequal distribution of income in favour of the upper strata encourages them to imitate the forms of consumption of the centres, an imitation which tends to spread to the middle strata. The privileged-consumer society which thus develops represents a considerable waste of capital accumulation potential.

This waste concerns not merely the amount but also the composition of capital. Closely linked with the technology which increases productivity and income, use is made of technology which constantly diversifies production of goods and services. As this change occurs in the production structure, together with other forms of investment, the proportion of non-reproductive capital increases without any growth of productivity or multiplication of employment, to the detriment of the reproductive capital necessary for fostering development.

These trends inherent in the internal logic of capitalism in the centres appear prematurely in the periphery on account of the great inequality in distribution.

In addition to all this, again at the expense of accumulation, there is the exorbitant siphoning-off of income by the centres, especially through the transnationals, as a result of their technical and economic superiority and hegemonic power.

This insufficient, stunted accumulation of reproductive capital, aggravated by the trend towards hypertrophy of the State and the extraordinary growth of the population, is the main reason why the system cannot intensively absorb the lower strata of the social structure and cope with other manifestations of redundancy of labour. This is the system's exclusive tendency.

These lower strata abound in agriculture, and as the demand for agricultural goods scarcely becomes diversified, labour tends to shift towards other activities. However, given the system's inadequate capacity to absorb labour, a serious redundancy arises which explains the relative deterioration of labour income in agriculture.

As long as this insufficient capacity to absorb labour lasts, technical progress in agriculture will not raise those incomes and correct their relative decline. Instead, it tends to harm relative prices when production outstrips demand. This is usually true of agricultural exports in particular, and has the effect of checking their growth to the detriment of development.

## 2. Changes in the power structure and crisis of the system

As technology penetrates the social structure, changes take place which are reflected in the power structure. The middle strata expand, and as the process of democratization advances their trade-union and political power develops and increasingly forms a counterweight to the economic power of those, especially in the

upper strata, in whose hands most of the means of production are concentrated. It is therefore in these strata that the labour force possessing social power is mainly found. These power relations between upper and middle strata exist both in the market and in the State spheres. In this way ever-increasing pressure develops for sharing out the fruits of the growth of productivity.

This twofold pressure is largely manifested through a rise in the remuneration of the labour force, either to increase its share in the fruits of productivity or to offset the unfavourable effects of certain factors, above all the tax burden which it bears directly or indirectly and through which the State copes with the trend towards its own hypertrophy.

Bureaucratic power and military power have their own dynamics in the State apparatus, supported by the political power of the middle strata in particular, as a result of which State activities develop beyond considerations of economic efficiency, both as concerns the amount and diversification of State services and in terms of the spurious absorption of labour.

In this way, through the growth of employment and social services the State seeks to correct the system's insufficient absorption of labour and its distributive unfairness; and this is a major factor in its hypertrophy.

To express the foregoing in a nutshell: the distribution of the fruits of the system's rising productivity is fundamentally the result of the changing play of power relations, in addition, of course, to individual differences in ability and dynamism.

As the labour force's sharing capacity increases and it acquires the ability to recoup its tax burden and compensate for the effects of other factors, the rise in remuneration tends to overtake the drop in the costs of enterprises resulting from successive rises in productivity. The excess then tends to be transferred to prices, and this is followed by fresh rises in remuneration in the familiar inflationary spiral.

In these circumstances, for it to be possible to absorb supply, increased by higher costs, it is essential that demand, and the incomes underpinning it, should increase in a correlative manner.

If the monetary authority resists the necessary creation of money in order to avoid or check the spiral, the growth of demand will be insufficient to meet the growth of final production, leading to economic recession which will continue until the authority changes its attitude and prices can rise in line with the higher costs. The rise in prices means that the surplus may once again increase through new rises in productivity, but only temporarily since it is once again compressed by the subsequent rise in remuneration. Thus accumulation declines with adverse consequences for development, besides the disturbances which accompany the heightening of the distributive quarrel.

It should be noted, however, that these phenomena occur when, thanks to the process of democratization, the labour force's trade-union and political power becomes ever greater in both the market and the State spheres, and the latter's expenditure steadily expands through its own dynamics.

In these circumstances, the spiral becomes inherent in peripheral development; and the conventional rules of the monetary game are powerless to avert or suppress it.

These rules are highly valid when distributive power (for sharing out and recouping) is non-existent or very incipient. This is the case when the democratization process is very weak or obstructed or manipulated by the dominant groups: democracy in appearance but not in substance.

Such, then, is the crisis of the system when the arbitrary play of power relations becomes very strong, which is what occurs in the advanced stage of peripheral development. The crisis of the system may be postponed for some time, particularly when plentiful resources are available from the exploitation of non-renewable natural wealth.

The political power of the upper strata, apparently on the wane with the advance of democracy, surges up again when the disturbances brought about by the inflationary crisis give rise to economic disorder and social disintegration. At that point the use of force is introduced, which makes it possible to break the trade-union and political power of the disadvantaged strata.

If the holders of military power are not necessarily under the sway of the economic and political power of the upper strata, one is tempted to ask why they intervene to serve the privileged-consumer society. Here undoubtedly a complex set of factors comes into play. The fundamental explanation, however, is that since the upper strata hold the dynamic key of the system, i.e., the capacity for capital accumulation, they must be left to get on with it from a desire to restore smooth development; but the social cost is tremendous, not to mention the political cost.

What in fact happens is that democratic liberalism breaks down, while the ideas of economic liberalism flourish: a fake liberalism which, far from leading to the dissemination of the benefits of development, flagrantly consolidates social inequity.

Democratic liberalism has not yet managed to become firmly rooted in the Latin American periphery. We are all too familiar with its vicissitudes, its promising advances and painful setbacks. But the past cannot account for everything: new, complex elements spring up as changes occur in the social structure. And the significance of the use of force is not what is was in the past: the creation of that total split between democratic liberalism and economic liberalism, despite the fact that both sprang from the same philosophical source.

## 3. The great paradox of the surplus

The foregoing considerations lead to very important conclusions, perhaps the most important in our interpretation of peripheral capitalism.

The surplus is subject to two contrary movements. On the one hand, it grows through successive increases in productivity. On the other, it shrinks through the pressure for sharing which stems from the market and from the State. The system functions smoothly as long as the surplus grows continuously as a result of those two movements.

Consequently, the upper strata, in whose hands most of the means of production are concentrated, can increase capital accumulation and at the same time their privileged consumption: they possess the dynamic key of the system.

This essential condition is satisfied so long as the sharing out of the surplus, both in the market and the State spheres through the play of power relations, occurs at the expense of successive rises in productivity. The surplus will continue to expand, although at a dwindling rate. However, the sharing out cannot go beyond the threshold at which the surplus would begin to shrink.

At that limit, however, the surplus will have become proportionately greatest in relation to the total product. Why is it impossible to continue improving the sharing, when there would be plenty of room for doing so by reducing the surplus? This is the weak point of the system of distribution and accumulation, because if the pressure for sharing outstrips the increase in productivity, the rise in the cost of goods will cause enterprises to raise prices.

The total surplus would undoubtedly allow much more sharing out at the expense of size, but there is nothing in the system to make this happen. It is conceivable that enterprises might take part of the surplus and transfer it to the labour force without raising costs; this would be direct participation in the surplus. But the system does not work like that. Any rise in remuneration over the increment in productivity raises costs, with the consequences described above.

Not all the pressure for sharing, however, takes the form of higher remuneration. As was pointed out earlier, in order to share out the surplus the State resorts to taxes falling on the labour force, which the latter seeks to recoup through higher remuneration; but the State also has the possibility of directly taxing the surplus or the incomes of the social groups in the upper strata who have no capacity for recouping such taxes. These taxes are not transferred to costs, but if their amount squeezes the surplus the rate of accumulation and of growth is weakened, thus accentuating the exclusive and conflictive tendencies.

Whatever the angle from which it is approached, there is no solution to the problem within the system, so long as the capacity for redistribution is strengthened in the advanced stage of the democratization process. Either the result is the inflationary spiral, if sharing leads to higher production costs —which, in addition to the upheaval caused by the spiral, undermines the dynamics of the surplus— or else some of the surplus is taken directly, again with adverse consequences for its dynamics, which sooner or later must be resolved using inflationary means.

However much thought one devotes to the question, it appears that the rules of the game of peripheral capitalism do not allow for an attack on its two major flaws: its exclusive tendency, which may only be remedied by a more intense accumulation of capital at the expense of the privileged strata and of the income transferred to the centres; and its conflictive tendency, unrelentingly heightened in the unrestricted play of power relations.

There is a great paradox in all this. When the surplus grows so far as to reach its ceiling and the pressure for sharing continues, the system reacts by seeking to achieve continued growth of the surplus. In order to attain this objective, it resorts to the use of force. However, the use of force is not a solution; the only solution is to change the system.

### 4. Crisis of the system and the use of force

Given the nature of the system, at the advanced stage of peripheral development and of the democratization process it is impossible to avert the tendency towards crisis. In the system's internal logic there is no lasting way of ensuring that the pressure for sharing does not jeopardize the dynamic role of the surplus and lead inevitably to the inflationary spiral.

The attempt to restore the dynamics of the system through the use of force entails the risk of serious disruption, usually involving a combination of theoretical inconsistency and practical incongruity.

If the system is handled skillfully, however, particularly in favourable external conditions, high rates of accumulation and of development may be achieved with striking prosperity for the privileged social strata, but at the cost of severe compression of the income of a considerable part of the labour force.

This solution, however, by no means strikes at the roots of the system's exclusive and conflictive nature. When the democratization process is resumed sooner or later, the pressure for sharing will tend to lead the system into a new political cycle, aggravated by the deformation which has taken place in the production structure to satisfy the exaltation of the privileged-consumer society.

# III

# Towards a theory of change

## 1. *The two options for a change*

The system of accumulation and distribution of the benefits of technical progress is not subject to any regulating principle from the standpoint of the collective interest. If appropriation is arbitrary when market laws prevail, so is redistribution when political and trade-union power becomes a counterweight to those laws.

It is therefore essential for the State to regulate the social use of the surplus, in order to step up the rate of accumulation and progressively correct distributive disparities of a structural nature, which are quite distinct from functional disparities.

At bottom, there are only two ways in which the State can undertake this regulatory activity: by taking into its own hands the ownership and management of the means of production which give rise to the surplus; or by using the surplus in a spirit of collective rationality without concentrating ownership in its own hands.

The political and economic significance of these two options is essentially different. I lean towards the second on account of two fundamental considerations. In the first place, because the major flaws of the system do not lie in private property itself but rather in the private appropriation of the surplus and the harmful consequences of the concentration of the means of production. Secondly, because the first option is incompatible with the paramount concept of democracy and the human rights inherent in it, while in the second that concept becomes fully compatible, both in theory and in practice, with vigorous development and distributive equity.

## 2. *The dissemination of capital and self-management*

The transformation of the system necessarily calls for raising the rate of accumulation of reproductive capital, particularly at the expense of the consumption of the upper strata. The social use of the surplus enables this to be done by disseminating ownership of capital among the labour force thanks to the surplus of the large enterprises in whose hands most of the means of production are concentrated.

In the remaining enterprises, greater accumulation would be undertaken by the owners themselves, but as they rose in the capital scale an increasingly proportion would have to go to the labour force in order to avoid concentration.

The change in the social composition of capital thus occurring in the large enterprises would have to be accompanied by gradual participation in capital until reaching self-management. Some principles of this type of management could also be followed in State enterprises, in special conditions which justified doing so.

These guidelines refer to countries which have attained advanced stages in their development; at less advanced stages, the social use of the surplus could take different forms. In any event, in either case it would be necessary to establish suitable incentives so that the transformations could take place without major upheavals.

This latter concern could lead to intermediate solutions, one of which might be to encourage greater accumulation, even in the large enterprises, in the same hands as at present, together with measures for the redistribution of some of the surplus.

### 3. The market and planning

In the new system all enterprises, whatever their nature, could develop freely in the market, in conformity with some basic, impersonal conditions established by the regulatory action of the State concerning both the social use of the surplus and other responsibilities pertaining to the State.

This regulatory activity has to fulfil objectives which the market itself cannot attain, but which would enable it to achieve great economic, social and ecological efficiency.

The criteria guiding the State's regulatory activity should be established through democratic planning. Planning means collective rationality, and that rationality requires that the surplus should be devoted to accumulation and redistribution, as well as to State expenditure and investment. Accumulation and redistribution are closely linked, since productivity and income should gradually rise as the labour force in the lower strata, as well as the labour employed spuriously by the system, are absorbed more and more productively. This is a dynamic redistribution, accompanied by other direct forms of social advancement responding to pressing needs.

Planning involves technical work of the utmost importance, which cannot be undertaken without a high degree of functional independence; it is, however, a technical and not a technocratic task, as it must be subordinated to democratically-adopted political decisions.

All this requires constitutional changes in the State machinery and new rules of the game ensuring both stability in the social use of the surplus and flexibility in responding to major changes in prevailing circumstances.

### 4. Synthesis of socialism and liberalism and power structure

The option for change outlined here represents a synthesis of socialism and liberalism. Socialism in that the State democratically regulates accumulation and distribution; liberalism in that it enshrines the essence of economic freedom, closely linked to political freedom in its original philosophical version.

This option calls for very important changes in the structure of political power, as does the option of concentrating ownership and regulatory activity in the hands of the State. In the course of the alterations of the social structure, the power of the upper strata is counterbalanced by the redistributive power of the middle and, possibly, lower strata. The latter, however, eventually shatters itself against the former in the dynamics of the system. Nevertheless, the crisis of the system opens the way for changing it, as it opens the possibility of reducing the power of the upper strata.

These changes in the power structure would perforce be confined to the periphery, as the power relations between the periphery and the centres, under the hegemony of the latter, especially the leading dynamic centre of capitalism, could not be radically changed by the action of the periphery alone. The power of the centres is considerable, and furthermore it lacks a sense of foresight, as is evidenced by its serious disruptions of the biosphere. This crisis may perhaps have the virtue —as has often been true of major crises in the past— of making the centres aware of the need for great foresight in their relations with the periphery and for containing their own power. I am inclined to think that if the main dynamic centre of capitalism had had this awareness, the breakdown of the international monetary system might have been avoided.

The myth of the worldwide expansion of capitalism has been exploded, as has that of the development of the periphery in the image and likeness of the centres. The myth of the regulatory virtue of market laws is also being dispelled.

Major changes are needed; but it is necessary to know why, how and for whom the changes are made. A theory of change is also needed; these pages, called forth by the pressing need for debate and enlightenment, seek to contribute to the formulation of such a theory.

# EFFECTS OF THE WORLD ECONOMIC CRISIS ON THE EAST EUROPEAN ECONOMIES[1]

RICHARD PORTES

The economies of the centrally planned countries of Eastern Europe played little part in the sequence of events leading to the world economic crisis of the 1970s.[2] Even so, once the crisis became generalized in powerful disturbances of world commodity and capital markets, Eastern Europe could not remain unaffected. Conscious economic planning and centralized allocation of resources *in natura* may replace or supplement the market within a country. Planning can extend further, to groups of countries, as it does to some extent among the member countries of the Council for Mutual Economic Assistance[3] (CMEA or Comecon), the "economic community" of the centrally planned economies. But national and even supranational boundaries are not natural economic boundaries; and central planners cannot, and do not, ignore the gains to be derived from trade with the market-oriented economies.

This chapter aims to provide the theoretical background necessary to understand and interpret the ways in which the world economic crisis was transmitted to the centrally planned economies of Eastern Europe and how they have responded. It specifies and quantifies the impacts of the crisis on the seven East European members of Comecon, (hereafter, referred to as "the Seven"), discusses the consequences for their economies, their policy responses and longer-run implications, and concludes with an overall assessment of the effects of the crisis on the Seven.

## EXTERNAL DISTURBANCES AND MACROECONOMIC ADJUSTMENT IN CENTRALLY PLANNED ECONOMIES

This section analyzes the interconnections between the foreign sector and the domestic macroeconomy of a centrally planned economy within a theoretical

[1]All tables and certain passages were omitted with the approval of the author.

[2]I am grateful for support and hospitality from the Maison des Sciences de l'Homme and the Ecole des Hautes Etudes en Sciences Sociales; and assistance from the French Programme of the Social Science Research Council (SSRC). This chapter arises out of work done under projects on "Macro-economic Adjustment and Foreign Trade in Eastern Europe, supported by a grant from the SSRC, and on "Western Private Financing and East European Adjustment Policies," supported by the Leverhulme Trust and based at the Atlantic Institute. John Burkett, Simon Price, Stan Rudcenko, David Winter, and Jenny Woods have been very helpful indeed at various stages of this work. An early version of the chapter was published in *Revue Economique*, Paris, and I am grateful for permission to use this material as well as for very thoughtful comments on it from Roger Bennett, Ralph Bryant, Jerry Cohen, Ed Hewett, Jacques de Miramon, and Tom Wolf. Peter Oppenheimer's editing was especially detailed and helpful. I am, however, responsible for any errors of fact, analysis, or opinions here.

[3]Founded in 1949, the CMEA, or Comecon, now consists of Bulgaria, Cuba, Czechoslovakia, East Germany, Hungary, Mongolia, Poland, Romania, Vietnam, and the Soviet Union. Albania has played no part since 1963, but apparently remains formally a member.

Reprinted (with author approved deletions) with permission from *The World Economy* 3(1), 13–52, Copyright 1980.

framework tied as closely as possible to a widely accepted theory for market economies. It allows us to specify the precise respects in which the centrally planned economies differ.[4]

All the standard national income identities apply also to the centrally planned economies if the definitions are interpreted appropriately. Hence it is possible to present the links between trade flows and internal macroeconomic variables in terms of the relation between domestic production (output) and internal demand (utilization of resources, expenditures, "absorption"). Let us write:

$Y$ = output            $B = X - M$ = balance of trade
$A$ = domestic utilization   $C$ = consumption
$X$ = exports           $I$ = investment
$M$ = imports           $G$ = government current expenditure

The resources available for domestic use are given by output plus imports less the amount exported; while domestic use goes to one of the mutually exclusive and exhaustive categories of private (marketed) consumption, investment in fixed or working-capital assets (whether private or state owned) and government (state) expenditure on current period objectives (defense, collective or nonmarketed consumption, etc.). Thus, we have:

$$A = Y + M - X$$
$$A = C + I + G,$$

hence,

$$Y - A = X - M = B$$
$$X - M = Y - (C + I + G)$$

Any excess of output over domestic use is available for net exports; and any excess of domestic use over production must be supplied by net imports. *Ex post*, this is an identity.[5] The two sides present quite different economic problems when seen as a need for equilibrium, whether for market decision makers or for central planners. The excess (possibly negative) of desired or planned output over desired or planned domestic use may be greater than what the economy is able to transfer abroad in net exports—for various possible reasons. The equation does not distinguish between tradable and nontradable goods and the "surplus" output may be neither exportables nor import substitutes. The foreign demand conditions for ex-

[4]For further theoretical background, see F. Holzman, "Soviet Central Planning and its Impact on Foreign Trade and Adjustment Mechanisms," in A. Brown and E. Neuberger (eds), *International Trade and Central Planning* (Berkeley: University of California Press, 1968) pp. 280–305; Peter Wiles, *Communist International Economics* (Oxford: Blackwell, 1968); Peter Kenen and L. Tyson, "The International Transmission of Economic Disturbances: a Theoretical Framework," in E. Neuberger and L. Tyson (eds.), *The Transmission of External Disturbances to Eastern Europe and the Soviet Union* (New York: Pergamon Press, 1981); and Portes, "Internal and External Balance in a Centrally-planned Economy," *Journal of Comparative Economics*, New York, No. 3, 1979, pp. 325–45.

[5]As a statement about aggregates, however, it ignores the effects of any differences between the domestic and foreign relative price structures, which are especially important in centrally planned economies; see T. Wolf, "On the Adjustment of Centrally-planned Economies to External Economic Disturbances," a paper presented at a conference on East European Integration and East–West Trade, Bloomington, Indiana, 1976.

portables may not allow exports to be raised sufficiently, or at acceptable terms of trade, or at prices that do not provoke foreign retaliation against "unfair competition." Complementarity in production or use between domestic output and imports may be so strong that cutting imports would strongly affect $Y$ and $A$, which therefore could not stay at their planned levels. In any of these circumstances, although the identity is still satisfied, we have[6] $(Y - A)^P > X - M$, so there is excess supply domestically. The normal consequences are an accumulation of undesired inventories and a revision of plans, leading to adjustments of all variables.

Perhaps less common in market economies, but frequent enough in centrally planned economies, is the imbalance $(Y - A)^P < X - M$, or $(A - Y)^P > M - X$. Here, planned domestic use cannot be met by planned output plus net imports, so there is excess demand domestically. It may be impossible to obtain (or finance) desired imports, at least in the short run, or exports may be unsuitable for diversion to domestic use, or impossible to withdraw from commitments to trading partners. Again, the identity will still be satisfied: for example, through informal rationing of consumer goods by queues or through delays to investment projects because machinery is not delivered (we disregard here and in the preceding the possibility that domestic prices might change). But plans will not be satisfied and will therefore eventually change. In a centrally planned economy, adjustment is likely to be slower and the short-run response mechanism less likely to involve reduction of net exports than in a market economy.

Whatever may be the *ex ante* relation between them, the two "gaps" $(Y - A)$ and $(X - M = B)$ have different economic content. Moreover, policymakers are concerned and markets influenced by the *levels* of the different variables at which the equation is satisfied. In particular, even if planned (desired), any "gap" $B = O$ must be *financed* unless and until there is *adjustment* to eliminate it. For the centrally planned economy the question of finance arises only on the international side. There is no domestic financial problem (such as financing the government deficit) created by an excess of expenditure over output: the banking system creates credit and emits currency in the required amounts. But the planners must provide for the borrowing or lending abroad that corresponds to $B \neq O$ and, insofar as they find this capital flow undesirable or infeasible, theymust adjust.

As in market-oriented economies, adjustment will require some combination of policies affecting both gaps, and, indeed, any single policy will typically affect both. In market-oriented economies, however, policies are usually directed toward domestic and foreign *demand* and their components. Thus we have "expenditure-reducing" and "expenditure-switching" policies for cutting balance-of-trade deficits, although we know that a measure which reduces $A$, and thus directly reduces expenditure on imports and exportables, will also start a process that somewhat reduces $Y$ (supply) as well; and a policy switching expenditure from foreign toward domestically produced goods will normally induce some increase in domestic output—and consequently in $A$. In a centrally planned economy, the planners are

[6]The superscript$^P$ denotes planned or desired values.

more capable of acting directly on the supply sides of both gaps *(Y* and *X)* and the system is better able to contain the indirect (and often undesired) effects of any policy.

Even supposing that the central planners in any centrally planned economy form a single decision-making authority with no internal conflicts, they still seek to attain separate, sometimes conflicting, policy objectives. We may call these targets "external balance" and "internal balance." The former is familiar: not necessarily $B = O$ continuously, but maintaining $B$ on a longrun path of optimal foreign investment or borrowing, while in the short run avoiding liquidity crises and, throughout, protecting the country from any undesired political consequences of being a debtor or creditor. Note that the centrally planned economies have never shown any desire to hold foreign-exchange reserves or run trade surpluses for their own sake.[7] Neomercantilism is less attractive when exporting is not perceived as a way to maintain full employment (see the following).

In a developed market-oriented economy, conventionally we take "internal balance" to mean full employment with (approximately) stable prices. In the centrally planned economies, both these targets have exceptionally high priority. Virtually all consumer prices are fixed centrally and held constant for long periods, while full employment is ensured by "over-full employment planning" ("planner's tension" or, in other words, excess real demand in the productive sector, enforced by over-ambitious plans) and restrictions against discharging workers. Thus, for our purposes, the full-employment target must be seen as the planner's objective of sustained rapid growth of output.

It is important, however, to disaggregate further. The only markets in these economies are for consumer goods, where households are free to spend their incomes on the supplies offered by the planners at the prices they fix; and for labor, where households are free to offer their services for the jobs and associated wage rates provided by the planners. Investment goods, nonmarketed consumption, and exports are allocated directly in physical terms by the planners. Hence, internal balance may be taken to include (1) avoiding excess demand or supply on the consumer goods market by keeping incomes and supplies balanced at the given prices, (2) avoiding excess demand for labor, which can generate excessive wage payments and thus threaten the equilibrium of the consumer goods market, and (3) avoiding overinvestment, caused by pressures for expansion both at the central and at the enterprise levels. A final, but essential, characteristic of internal balance in these economies is a politically acceptable distribution of income, for which the planners have always consciously accepted responsibility.

One consequence of these goals is the absence of the "demand multiplier" effects of Keynesian macroeconomics. An increase in demand cannot elicit an increase in employment and aggregate supply, since resources are already fully employed, which means it cannot generate further incomes and expenditure. In

[7]In fact, we find empirically for four centrally planned economies (mid-1950s to 1975) that the higher is $B_{t-1}$, *ceteris paribus,* the lower is $X_t$ and the higher is $M_t$. See Portes, Winter, and Burkett, *op. cit.*

particular, exports have no multiplier effects and no role as a component of the "autonomous" final demand expenditure which, in a market-oriented economy, creates employment and growth. On the whole, exports are a "necessary evil" that permit the imports essential to production (although exporting may also serve to dispose of surplus output resulting from planners' errors). Consumer-goods imports are kept relatively low and imports are dominated by intermediate goods. Thus, we have the "bottle-neck multiplier," referring to the possibility that a fall in exports might force cuts in imports, hence in production, hence in the supply of goods for export, etc.[8] Note finally that, in these circumstances, both the aggregate demand for imports and supply of exports will be price inelastic, but the volume of imports will be closely related to aggregate output (it is also observed[9] that aggregate import and export volumes are, respectively, increasing and decreasing functions of the previous period's aggregate excess demand for consumer goods).

To seek their objectives, the planners dispose of a rather different range of policy instruments than those used by policymakers in market-oriented economies. How these instruments work and why others do not are explained by other special features of centrally planned economies that affect the manner in which external disturbances are transmitted to and propagated in them.

First, the "price equalization account" completely separates domestic prices from foreign prices. The foreign-trade enterprises buy imports and sell exports abroad at foreign-currency prices, which for their internal accounts (and for the official foreign trade statistics) are converted into units of the domestic currency at an accounting exchange rate bearing no relation whatsoever to the domestic price level. The enterprises sell the imports to, and buy the exports from, domestic users and producers, respectively, at the ruling domestic prices. The accounting profit or loss they show on the aggregate of these transactions is directly absorbed into, or made good by, the state budget. Thus, there is no functioning exchange rate, no link between domestic and foreign prices, and no link between trade and foreign-exchange reserve flows, on the one hand, and the monetary assets held by households and firms, on the other. The currency is completely inconvertible. There are no "autonomous" international capital flows (and those accommodating trade are directly controlled by the planners) and no real balance effects arising from the foreign sector. Thus the exchange rate cannot be used as a policy instrument and the planners cannot devalue as conventionally understood. They can, of course, take some related policy measures, such as lowering their export-supply prices in foreign currency, or raising the domestic prices of consumer-goods imports; but these do not have the "automatic" across-the-board character of currency devaluation.

Furthermore, in the centrally planned economies, the endogenous feedback mechanisms characteristic of macroeconomic relationships in market-oriented economies are broken. There are no domestic markets for government securities,

[8]This is similar to another problem that the planners must take care to avoid, the "supply multiplier" process of repressed inflation, in which excess demand for consumption goods discourages labor supply, thus reduces output, and thus the supply of consumption goods.

[9]Portes, Winter, and Burkett, *op. cit.*

nor any private debt or equity instruments. Hence, the planners need not be concerned with domestic interest rates (except as incentives for household savings) or the effects of government financial policy on domestic investment or international capital flows. Changes in wages and other costs do not affect prices—at least in the short- and medium-term. Conversely, changes in prices have no direct effect on factor incomes, hence none on demand. Moreover, changes in the demand for goods need not affect their supply and therefore do not affect the demand for labor. Under these conditions, neither external nor internal shocks are likely to start a "wage-price spiral."

As a result, the centrally planned economy is, to a considerable extent, insulated from external shocks. It cannot, however, be insulated from shifts in nonhorizontal foreign demand and supply curves, nor from the real income effects of changes in its terms of trade. On the other hand, the planners can control the *incidence* of such shocks: for example, how the burden of a deterioration in the terms of trade is shared (between $C$, $I$, and $G$, between households, etc.) but their economic power carries a corresponding responsibility. It is much harder for central planners than for governments in developed market-oriented economies of the West to use external events as justification for unpopular domestic policies, or to shift blame onto impersonal market forces that would be neither feasible nor wise to oppose. Their burden is not only their control over the economy but also their claim to protect it from the maladies of the capitalist market-oriented economies and international system.

The policy instruments they use, then, are typically direct controls. They can decree changes in the allocation of resources to investment, government expenditures, and exports. They can directly control the volume of imports—recalling that this may constrain output. They can vary the supply of goods to households for consumption—bearing in mind that this must be related to wages so as to avoid excess demand, disincentive to labor supply, and, again, effects on output.

Some examples will indicate more concretely how the centrally planned economy and its planners may respond to external disturbances. Suppose that all foreign prices rise by a uniform percentage, leaving relative prices of traded goods constant. In a market-oriented economy, this could be offset by a corresponding appreciation of the exchange rate, but either the authorities or the foreign-exchange market would have to follow the world price movement continuously and precisely, meanwhile avoiding any speculative phenomena—an unlikely outcome. In a centrally planned economy, with trade initially balanced and no net foreign-currency assets or liabilities, uniform worldwide inflation has no effect whatsoever. The price equalization system automatically does the equivalent of continuous revaluation, and thus there are no real balance effects and no effects on the domestic relative prices of tradables and nontradables.[10]

A more interesting case is that of a deterioration in the terms of trade. Suppose,

---

[10]For a detailed explanation of how the "price equalization account" system perfectly insulates the domestic monetary system and internal prices of a centrally planned economy from uniform world inflation, provided that real trade flows are not adjusted, see Wolf, *op. cit.*

for realism, there is an increase in the price of oil for a substantial net importer (only Poland, Romania, and the Soviet Union are major producers of fuels) and that initially there is balance-of-trade equilibrium, capital-account equilibrium, and full employment everywhere.

For an individual oil-importing market-oriented economy, the oil price increase means a worsening of the terms of trade and, at the same time, is equivalent to an excise tax. Income falls, therefore expenditure falls; and a multiplier process gives a fall in output and employment. This reduces imports, but probably not as much as the initial increase. Fiscal or monetary policy may be used to restore full employment by raising absorption. The consequence is a trade deficit at full employment, typically financed directly or indirectly by an inflow of financial capital from oil exporters. If the policy-induced increase in absorption involves an increase in investment, this at least is creating assets, which will eventually raise output and thereby make it easier ultimately to repay the debts accumulating abroad. In any case, the country is postponing the transfer of real resources abroad, which is required by the deterioration in the terms of trade. An alternative policy is to try to raise net exports immediately. For all oil importers taken together, this requires raising the net imports of the oil exporters; otherwise, some of the former will simply export their unemployment and trade deficits to others. (Recall that, initially, total world savings rose; *some* countries must subsequently increase absorption if full employment is to be maintained.)

The story is quite different for the oil-importing centrally planned economy because domestic users do not automatically pay the higher oil prices. Thus there is no "tax" effect and no fall in domestic incomes, employment, or output. Because of the change in the terms of trade, however, real output at full employment does fall when measured in units of foreign goods; alternatively, real absorption rises when measured in units of domestic output. The problem of a trade deficit is in a sense more acute because there is no automatic reduction in domestic spending and production and thus no induced fall in imports. As long as imports and full employment can be maintained by borrowing abroad, the centrally planned economy makes no contribution to the fall in world demand.

Here too, however, the question arises whether there will be a shift within $A$ from $(C + G)$ to $I$ in order to accumulate additional real assets that will assist in repaying the debts to foreigners. The alternative policy is to reduce $A$ immediately in order to shift resources from $(C + I + G)$ to $(X - M)$. Note that there is no reason to reduce output or its growth rate, as a means of reducing expenditure, because the planners can act directly on the components of absorption. Admittedly, a cut in investment will ultimately reduce the growth rate; a cut in the supply of consumption goods may depress output through the "supply multiplier" (unless wages can be held back or consumer prices raised correspondingly, but this too has incentive effects); an increase in exports may require reducing their prices, thereby increasing the burden of a deterioration in terms of trade; and both increases in exports and reductions in imports may create domestic excess demand initially, with eventual "bottle-neck" effects restricting output growth. At *no* stage, however, does the problem of adjustment appear to the planners as one of avoiding deflation and unemployment.

Some further contrasts appear for the Seven taken as a group. Within the Comecon area trade flows are balanced bilaterally. The "transferable rubles" in which they are measured are an inconvertible accounting currency (a small percentage of intra-Comecon trade is accounted in hard currency, but there are reportedly even restrictions on the use of balances so accumulated). Until 1975, intra-Comecon trade prices were set at the beginning of each 5-year-plan period on the basis of average world-market prices of the preceding few years, then held essentially constant for 5 years (the new "moving average rule" and its effects are discussed in the next part of this chapter). Thus, with minor exceptions, the Seven do not use (hard currency) earnings outside the Comecon area to finance trade imbalances between themselves, nor can one Comecon country's surplus with another be used to finance a deficit with a third. Internally, deficits are in effect met by direct intercountry lending. Until recently, this was done entirely on a year-to-year basis, but there is now evidence that the Soviet Union is extending some long-term loans.

Within the Comecon area, there is never a "beggar-thy-neighbor" problem in the usual sense. Raising net exports to Comecon partners does not export unemployment to them, but simply transfers real resources from those who lose on changes in the terms of trade to those who gain (returning to our example in the preceding). Indeed, the employment and output effects will, if anything, be the opposite, with bottlenecks created in countries that raise net exports and resolved in those whose net imports increase. Thus, in the West, where the problem is unemployment, the "beggar-thy-neighbor" policy is to export it, while in the East, where the usual problem is excess demand, the corresponding policy is to export it by running an import surplus.

Suppose, however, that the Seven as a group seek to raise their net exports to (oil-importing) market-oriented economies in order to cover an increase in their import prices. Although such a policy is not meant to export unemployment, it may indeed have the effect of creating further unemployment in the market-oriented economies and possibly (if it creates bottlenecks) *even within the Comecon area.* Conversely, if the market-oriented economies raise their net exports to Comecon countries, it will *raise employment* in *both* groups, so long as someone is willing to finance the resulting Comecon deficits (in the first instance, say, Western banks, acting as intermediaries for the oil-exporters' surpluses).[11]

---

[11]Hungary and Poland present some exceptions to these generalizations. The Hungarian reforms of 1968 did eliminate, in principle, the price-equalization account and introduced the equivalent of exchange rates to convert intra-Comecon transferable ruble prices and world-market convertible-currency prices to domestic prices paid by importers and received by exporters. Some differentiated subsidies and taxes on imports and exports were eliminated, investment finance was significantly decentralized, many consumer prices were made flexible and money incomes became substantially dependant on enterprise performance (profits). The Hungarian authorities were, however, still inexperienced at managing exchange-rate, credit, and fiscal policies when the crisis developed. In Poland, too, domestic prices and incentives became dependent on foreign prices and the money wage control system was relaxed to promote incentives after 1971. In both countries, however, domestic capital markets remained absent and

## IMPACTS OF THE CRISIS ON EASTERN EUROPE

The direct impacts[12] on the Seven of the crisis in the world economy have come through goods markets and capital markets. Given these countries' severe restrictions on labor migration across national boundaries, effects on the labor markets may be ignored. It is convenient to consider quantity and price effects separately—as far as possible. It will also be important to distinguish among the Seven, according to the composition of their trade.

Eastern Europe had not been affected significantly by any supply restrictions on its imports from the rest of the world until the disturbances of 1978–1979 in world oil and gas markets (in particular, the Iranian default on supply commitments). On the other hand, the Western recession brought a fall across the board in demand for Eastern exports (except fuels). The slowdown in industrial activity hit exports of materials and imtermediate goods; and excess supply of manufactures in the West obstructed Eastern efforts to expand their exports of manufactures. The Organization for Economic Cooperation and Development (OECD) has calculated that the *volume* of OECD imports from the Seven fell 10% below trend in 1974, then 15% below trend in 1975.

After the initial shock, the character of the problem changed somewhat. It has become evident that the composition of East European exports of manufactures to the West is too heavily slanted toward goods regarded in the West as "sensitive" because of the excess supply created by the crisis, by structural rigidities in the West, and by competition from newly industrializing developing countries. Thus, the crisis has brought the Eastern countries up against *protectionist measures* and threats of further restrictions in areas such as textiles, clothing, footwear, steel, electric motors, meat products, and basic chemicals.

At the same time, the excess supply of capital goods created by the fall of investment demand in the West pushed exporters and their governments into major sales efforts in Eastern Europe. OECD data indicate that, from 1973 onward, officially supported export credits to Eastern Europe have accounted for about 20% of total OECD export credits to all countries (although Eastern Europe takes only 4% of total OECD exports). About 75% of capital-goods exports to the Seven in 1973-1977 were financed by official export credits with terms longer than a year. New officially supported export credit commitments by OECD exporters on signed contracts with Eastern Europe totalled $4000 million in 1973, $6600 million in 1974, $5700 million in 1975, $7200 million in 1976, and about the same in 1977. None of these data include the substantial amount of Western bank lending without official support or guarantees that has accompanied export contracts won by the

currencies inconvertible, reserve flows continued to be sterilized, and overriding emphasis was still placed on full employment (with a correspondingly weak "demand multiplier"). For further theoretical discussion of the effects of external disturbances on these "modified" centrally planned economies, see Wolf, "Balance of Payments Adjustment in Modified Centrally-planned Economies," in Neuberger and Tyson, *op. cit.*

[12]Several papers relevant to this and the next part of this chapter, including case studies of particular countries, are contained in Neuberger and Tyson, *op. cit.*

banks' domestic customers. All this testifies to a strong and highly competitive Western effort to find Eastern markets for capital goods in response to weakened demand in the West itself.

The crisis has also been marked by great changes in the world-market prices facing the Comecon countries. Available data are not entirely consistent (the indices may be calculated by different methods, from different samples of goods and trading partners, etc.), but the overall picture is fairly clear. Price levels facing the Eastern countries from outside the Comecon area approximately doubled from 1972 to 1976. Bulgaria, Poland, the Soviet Union, and probably Romania experienced improvements in their terms of trade with the West from 1972 to 1976, while for Hungary, East Germany, and Czechoslovakia, there was a substantial deterioration. This reflects the composition of their trade: The first group exporting predominantly primary products to the West (food for Bulgaria and mainly fuels for the others); and the second group exporting mainly manufactures. Overall, the gainers outweigh the losers and the Comecon areas's aggregate terms of trade with the West have risen significantly. The main gains came in 1974 and 1976 with losses in 1972–1973 and 1975.

Intra-Comecon prices, under the old rules, remained essentially unaffected by the world price changes from 1971 onward until the end of 1974. But, during 1974, the Soviet Union made it clear to the Six that it would not be willing to bear for long the full cost of foregoing increases in its raw-material (especially oil) export prices to other Comecon countries, when it could get much higher prices in hard currency on world markets. Thus the rules were changed. For 1975, Comecon trade prices were set on the basis of the average of world-market prices for 1972–1974; for 1976, the average covered 1972–1975; and for each year from 1977 onward, intra-Comecon trade prices are based on a moving average, covering the 5 preceding years.

Data show the Soviet Union to be the major gainer and the heavy raw-material importers to be the hardest hit by the changes of 1975. Note that the positions of Hungary and East Germany had been deteriorating fairly continuously from 1970 onward, presumably because of unfavorable shifts in the structure of their trade. The new rules improved the Soviet Union's terms of trade with all the Six except Romania, which does not import Soviet oil. But the changes within the Comecon area were still very moderate compared with what was happening outside. It is estimated[13] that had it forced an immediate and full shift to world market prices, the Soviet Union could have obtained an improvement in the terms of trade with the Six of as much as 40% by 1976, in comparison with the 14% it actually received. Moreover, the Soviet Union further cushioned the effects of the price changes by allowing its partners to go into deficit with itself and these deficits rose continuously from 1974 to peak in 1977 at about $2000 million, in aggregate. They were covered by 10-year loans from the Soviet Union. The relation of the Soviet Union's import quantity to its export quantity did not (except *vis-à-vis* Hungary) improve as much

---

[13]E. Hewett, "The Impact of the World Economic Crisis on Intra-CMEA Trade," in Neuberger and Tyson, *op. cit.*

as its terms of trade—in other words, it did not take all the real resources that the new rules would have allowed. Overall, Soviet moderation considerably eased the short-run burden of adjustment for the Six.

The geographical structure of Bulgaria's trade must have altered quite unfavorably (in other words, toward other Comecon countries and especially the Soviet Union), because it now emerges as a significant loser, joining Hungary, East Germany, and Czechoslovakia. And the losses were very significant indeed: For Hungary, they have been assessed at 5–8% of national income in each of the years 1974–1977.[14] Thus, although foreign price changes lessened the strain of adjustment to the crisis for Comecon *as a whole,* this benefit was very unequally distributed. It seems, moreover, that Comecon countries faced large and wide-ranging changes in foreign *relative* prices within quite disaggregated commodity groups. The question here is not so much adjustment to broad terms-of-trade effects as flexible adaptation of the detailed structure of trade to these new foreign price ratios. It has also been suggested that Hungary was not very successful in this regard—and one might have expected its economic system to be more responsive than the others.[15]

The impact of the crisis on the goods markets went hand in hand with major effects on world capital markets. Just as the centrally planned economies were basically unaffected by the changes in exchange-rate regimes, neither were they vulnerable to speculative capital flows associated with exchange-rate variation. On the other hand, the Seven were very much a part of the developments in international lending and borrowing that accompanied the crisis. The commodity (especially oil) price increases—and the associated trade deficits and surpluses—not only generated tremendous increases in the supply of funds to be lent by international financial institutions, and the demand for them by countries needing balance-of-payments financing, but also shifted the process of financing and international liquidity creation from official to private sources and, in particular, to the Eurocurrency market.

The new pattern of international financial intermediation opened new opportunities to the Seven, who (except Romania) had no access to lending from the International Monetary Fund (IMF), but were welcomed as major customers by the international banks.[16] The supply of medium-term funds to Eastern Europe at market terms was effectively unconstrained, at least until mid-1976, and even since then it has been plentiful, especially in the renewed borrowers' market conditions from

[14]A. Brown and M. Tardos, "Global Stagflation and the Hungarian Economy," in Neuberger and Tyson, *op. cit.* Brown and Tardos also estimate that Hungary's total overall (ruble and dollar) trade deficit in 1974–1976 was approximately equal to the loss from the deterioration in terms of trade after 1972. In *Les Conditions de l'Endettement des Pays de l'Est* (Paris: Centre d'Etudes Prospectives et d'Informations Internationales [CEPII], 1979) Tables 7–8, it is calculated that the improvement in terms of trade for the Soviet Union substantially reduced her hard-currency deficit, although it was of only limited importance for that of Poland.

[15]Brown and Tardos, *op. cit.*

[16]Portes, "East Europe's Debt to the West," *op. cit.*

the autumn of 1977 onward. Only Hungary had, even on a limited scale, carried out any medium-term Euromarket borrowing before 1973. Since then, all the Seven have at various times, and some continuously, been major borrowers. The margins ("spreads") applied to Eurocurrency borrowings by the Seven have been typically only slightly higher than those for OECD borrowers as a group, substantially lower than for developing-country borrowers and even somewhat better than those available to the members of the Organization of Petroleum Exporting Countries (OPEC).[17]

It is important to realize that the goods-market and capital-market manifestations of the crisis were not independent. It has been said, for example, that Poland was "unlucky" insofar as the crisis brought a fall in demand for her exports when she needed to mount a major export drive to service the debt she was accumulating. But Poland could not have borrowed so freely during 1974–1976 to finance her investment drive (and thereafter to continue her deficits and service her debt) if the crisis had not simultaneously made Western banks and exporters seek borrowers and buyers. The changes in terms of trade, which imposed so great a burden on Hungary, East Germany, and Czechoslovakia, were associated with the events in capital markets which permitted borrowing to support this burden in the short and medium run. The *external* problems creating the need for adjustment also created the opportunity to postpone it through financing. This applies rather generally for the crisis, although not of course to the *internal* problems such as harvest failures or labor shortages. We shall try to sort out these influences in the next section of this chapter.

A Marxist interpretation of Western behavior would also stress the relation between goods and capital-market phenomena. With the crisis, the mature capitalist economies required a vent for both financial and real surplus. Although not an obvious field for imperialist exploitation, Eastern Europe in many ways offered more attractive markets than the developing countries, and both finance and industrial capitals have competed fiercely for these markets. The one twist in the argument is that (as already stressed) so long as the lending and debt burden are sustainable, the process promotes employment and growth *in the East* as well as in the West. If a crash should follow the crisis, the East would probably suffer less than the West and, in addition, its debt might then vanish either in a wave of inflation or, if repudiation became respectable, in a wave of defaults. Lenin would doubtless have perceived all this and so may have his successors.

## EAST EUROPEAN POLICY RESPONSES

The crisis in the world economy has deeply affected foreign commodity price levels and relative prices (terms of trade), the demand for exports, and the availability of external finance. The macroeconomic data show the joint effects of these

[17]See CEPII, *Les Conditions de l'Endettement des Pays de l'Est, op. cit.,* table 12; OECD, *Financial Market Trends,* Paris, June 1979, No. 10; and various issues of *World Financial Markets,* Morgan Guaranty Trust Company, New York.

shocks and the policies responding to them. Our task now is to separate the two as
far as possible, analyze the policies, and then assess the overall effects of the crisis
on the Eastern countries.

The aggregate net material product *(Y)* figures[18] show no disturbance at all in
1974 (although growth in the West came to a halt) and then a general fall in growth
rates in 1975–1978, with 1978 the worst year yet. It was clearly impossible to meet
the 1976–1980 plans, although they were uniformly lower than the actual results of
1971–1975. Moreover, from 1976 onward, output growth fell short of the *annual*
plans in almost every case. On the other hand, this is only a "growth recession"; no
country's aggregate output has fallen in any year and the lowest growth rate recorded is Poland's 2.8% in 1978.[19] For the Six, aggregate utilization (our *A*) grew
faster than output in 1974, as the shifts into overall trade deficit testify. But only
East Germany and Czechoslovakia then kept the growth of *A* below that of *Y* in
1975; Bulgaria and Hungary reacted to their deficits in this way in 1976; and Poland
could not manage to do so until 1977. Meanwhile, East Germany relaxed somewhat
in 1976 and Hungary rather more in 1977–1978. The only case when total absorption actually fell was in Hungary (and probably Poland) in 1979.

Looking at components of total output, we see that the growth rates of industrial
output have also slowed, especially from 1976 onwards, although they still remain
at quite respectable levels. It appears that the initial slowdown was intended to
reduce the growth of imports.

In some cases (e.g., Poland and Czechoslovakia in 1978), shortages of imports
themselves began to constrain the growth of output. Agricultural performance,
which is independent of the crisis but influences responses to it, has of course been
much more variable, with actual falls in output experienced in several instances.

On the side of absorption, there is a strong contrast between countries in which
investment and incomes have followed fairly steady growth paths—East Germany,
Czechoslovakia, and the Soviet Union—and those where both these variables, but
especially investment, have been more variable—Bulgaria, Hungary, Poland, and
Romania. In East Germany, and to a lesser extent in Czechoslovakia, steady growth
has been matched by steadily increasing trade deficits, both overall and in hard
currency although Czechoslovakia has done much less badly than East Germany.
This association is not valid for the Soviet Union mainly because of the impact of
varying grain harvests. On the other hand, the converse relation holds: In the
generally dismal trade performances of 1974–1978, there are wide year-to-year
swings for Bulgaria and Hungary; Poland's deficits grew to a peak in 1976, then

[18]It may be useful to keep in mind the relative sizes of the economies of the Seven. Western
recalculations of GNP by T. Alton and Associates, L. W. International Financial Research Inc., New
York, give the following for 1977 in current prices: Bulgaria $22,100 million; Hungary $29,500 million;
Poland $100,700 million; East Germany $73,200 million; Romania $60,300 million; Czechoslovakia
$63,400 million; and the Soviet Union $1,018,100 million. These are probably rather conservative
estimates.

[19]Preliminary data now suggest growth in 1979 was less than in 1978 everywhere except Bulgaria and
East Germany—and that net material product did actually fall in Poland in 1979.

declined somewhat; and Romania's deficits behaved conversely, improving up to 1976 and deteriorating thereafter.[20]

There appears to be a mild downward trend in the growth rate of *per capita* real incomes in East Germany, Czechoslovakia, and the Soviet Union. Hungary, Poland, and, probably, Romania failed to control money incomes, with consequent increases in consumer prices and (in Poland and Romania) some excess demand for consumer goods. Nevertheless, even Poland's price increases have been fairly modest by Western standards, and the official consumer-price indices for the other countries are effectively flat until 1979, when finally some major adjustments have been reported. Nowhere in the East have consumer prices shown any significant relation to world price increases, in either the timing or the extent of changes. The official price indices are, of course, suspect.[21] There has doubtless been some hidden and repressed inflation throughout the region in the past few years, as both internal and external pressures have restricted the possibilities for increasing the supply of consumption goods more than the planners have been able to restrain wages.

The changes in the volume of foreign trade are of some interest. Expansion of intra-Comecon trade initially slowed up in 1975–1976, but then re-accelerated as trade with the rest of the world weakened. Trade between the Soviet Union and the Six slowed especially in 1975–1976 (volumes may even have declined)[22] as the Six sought to economize on their imports of primary products and the Soviet Union tried to shift some exports of these goods to lucrative hard-currency markets. Overall, intra-Comecon trade accounts for most growth of export volumes in 1974–1975 and again in 1977–1978; the Soviet Union raised its primary product exports once more, while its partners had to export substantially more in return.

While observing the hard-currency merchandise trade balances, it is important to distinguish between the Soviet Union and the Six. The Soviet Union's deficits are relatively modest and largely attributable to the bad grain harvests of 1972, 1975, and 1977. On the other hand, the hard-currency deficits of the Six had already, in 1972–1973, begun to show the effects of rising commodity prices (and Poland's imports of Western machinery) and in the 5 years of 1974–1978 their aggregate deficits with the West averaged $6000 million per annum. Contrary to expectations of some observers, there has been *no significant improvement* since the figure of $6600 million in 1976. Indeed, excluding Poland, the overall situation was worst in 1978. Nor does the picture for the Six change much if we consider their trade with developing countries: Their surplus of $550 million in 1976 rose to $1000 million in 1977, but fell back to $800 million in 1978 (most of which was earned by Bulgaria). On the other hand, the Soviet Union's surplus with the developing countries rose

[20]Preliminary figures for 1979 trade of the Seven with industrialized Western countries show little change from 1978: some improvement for Bulgaria, Hungary, Poland, and the Soviet Union, deterioration for East Germany and Romania.

[21]Portes, "The Control of Inflation: Lessons from East European Experience," *Economica,* London, No. 44, 1977, pp. 109–130.

[22]Hewett, *op. cit.*

from $1200 million in 1976 to $3200 million in 1977 and $4300 million in 1978, mainly (it seems) on the strength of exports of armaments.

The consequences are evident in the hard-currency debt figures. The gross total at the end of 1978 was $69,000 million, of which the Six individually accounted for $46,000 million. The burden of interest and amortization payments is now considerable, as indicated by the debt-export and debt-service ratios. Hungary and Poland, for example, must have had to use 30–35% of their earnings from exports to the West in 1978 simply to pay the interest on their debt. The relatively short-dated maturity structure of the debt also presents short-run liquidity problems.[23] None of the debt is risk capital and much of it carries variable interst rates. This is, indeed, another channel of transmission of Western economic conditions. The 6-month Eurodollar rate (LIBOR) rose from 5.4% at end-December 1976 to 12.3% at end-December 1978. Adding an average spread of around 1%, say, one still finds that the interest burden on outstanding Eurodollar loans more than doubled in that period.[24] Although the corresponding Western inflation reduces the longer-run real burden of repayments of principal, the size of debt relative to GNP is now considerable in some countries. Overall, Poland's debt problem is most serious with Bulgaria and Hungary, with East Germany next in line.

Common to the experience of all the Seven is the unimportance of rises in the overall world price level. Even in Hungary relatively little inflation has been imported.[25] Moreover, I would maintain that in Hungary and, even more so, in Poland, consumer-price increases have primarily reflected the need to limit real consumption and excess demand while money-wage growth continued. The money-wage increases were partly intended for incentives, to widen differentials and to relieve social tensions, but they also went beyond the planners' intentions.[26]

All the Seven have introduced large price increases for certain specific commodities. These have been well-publicized in the West and are often interpreted as evidence of the inability of central planning (and Comecon), to protect the Eastern countries from the crisis and worldwide inflation. The truth, however, is that the increases are basically responses to *relative* price distortions aggravated by both external and internal pressures. The increases for petrol, fuels, and energy are overdue and perhaps still inadequate. The usual objective has been to cut some specific imports, or release supplies of certain goods for export, as well as to reduce subsidies whose justification (for income redistribution) has become weaker as overall real incomes have risen. The price increases will have relatively little effect on

[23]Sixty-nine percent of Eastern Europe's gross debt at end-1978 was in form of liabilities to Western banks; of this $47,500 million, 43% carried maturities of no more than a year, and a further 11% maturities from 1 to 2 years. *Bank of International Settlements* data.

[24]Note also that the Eurodollar rate (LIBOR) rose substantially more through 1979 and stood at 17.0% at end-February 1980.

[25]While import prices in non-ruble trade rose 138% from 1970 to 1977, consumer prices rose 25%. See *Foreign Trade Statistical Yearbook*, Budapest, 1977, p. 408.

[26]How this can happen in a centrally planned economy is explained in Portes, "The Control of Inflation: Lessons from East European Experience," *op. cit.*

aggregate demand—nor was this the intention. In principle, the planners could have achieved increases in the relative prices of the goods in question by reducing the absolute prices of all others and temporarily stopping increases in money wages. But merely to put the problem this way reveals that solution to be administratively, and even politically, unfeasible when the number of "problem" goods, and their weights in household budgets, is not very large. There are nevertheless significant political costs to the price increases, precisely because households are unaccustomed to any price rises and because the increases are abrupt rather than continuous. The planners cannot easily shift responsibility onto external forces, although they have naturally pointed to conditions in the world market and the harsh winter of 1978–1979.

To consider policy responses in *real* variables, we must divide the period into two subperiods, 1974–1975 and 1976 onward; and also divide the Seven into those which gained from changes in terms of trade and those which lost. The former group includes the Soviet Union, Romania, Poland, while the evidence is ambiguous for Bulgaria; and Czechoslovakia, Hungary, and East Germany all lost. The Soviet Union is a special case because of its size, low trade dependence, and key role in the bloc. Its responses to the crisis have taken the form of changes less in macroeconomic policy than in its economic and political relations within and outside the Comecon area.

In 1974–1975, none of the Seven, except Romania, reacted quickly to their increased trade deficits, whether these were due to terms-of-trade or export-demand shocks, investment and machinery import drives, or harvest failure.[27] Output grew rapidly (except in Soviet agriculture), but absorption grew faster. During this period, the crisis affected virtually nothing except trade balances and foreign borrowing—but specifically *not* internal policies.

The 1976–1980 plans all proposed that absorption should grow considerably slower than output, which itself was to grow somewhat less than in 1971–1975. Adjustment, however, has not been achieved. First, even the more moderate output plans were much too ambitious, given the internal pressures already outlined as well as the continuing losses in terms of trade for the losing group, as intra-Comecon trade prices followed the earlier change in world prices. While the output plans were not fulfilled, the attempts to reach them ("overfull employment planning") led to faster increases in absorption than had been intended. This showed in investment outlays, in excess demand for labor and wage increases that (together with diversion of goods from consumption supply) generated excess demand for goods, and directly in import demand and the supply of exports. When imports were restricted, as in most countries in 1978, shortages of intermediate inputs resulted and constrained output growth.

The original plans for limiting the growth of absorption in 1976–1980 were also too ambitious. It proved difficult to control wages in Hungary, Poland, and even

---

[27]Note that *import volume* exceeded plans in 1971–1975 only in Bulgaria, Poland, and the Soviet Union. The internal causes are clear in the latter two.

East Germany, and difficult to control investment in these countries, as well as in
Romania and Bulgaria. Investment exceeded the plans over the period 1976–1978
in the first four and, as late as 1978, in all the Seven except Bulgaria. Where import
volumes were controlled, resources were insufficient to meet domestic claims and
raise export supply adequately, especially to Western markets (given that all of the
Six except Romania had to export more to the Soviet Union). Primary products that
could have been sold were unavailable (e.g., Poland was unable to divert enough of
her coal output to exports). Where there were supplies for export, the goods were
typically those not required to meet internal needs, but also not desired in either
western or Eastern markets (textiles, clothing, and some types of machinery).

Typically, then, one does *not* observe in the East internal imbalances resulting
from attempts to correct external imbalances caused by the crisis. Rather, one has
seen domestically and independently generated *internal imbalances creating external imbalances,* which in some cases were eased and in others exacerbated by the
crisis in the West. Long-run pressures have left the planners very little room to
maneuver, insofar as they have little scope for shifting resources into net exports
without restricting consumption and output-growth rates more than they believe to
be desirable or feasible. This part of the story seems to hold for the Soviet Union as
well, despite its special position.

Thus, the Eastern countries have for the most part chosen *financing rather than
adjustment.* Adjustment is *painful.* For four of the Seven, just to eliminate their
trade deficits with the West would require shifting 2–3% of GNP into net exports
and to this must be added the interest burden of accumulated debt. And with no
unutilized resources (ignoring potential short-run reallocations that would improve
efficiency), output growth cannot be accelerated; so this adjustment means reducing
the growth of absorption (and consequently that of output, to some extent).

Adjustment is *difficult.* It is difficult to convince politicians or consumers to
lower their aspirations and expectations. Even in a centrally planned economy, it is
hard to control lower-level pressures for more investment, more wage funds and
more imported materials and equipment. It is hard to sell more in the West, especially in a generalized crisis, when you have a bad image (partly justified) for
quality and after-sales service and when your attempts to compensate by cutting
prices provoke charges of dumping (and you cannot achieve the same result by a
competitive exchange-rate devaluation).

Adjustment is, of course, *not necessary* if finance is freely available at reasonable
terms. There has been and continues to be a positive *push* from Western exporters
and capital markets for the Eastern countries to accept finance. More export credits
are offered and loans are oversubscribed. The borrowing may have been an appropriate short- and medium-run response at the outset, if it appeared that adjustment
could be achieved fairly soon, and that trying to adjust faster would have been too
costly. Borrowing might also have been seen as part of a long-run strategy, as in the
Polish case; but then it is justified only if the borrowed resources go into increasing
investment or if there is an explicit decision to consume more in the present at the
cost of consumption in the future.

In most cases, however, these points have long since been passed. We must view financing as a policy response to internal problems and shocks to the goods market, a response predicated on the availability of funds in Western capital markets and to some extent responding to an aggressive supply from those markets. If absorption exceeds output, at the level of the Seven taken together, the only possible outlet is hard-currency borrowing to finance import surpluses from outside the bloc. It would seem, however, that this hard-currency financing has allowed the continuation of policies that give little promise of improved economic performance and external balance in the long run.

If the problem of continuing East European hard-currency trade deficits is indeed an expression of the planners' preferences for financing over adjustment, then there can be no single-cause explanation. Neither bad harvests nor losses on the terms of trade can account for the problem. Only the Soviet Union's hard-currency trade balance seems closely related to its agricultural performance. And with Romania back in substantial deficit after 1977, the countries that have gained on the terms of trade have trade balances no healthier than those which have lost (of which there are only three clear cases out of the Seven).

One popular explanation is the "technology gap." Supposedly, the large import surpluses since 1973 are due to imports of Western machinery intended to remedy the East European lag behind the West in developing new, more efficient techniques of production. This has obviously been important in Poland and significant elsewhere; but it cannot be taken to be the primary cause of the deficits.

It will be seen that the overall importance of technology exports in total exports to the Seven has slightly *declined* from the late 1960s, when East European hard-currency trade deficits were small in aggregate, to the period 1973–1977, when such deficits totalled $41,000 million. This decline is due almost entirely to the Soviet Union; for the Six, the shares are remarkably stable. Clearly, the pressures for increased imports of machinery, equipment, and other high-technology goods have coincided with equally powerful pressures for imports in other categories. The "structural disequilibrium" in trade with the West is primarily a manifestation of aggregate level and macroeconomic pressures: The hard-currency imports the planners require cannot be covered by hard-currency earnings because of a shortage of exports that could be sold at acceptable prices. But the resulting deficits, and the assistance they provide to Western trade balances and to employment in both East and West, can continue only as long as Western lenders allow. And enforced cutbacks are more likely, in the medium run, to hit imports of Western capital goods than of materials and intermediate goods.

## CONCLUSIONS

The future expansion of Eastern trade with countries outside the Comecon area faces severe constraints. The long-run potential of the Seven to increase their exports of primary products outside the bloc is limited and their manufactured exports to the West will face growing competition from newly industrializing developing countries. Limits on export earnings, obligations to service existing hard-currency debts, and eventual restrictions on further borrowing will constrain their hard-currency imports more severely than the planners have been willing or able to do so far.

The repercussions of the crisis have undoubtedly contributed to the centripetal forces that are already gathering strength in Comecon. The new obstacles to expanding trade outside the bloc have been superimposed on the movement toward more specialization and integration in Comecon manufacturing production and trade under the Complex Program. Perhaps the most powerful effect, however, is the increased dependence of the Six on the Soviet Union for energy and raw-material supplies and the shift in economic influence accompanying the improvement in her intra-Comecon terms of trade. This reinforcement of intrabloc ties makes even more unlikely any move towards convertibility of the transferable ruble or participation in the IMF of others besides Romania. Thus, contrary to some commentators, the small amount of intrabloc trade now done in hard currency is no indication of any weakening of intra-Comecon cohesion.[28]

The effects of the crisis on domestic economic institutions are not quite so clear. Certainly the external pressures of 1973–1975 contributed to the forces that blocked further development of the Hungarian economic reforms (although they were not as important in this regard as domestic tensions which preceded the crisis). The limited Polish decentralization of 1973 was also being eroded by 1975—but again the internally generated pressures were at least as important as those from abroad. The external instability and imbalances of the mid-1970s clearly presented an environment that was not conducive to decentralization. On the other hand, both Hungary and East Germany now seem to be moving in that direction (the latter starting from a much more centralized system), partly in response to their hard-currency problems. One may expect these new efforts to be relatively unsuccessful, and probably short-lived, but it is too early even to conjecture how far they will go.

I judge the most significant impact of the crisis on the Seven to be the unprecedented availability of external finance in hard currency. This, in some cases, positively stimulated excessive imports of Western capital goods, permitted some countries to lose control of macroeconomic policy for more or less extended periods and allowed all the Seven to postpone adjustment to internal pressures as well as to other impacts of the crisis.

Next in order of importance come the changes in terms of trade: initially with the rest of the world, then in consequence with Comecon countries. The Seven as a group benefited, but this observation is almost meaningless because of the great disparities between countries in the effects on real income, trade balances, and economic power. For the three countries that suffered substantial losses, as well as for the Soviet Union, the impact on the terms of trade has been very strong indeed. The fall in Western demand for Eastern exports has in certain cases been significant, but perhaps overemphasized by some who neglect the strong internal constraints on

---

[28]Portes, "Est, Ouest et Sud: le Rôle des Economies Centralement Planifiées dans l'Economie Internationale," *op. cit.*

export supply. Except in Hungary, the general worldwide inflation has been basically neutralized and has had little to do with those price rises that have occurred in Eastern Europe.

It should also be evident that through the mechanisms previously outlined, central planning has, to a considerable extent, insulated the economies of the Seven from the effects of the crisis. Other countries, too, have had access to abundant external finance, but have not thereby avoided widespread unemployment, substantial shortfalls of actual below-potential output, and double-digit inflation. Comecon as a trading group, with its own more stable prices and long-term commitments, has also protected its members from some of the short-run instability of the world economy, although it is by no means independent of influences from world markets. The Soviet Union has benefited from those influences, but apparently not to the extent that it could have done under Comecon rules—and those rules were themselves less advantageous to it than trading outside the bloc would have been—and has significantly assisted its partners in meeting the burdens imposed by recent world economic events. It may now feel that the period of intra-Comecon adjustment to the crisis should be coming to an end. On the other hand, for the next 5-year-plan, it is apparently putting less emphasis on contributions by the Six to investment in projects for extraction of Soviet raw materials, which has put a burden of several billion dollars on its partners during 1976–1980 (in return for future raw-material supplies, of course).

To some extent, this protection has merely postponed problems and burdens—and insulation at the macroeconomic level has delayed necessary structural changes. The economies of the Seven are by no means healthy and trouble-free. But their current difficulties are consequences more of pressures originating internally, and of past policy choices, than of the world crisis from whose full shocks they have been cushioned.

# The Tokyo Round of multilateral trade negotiations (1973-79)

The GATT Secretariat

The Tokyo Round of multilateral trade negotiations, the most comprehensive of all the seven rounds of negotiations held within the GATT since its founding in 1948, was concluded in 1979. All the agreements resulting from the Tokyo Round, covering not only tariffs but also certain non-tariff measures, bovine meat, dairy products, civil aircraft, and an improved legal framework for the conduct of world trade, took effect on 1 January 1980, except for those covering government procurement and customs valuation which will take effect on 1 January 1981.

Nearly all the Tokyo Round agreements were concluded in early 1979 [1]; on 12 April, a Procès-Verbal, in which were listed texts embodying the Tokyo Round results, had been approved by the Trade Negotiations Committee—the body responsible for overall supervision of the Tokyo Round—and opened for authentication of the texts by signature of the participating governments. Most of the agreements listed in the Procès-Verbal were definitive and ready for incorporation in the final Tokyo Round package. There were, however, some notable exceptions, and negotiations on these items continued over the following months, with attention focussed particularly on the major unresolved problem of whether, and in what way, to revise GATT rules on safeguard action against disruptive imports. Also, divergent views on certain aspects of the agreements on customs valuation and anti-dumping had to be reconciled; and a number of bilateral negotiations on tariffs and tropical products had to be completed. The further

---

[1] A detailed account of the position reached in the negotiations up to April 1979 will be found in *GATT Activities in 1978*.

Reprinted from "GATT Activities in 1979 and Conclusion of the Tokyo Round Multilateral Trade Negotiations (1973–1979)," pp. 9–35, with permission of the GATT Secretariat, Geneva, Copyright 1980.

development of active co-operation in the agricultural sector within an appropriate consultative framework was among other matters still outstanding in April 1979.

Although the Tokyo Round was conducted within GATT, it was initiated not by GATT's member states as such, but by a conference of Ministers, including Ministers from non-member countries of GATT, at Tokyo in September 1973. Consequently, a principal task of the GATT member states at their November 1979 session was to take the necessary action to bring the Tokyo Round results within the framework of GATT, both as an institution and as a treaty.

This action was on the following lines. First, the member states took note of the two Tariff Protocols opened for acceptance during 1979, which, with their annexed Schedules of Concessions, comprise the Tokyo Round results in the field of tariff-cutting commitments.

Next, they adopted a consensus decision establishing the relationship with GATT of the agreements covering non-tariff measures, dairy products, meat, and civil aircraft. This decision reaffirms the responsibility of GATT's member states in overseeing the operation of the GATT system as a whole; it defines the rights of GATT members which have not signed particular Tokyo Round agreements, and stipulates that each of the Councils or Committees set up to supervise the various Tokyo Round agreements will report regularly to GATT's member states as a whole on developments relating to operation of the agreements. Furthermore, the decision makes clear that GATT member states which have not signed a particular agreement " will be able to follow the proceedings of the Committees or Councils in an observer capacity, and that satisfactory procedures for such participation will be worked out by the Committees or Councils ".

Third, they adopted another consensus decision which makes an integral part of the GATT rules the so-called " Framework " agreements on differential and more favourable treatment, reciprocity and fuller participation of developing countries, safeguard action for development purposes, trade measures for balance-of-payments purposes, and improved dispute settlement procedures.

Fourth, the member states established a new committee within GATT to carry forward the negotiations on safeguards, which was the major piece of unfinished business left over from the Tokyo Round. Another decision, linked to that on safeguards, set up a sub-committee of GATT's Committee on Trade and Development to examine protective measures taken against imports from developing countries.

## Background to the negotiations

Agreement to open the Tokyo Round was embodied in a document known as the Tokyo Declaration [1] adopted at the Ministerial meeting in the Japanese capital in September 1973. While developed countries were expected to negotiate on a basis of reciprocity, i.e. to make trade concessions balancing those that they received, they agreed not to expect from developing countries contributions inconsistent with their individual development, financial and trade needs.

---

[1] For text, see *GATT Activities in 1973*.

Ninety-nine countries of widely differing levels of development and economic systems, both GATT and non-GATT members, and accounting for nine-tenths of world exports, were involved in the Tokyo Round: the industrialized countries of Western Europe and North America as well as Japan; less-industrialized countries, such as Australia and New Zealand; countries of Eastern Europe; and the whole range of developing countries from the least developed to the most advanced. (The participating countries are listed on the final page of this booklet.)

The Tokyo Round, more extensive and comprehensive than any undertaken before, was designed not only to reduce or eliminate tariff and non-tariff barriers to trade in agricultural as well as industrial products, but also to shape the multilateral trading system and international trade relations into the 1980's and beyond.

Also distinguishing the Tokyo Round from all earlier rounds of GATT trade negotiations was the part played in it by the developing countries. For the first time in GATT multilateral trade negotiations, the problems of these countries assumed a prominent place, reflecting their increased economic and political significance in international affairs, and the importance and weight of their participation in the negotiations themselves.

The Trade Negotiations Committee, consisting of representatives of all the 99 countries engaged in the Tokyo Round, guided the negotiations. The main responsibility for the actual bargaining fell on seven negotiating groups, each open to all interested participating countries, which together covered the main areas laid down in the Tokyo Declaration: tariffs, non-tariff measures, the sectoral approach to negotiations, safe-guards, agriculture, tropical products, and improvements in the international framework for the conduct of world trade.

Further information on the background, main aims and provisions of all the Tokyo Round agreements can be found in Chapter II of *GATT Activities in 1978*.

More comprehensive information on the Tokyo Round can be found in a two-volume report [1] by the Director-General of GATT which systematically surveys the issues in the Tokyo Round and how they were dealt with, and provides a detailed assessment of the significance of the results achieved.

## Results for developing countries

A major objective of the developing countries in the Tokyo Round was to seek improved and predictable conditions of access for their increasingly diversified range of exports, an improved legal framework for the future conduct of international trade taking into account their development, financial and trade needs, and special and differential treatment where this was feasible and appropriate, including special treatment for the least-developed countries. They were also concerned to ensure that any liberalization achieved would be placed on a secure footing.

---

[1] " The Tokyo Round of Multilateral Trade Negotiations ", available in English, French and Spanish language editions from the GATT Secretariat, Centre William Rappard, 154, rue de Lausanne, 1211 Geneva 21, Switzerland. (Volume I, April 1979, Price SF. 17.— or US$ 10.—; Volume II, January 1980, Price SF. 8.— or US$ 5.—).

Developing countries, which comprised three-quarters of the Tokyo Round participants, won more flexible provisions for their international trading interests in the agreements covering tariffs, non-tariff measures, agriculture, tropical products, and an improved framework for the conduct of world trade.

A major result of the Tokyo Round was the provision of a permanent legal basis within the GATT for preferential trade treatment on behalf of, and between, developing countries.

Taking into account the uncertain financial and economic conditions in which the Tokyo Round took place, the results achieved represent a significant improvement of trading conditions for developing countries, including the tariff treatment of their exports. It should be noted that although the tariff concessions agreed upon in the Tokyo Round were negotiated bilaterally or plurilaterally, their benefits are, under the most-favoured-nation rule, automatically extended to all GATT member countries.

Developing countries' exports of tropical products have already benefited from the tariff and non-tariff concessions and contributions implemented by most developed countries more than three years ago as part of the Tokyo Round negotiations, in accordance with the priority given to negotiations on tropical products by the Tokyo Declaration. More concessions and contributions, offered during the later stages of the Tokyo Round, are taking effect this year.

The multilateral agreements on non-tariff measures and agriculture that have been negotiated should produce greater transparency in trade, reduce the scope for arbitrary use of non-tariff measures in a number of areas and provide mechanisms for consultation and dispute settlement, aimed at greater international co-operation and more effective monitoring of trade practices. The effective working of agreed rules and procedures will help developing countries protect their commercial interests as they participate increasingly in world trade.

However, as many of the participating countries—both developed and developing—have made clear, progress in certain areas of the Tokyo Round fell below their aims. The safeguards question still remains to be settled. Developing countries have also stressed the need for further efforts in such areas as tariff escalation, quantitative restrictions, and tropical products in order to provide greater opportunities for their exports in developed country markets.

## Framework for the conduct of world trade

The Tokyo Round presented an opportunity to review and improve the working of some of the fundamental provisions of the General Agreement, notably Article I (the most-favoured-nation clause) which guarantees non-discriminatory trade between GATT's member countries. The Tokyo Declaration called for consideration to be given " to improvements in the international framework for the conduct of world trade ", a phrase which thus also provided the name for the " Framework " negotiating group entrusted with the task.

Four important agreements emerged from the Framework group, and were adopted by GATT's member states at their annual session in November 1979, thus taking effect immediately. They are listed below. As part of GATT's future work programme, the member states also agreed to examine the forum and modalities for carrying out future work in the area of export restrictions.

## (i) Differential and more favourable treatment, reciprocity and fuller participation of developing countries

This agreement marks a turning point in international trade relations by recognizing tariff and non-tariff preferential treatment in favour of and among developing countries as a permanent legal feature of the world trading system.

The agreement has sometimes been referred to as the "enabling clause", since a key provision permits developed country members of GATT to give more favourable treatment only to developing countries, and special treatment to the least-developed countries, notwithstanding the most-favoured-nation provisions of Article I of the General Agreement.

## (ii) Trade measures taken for balance-of-payments purposes

This agreement states principles and codifies practices and procedures regarding the use of trade measures by governments to safeguard their external financial position and their balance of payments.

GATT recognizes the particular difficulties that developing countries may have with regard to their balance-of-payments. This agreement improves the procedures for review of these difficulties and provides an equitable and secure basis for the participation of developing countries in the process of consultation on the use of trade measures for balance-of-payments purposes provided for in the General Agreement.

## (iii) Safeguard action for development purposes

This agreement concerns the derogations from other GATT provisions which are accorded to developing countries under Article XVIII of the General Agreement, giving them greater flexibility in applying trade measures to meet their essential development needs. The new provisions regarding application of Article XVIII are expected to make it easier for developing countries to adapt their import policies to the changing needs of their economic development.

## (iv) Understanding on notification, consultation, dispute settlement and surveillance in GATT

This Understanding provides for improvements in the existing mechanisms concerning notification of trade measures, consultations, resolution of disputes and surveillance of developments in the international trading system. It also contains an "agreed description" of customary GATT practices in the field of dispute settlement.

The adoption in the Tokyo Round of these commonly agreed and strengthened rules for resolution of trade disputes is an important contribution to the maintenance of an open and balanced international trading system. Such rules are valuable for safeguarding the trade interests of all countries, especially of the developing and smaller developed countries.

# Tariffs

In the second half of 1979, two Protocols embodying results of the Tokyo Round tariff negotiations were opened for acceptance by governments. By annexing to the Protocols their Schedules of Concessions, and by accepting the Protocols, governments made their tariff-cutting commitments legally binding within the GATT. By the end of 1979, nearly 50 countries, both developed and developing, had accepted the Protocols.[1]

Most of the tariff reductions began on 1 January 1980, to continue with equal annual cuts, the total reduction to become effective not later than 1 January 1987.

The total value of trade affected by Tokyo Round most-favoured-nation (m.f.n.) tariff reductions, and by bindings of prevailing tariff rates, amounts to more than $155 billion, measured on m.f.n. imports in 1977.

As a result of these cuts, the weighted average tariff (that is, the average tariff measured against actual trade flows) on

---

[1] *Geneva (1979) Protocol to the General Agreement on Tariffs and Trade* (Four volumes: *Vol. I*, Protocol plus Schedules of Canada, Czechoslovakia, New Zealand, Norway and South Africa. *Vol. II*, Schedules of United States, Finland and Sweden. *Vol. III*, Schedules of Austria, Japan, Spain, Yugoslavia and Switzerland. *Vol. IV*, Schedules of Iceland, Argentina, Jamaica, Romania, Hungary and European Communities) June 1979. Price US$ 120.00 or SF. 200.00 per set, or US$ 30.00 or SF. 50.00 per single volume.

*Supplementary Protocol to the Geneva (1979) Protocol to the General Agreement on Tariffs and Trade.* (Supplementary Protocol plus Schedules of Australia, Brazil, Canada (concessions additional to those annexed to the Geneva (1979) Protocol), Chile, Dominican Republic, Egypt, European Economic Community (a further Schedule of Concessions containing special commitments affecting developing country exports), India, Indonesia, Israel, Ivory Coast, Korea, Malaysia, Pakistan, Peru, Singapore, Spain (text of the Schedule in French; the text in Spanish was annexed to the Geneva (1979) Protocol), Uruguay, Zaire. *To be published mid-1980.*

Bulgaria has also established its Schedule of Concessions, but since it is not a GATT member, its Schedule is not annexed to either Protocol, but is contained in a separate instrument.

manufactured products in the world's nine major industrial markets will decline from 7.0 to 4.7 per cent, representing a 34 per cent reduction of customs collection.

Since the tariff-cutting formula adopted by most industrialized countries results in the largest reductions generally being made in the highest duties, the customs duties of different countries will be brought closer together or " harmonized ".

Concessions by the European Communities and eight industrial countries (Austria, Canada, Finland, Japan, Norway, Sweden, Switzerland and the United States) covered imports valued at $141 billion ($14 billion in agriculture, and $127 billion in industry). Concessions by other developed countries affected imports valued at $0.4 billion in agriculture and $2.7 billion in industry.

Developing countries made tariff-cutting commitments, in the form of tariff bindings or reductions, on $3.9 billion of their imports in 1977.

Imports into the nine above-mentioned industrialized markets from developing countries affected by m.f.n. tariff concessions amounted to $39 billion: $11 billion in agriculture and $28 billion in industry.

The nine industrialized markets' m.f.n. tariffs facing developing countries' exports of manufactured products will be reduced by 27 per cent based on the weighted average tariff, and by 38 per cent based on the simple average tariff. Concessions on an m.f.n. basis in agriculture resulted in the decline of the weighted average tariff for developing countries from 8.1 to 7.1 per cent: most tariff action on agricultural products of interest to developing countries was however taken in the form of improvements to the Generalized System of Preferences (GSP), or in the framework of the Tokyo Round tropical products negotiations.

The effect of Tokyo Round m.f.n. concessions on the GSP is difficult to assess because of the imprecision underlying GSP statistics. Products entitled to GSP represented $4.6 billion or 23 per cent of dutiable imports of agricultural products, and $22.5 billion or 65 per cent of dutiable imports of industrial products. GSP contributions as a result of the Tokyo Round would increase the GSP product coverage by $0.9 billion in agriculture, and in industry they would nearly compensate for elimination of GSP preference resulting from m.f.n. concessions at zero rates. The GSP preferential margin would be slightly increased in agriculture as m.f.n. concessions were more important on non-GSP products. In industry, where the GSP coverage is more extensive especially in processed goods, the GSP preferential margin shows an expected decrease as a result of application of the tariff-cutting formula to items where GSP admission was free of duty.

## Non-tariff measures

As the general level of tariff protection declined in the post-World War II period, so the distorting effects on world trade of non-tariff measures became more pervasive. The complex and often very difficult negotiations to counter the negative effects of these measures constituted one of the major features that distinguished the Tokyo Round from earlier rounds of GATT trade negotiations.

The core of the Tokyo Round results consists of the agreements, or codes, aimed at reducing, and bringing under more effective international discipline, these non-tariff barriers.

Standing committees were set up in early 1980 to administer the agreements covering non-tariff measures which took effect on 1 January 1980. All the agreements contain provisions for consultation and dispute settlement; they also provide for special and more favourable treatment for developing countries.

Below are brief descriptions of the non-tariff measure agreements, and in the case of the Customs Valuation Code and the revised GATT Anti-Dumping Code, brief accounts of the negotiations in 1979 which led to final agreement on outstanding issues concerning these Codes. More detailed information on the agreements can be obtained from *GATT Activities in 1978*, or from the two-volume report on the Tokyo Round by the Director-General referred to in the footnote on page 13.

### (i)  Subsidies and countervailing duties

Production and export subsidies have in recent years had a growing and distorting influence on international trade, often protecting inefficient production at the expense of competitive industries; the use of countervailing duties has grown proportionately, and resort to both measures has been encouraged by increasing protectionist pressures. For this reason, the *Code on Subsidies and Countervailing Duties* reached in the Tokyo Round is one of the most important results of the negotiations.

The Code (formally known as the *Agreement on Interpretation and Application of Articles VI, XVI and XXIII of the General Agreement on Tariffs and Trade*) aims to ensure that the use of subsidies by any signatory does not harm the trading interests of another, and that countervailing measures do not unjustifiably impede international trade. The Code, which entered into force on 1 January 1980, establishes an agreed framework of rights and obligations covering these measures, and a mechanism for international surveillance and dispute settlement.

## (ii) Technical barriers to trade

The *Agreement on Technical Barriers to Trade* (also known as the *Standards Code*) aims to ensure that when governments or other bodies adopt technical regulations or standards, for reasons of safety, health, consumer or environmental protection, or other purposes, these regulations or standards, and the testing and certification schemes related to them, should not create unnecessary obstacles to trade. It provides for notification and consultation on these measures and contains provisions for granting technical assistance and special and more favourable treatment to developing countries. For the first time in the field of standardization, there will be worldwide legally binding rules between governments, enabling them to complain about, and obtain redress for, code violations by other signatories. The Code entered into force on 1 January 1980.

## (iii) Import licensing procedures

The *Agreement on Import Licensing Procedures* recognizes that these procedures can have acceptable uses, but also that their inappropriate use may hamper international trade; it aims at ensuring that they do not in themselves act as restrictions on imports. By becoming parties to the Agreement, which took effect on 1 January 1980, governments commit themselves to simplifying their import licensing procedures and to administering them in a neutral and fair way.

## (iv) Government procurement

The *Agreement on Government Procurement* aims to secure greater international competition in the government procurement market. Increased competition, besides benefiting exporters, should also make more effective use of taxpayers' money in government purchases of goods.

The Agreement, which will enter into force on 1 January 1981, contains detailed rules on the way in which tenders for government purchasing contracts should be invited and awarded. It is designed to make laws, regulations, procedures and practices regarding government procurement more transparent, and to ensure that they do not protect domestic products or suppliers, or discriminate among foreign products or suppliers. The Agreement's provisions will apply to individual government contracts worth more than SDR 150,000 (about US$ 197,000).

The Agreement applies to products rather than services (which are covered only to the extent that they are incidental to the supply of products and cost less than the products). For the purposes of the Agreement, the buyer is a government entity or agency which has been listed in an annex. This list of entities resulted from negotiations among the signatories of the Agreement, which also envisages future rounds of negotiations to bring further entities within its coverage. To join the Agreement, a country is required to make a contribution in the form of a list of its purchasing entities which are covered.

## (v)  Customs Valuation

The *Agreement on Implementation of Article VII of the General Agreement on Tariffs and Trade* (known as the *Customs Valuation Code*) is intended to provide a fair, uniform and neutral system for the valuation of goods for customs purposes: a system that conforms to commercial realities, and which outlaws the use of arbitrary or fictitious customs values.

The code provides a revised set of valuation rules, expanding and giving greater precision to the provisions on customs valuation already found in the GATT. It allows developing countries to delay applying it for five years from the date of its entry into force. The Code will enter into force on 1 January 1981; the United States and the European Communities will apply its provisions as from 1 July 1980.

A *Protocol to the Customs Valuation Code*, which is deemed to be part of the Code, gives greater powers to developing country customs authorities to counter potentially unfair advantages to exporters and importers who are related, and also to combat what they might judge to be fraudulent invoicing. It also provides that signatories will give sympathetic consideration to any developing country which asks for an extension of the five-year delay in applying the Code, and provides for technical assistance to developing countries to help them set up new valuation systems based on the Code's provisions.

Two texts relating to the *Customs Valuation Code* were listed in the Procès-Verbal opened for signature by governments on 12 April 1979. The first was a complete version of the Code, to which both developed and developing countries subscribed; the second contained additional provisions which many developing countries wanted to see incorporated in the Code.

Some of these additional provisions presented considerable difficulties for other participants in the Tokyo Round, but following negotiations during the summer and autumn of 1979, the participants agreed on the Protocol which provided for much of what the developing countries had been pressing for. The Protocol was in effect a compromise between the developed and developing countries, with neither side securing everything it might have wished. Subsequently, both the Code and the Protocol were opened for signature by governments.

## (vi) Revised GATT Anti-Dumping Code

Participants in the Tokyo Round agreed on a revision of the GATT Anti-Dumping Code (formally known as the *Agreement on Implementation of Article VI of the General Agreement on Tariffs and Trade*), which was negotiated by a group of major industrialized countries during the Kennedy Round (1964-1967).

This revised version of the Anti-Dumping Code brings certain of its provisions (notably those concerning determination of injury; price undertakings between exporters and the importing country; imposition and collection of anti-dumping duties) into line with the relevant provisions of the Code on Subsidies and Countervailing Duties.

Following negotiations over the summer and autumn months of 1979 on outstanding differences between developed and developing countries, certain understandings were reached which enabled the revised GATT Anti-Dumping Code to be opened in November 1979 for acceptance by governments.

The most important of these understandings recognizes that, as special economic conditions in developing countries affect prices in the domestic market, these prices do not provide a commercially realistic basis for dumping calculations; and the fact that a developing country's export price is lower than the comparable domestic price shall not in itself justify an investigation or the determination of dumping. In such cases, the normal value, for the purposes of ascertaining whether the goods are being dumped, is to be determined by methods such as a comparison of the export price with the comparable price of the like product when exported to any third country, or with the cost of production of the exported goods in the country of origin, plus a reasonable amount for administrative, selling and any other costs and for profits.

A second understanding accepts that developing countries may have difficulties in adapting their legislation to the requirements of the Code as regards anti-dumping investigations initiated by them. It provides for granting, on a case-by-case basis, time-limited exceptions from the relevant provisions of the Code. It also provides for technical assistance to developing countries which have accepted the Code. This covers implementation of the Code, training of personnel, and the supply of information on methods, techniques, and other aspects of investigating dumping practices.

The revised Anti-Dumping Code took effect on 1 January 1980.

# Agriculture

The agreements on tariff and non-tariff concessions, and all the multilateral agreements reached in the Tokyo Round apply to world trade in farm products, as well as to industrial products.

Participating countries in the Tokyo Round also drew up multilateral agreements on Bovine Meat and on Dairy Products.

The Trade Negotiations Committee in April 1979 recommended GATT's member states to further develop active cooperation in the agricultural sector. The GATT Council later requested the Director-General to consult with interested delegations on this matter and report to the next annual session of GATT's member states in November 1980.

## (i)  Bovine Meat

The *Arrangement Regarding Bovine Meat* aims to promote expansion, liberalization and stabilization of international trade in meat and livestock as well as to improve international co-operation in this sector. The Arrangement covers beef and veal, and live cattle.

Signatory governments have established within GATT an *International Meat Council* which will review the functioning of the Arrangement, evaluate the world supply and demand situation for meat, and provide a forum for regular consultation on all matters affecting international trade in bovine meat, including bilateral agreements on trade in this sector reached during the Tokyo Round.

Entry into force of the Arrangement on 1 January 1980 entailed abolition of the International Meat Consultative Group established by GATT in 1975.[1]

## (ii) ` Dairy products

*Background*

In May 1970, most of the main participants in international trade in dairy products reached an Arrangement in GATT to fix minimum export prices for skimmed milk powder, in order to restore stability to the world market for this product. In 1973 they reached agreement upon, and brought into effect, a Protocol which made similar arrangements for milk fats such as butter oil. These arrangements (formally entitled The Arrangement Concerning Certain Dairy Products and the Protocol Relating to Milk Fat) worked satisfactorily throughout the 1970's, as did the " Gentleman's Agreement " on minimum prices for whole milk powder which operated from 1963 onwards under the auspices of the OECD.

*International Dairy Arrangement*

All three arrangements referred to above lapsed at the end of 1979, when they were superseded by an *International Dairy Arrangement*, resulting from the Tokyo Round, which entered into force on 1 January 1980.

The aims of the new Arrangement are to expand and liberalize world trade in dairy products; to achieve greater stability in this trade and therefore, in the interests of exporters

---

[1] See page 62.

and importers, to avoid surpluses and shortages, undue fluctuations in prices and serious disturbances in international trade; to provide better possibilities for developing countries to participate in the expansion of world trade in dairy products so as to further their economic and social development; and to improve international cooperation in these areas.

The Arrangement in general covers all dairy products. More specifically, as Annexes to the Arrangement, there are *three Protocols* which set specific provisions, including minimum prices, for international trade in: (i) certain milk powders, (ii) milk fats including butter, (iii) certain cheeses.

The participants have established within GATT an *International Dairy Products Council* which will review the functioning of the Arrangement and evaluate the situation in, and future outlook for, the world dairy market.

# Tropical products

Many developing countries depend largely on tropical products for their export earnings, and the Tokyo Declaration stipulated that this should be treated as a special and priority sector.

The concessions and contributions by industrialized countries on exports of tropical products from developing countries were the first concrete results of the Tokyo Round. Most of the industrialized countries implemented their concessions and contributions in 1976 and 1977. Further concessions offered during the later stages of the negotiations, including the concessions offered by the United States, are taking effect from 1980 onwards.

Some forty-six developing countries submitted requests to eleven developed participants for tariff and non-tariff concessions on exports of tropical products. These requests covered not only typical tropical products, but also a broad range of other products—agricultural, raw materials and minerals, semi-manufactures and manufactures—of which developing countries are producers and actual or potential exporters.

Of the 4,300 dutiable items at the tariff line level subject to requests, m.f.n. concessions and GSP contributions were granted with respect to some 2,855 tariff lines. Within the latter figure, approximately 940 m.f.n. concessions and GSP contributions were implemented in the initial phase of the tropical products negotiations during the period 1976-1977. (For details, see *GATT Activities in 1976*, pp. 29-33).

## Agreement on Trade in Civil Aircraft

Several major industrialized participants in the Tokyo Round reached an *Agreement on Trade in Civil Aircraft* which committed signatory governments to eliminate, by 1 January 1980, all customs duties and any similar charges of any kind on civil aircraft, aircraft parts, and repairs on civil aircraft. These zero duties are legally " bound " under the GATT, and thus, in accordance with the most-favoured-nation rule, apply to all GATT member countries.

The Agreement contains an Annex listing all the products covered, ranging from passenger airliners, helicopters, gliders and ground flight simulators to food warmers and oxygen masks.

The signatories have established within GATT a *Committee on Trade in Civil Aircraft* to review operation of the Agreement, and to provide a forum for consultation and dispute settlement.

# The multilateral safeguard system

The Tokyo Declaration called for " an examination of the adequacy of the multilateral safeguard system, in particular of the way in which Article XIX is applied ".

Against the background of increasing world economic difficulties and strong protectionist pressures, several participants in the Tokyo Round hardened their negotiating positions on this key issue during the negotiations, and it had still not been resolved by the time that GATT's member states met for their annual session in November 1979.

The right to impose import controls or other temporary trade restrictions to prevent commercial injury to a domestic industry, and the corresponding right of exporters not to be lightly deprived of access to markets are, broadly speaking, what is meant by " the multilateral safeguard system ". Provisions to this effect exist in GATT, and in particular in Article XIX, which is entitled: " Emergency Action on Imports of Particular Products ". However, experience has shown that GATT member countries have tended to turn to other articles of the Agreement for justification of their action or even to remedies outside GATT, such as " voluntary " restraint agreements and " orderly marketing arrangements ".

Under Article XIX, which is the main GATT " escape clause ", member countries have to show that imports of a certain product take place in such increased quantities and under such conditions as to cause or threaten serious injury to domestic producers of like or directly competitive products before they can be allowed to take emergency action. The article thus limits their freedom of action by specifying the circumstances in which such action can be taken and defining the GATT obligation or concession that may be suspended or withdrawn. It also calls for consultations with the supplying countries affected and allows the latter to take retaliatory measures if the consultations do not lead to an agreement.

Action taken under Article XIX has traditionally been non-discriminatory, i.e. applied to all GATT members, and not only against the country whose exports are causing injury. One of the major differences of opinion in the Tokyo Round negotiations concerned this point. It was argued that governments would more willingly accept wide-ranging trade liberalization if safeguard measures could be taken selectively. Against this it was argued that the principle of non-discriminatory application of safeguard measures should be maintained, as selective applications might allow easy or arbitrary establishment of trade barriers.

By the time that the Trade Negotiations Committee met on 11-12 April 1979, agreements had been reached covering nearly all areas of the Tokyo Round. The lack of agreement on safeguards was the major piece of unfinished business. The Committee agreed that further negotiations on this issue should be continued as a matter of urgency with the aim of reaching agreement by mid-July. Negotiations resumed quickly and by the beginning of July attention was almost exclusively focussed on the problem of selectivity. Earlier in the year, both developed and developing country participants had already accepted selectivity as a working hypothesis. However, considering they had made a major concession in the interest of reaching a mutually acceptable compromise, many countries, developed and developing, insisted on the need for the application of strict rules, criteria and surveillance arrangements relating to the use of selective measures. For these countries it was essential that any country taking selective safeguard action should gain agreement for such action by the affected exporting country or countries concerned; or, in the absence of such agreement, that there should be a prior determination by the Committee that would be established to implement any Safeguards Code to the effect that the conditions and criteria for selective action had been fulfilled. These countries also insisted that actual, and not only potential material injury to domestic production in the importing country had to be proven; that account had to be taken of damage that might be caused to export industries in developing countries; and that safeguard measures should not be used as a substitute for structural economic adjustment.

For certain industrialized countries, some of these conditions were unacceptable; in particular, they opposed the call for prior determinations by the Committee which would be set up to supervise the Safeguards Code. They favoured an approach that would permit unilateral, selective action with subsequent review by the Committee, especially in critical circumstances where delay in restricting imports would cause damage difficult to repair.

This difference in positions proved to be a major sticking point, and despite intensive negotiations during July, it proved impossible to close the gap on the fundamental issue of selectivity.

In July, the Director-General proposed to the Council that GATT's member states should establish a committee in GATT to continue the negotiations on safeguards. This initiative, which brought the safeguards issue squarely into the GATT, was the subject of continuous consultations from mid-September onwards, and in early November the GATT Council unanimously supported establishment of a Committee in GATT (rather than in the framework of the Tokyo Round) to continue the safeguards negotiations.

Later in November, at their annual session, GATT's member states formally adopted a decision to establish the Committee, whose membership is open to all the countries which participated in the Tokyo Round. In the same decision, the GATT member states " reaffirm their intention to continue to abide by the disciplines and obligations of Article XIX of the General Agreement ".

Another decision, linked to that on safeguards, set up a sub-committee of GATT's Committee on Trade and Development to examine protective measures taken against imports from developing countries; this fulfilled a commitment embodied in resolution 131 of the UNCTAD conference at Manila in May 1979 inviting GATT to set up such a body.

Failure to negotiate an agreement on safeguards in the Tokyo Round was regarded as a setback by a large number of participating countries, especially the developing countries. But fortunately, no country has closed the door on these negotiations. The way is still open for a solution to this difficult and politically sensitive problem. Agreement on a strengthened multilateral safeguard system within the GATT is important to both developed and developing countries for the future conduct of their international trade. Given present protectionist pressures, it is essential that all nations should be able to refer to, and rely on, a commonly agreed and realistic set of rules for safeguard action against imports.

# INTERNATIONAL ASSISTANCE POLICIES

## H. B. CHENERY AND A. M. STROUT

Our analysis has shown the conditions under which external assistance may make possible a substantial acceleration in the process of economic development. It has focused on the interrelations among external resource requirements and the development policies of recipient countries. Analysis of these interrelations leads to several principles of general applicability to international assistance policy.

The central questions for assistance policy are the measurement of the effectiveness of external assistance, the policies that recipient countries should follow to make best use of external resources, and the basis for allocating assistance among countries. This chapter summarizes the main implications of our analysis for each of these questions and adds some qualitative elements which have been omitted from the formal analysis.

## THE EFFECTIVENESS OF ASSISTANCE

In the short run the effectiveness of external resources depends on their use to relieve shortages of skills, saving, and imported entities. The productivity of additional amounts of assistance over longer periods can be measured by the increase in output resulting in the fuller use of domestic resources which they make possible.

Over longer periods, the use that is made of the initial increase in output becomes more important. Even if the short-run productive aid is high, the economy may continue to be dependent on outside assistance indefinitely unless the additional output is allocated so as to increase saving and reduce the trade gap. Over the whole period of transition to self-sustaining growth, the use that is made of the successive increments in GNP is likely to be more important than the efficiency with which external assistance was utilized in the first instance. To emphasize this point, let us assume that the productivity investment in the first 5 years of the upper-limit development sequence outlined above for Pakistan had been one-third lower, requiring a correspondingly larger amount of investment and external aid to achieve the same increase in GNP. The effect would be to increase the total aid required over the 17-year period to achieve self-sufficiency by some 45%. This, however, is less than the effect on aid requirements of a reduction in the marginal saving rate from .24 to .22% if critical elements in the development sequence are getting the increase in the rate of growth, channeling the increments in income into increased saving, and allocating investment so as to avoid balance of payments bottlenecks. These long-run aspects are likely to be considerably more important than the efficiency with which external capital used in the short run.[1]

---

[1] This conclusion is demonstrated in the evaluation of the effectiveness of aid to Greece in Organization for Economic Cooperation and Development, *Development Assistance Efforts and Policies:* 1965 Review, Report of the Chairman of the Development Assistance Committee, Paris, 1965.

Reprinted with permission from *The American Economic Review*
**56**(4), 723–729, Copyright 1966

The long-run effectiveness of assistance is also likely to be increased by supporting as high a growth rate as the economy can achieve without a substantial deterioration in the efficiency of use of capital. There are also several factors omitted from the formal models that argue for more rapid growth:

1. the fact that a smaller portion of the increase in GNP is offset by population growth;

2. the gain in political stability and governmental effectiveness that is likely to result;

3. the greater likelihood of being able to raise marginal saving rates and export growth when GNP is growing more rapidly;[2]

4. the greater likelihood of attracting foreign private investment to finance the needs for external capital.

While the last three factors cannot be measured with any accuracy, they appear to have been important in most countries that are successfully completing the transition, such as Israel, Greece, Taiwan, Mexico, Peru, and the Philippines. These examples support the theoretical conclusion that the achievement of a high rate of growth, even if it has to be initially supported by large amounts of external capital, is likely to be the most important element in the long-term effectiveness of assistance. The substantial increases in internal saving ratios that have been achieved in a decade of strong growth—from 7% to 12% in the Philippines, 11% to 16% in Taiwan, 6% to 14% in Greece, and 9% to 12% in Israel—demonstrate the speed with which aid-sustained growth can be transformed into self-sustained growth once rapid development has taken hold.

## POLICIES FOR RECIPIENT COUNTRIES

While the receipt of external assistance may greatly reduce the time required for a country to achieve a satisfactory rate of growth, dependence on substantial amounts of external resources creates some special policy problems. One lesson from the preceding analysis is that the focus of policy should vary according to the principal limitations to growth. Just as optimal countercyclical policy implies different responses in different phases of the business cycle, optimal growth policy requires different "self-help" measures in different phases of the transition.

In Phase I, where the growth rate is below a reasonable target rate, the focus of policy should be on increasing output, implying an increase in the quality and quantity of both physical capital and human resource inputs. Our statistical comparisons suggest that a rate of growth of investment of 10–12% is a reasonable target for countries whose initial investment level is substantially below the required level. Phase I can be completed by most countries in a decade if this increase in investment is accompanied by sufficient improvement in skills and organization to

---

[2]The advantages of more rapid growth with constant per capita marginal savings rates are demonstrated by J.C.H. Fei and D.S. Paauw, "Foreign Assistance and Self-Help: Reappraisal of Development Finance," *Rev. Econ. Stat.,* Aug. 1965, 67, 251–67.

make effective use of the additional capital that becomes available. Although it is probably more important in this phase to focus on securing increases in production and income, a start must also be made on raising taxes and saving if international financing is to be justified by performance.

As Phase I is completed, the rate of increase in investment can be allowed to fall toward a feasible target rate of GNP growth, which is unlikely to be more than 6–7%. The focus of development should then be increasingly on (1) bringing about the changes in productive structure needed to prevent further increases in the balance of payments deficit, and (2) channeling an adequate fraction of increased income into saving. Although theoretical discussion has tended to stress the second requirement, the first appears to have been more difficult in practice for many countries. Since substantial import substitution is required just to prevent the ratio of imports to GNP from rising, export growth at least equal to the target growth of GNP is seen to be necessary in order to reduce external aid.

As the focus of development policy changes, the instruments of policy must change accordingly. Somewhat paradoxically, successful performance in Phase I, which would justify a substantial and rising flow of foreign assistance, may make success in Phase III more difficult. If investment and other allocation decisions are based on the exchange rate that is appropriate for a substantial flow of aid, they are not likely to induce sufficient import substitution or increased exports to make possible a future reduction in the capital inflow. Planning should be based on the higher equilibrium exchange rate that would be appropriate for a declining flow of aid in order for the necessary changes in the production structure to be brought about in time.

It is the need for rapid structural change that sets the lower limit for the time required to complete the transition to self-sustaining growth. Although there is the possibility of completing this transition in less than 20 years starting from typical Asian or African conditions, it is very unlikely that any such country can meet all the requirements of skill formation, institutional building, investment allocation, etc. in less than one generation.

## POLICIES FOR DONOR COUNTRIES

Donors are concerned with criteria for the allocation of aid among recipients, and the means for controlling its use. Allocation policies are complicated by the mixture of objectives that motivate international assistance, the most important of which are (1) the economic and social development of the recipient, (2) the maintenance of political stability in countries having special ties to the donor, and (3) export promotion. This mixture of motives has led to a complex system of aid administration in all countries.

The predominant basis for development loans is the individual investment project, for which external financing is provided to procure capital goods from the donor country. Loans not limited to equipment for specific projects are provided to a few selected countries against the balance-of-payments needs of development programs.[3] Substantial but declining amounts of grants are also furnished for budgetary support of ex-colonies and other dependent areas.

Our analysis suggests some directions in which improvements can be sought in the present methods of supporting economic development, which is the objective on which all parties agree. We first consider methods of transferring resources to individual countries and then allocation of assistance among countries.

## The Transfer of Assistance

Any system for transferring resources must include: (1) a basis for determining the amount of the transfer, (2) specification of the form of resources to be furnished, and (3) a basis for controlling their use. On all these counts the project system has the virtue of simplicity. It also provides for detailed evaluation of the investments that are directly financed from external aid—which may be 10% or so of total investment—and for increasing their productivity through technical review.

While the project system has much to commend it when the main focus is on increasing the country's ability to invest, it becomes increasingly inappropriate as the development process gets under way. As the rate of growth increases, we have shown that the effectiveness of aid depends more on the use that is made of the additional output than on the efficiency with which a limited fraction of investment is carried out. Furthermore, an attempt to finance the amount of external resources needed during the peak period of an optimal growth path—which may imply aid equal to 30–40% of total investment—by the project mechanism alone may greatly lower the efficiency of use of total resources. Limiting the form of assistance to the machinery and equipment needed by substantial investment projects is likely either to lower the rate of growth or to distort the pattern of investment.

In these circumstances, assistance would be more effective if the range of commodities supplied could be broadened to permit the recipient's pattern of investment and production to evolve in accordance with the principle of comparative advantage.[4] While domestic supply can—and indeed must—lag behind demand in some sectors to accommodate the needed resource transfer, the country should also be preparing to balance its international accounts by the end of a specified transitional period.

Since donors fear that uncontrolled imports may be wasted in increased consumption without the restraints imposed by the project mechanism, an alternative means of control is needed. Part of the solution lies in relating the amount of aid supplied to the recipient's effectiveness in increasing the rate of domestic saving, so that the added aid necessarily increase saving and investment as income grows. As development planning and statistics on overall performance improve, this kind of

---

[3]In the terminology of AID, the latter are called program loans. About half of United States development lending is on a program basis in contrast to a much smaller proportion for other Organization for Economic Cooperation and Development (OECD) Development Assistance Committee members, or the World Bank.

[4]This observation applies to aid in the form of agricultural commodities as well as to aid in the form of machinery or any other specified goods.

"program approach" is becoming increasingly feasible, both from the point of view of determining the amounts of assistance needed and assessing the results.[5]

The strongest argument for the program approach arises for countries in Phase III where the balance of payments is the main factor limiting growth and where there is typically excess capacity in a number of production sectors. In this situation, the highest priority use of imports is for raw materials and spare parts to make more effective use of existing capacity; project priorities should give primary weight to import substitutes and increased exports. In this situation donor controls should be primarily concerned with the efficient use of total foreign exchange resources, which can only be assessed adequately in the framework of a development program.

## Allocation of Assistance

If the objectives of the donor countries could be expressed as some function of the growth of each recipient, it would be possible to allocate aid primarily on the basis of expected development performance. The varying political objectives of the donors complicate the problem because each would give somewhat different weights to a unit of increase in income as among recipients. Even with this limitation, however, there may be considerable scope for reallocating a given amount of aid or for selective increases in individual country totals in accordance with criteria of self-help.

The predominant project approach now in use favors countries whose project preparation is relatively efficient. Other qualities that are equally important to successful development—tax collection, private thriftiness, small-scale investment activity, export promotion—are ignored in focusing on this one among many aspects of better resource use.[6]

Where fairly reliable statistics are available, an alternative procedure would be to establish minimum overall performance standards for each country and to share the aid burden among interested donors through a consortium or other coordinating mechanism. For example, a country starting in Phase I might have as its principal performance criteria: (1) growth of investment at 10% per year at a minimum standard of productivity, and (2) the maintenance of a marginal saving rate of .20 (or alternatively a specified marginal tax rate). There would be little possibility to waste aid on these terms, since the required increase in savings would finance a large proportion of total investment. Appropriate overall standards for saving rates and balance-of-payments policies for countries in Phase II and Phase III could also be established without great difficulty. A country maintaining high standards—say a marginal savings rate of .25 and a marginal capital-output ratio of less than 3.3—could safely be allotted whatever amount of aid it requested in the knowledge that the larger the amount of aid utilized, the higher would be its growth rate, and the more rapid its approach to self-sufficiency.

[5]The United States government has been using the program approach in India, Pakistan, Turkey, Tunisia, Chile, Colombia, and Brazil.

[6]It is perhaps more than coincidence that most of the striking successes in development through aid—Greece, Israel, Taiwan, etc.—were financed largely on a nonproject basis.

# Part VIII

## International Payments

Meade
Johnson
Cooper
Karlik
de Strihou
Dornbusch
Artus and Crockett
International Monetary Fund

*The Economic Journal,* **88** *(September* 1978), 423-435      JAMES MEADE

*Printed in Great Britain*

## THE MEANING OF "INTERNAL BALANCE"*

### I

It is a special privilege for me on this occasion to have my name associated with that of Professor Bertil Ohlin. By the younger generation of economists we are no doubt both regarded as what in my country and now in his own are now termed senior citizens; but I am just that much younger than Professor Ohlin to have regarded him as one of the already established figures when I was first trying to understand international economics. His great work (Ohlin, 1933) on *International and Inter-regional Trade* opened up new insights into the complex of relationships between factor supplies, costs of movement of products and factors, price relationships, and the actual international trade in products, migration of persons, and flows of capital. Of the two volumes which I later wrote on International Economic Policy – namely, *The Balance of Payments* and *Trade and Welfare –* (Meade, 1951, 1955) – it is in the latter that the influence of this work by Professor Ohlin is most clearly marked.

Professor Ohlin also made an important contribution to what now might be called the macro-economic aspects of a country's balance of payments. In 1929 in the ECONOMIC JOURNAL he engaged in a famous controversy with Keynes on the problem of transferring payments from one country to another across the foreign exchanges. In this he laid stress upon the income–expenditure effects of the reduced spending power in the paying country and of the increased spending power in the recipient country. In doing so he made use of the usual distinction between a country's imports and exports; but in addition he emphasised the importance of the less usual distinction between a country's domestic non-tradeable goods and services and its tradeable, exportable and importable, goods. I made some use of this latter distinction in my *Balance of Payments*; but looking back I regret that I did not let it play a much more central role in that book.

### II

Indeed I realise now, looking back with the advantage of hindsight, that my two books were deficient in many respects. From this rich field of deficiencies I have selected one as the subject for today's lecture, because it raises an issue which in my opinion is at the present time perhaps the most pressing of all for the maintenance of a decent international economic order.

The basic analysis in *The Balance of Payments* was conducted in terms of static equilibrium models rather than in terms of dynamic growing or disequilibrium models. The use of this method of comparative statics was a result of Keynes's work.

Keynes (1936) in *The General Theory* applied Marshall's short-period analysis to the whole macro-economic system instead of to one single firm or industry. In this model additions to capital stocks are taking place; but we deal with a period of time over which the addition to the stock bears a negligible ratio to the total existing stock. Variable factors, and in particular labour, are applied to this stock with a rising marginal cost until marginal cost is equal to selling price – an assumption which can be modified to accommodate micro-economic theories about determinants of output and prices in conditions of imperfect competition. The rest of the Keynesian analysis with its consumption function, liquidity preference, and investment function can be used to determine the short-period, static, stable equilibrium levels of total national income, output, employment, interest, and so on, in terms of such parameters as the money wage rate, the supply of money, entrepreneurs' expectations, rates of tax, levels of government expenditure, and the foreign demand for the country's exports. The model can then be used to show how changes in these parameters would affect the short-period equilibrium levels of the various macro-economic variables. Keynes was not, I think, interested in the process of change from one short-period equilibrium to another, though he was very interested in the way in which expectations in a milieu of uncertainty would affect the short-period equilibrium, in particular through their effect upon investment. If my interpretation is correct, he judged intuitively that the short-period mechanisms of adjustment were in fact such that at any one time the macro-elements of the system would not be far different from their short-period equilibrium values; and he may well have been correct in this judgement in the 1930s.

*The Balance of Payments* was essentially based on macro-economic models of this kind. What I tried to elaborate was the international interplay between a number of national economies of this Keynesian type. For this purpose I discussed the different combinations of policy variables which would serve to reconcile what I called "external balance" with what I called "internal balance". By "external balance" was meant a balance in the country's international payments; and although this idea presents, and indeed at the time was realised to present, considerable conceptual difficulties, nevertheless I still instinctively feel that it is not a foolish one. But can the same be said of the idea of "internal balance"? Does it mean full employment or does it mean price stability?

I don't believe that I was quite so stupid as not to realise that full employment and price stability are two quite different things. But one treated them under the same single umbrella of "internal balance" because of a belief or an assumption that if one maintained a level of effective demand which preserved full employment one would also find that the money price level was reasonably stable. The reason for making this tacit or open assumption was, of course, due to a tacit or open assumption that the money wage rate was normally either constant or at least very sluggish in its movements. In this case with the Keynesian model the absolute level of money prices would be rather higher or lower according as the level of effective demand moved the economy to a higher or lower point on the upward-sloping short-period marginal cost curve. But there would be no reason to expect a rapidly rising or falling general level of money prices in any given short-period equilibrium position.

This may have been a very sensible assumption to make in the 1930s. It is more doubtful whether it was a sensible assumption to make in the immediate post-war years when *The Balance of Payments* was being written. In any case if I were now rewriting that book I would do the underlying analysis not in terms of the reconciliation of the two objectives of external balance and internal balance, but in terms of the reconciliation of the three objectives of equilibrium in the balance of payments, full employment, and price stability.

Why did I not proceed in this way in the first place?

I was certainly aware of the danger that trade union and other wage-fixing institutions might not permit the maintenance of full employment without a money cost-price inflation. But I suppose that writing immediately after the war I adopted the basic model which was so useful before the war and simply hoped that somehow or another it would be possible to avoid full employment leading to a wage-price inflation. Having done so I found that there remained quite enough important international relationships to examine even on that simplifying assumption. That is not perhaps a very strong defence of my position, but I suspect that it is the truth of the matter.

I am well aware that I could now adopt a more sophisticated line of defence of my past behaviour. It is quite possible to define as the natural level of employment, that level which – given the existing relevant institutions affecting wage-fixing arrangements – would lead to a demand for real wage rates rising at a rate equal to the rate of increase of labour productivity. One has only to add to this the assumption that one starts from a position in which there is no general expectation of future inflation or deflation of money prices to reach the position in which the maintenance of this natural level of employment is compatible with price stability. If this natural level of employment is treated as "full employment" one has succeeded in defining a situation of "internal balance" in which "full employment" and "price stability" can be simultaneously achieved.

One could then go on to discuss the many institutions which affect this so-called full employment level. Decent support of the living standards of those who are out of work may mean that unemployed persons are legitimately rather more choosey about the first alternative job which is offered to them, quite apart from the existence of a limited number of confirmed "sturdy beggars" who prefer living on social benefits to an honest day's work. The obligation to make compulsory severance or redundancy payments when employees are dismissed may make some employers less willing to expand their labour force in conditions in which future developments are uncertain. Monopolistic trade union action may put an extra upward pressure on money wage demands which means that unemployment must be maintained at a higher level in order to exert an equivalent countervailing downward pressure. Some statutory wage-fixing bodies in particular occupations may exert a similar influence.

It is not very helpful to squabble about definitions. There is, however, a very real difference of substance between those who do, and those who do not, consider these labour market institutions to cause very real difficulties. Is it necessary to achieve some radical reform of these institutions in order to make reasonable price stability compatible with reasonably low levels of unemployment? Or is it a fact that, if affairs could for a time be so conducted as to remove the expectation of any marked future inflation of the money cost of living, we would find that even with present institutions the natural level of unemployment would not be at all excessive? I myself would expect that in many countries including the United Kingdom the recasting of labour market institutions would still be found to be of crucial importance.

As far as the less important question of definition is concerned, I prefer to think of "full employment" and "price stability" as being two separate and often conflicting objectives of macro-economic policies. Anyone who has this preference can, of course, be legitimately challenged to define what is meant by full employment. Perhaps I would be driven to the extreme of defining full employment as that level of employment at which the supply-demand conditions would not lead to attempts to push up the real wage rate more rapidly than the rate of increase in labour productivity if there were perfect competition in the labour market – no monopsonistic employers, no monopolistic trade unions, no social benefits to the unemployed, no obligations on employers to make compulsory severance or redundancy payments to dismissed workers, and so on – though I am not at all sure whether this extreme form of definition has much meaning. However, in so far as full employment could be defined somewhere along these lines, one would end up with price stability and full employment as separate macro-economic objectives in any real world situation with labour market institutions as one of the instruments of policy. This is the way in which I like to think of macro-economic problems.

If one adopted this approach, how should *The Balance of Payments* be recast? In the basic model we would have the three targets of external balance, full employment, and price stability. If one continued to think in terms of matching to each "target" a relevant policy "weapon", one could divide the weapons into three main armouries: the first containing the weapons which directly affect the level of money demands (e.g. monetary and budgetary policies); the second containing the weapons which directly affect the fixing of money wage rates; and the third containing the weapons which directly affect the foreign exchanges, such as the fixing of rates of exchange, measures of exchange control, and commerical policy measures designed directly to affect the total value of imports and exports.

My subsequent education in the rudiments of the theory of the control of dynamic systems suggested to me that this was not the best way to have proceeded. One should not pair each particular weapon off with a particular target as its partner, using weapon $A$ to hit target $A$, weapon $B$ to hit target $B$, and so on. Rather one should seek to discover what pattern of combination of simultaneous use of all available weapons would produce the most preferred pattern of combination of simultaneous hits on all the desirable targets. With this way of looking at things no particular weapon is concentrated on any particular target; it is the joint effect of all the weapons on all the targets which is relevant.

There is no doubt that this is the way in which a control engineer will look at the problem and that in a technical sense it is the correct way to find the most preferred pattern of hits on a number of targets simultaneously. For a considerable period between the writing of *The Balance of Payments* and the present time I was fully enamoured of this method.

I am now, however, in the process of having second thoughts and of asking myself whether the idea of trying to hit each particular target by use of a particular weapon or clearly defined single armoury of weapons is really to be ruled out. This onset of second childhood is due to a consideration of the political conditions in which economic policies must be operated. It is most desirable in a modern democratic community that the ordinary man or woman in the streets should as far as possible realise what is going on, with responsibilities for success or failure in the different fields of endeavour being dispersed but clearly defined and allocated. To treat the whole of macro-economic control as a single subject for the mysterious art of the control engineer is likely to appear at the best magical and at the worst totally arbitrary and unacceptable to the ordinary citizen. To put each clearly defined weapon or armoury of weapons in the charge of one particular authority or set of decision makers with the responsibility of hitting as nearly as possible one well defined target is a much more intelligible arrangement.

Of course there are obvious disadvantages in any such proposal. Thus the best way for authority $A$ to use weapon $A$ to achieve objective $A$ will undoubtedly be affected by what authorities $B$ and $C$ are doing with weapons $B$ and $C$. It depends upon the structure of relationships within the economic system how far these repercussions are of major importance. Perhaps a mysterious dynamic model operated inconspicuously in some back room by control experts for silent information of the authorities concerned might be useful; and in any case in the real world it would be desirable for the different authorities at least to communicate their plans to each other so that, by what one hopes would be a convergent process of mutual accommodation, some account could be taken of their interaction. But in the modern community there is, I think, merit in arrangements in which each authority or set of decision makers has a clear ultimate responsibility for success or failure in the attainment of a clearly defined objective.

### III

There are six ways in which each of three weapons can be separately aimed at each of three targets. Some of these patterns make more sense than others. In this lecture I can do no more than give a brief account of that particular pattern which, as it seems to me at present, would make the best sense if one takes into account both economic effectiveness and also comprehensibility of responsibilities in a free democratic society.

With this pattern:

(1) the instruments of demand management, fiscal and monetary, would be

used so to control total money expenditures as to prevent excessive inflations or deflations of total money incomes;

(2) wage-fixing institutions would be modelled so as to restrain upward movements of money wage rates in those particular sectors where there were no shortages of manpower and to allow upward movements where these were needed to attract or retain labour to meet actual or expected manpower shortages. This should result in the preservation of full employment with some moderate average rise in money wage rates in conditions in which demand management policies were ensuring a steadily growing money demand for labour as a whole; and

(3) foreign exchange policies would be used to keep the balance of payments in equilibrium.

This pattern implies the use of the weapons of demand management to restrain *monetary* inflation and of wage-fixing to influence the *real* level of employment and output. Many of my friends and colleagues who share my admiration for Keynes will at this point part company from me. "Surely", they will say, "you have got it the wrong way round. Did not Keynes suggest that the control of demand should be used to influence the total amount of real output and employment which it was profitable to maintain, while the money wage rate was left simply to determine the absolute level of money prices and costs at which this level of real activity would take place?" I agree that this is in fact the way in which Keynes looked at things in the late 1930s when it could be assumed that the money wage rate was in any case constant or rather sluggish in its movements. What he would be saying today is anybody's guess; and I do not propose to take part in that guessing game except to say that he would be appalled at the current rates of price inflation. It is a complete misrepresentation of the views of a great and wise man to suggest that in present conditions he would have been concerned only with the maintenance of full employment, and not at all with the avoidance of money price and wage inflation.

Whatever Keynes's policy recommendations would be in present circumstances, I would maintain that the way in which I have distributed the weapons among the targets is in no way incompatible with Keynes's analysis. In the 1930s Keynes argued, rightly or wrongly, that cutting money wage rates would have little effect in expanding employment because its main effect would be simply to reduce the absolute level of the relevant money prices, money costs, money incomes, and money expenditures, leaving the levels of real output and employment much unchanged. It is a totally different matter, wholly consistent with that Keynesian analysis, to suggest that the money wage rate might be used to influence the level of employment in conditions in which the money demand was being successfully managed in such a way as to prevent changes in wage rates from causing any offsetting rise or fall in total money incomes and expenditures. If one is going to aim particular weapons at particular targets in the interests of democratic understanding and responsibility, it is, in my opinion, most appropriate that the Central Bank which creates money and the Treasury which pours it out should be responsible for preventing monetary inflations and deflations, while those who fix the wage rates in various sectors of the economy should take responsibility for the effect of their action on the resulting levels of employment.

Earlier I spoke of "price stability" as being one of the components of "internal balance". Yet in the outline which I have just given of a possible distribution of responsibilities no one is directly responsible for price stability. To make price stability itself the objective of demand management would be very dangerous. If there were an upward pressure on prices because the prices of imports had risen or indirect taxes had been raised, the maintenance of price stability would require an offsetting absolute reduction in domestic money wage costs; and who knows what levels of depression and unemployment it might be necessary consciously to engineer in order to achieve such a result? This particular danger might be avoided by choice of a price index for stabilisation which excluded both indirect taxes and the price of imports; but even so, the stabilisation of such a price index would be very dangerous. If any remodelled wage-fixing arrangements were not working perfectly – and it would be foolhardy to assume a perfect performance – a very moderate excessive upward pressure on money wage rates and so on costs might cause a very great reduction in output and employment if there were no rise in selling prices so that the whole of the impact of the increased money costs was taken on profit margins. If, however, it was total money incomes which were stabilised, a much more moderate decline in employment combined with a moderate rise in prices would serve to maintain the uninflated total of money incomes.

The effectiveness of the pattern of responsibilities which I have outlined rests upon the assumption that there is a reasonably high elasticity of demand for labour in terms of the real wage rate, since success is to be achieved by setting a money wage rate relatively to money demand prices which gives a full employment demand for labour by employers. I have no doubt myself that in the longer-run the elasticity of demand is great enough. But what of the short-run? What if in every industry there is a stock of fixed capital in a form which sets an absolute limit to the amount of labour which can be usefully employed, while, for some reason or another of past history there is more labour seeking work than can be usefully employed? There will be unemployment in every industry; and any resulting reduction in money wage rates combined with the maintenance of total money incomes would merely redistribute income from wages to profits.

I have explained the danger in its most exaggerated form; but it would remain a real one even in a much moderated form. There should, of course, never be any question of the wholesale immediate slashing of wage rates in every sector in which there was any unemployment. Any such arrangement would, for the reasons which I have outlined, be economically most undesirable even if it were politically possible. What one has in mind is simply that in a milieu in which total money incomes are steadily rising at a moderate rate, money wage rates should be rising rather more rapidly in some sectors and less rapidly or not at all in other sectors according to the supplies of available labour and the prospects of future demands for labour in those sectors. There would be no requirement that any money wage rates must be actually reduced.

Putting more emphasis on supply-demand conditions in the settlement of particular wage claims could only work if there were general acceptance of the idea by the ordinary citizen; and such acceptance would depend *inter alia* upon a marked change of emphasis about policies for influencing the redistribution of the national income. I have for long believed that it is only if, somehow or another, the ordinary citizen can be persuaded to put less emphasis on wage bargaining and more emphasis on fiscal policies of taxation and social security for influencing the personal distribution of income and wealth that we have any hope of building the sort of free, efficient, and humanely just society in which I would like to live. But that raises a host of issues which I cannot discuss today.

There is, however, one feature of this connection between the supply–demand criterion for fixing wage rates and the attitude of the wage earner to his real standard of living on which I do wish to comment. Suppose, for example because of a rise in the world price of oil or of other imported foodstuffs or raw materials, that the international terms of trade turn against an industrialised country. This is equivalent to a reduction in the productivity of labour and of other factors employed in the country in question. If money wage rates are pushed up as the prices of imported goods go up in order to preserve the real purchasing power of wage incomes, money wage costs are raised for the domestic producer without any automatic rise in the selling price of the domestic components of their outputs. Profit margins are squeezed. The demand for labour will fall unless and until profit margins are restored by a corresponding rise in the selling prices of domestic products. But such a rise would in turn cause a further rise in the cost of living, followed perhaps by a further offsetting rise in money wage rates, with a further round of pressure on profit margins. In fact workers are attempting to establish a real wage rate which, because of the adverse effect of the terms of international trade, is no longer compatible with full employment. The resulting rounds of pressure on profit margins, rises in domestic selling prices, further rises in money wage rates, further pressure on profit margins, and so on, may result in stagflation – a level of employment below full employment with a continuing inflation of money prices.

This story may in fact help to explain what has happened recently in some industrialised countries, but my purpose in telling it is merely to give a vivid illustration of the fact that an effective combination of full employment with the avoidance of inflation necessarily requires that wage-fixing should take as its main criterion the supply–demand conditions in the labour market without undue insistence on the attainment and defence of any particular real wage income. The latter must be the combined result of domestic productivity, the terms of international trade, and tax and other measures taken to affect the distribution of income between net-of-tax spendable wages and other net-of-tax incomes.

## IV

So much for the specification of targets and for the distribution of weapons among targets, we must now ask "What about the detailed specification of the weapons themselves?"

If the velocity of circulation of money were constant, a steady rate of growth in the total money demand for goods and services could be achieved by a steady rate of growth in the supply of money, and this in turn could be the task of an independent Central Bank with the express responsibility for ensuring a steady rate of growth of the money supply of, say, 5 % per annum. It is a most attractive and straightforward solution; but, alas, I am still not persuaded to be an out-and-out monetarist of this kind. It is difficult to define precisely what is to be treated as money in a modern economy. At the borderline of the definition substitutes for money can and do readily increase and decrease in amount and within the borders of the definition velocities of circulation can and do change substantially. Can we not use monetary policy more directly for the attainment of the objective of a steady rate of growth of, say, 5 % per annum in total money incomes, and supplement this monetary policy with some form of fiscal regulator in order to achieve a more prompt and effective response? For this purpose one would, of course, be well advised to call in aid the skills of the control engineer, in order to cope with the dynamic problems of keeping the total national money income on its target path. Am I to be regarded as a member of the lunatic fringe or as an unconscious ally of authoritarian tyranny if I express this remaining degree of belief in the possibilities of rational social engineering?

I find very attractive the idea that this monetary control should be the responsibility of some body which was not directly dependent upon the government for its day-to-day decisions but which was charged by its constitution independently to achieve this stable but moderate growth of money incomes. But there is real difficulty in endowing any such independent body with powers to use fiscal policy as well as monetary policy to achieve its objective.

Let me take an example. Suppose that overseas producers of oil raise abruptly the price charged to an importing country; and, to isolate the point which I want to make, suppose further that the oil producers invest in the importing country any excess funds which they receive from the sale of an unchanged supply of oil, so that there is no immediate need to cut imports or to expand exports in order to protect the foreign exchanges. The abrupt price change will, however, tend to cause a deflation of money incomes in the importing country whose citizens will, out of any given income, spend less on home produced goods in order to spend more on imported oil, the receipts from which are saved by the oil producers. With the scheme of responsibilities which I have outlined it is now the duty of the demand managers to reflate the demand for goods and services in the importing country in order to prevent a fall in money incomes in that country.

There are at least two alternative strategies for such reflation.

In the first place, the taxation of the citizens of the importing country might be reduced so that, while they have to spend more on imported oil, they have just so much more spendable income to maintain their demands for their own products. In this case the government directly or indirectly borrows funds saved by the oil producers to finance the larger budget deficit due to the reduced tax payments by the domestic consumers. No one's standard of living is immediately affected.

If this solution is adopted, the importing country faces an ever-growing debt to the foreign oil producers with no corresponding growth of domestic or foreign capital to set against it. If this is considered undesirable, the private citizens must not be relieved of tax; their current consumption standards must be allowed to fall as a result of the rise in the price of oil; and the reflation of domestic income must be brought about by measures which stimulate expenditure on extra real capital development at home, the finance of which will mop up the savings of the oil producers. Such action will depend upon monetary policies rather than, or at least as well as, upon fiscal action.

I have told this particular story simply to make the point that the choice between fiscal action and monetary action must often depend upon basic policy issues which should certainly be the responsibility of the government rather than of any independent monetary authority. Perhaps the best compromise is an independent monetary authority charged so to manage the money supply and the market rate of interest as to maintain the growth of total money income on its 5 % per annum target path, after taking into account whatever fiscal policies the government may adopt. One would hope, of course, that there would be a suitable discussion of their plans and policies between the government and the monetary authority; but the latter would be given an ultimately independent duty and independent choice of monetary policy for keeping total money incomes on their target path.

The difficulties involved in the specification of the weapons of demand management are real enough; but they fade into insignificance when they are compared with the problems of remodelling wage-fixing arrangements in such a way as to ensure a greater emphasis on supply–demand conditions in each sector of the labour market.

I can think of five broad lines of approach.

First, one can conceive of wage fixing in each sector of the labour market by the edict of some government authority. An efficient use of this method would be extremely difficult in a modern economy with its innumerable different forms and skills of labour in so many different and diverse regions, occupations, and industries. It would, I think, in any case ultimately involve a degree of governmental authoritarian control which I personally would find very distasteful.

Secondly, there is the corporate state solution in which a monopoly of employer monopolies agrees with a monopoly of labour monopolies on a central bargain for the distribution between wages and profits in the various sectors of the economy of the total national money income which the demand managers are going to provide. I suspect that, in the United Kingdom at any rate, any such bargain would be very difficult to attain without leaving some important, but relatively powerless, sectors out in the cold of unemployment or of very low wages. In any case I ought to reveal my prejudice against being ruled by a monopoly of uncontrolled private monopolists.

Thirdly, the restoration of competitive conditions in the labour market would in theory do the trick, since the competitive search for jobs would restrain the wage rate in any sector in which there was unemployed labour and the competitive search for hands by competitive employers would raise the wage rate in any sector in which manpower was scarce. There is little doubt in my own mind that in some cases trade unions have attained an excessively privileged position and that some reduction of their monopoly powers might help towards a solution.

But I do not believe that any full solution is to be found along this competitive road. On the employers' side it may be impossible to ensure effective competition where economies of large scale severely restrict the number of employers. On the employees' side the whole of history suggests the powerful psychological need for workers with common concerns to get together in the formation of associations to represent their common interests. Moreover, reliance on individual competition might well involve the reduction, if not elimination, of support for workers who were unemployed and of compulsory severance or redundancy payments to workers whose jobs disappeared. But what one wants to find is some effective, but compassionate and humane, method which applies supply–demand criteria for the fixing of wage rates for those in employment without inflicting needless hardship and anxiety on those particular individuals who are inevitably adversely affected by economic change.

Fourthly, there are those who see the solution in the labour-managed economy in which workers hire capital rather than capital hiring workers. In such circumstances there would be no wage rate to fix. Workers would share among themselves whatever income they could earn in their concerns after payment of whatever fixed interest or rent was necessary to hire their instruments of production. These ideas are very attractive; but, alas, there is, I think, good reason to believe that satisfactory outcome on these lines is possible only in those sectors of the economy where small scale enterprises are appropriate and where conditions make it fairly easy to set up new competing co-operative concerns.

Finally, there remains the possibility of the replacement in wage bargaining of the untamed use of monopolistic power through the threat of strikes, lock-outs, and similar industrial action by the acceptance of arbitral awards made by trusted and impartial outside tribunals – awards which would, however, have to be heavily weighted by considerations of the supply–demand conditions of each particular case, if they were to achieve what I have suggested should be the basic objective of wage-fixing arrangements.

This is the civilised approach; but I am under no illusion that it is an easy one. It relies upon a widespread acceptance of the idea that some such approach is necessary for everyone's ultimate welfare and, in particular, as I have already indicated, upon the belief that there are alternative fiscal and similar policies to ensure social justice in the ultimate distribution of income and wealth. But even if in the course of time such a general acceptance could be achieved, some form of sanction for its application in some particular cases would almost certainly be needed. The punishment of individuals as criminals for taking monopolistic action to disturb a wage award does not hold out much hope of success, but is it pure dreaming to conceive ultimately of a state of affairs in which (1) in the case of any dispute about wages either party to the dispute or the government itself could apply for an award of the kind which I have indicated and (2) in which certain financial privileges and legal immunities otherwise enjoyed by the parties to a trade dispute would not be available in the case of industrial action taken in defiance of such an award?

Perhaps this is merely an optimist's utopian fantasy; but I can think of nothing better.

## V

So much for the attainment of price stability and full employment through the instrumentalities of demand management and wage-fixing, we must now ask "What about the attainment of external balance through foreign exchange policies?"

In my view the appropriate division of powers and obligations between national governments and international institutions is that the national governments should be responsible for national monetary, fiscal, and wage policies which combine full employment with price stability and that external balance should be maintained by foreign exchange policies under the supervision of international institutions.

Variations in the rate of exchange between the national currencies combined with freedom of trade and payments should in my view be the normal instrument of such foreign exchange policies, but this is not to say that there will never be occasion for the use of other instruments of foreign exchange policy. Special control arrangements may be appropriate where the removal of an international imbalance requires wholesale industrial development or structural change, or where abrupt changes in the international flow of capital funds require special offsetting measures, or where differences in national tax regimes would distort international transactions in the absence of offsetting measures. But where such exceptions to the free movement of goods and funds arise, these should be under the rules and supervision of appropriate international institutions.

After the war we managed to lay the foundations of an international system of this kind with the pivotal institutions of the International Monetary Fund, the International Bank for Reconstruction and Development, and the General Agreement on Tariffs and Trade, a system which for a quarter of a century resulted in a most remarkable expansion of international trade. In my opinion there was one important original flaw in this system, namely the insistence on the International Monetary Fund's very sticky adjustable peg mechanism for the correction of inappropriate exchange rates. But even this flaw has now gone as the International Monetary Fund seeks to find the most appropriate rules for running a system of international flexible exchange rates.

Yet we seem now to be faced with the possibility of a gigantic tragedy, with this initial success being fated unnecessarily to end in calamity. Why is this so? In my view the answer is obvious; it is simply because so many of the national governments of the developed industrialised countries have failed to find appropriate national institutional ways of combining full employment with price stability.

If they could do so, not only would the domestic tragic waste and social discontent of heavy unemployment in such countries be removed, but the international scene would be transformed. The pressure for the use by developed countries of massive import restrictions rather than of gradual and moderate changes in exchange rates to look after their balances of payments would, I suspect, be very greatly reduced. It is the spectacle of imports competing with the products of domestic industries in which there is already serious unemployment which is the greatest threat to the freedom of imports into the developed countries. With full employment and price stability at home the balance of payments could with much more confidence be left to the mechanism of flexible foreign exchange rates. The developed countries would then have less difficulty in giving financial aid to the third world; and, what in my opinion is even more important, they could much more readily accept the inflow from the third world of their labour-intensive products.

In this lecture I have marked an occasion which is concerned with international economics with a lecture on internal balance. But I suggest that in present conditions this is not anomalous. I do not, I think, exaggerate wildly when I conclude by saying that one – though, of course, only one – of the really important factors on which the health of the world now depends is the recasting of wage-fixing arrangements in a limited number of developed countries.

*Little Shelford, Cambridge*

*Date of receipt of final typescript: February 1978*

REFERENCES

Keynes, J. M. (1936). *The General Theory of Employment, Interest and Money*. London: Macmillan.
Meade, J. E. (1951). *The Theory of International Economic Policy*. Vol. I. *The Balance of Payments*. London: Oxford University Press for the Royal Institute of International Affairs.
—— (1955). *The Theory of International Economic Policy*. Vol. II. *Trade and Welfare*. London: Oxford University Press for the Royal Institute of International Affairs.
Ohlin, B. (1933). *International and Interregional Trade*. Harvard University Press.

# MONEY, BALANCE OF PAYMENTS THEORY, AND THE INTERNATIONAL MONETARY PROBLEM

## Harry G. Johnson

This Essay is adapted with small modifications from the David Horowitz Lectures that I delivered in Israel in 1975. I was honored by and grateful for the invitation to give them. My only previous visit to Israel had been ten years earlier, for a Rehovoth Conference graced by an address by Governor Horowitz himself. At that time, he was extremely active in two major world economic-policy debates—the reform of the international monetary system, which already appeared as a necessary task but one that could be tackled with due deliberation by economic statesmen of good will, and the problem of devising new ways to transfer resources for development to the less developed countries, a problem to which world attention had been dramatically called by the First United Nations Conference on Trade and Development in Geneva in 1964. Governor Horowitz, in common with many other international monetary experts, sought to solve the two problems simultaneously by linking reserve creation in some way to development assistance. That general class of proposals, I must admit, did not appeal to me then and has never appealed to me any more strongly since. After years of learning to appreciate the necessity and the difficulty of distinguishing between monetary and real phenomena, I find intellectually obscurantist any analysis or proposition that unwittingly or willfully confuses the creation of money with the liberation of real resources, however noble the intention. The world inflation of recent years bears ample evidence to the dangers of the politically popular belief that desirable real results can be achieved by manipulations of monetary magnitudes and maneuvers with monetary mystique. Nevertheless, I admired the combination of ingenuity and economic statesmanship that distinguished the contributions of Governor Horowitz to the debate—a combination that is in some ways reminiscent of John Maynard Keynes at his best as a policy advisor.

There is a second reason why I was glad to return to Israel: so many Israeli economists have contributed to the development of the two main fields of economic theory I am interested in, monetary theory and international economic theory. In particular, at the time of my previous visit the second edition of Don Patinkin's *Money, Interest and Prices* (1965) was just about to appear—in fact, he showed me an advance copy. His book had only begun to establish the classic position in monetary theory that it has since come to enjoy, and neither of us, I am sure, had any thought that it would become a major source of ideas and techniques for the analysis of a problem it did not deal with at all, the theory of the balance of payments and the international monetary system. In brief, as I shall show later, Patinkin's work on the integration of money and value theory through the real balance effect, and on the interaction of stock and flow adjustments on the establishment and the stability requirements of full economic equilibrium, provided the key to understanding classical income-expenditure, monetary balance-of-payments theory. That key was necessary for an effective return from the post-Keynesian tradition to the classical tradition of analysis of international disequilibrium problems as monetary phenomena.

My two Horowitz Lectures were entitled "Money and the Balance of Payments" and "The International Monetary Problem." That my selection of topics may seem on the one hand an arbitrary linking of two largely unrelated subjects, and on the other hand a choice catering to the widely different interests of two eminent Israeli economists, Governor Horowitz and Don Patinkin, is admitted. But in my own mind the two topics are firmly interwoven: both areas of economic concern illustrate the difficulties that professional economists and policy-makers get into when they forget the fundamental truism that a monetary economy is different from a barter economy because it is monetary, and the corollary that in some broad sense the demand for and supply of money are relevant to what happens to monetary magnitudes in such an economy. One cannot hope to reason effectively about a monetary economy in the terms appropriate to the barter economy of value theory.

Both topic areas exemplify the pitfalls of attempting to analyze a monetary world with the tools of "real" theory. I include under this term theories like sophisticated Keynesianism that attempt to create a simulacrum of a monetary economy by treating money and monetary policy as a determinant of a real quantity or real price in the shape of a real quantity of money or "liquidity" or a rate of interest. Notable examples are the "Yale School's" "portfolio balance" approach and the alternative liquidity approach presented in recent writings of J. R. Hicks. My own experience as both a pure theorist and a minor participant in over fifteen years of discussion of the international monetary system and its possible reform has led me increasingly to ask myself, as a social scientist: Why do policy-makers and their professional economic advisors, who should know better, consistently retreat into "real" analyses of and solutions for monetary problems? I can offer only a brief sketch of an answer here: The "real" world is familiar, and identical with the "monetary" world as long as the price level is reasonably stable; everyone lives his normal life in a partial-equilibrium

2

context in which money price changes are also real price changes. The "money" world of monetary macro-equilibrium and disequilibrium is by contrast unfamiliar and strange. Few people indeed possess either a systemic concept of the economy as a whole, as distinct from their own small corner of it, or the imagination to recognize what seem like "real" changes with "real" causes as being in reality monetary changes with monetary causes. Hence they do "what comes (intellectually) naturally," treating unemployment as due to business pessimism, automation, inadequate training of the labor force, and so on, and inflation as due to the monopoly power and greediness of big business or big labor, or to excessive and wasteful public spending on warfare or welfare, according to political taste.

## Money and the Balance of Payments

The new, so-called "monetary approach" to the theory of the balance of payments has been developing and gaining popularity in recent years as an alternative to the "elasticity approach," the "absorption approach," and various other Keynesian approaches which may be termed "the foreign-income multiplier approach" and "the Meade-Tinbergen-Keynesian economic-policy approach." (The meaning of these approaches will be explained more fully later.)

At the outset, it is important to note that the monetary approach is new only in the context of balance-of-payments theory as it has developed since the 1930s, when the collapse of the liberal international economic order based on the gold-standard system was accompanied by the Keynesian revolution in economic theory. The monetary approach actually represents a return to the classical tradition of international monetary theory established by the work of David Hume, summarized in the classical price-specie flow mechanism of adjustment to international monetary disequilibria, and foreshadowed in the work of Isaac Gervaise (1720). This tradition has dominated international economics for most of the two centuries during which economics has existed as a scientific system of thought. It is important to emphasize this point, because the development of the monetary approach to the balance of payments has been confused in many so-called minds with something called "monetarism," which is one side of an argument about domestic macroeconomic policy management that has been conducted mainly in the United States, though with a subsequent and derivative manifestation in the United Kingdom. The argument is between those who place their faith in fiscal policy, following the Hansenian American version of the Keynesian revolution,

3

and those who emphasize the necessity of proper monetary policy for the stabilization of the economy, led but by no means dragooned by Milton Friedman.

The issue has been further confused by the fact that Robert Mundell and I, as the two most visible exponents of the new approach, were associated with the University of Chicago during the crucial period when the approach was developed and hence are easily identified by the unthinking as lesser lights in the contemporary "Chicago School" led by Milton Friedman. Yet we are Canadians by birth and citizenship, did our graduate work at M.I.T. and Harvard respectively, and were strongly influenced in our early professional years by the balance-of-payments theorizing of James Meade of the London School of Economics. Mundell worked out his central ideas at the Johns Hopkins Bologna Center and the International Monetary Fund before he joined the Chicago department. Most of my own work on the subject was done during my periods at the London School of Economics, in response to international monetary developments as seen from—more accurately, not understood by—the United Kingdom. Unfortunately, however, the description of scientific activity as a debate between risible and reasonable schools of thought and the assignment of skeptics about prevailing orthodoxy to a ludicrous school through guilt by association, however tenuous geographically and intellectually, has become a hallmark of post-Keynesian discussion of monetary economics. The ability to do so with fluency and plausibility has been assumed by many to be more than adequate to justify the earning of a Ph. D. at public expense.

In order to explain the nature of the new approach, I find it convenient to begin with a brief history of the development of balance-of-payments theory, an *excursus* that will allow me to make some incidental digressions on the inherent difficulty of monetary theory and the shackles imposed on free theoretical inquiry in economics by the limitations of the tools—particularly the mathematical tools—of theoretical analysis.

Hume's price-specie flow analysis was developed as an answer to the mercantilist contention that the path to augmentation of national wealth and power lay in the development and maintenance of a balance-of-payments surplus through measures to increase exports and decrease imports ("a policy of import substitution," in the modern phrase) so as to produce a continuing inflow of precious metals ("treasure") into the country. Hume's analysis demonstrated that such a policy would inevitably be self-defeating, since the accumulation of money stocks would satiate the demand for them and any ex-

4

cess stocks would "leak out" through a balance-of-payments deficit. (Remember that, in an open system, actual stocks of real balances are adjusted to desired levels not by price inflation or deflation, as in a Patinkinian closed economy, but by outflow or inflow of nominal money through the balance of payments.)[1] In illustrating this proposition, Hume showed that the expansion of issue of paper-currency substitutes for precious metals would lead merely to an outflow of precious metals. The parallel in the contemporary monetary approach is the proposition that excessive expansion of "domestic credit" by a country's banking system will lead to a balance-of-payments deficit under fixed exchange rates and a loss of international reserves. A corollary of Hume's analysis is the assertion that there is a "natural distribution" of the world money or reserve stock among the member countries of the world system toward which the actual distribution will gravitate. (Note the parallel with the Archibald and Lipsey [1958] critique of Patinkin's first-edition analysis of the effects of a disproportionately distributed increase in nominal money.)

Hume's analysis was related to the economic world of his time, but such is the propensity of economists to live with archaic old facts rather than open their eyes to new facts that the work of Hume and his immediate successors left a permanent mark on balance-of-payments theory. The most important and pervasive point of influence was his concentration on the trade account—exports and imports of goods—as the locus of adjustment to international monetary disturbances. This concentration has remained a valid point of complaint by practical men against academic balance-of-payments theory, especially as it has been carried over to, and accentuated in, the post-Keynesian "elasticities," multiplier, and policy approaches to the balance of payments. A second influence, which—apart from some work by Ohlin in the 1920s (see, e.g., Ohlin, 1929)—has only in the last four years begun to be questioned, was the assumption incorporated in the phrase "price-specie flow mechanism" that the domestic price level of a country possessing excess money stocks must rise relative to other countries' price levels before trade flows are affected and a balance-of-payments deficit emerges. This view assumes limited holdings of commodity stocks and long lags in transportation and in the dissemination of information about markets, assumptions appropriate to Hume's time but a decreasingly realistic approximation for contemporary integrated world markets. Furthermore, Hume's

---

[1] This proposition cleared monetary phenomena from policy discussion and permitted the advocacy of free trade as the way to maximize output from national resources.

5

account predated the development of large-scale commercial banking subject to control by a central bank. By this century, however, the theory had been extended by the addition of the standard textbook analysis of the role of bank-rate adjustments in stimulating short-term capital movements as substitutes for actual specie movements.

The classical Humean tradition of international monetary analysis, like so much else of value in the classical and neoclassical traditions of monetary theory, was swept aside and, at least transiently, completely suppressed in the wake of the Great Depression and the Keynesian revolution. I attribute the fragility of that tradition, and its vulnerability to attack by what purported to be "common sense," to the inherent difficulty of monetary theory. "Real" theory began with the notion that value is created by the expenditure of human effort over time, a notion that raises no real questions of understanding or ethical justification. But it ran into problems once it became necessary to explain the productive role of material capital and the existence of a return on it, problems that still and needlessly confuse, or are confused by, the present-day Cambridge successors of the English classical tradition. But real capital at least requires sacrifice to accumulate, and it contributes in tangible form to total output. Money, on the other hand, is a stock that ultimately requires confidence, not tangible effort, to create, appears to have no inherent usefulness in its medium-of-exchange function, and yields no explicit return identifiable with an easily measurable contribution to production. Hence it requires a great deal of sophistication to treat money as a stock requiring application of stock-flow adjustment mechanisms. It is not surprising that even great monetary theorists like Wicksell and Keynes have found it more congenial to treat monetary adjustments proximately in terms of income-expenditure flow relationships motivated by the fixing of a disequilibrium relative price (the interest rate) through monetary policy, while politicians and the public prefer to attribute balance-of-payments deficits to prices being too high, businessmen and workers too lazy, or governments too spendthrift with the taxpayers' money.

Be that as it may, the classical approach to international monetary equilibrium and disequilibrium and balance-of-payments problems was swept away in the 1930s in favor of a succession of alternative approaches that attempted to treat balance-of-payments equilibria and disequilibria as flow equilibria. The implicit or explicit assumption on which these approaches were based was that flow trade deficits or surpluses (or, in some cases, surpluses or deficits on the balance of trade and services) entailed corresponding outward or inward flows of international reserves. This brief description, incidentally, encapsu-

6

lates the two main objections to these approaches made on behalf of the monetary approach. The first, which is one of those blindingly obvious elementary tautological points that economists are carefully trained to disregard in the process of formal model building, is that (in a fixed-exchange-rate system) an excess demand for money can be supplied *either* by the acquisition of international money through a balance-of-payments surplus *or* through the creation of money against domestic credit by the domestic monetary authority. This point has pervasive implications for international economic policy; they can be summarized in the proposition that no policy for improving the balance of payments can be successful unless supported by an appropriate restriction of domestic credit. The second objection, which requires sophisticated understanding of the basics of stock-flow relationships and adjustments subsumed in Patinkin's "real balance effect," is that a balance-of-payments deficit or surplus represents a transient stock-adjustment process evoked by an initial inequality of desired and actual money stocks. It cannot be treated as a continuing flow equilibrium. It is worth noting in passing that Keynes never made that mistake—he dealt entirely with a closed economy and a full stock and flow equilibrium in the goods, money, and bonds market. It was entirely a creation of others, who committed the error of analyzing a disaggregated monetary economy as if overall stock-flow equilibrium was enough and continuing net cash flows between its national parts would leave other flow equilibria unchanged.

The first popular successor to the traditional framework of analysis, one that still prevails in official and public policy discussions, was the so-called "elasticities approach," attributable to a classic essay by Joan Robinson (1950), though traceable to early work by the eccentric Bickerdike. That approach was pre-Keynesian, in the direct tradition of Marshallian partial-equilibrium analysis, which ignored repercussions of changes in production and expenditure in one sector on the equilibrium of the rest of the economy. Specifically, the approach regarded exports and imports as separate small sectors whose equilibria were determined by sector demand-and-supply functions in terms of domestic money prices (proxying for relative prices of traded goods in terms of domestic nontraded goods in general) as affected by the exchange rate applicable to conversion of domestic into foreign prices and vice versa.

The elasticities approach had the advantage of apparently shedding light, mistakenly it now appears, on two questions of contemporary concern apart from the effect of exchange-rate changes on the balance of payments itself: the effect on domestic employment, where im-

7

provement in the balance of payments *in domestic currency* (as distinct from foreign) increases demand for domestic output and the amount of employment, and the effect on the terms of trade, assumed to constitute an index of economic welfare. (In the latter connection, Robinson and others attempted to establish a presumption where none can exist, to the effect that the terms of trade will tend to move against the devaluing country.)

The approach had three major drawbacks, however. First, it expressed the criteria for a devaluation to improve the trade balance in terms of separate elasticities for exports and for imports, in an unfamiliar formula making improvement depend in the simplest case on whether the demand elasticities summed to more or less than unity. This formulation concealed the point that the question was one of market stability and concentrated attention on empirical guesswork as to what the magnitudes of the elasticities were likely to be in particular cases. Second, the analysis involved both the minor inelegance of ignoring cross-elasticity relations among exports and imports and the major theoretical error of ignoring the multiplier implications of the increase in demand for domestic output that was the counterpart of an improvement in the balance of payments. (This was *not*, it should be recognized, an error committed by Robinson, who clearly stated the multiplier implications, but an important error in popular interpretation of her analysis.) In consequence of this error, less skilled theorists took the elasticities formulation as the total of the analysis. In the early postwar controversy over elasticity pessimism versus elasticity optimism, they ignored the question of availability of unemployed domestic resources to supply the devaluation-induced increase in demand and attempted to cram this consideration into the determination of the likely magnitudes of the elasticities themselves. Finally, as already mentioned, the model identified an excess flow demand for money with an excess flow demand for international reserves, thereby treating balance-of-payments disequilibria as continuing flow phenomena and ignoring the importance of domestic monetary policy in determining the effect of the presumed cash-flow demand on the flow of international reserves.

In the 1930s, the international-economics application of Keynesian theory proper was primarily concerned with the international extension of the multiplier concept. Initially, there was a controversy over whether the trade balance or total exports should be used as the multiplicand, and whether the marginal propensity to save or the sum of the marginal propensities to save and to import should be used as the multiplier. The controversy about the multiplicand was soon re-

8

solved, correctly, in favor of total exports. Later work by Metzler (1942) and Machlup (1943) was concerned with the question of whether quantity adjustments through the multiplier could fully replace the classical relative-price adjustments, the answer being in the negative so long as both countries in the world system had positive marginal propensities to save internationally.

A fully consistent multiplier analysis of the effects of devaluation, however, had to wait until the postwar period. The analysis was simplified by assuming perfectly elastic supplies of exports and imports and no nontraded goods, so that devaluation involved essentially a change in the real relative price of the two goods. This relative price change triggered multiplier expansion and contraction of income depending on whether the demand elasticities summed to more or less than unity *and* whether or not both marginal propensities to save internationally were positive. The two requirements appeared multiplicatively in the overall formula for the effects of a devaluation, and the concept of the marginal propensity to save disguised the fact that what was really represented was a bastard stock-flow concept of a marginal propensity to accumulate international reserves. It was thus easy to interpret the concept as making the effect of a devaluation depend essentially on the standard stability criterion that the sum of the price elasticities of import demand be greater than unity, the result of the devaluation, if successful, being a continuing inflow of international reserves (domestic credit policy being ignored).

Meanwhile, the fact that the immediate postwar situation was one of inflationary pressure rather than mass unemployment led most policy economists to attempt to torture the elasticities approach into conformity with inflationary conditions. Alexander (1952) responded by producing the rival "absorption approach" to devaluation. Alexander's essential contribution was to observe that, for a devaluation undertaken *by itself* under full-employment conditions, the resulting extra inflationary pressure would make the elasticities of export and import demand and supply irrelevant and the effect of the devaluation in reducing the deficit depend on the consequences of the inflation itself in reducing aggregate domestic demand for output. These consequences were of two kinds: Keynesian effects, of theoretically doubtful reliability, working via various kinds of income redistribution, and the monetary-theoretic effect of inflation in reducing real balances.

The absorption analysis, while important in shifting attention from microeconomic elasticities to the macroeconomic balance of aggregate demand and supply, was itself defective in two important re-

9

spects and is best regarded as constituting a halfway house to a correct analysis. The first important defect lay in taking devaluation *by itself* as a policy for analysis, under circumstances in which a combination of devaluation *and* deflationary macroeconomic policy is clearly required. The second defect was that the absorption approach still concentrates on expenditure flows, not recognizing that a continuing deficit will eventually correct itself without devaluation by reducing the economy's real balances, unless real balances are continually renewed by domestic credit expansion to offset the effects of reserve losses. In such a case, devaluation will not improve the balance of payments by deflating real balances.

The fourth, and most theoretically satisfactory, stage of post-Keynesian development of Keynesian balance-of-payments theory came almost simultaneously with the publication in 1951 of James Meade's *The Theory of International Economic Policy*, Volume I, *The Balance of Payments*. Meade shifted the whole theory from the "positive" analysis of the effects of individual policies on the balance of payments to the "normative" analysis of the combination of policies the authorities must follow if they wish to implement policy objectives with respect to both domestic employment and the balance of payments ("internal balance" and "external balance," in Meade's terminology). (Tinbergen's [1952] contribution, incidentally, was to show that the government must have as many independent policy instruments as it has objectives.)

Basically, Meade's analysis showed that the authorities must have fiscal or monetary policy to control aggregate domestic expenditure, and devaluation or controls over international trade and payments to control the allocation of domestic and foreign expenditure between domestic and foreign output. Note that, insofar as the authorities maintained exact balance-of-payments equilibrium, actual and desired money holdings would balance and there would be no inconsistency with monetary-theoretic requirements. Inconsistency could arise only from the implication that if government policy erred, the result would be a continuing flow-equilibrium deficit or surplus whose elimination would require a change in governmental economic policy.

The final stage of development of Keynesian balance-of-payments theory came with the apparent foreclosure of the possibility of using exchange-rate change or trade and payments controls as policies for affecting the allocation of domestic and foreign demand between domestic and foreign output. Mundell (1962) pointed out that the need for as many policy instruments as policy objectives could be met

10

by recognizing that fiscal and monetary expansion have effects in the same direction on the current account but in opposite directions on the capital account of the balance of payments. This policy model—the so-called "theory of fiscal-monetary policy mix" (Mundell, 1962)—also involved no monetary-theoretic inconsistency in cases of preservation of balance-of-payments equilibrium. It was correctly criticized, however, for treating as continuing flow phenomena what are properly regarded as securities-portfolio stock adjustments in response to changes in international interest-rate differentials, and for neglecting the consequences of such portfolio adjustments on the services-account component of the current account. (Moreover, as a practical policy suggestion for the United States, it turned out to be a resounding failure to the extent that it was tried.)

The alternative "monetary approach" to the balance of payments starts from the proposition that balance-of-payments disequilibrium involves an inflow or outflow of international money and hence must be treated as a monetary phenomenon requiring the application of the tools and concepts of monetary theory (Frenkel and Johnson, eds., 1976). This approach, as mentioned earlier, involves two major changes in theoretical orientation and formulation. The first is the simple tautological point that domestic money can *either* be created or destroyed by domestic monetary policy operating on the volume of domestic credit extended by the banking system *or* be imported or exported by running a surplus or deficit on accounts of the balance of payments other than the money account. (The phrasing here is carefully chosen, for a reason that will appear shortly.) This change implies, most fundamentally, that balance-of-payments theory, analysis, and policy prescription must necessarily include exact specification of domestic monetary policy. The second and more subtle change is that international money flows are a consequence of stock disequilibria—differences between desired and actual stocks of international money—and as such are inherently transitory and self-correcting. This is, of course, nothing more than a contemporary restatement of the Humean price-specie flow mechanism, but one refined by modern understanding of the nature of money as a stock yielding either utility to consumers or productive services to producers and by recognition of the possibility that adjustment of desired to actual stocks of international money may occur through either the trade account (surpluses or deficits on exports relative to imports of goods and services) or the capital account (international capital flows of various descriptions) or both. In other words, the monetary approach frees balance-of-payments theory from its traditional concentration on bal-

11

ance-of-payments adjustment through changes in the trade balance (the modern equivalent of which is concern with the "basic balance" or combined balance on goods, services, and long-term capital accounts, the last account being assumed for some reason to be more predictable and amenable to economic explanation and policy influence than movements in the money and short-term capital accounts).

The monetary approach also has the attraction of clarifying the role of movements in the terms of trade—the prices of imports relative to the prices of exports—in the process of international adjustment. Prevailing theory has strongly implied that the purpose of devaluation is precisely to produce an adverse movement in the terms of trade (thereby making the devaluing government appear to be deliberately choosing to impose a national loss) and has further tended to suggest that the reduction of material welfare consequent on devaluation can be avoided by alternative interventionist balance-of-payments policies. By contrast, the monetary approach indicates that terms-of-trade changes, which in principle may go in either direction, are either a transient feature of a monetary stock-adjustment process or a necessary concomitant of movement from an unsustainable deficit situation (in which a country is "living beyond its income" with the help of distress monetary transfers from the rest of the world) to a sustainable position of balance-of-payments equilibrium.

Recognition of this point has led monetary balance-of-payments theorists to transfer their analytical interest away from models stressing imperfect substitution between foreign and domestic tradeable goods and the role of elasticities of demand for such goods in producing terms-of-trade changes that may go either way. They are concerned instead with models stressing the distinction between traded international and nontraded domestic goods, whose relative prices must change in a particular way as domestic expenditure varies relative to income in the process of international monetary stock adjustment. (For an early example of such a model, see Salter, 1959.) This clarification of the role of elasticities and relative price changes in the process of international adjustment is worth emphasizing. The concentration of analysis by monetary balance-of-payment theorists on the so-called "small country" assumption—that the country under analysis is so small that all goods prices and interest rates can be treated as internationally fixed—has often been mistakenly interpreted to mean that the monetary approach is confined to the analysis of such trivial cases. In fact, the procedure has been prompted by the opposite purpose of clearing secondary and essentially trivial analytical complications out of the way of understanding the essentials.

12

Insofar as theoretical development in international economics is promoted by the observed failure of existing theories to fit the facts of experience rather than by the refinement of professional standards of theoretical elegance and the instinct of scientific workmanship, the development of the monetary approach to balance-of-payments theory can be attributed to two recent historical events. The first is the failure of the prevalent elasticity approach to account for a mounting accumulation of awkward failures of prescription and prediction with respect to devaluations and revaluations and other balance-of-payments policies, most notably the initial failure of the British devaluation of 1967 and its short-lived success after the British authorities turned temporarily to tight control of domestic credit expansion. The other is the failure of Keynesian theory to explain and account for the world inflation that has been going on since 1965 or so, a phenomenon that is easily explainable on classical Humean lines by the generation of an excessive rate of growth of the world money supply initiated by U.S. monetary policy.

Correspondingly, the ultimate test of the monetary approach is its superiority in empirical explanation of balance-of-payments phenomena. Work on the empirical testing of the theory has been proceeding apace behind the scenes, though constantly impeded by the unfortunate institutional fact of life that the reputations of young economists can be made much more quickly and definitively by elegant and comprehensive mathematical theorizing than by the empirical testing of hypotheses. The main positive findings so far (see Frenkel and Johnson, eds., 1976) have provided underpinning for the proposition that balance-of-payments improvement is inversely connected with domestic credit expansion. The most robust specific proposition is that, contrary to Keynesian predictions, the fastest-growing countries will have the strongest (the surplus) balance-of-payments positions, because their demand for money will tend to grow faster than the supply of domestic credit. Empirical testing has, however, run into two major difficulties: First, there is a dangerous temptation to test and confirm the monetary approach spuriously, by verifying statistically the tautology that an increase in domestic money must be provided either by domestic credit creation or by reserve acquisition. Second, in devising a proper test of the theory, which involves testing the existence and stability of the domestic demand for money, one runs into all the problems previously encountered in domestically oriented research on the quantity theory of money, most noticeably the interdependence of demand and supply of money, lags in the adjustment of actual to desired quantities on both sides, and the division

13

of the effects of monetary changes between price changes and output changes.

In this section, I have dealt with the application of monetary theory to the theory of the balance of payments, criticizing the successive stages of development of balance-of-payments theory since the early 1930s for their attempt to analyze the monetary phenomena of balance-of-payments surpluses and deficits with theoretical constructs designed to deal with a "real" or "barter" system. I outlined a new "monetary" approach to these problems—actually a restatement of the main tradition of international monetary theory going back to David Hume's formulation of the price-specie flow mechanism but improved by the incorporation of modern concepts of stock-flow adjustments in monetary equilibration processes. The fundamentals of the monetary approach involve two central points: the tautology that changes in the domestic money supply may be brought about either through changes in the volume of domestic credit or through international exchanges of international reserve money for goods or securities, and the proposition that a balance-of-payments deficit or surplus is a monetary phenomenon representing a process of adjustment of actual to desired stocks of money and cannot therefore be appropriately treated as a continuing flow phenomenon representable as the residual of inflows and outflows of expenditure on goods (and possibly securities) governed by incomes and relative prices (and possibly relative interest rates).

## The International Monetary Problem

I turn now from pure theory to the application of the monetary approach to a practical problem in international economic policy, the international monetary problem. The problem, in simple terms, is that the international monetary system of fixed but "flexible" exchange rates, created at Bretton Woods after the international monetary collapse of the 1930s and centered on the International Monetary Fund, itself collapsed in February-March 1973 into a regime of "dirty" or "managed" floating exchange rates. At the time, the official international monetary experts were still arguing in a rather leisurely fashion about the precise institutional changes required and negotiable to strengthen the International Monetary Fund system against certain weaknesses that had become increasingly evident from the early 1960s on.

What have we learned from the experience of collapse and its aftermath, and where do we go from here? I find it useful, in examining

14

these questions, to concentrate on the international monetary system as a monetary *system*, that is, a system governing monetary relationships among the constituent national members of the international economy, and to visualize in a very long historical perspective the problems of an international monetary system based on the concept of fixed rates of exchange among national currencies. That means starting, though very briefly, with the traditional nineteenth-century gold standard—even though to speak of such a "traditional" system is in part mythological, since one of the safety valves of the nineteenth-century system was that, for most of the period, nations had a choice between the gold and the silver standards. They opted gradually for the gold standard at their own convenience, the United States coming firmly onto gold only in 1900. (By the same token, immediately after World War I, the international experts of the League of Nations saw as one of the chief problems of reestablishing the gold standard in a world of many more independent nations the danger of a shortage of gold relative to the demand for it, and set themselves to propagandizing, largely unsuccessfully, for the adoption of the gold-reserve standard.)

The gold standard, in common with any other fixed-exchange-rate system based either on a produced commodity or on an international credit instrument bearing a zero or uncommercially low rate of return, is subject to two major and interacting difficulties. The first is to provide for a rate of growth of the international-reserve-base money of the system approximately equal or closely related to the growth of demand for international reserves at stable prices. With a stable reserve growth at such a rate, broadly full-employment growth of the world economy can occur at a stable, or only mildly rising or falling, world price level. Without it, the fixed-rate system provides, for its member countries, not monetary stability but the obligation to experience roughly the same degree of price inflation or deflation as all the other members experience. The second difficulty arises from the fact that money derives its function not from its inherent value or characteristics but from confidence in its usability in exchange. Consequently, so long as the base money involves a functionally unnecessary investment of real resources or commitment to hold liquid assets in a zero or low-yielding form, there will be natural economic pressures to find higher-yielding substitutes for the holding of actual base money. This can be done with apparent safety so long as confidence is maintained in the ultimate convertibility of base-money substitutes into base money itself. The result of the process, however, is on the one hand greatly to complicate the problem of determining the rate of

15

expansion of world base money appropriate to the maintenance of reasonable price stability, and on the other hand to make the system vulnerable to waves of excessive confidence and loss of confidence in the convertibility of base-money substitutes into base money itself.

These two difficulties, it may be noticed in passing, are precisely analogous to the problems encountered in the historical evolution of national central banking. Those problems resulted in the conception of the central bank as having two not entirely consistent or easily combinable functions, that of controlling the growth of the money supply in the interests of monetary stability, and that of serving as lender of last resort to the commercial banking system in times of liquidity crisis. The solution in principle to the inconsistency was that in a liquidity crisis the central bank should lend *without stint* but lend *at a penalty rate*, in Bagehot's famous phrase, so that excess money created in a crisis would be returned to the central bank as soon as possible and not remain to overhang the market.

The nineteenth-century gold standard solved these problems surprisingly well, thereby maintaining an international monetary climate conducive to liberal policies of international trade and investment and peaceful world economic growth. But the gold-reserve standard reestablished after World War I very quickly fell victim to a collapse of confidence in national currency substitutes for nonexistent gold. The loss of confidence was triggered by the failure of the U.S. Federal Reserve System to prevent a collapse of the American money supply and was complicated by intra-European national rivalries and American isolationism, which prevented the salvaging of the system by international monetary cooperation. The international monetary system could have been rescued in three relatively painless alternative ways: coordinated national policies of domestic monetary expansion; international agreement to raise the national-currency prices of gold; and the invention by international agreement of a new international credit-reserve asset to replace gold. Failing the requisite willingness to solve the problem in one of these ways by international cooperation and invention, the only remaining alternative was the painful and socially disastrous one of lowering the national-money and gold prices of commodities through savage deflation and its accompanying mass unemployment. (In fact, the 1930s ended with exchange rates among national currencies more or less what they had been at the beginning, but with the national-currency prices some 75 per cent higher and national unemployment levels far higher on the average than the pre–World War I norm.)

16

The fundamental source of the international monetary collapse was only imperfectly understood at the time. In accordance with what I said in the preceding section about the strong temptations for both "practical" men and professional economists to retreat into the finding of "real" explanations and the proposal of "real" remedies for monetary problems, the monetary causation was increasingly dismissed or disregarded in favor of real explanations and remedies. Thus, the results of a failure of governmental monetary management were transmuted into evidence of the instability of capitalism and its alleged inherent tendencies to depression, and the failure to resort to expansionary monetary policy as evidence that monetary policy was powerless to make capitalism behave properly. The apotheosis of these ideas found expression in the American Hansenian version of Keynesianism, with its faith in the reality of "secular stagnation" and its emphasis on fiscal policy as the only effective tool available for macroeconomic management.

In the narrower context of international monetary organization, the experience gave rise to a number of ideas that constituted the ethos of opinion about the problems that were intended to be solved by the Bretton Woods system. Chief among these were the following: belief in the inherent "deflationary bias" of the gold standard, against which national full-employment policies had to be protected by the freedom to devalue in cases of "fundamental disequilibrium"; fear of a chronic shortage of international liquidity, to be made good by international provision of credit substitutes for gold—a fear that dominated International Monetary Fund thinking well into the 1970s, in spite of accelerating world inflation, and is still evident in the Fund's concern with providing additional short-term credit for consumer-country victims of the oil-price escalation; belief in the need to exercise surveillance to prevent "competitive devaluation," together with "elasticity pessimism," both derived from misinterpretation of a situation in which general devaluation was required to raise the price of gold as one in which devaluation was required to correct individual deviant behavior in an international monetary system in overall equilibrium; and belief that the chief threat to the stability of the system was another great depression in the United States economy.

The related set of preconceptions and problem orientations naturally meant that official opinion was unprepared when the chief problem of the system eventually turned out to be world inflation rather than world depression. The 1930s problem was turned on its head: excessive liquidity, excessive willingness to accept U.S. dollars as

17

credit substitutes for gold or Special Drawing Rights, "unfair" competition meaning reluctance to revalue rather than eagerness to devalue, and a chronic U.S. deficit as the engine of world inflation.

For a while, however, the Bretton Woods system worked well in providing a monetary framework for sustained economic growth and a trend toward a more liberal international trade and payments system. One must be careful, however, not to exaggerate how well it worked, and for how long. The European currencies became convertible only at the end of 1958, and tensions over the U.S. dollar glut and the adjustment of exchange rates began very soon thereafter. The system was buoyed up for a long time after the war's end by the belief that the United States had a disproportionate share of the world's gold and should, if anything, encourage a balance-of-payments deficit to relieve the postwar "dollar shortage" and redistribute the gold. Nevertheless, in the mid-1960s the view was gradually accepted that the reserve-currency position of the United States posed a special problem, and that the solution was gradual replacement of the dollar as international money by the creation and steady augmentation of new genuinely international credit-reserve assets in the form of Special Drawing Rights.

What forestalled this expected leisurely and deliberate progress toward a new fixed-exchange-rate system based on international credit reserves was the decision of President Johnson in 1965 to escalate the war in Vietnam *without* introducing the substantial increase in taxes required to finance that escalation. The result was necessarily inflation, which was compounded by later inflationary mistakes of American monetary policy. Owing to the fixed-exchange-rate system and the dominance of the United States as reserve-currency country and leading trading and investing country, the inflation permeated the world economy.

The period of inflation with fixed exchange rates raises two major problems for analysis. The first problem is why the major countries, primarily the European countries, were unable to cope with the American inflation. The international monetary system did, after all, permit exchange-rate changes in both directions, and it would have been possible in principle to confine most or all of the inflationary pressure to the United States by a series of revaluations of other countries' currencies against the dollar. This procedure would have been more disturbing than the adoption of floating rates against the dollar, since it would have amounted to a speeded-up version of the crawling peg. Nevertheless, there had been enough discussion of the need for smaller and more frequent changes in the adjustable pegs after the

18

1967 devaluation of sterling to permit not-too-startling innovations. The main reasons why European countries did not cope were two: (1) They had got used to the idea of the dollar as a currency with a fairly stable real purchasing power, in relation to which they could adjust their currency values to take account of more or less inflationary domestic price trends than those of their major trading partners. American inflation deprived the system of this cornerstone of stability and made it necessary for the European countries to learn to cooperate in concerted revaluations against the dollar and American inflation. This they were unwilling and unable to do. (2) The franc-mark realignment of 1969 dashed the hopes of the Common Market establishment that the Common Agricultural Policy of the EEC would make it impossible ever again to change the exchange rates of member countries against each other. The establishment reacted by pressing for the creation of a common European currency directly, rather than implicitly as an uncovenanted implication of the Common Agricultural Policy.

As a start, the "snake in the tunnel" concept narrowed the fluctuations of European currency rates in relation to each other, by comparison with fluctuations against the dollar. Given enough cooperation, the snake might eventually have produced a situation in which the European currencies could be revalued in common against the dollar, thus turning American inflation back onto the United States. But its main actual result was to freeze the exchange values of the European currencies and prevent individual action against world inflation. In the end, it was the United States, worried by its mounting balance-of-payments deficit and especially the adverse trend of its merchandise-trade balance, and not the Common Market countries that forced a revaluation of other currencies in terms of the dollar (or, if one prefers, a devaluation of the dollar) in 1971. The Smithsonian Agreement to this effect lasted barely long enough to let President Nixon win reelection. Thereafter, in 1973, an American decision to devalue by another 10 per cent precipitated the collapse of the fixed-exchange-rate system into a regime of "dirty floating."

The second problem concerning this period is why governments, and official and academic economists, were so determined to deny the existence of a world inflation sparked by U.S. inflation and communicated by the fixed-rate system. The requisite analysis was certainly obvious enough, in conformity with time-tested theory, and not entirely unknown from previous history, especially the well-known case of the effects of Spain's imports of precious metals from conquered Latin America. Yet all the official economists in Europe known to me

19

rushed to present alleged statistical disproof of the contention that I and various European colleagues advanced that the fundamental problem was a world inflation.

The reasons appear, at this juncture, to lie in two basic defects of Keynesian monetary theory (or, perhaps better, vulgar post-Keynesian macroeconomic policy theory), which those convinced of the rectitude of their position regard as a source of invincible strength against alternative "classical" monetary analysis. First, the *General Theory* provides no theory whatsoever about what determines money wages and changes in them. Consequently, when it comes to this question anything goes, and what goes best for the policy-maker accustomed to keep his brains sharp by reading the headlines and occasionally an editorial in his morning newspaper is a mishmash of *ad hoc* sociological analysis of union behavior, ending in the conclusion that what is required is an incomes policy. Second, Keynesian theory, like income-expediture theories in general, insists that monetary influences must affect aggregate output and prices through certain channels defined by the theory itself; if the influence cannot be seen moving in the specified channels, it does not exist. Thus, for a closed economy or one treated as closed, Keynesian theory asserts that monetary policy operates by influencing direct fixed investment. This remains an article of faith even though econometrics has been remarkably unsuccessful in finding such an influence and there is a growing body of evidence that monetary-policy changes have a fairly reliable influence on consumption expenditure (in typical "Keynesian," not sharply inflationary, circumstances). For an open economy, Keynesian theory similarly insists that world inflation must be communicated either through an inflow of reserves or through a sharp increase in the export surplus. Yet elementary monetary theory indicates that the money supply may increase either through reserve inflow or through domestic credit expansion designed to avert unwanted reserve inflow and that, through arbitrage, the prices of exports of closely substitutable goods will tend to stay in alignment in the various supplier countries rather than be forced up by a prior increase in demand.

There is a third and related problem: Why did the very official sources that denied the reality of world inflation during the closing stages of the fixed-exchange-rate system (when it was a necessary implication of the fixed-rate system) turn around within a few months of the switch to a floating-rate system (which made participation of foreign inflation in principle unnecessary) to the position that there was indeed a world inflation whose manifestation in domestic inflation

20

they were powerless to influence? It strains credulity—though not all the way to the breaking point—to hypothesize that it is a professional obligation of official economists to assert the exact opposite of prevailing economic truth in order to give maximum scope for ingenious policy recommendations. But another hypothesis seems plausible. Having tried unsuccessfully to hold back the tide of world inflation by sweeping vainly at it with the domestic brooms of fiscal, monetary, and incomes policy, only to see inflation become an endemic problem arrestable solely by thoroughgoing deflation and unemployment, governments and their economists found it easiest to blame the problem they had created for themselves on a foreign cause they could not be expected to overcome.

Leaving that issue aside, the floating-rate system, "dirty" or "managed" as it has been, especially in its early stages, has in my judgment worked very well. Contrary to the dire predictions of the adherents, defenders, and beneficiaries of fixed rates, and in accordance with the theoretical expectations of exponents of floating rates, the floating-rate system has not led to the fragmentation of the world economy and the cessation of growth of international trade and investment. Early fears of such fragmentation, and particularly of the proliferation of controls over international capital movements, were connected with the initial European effort to maintain a common float against the dollar and the belief of the French, since abandoned by them along with the common float, that this required a system of fixed rates for current-account transactions and a floating rate for capital-account transactions. Well before the oil crisis, the efficacy of the floating-rate system had removed any urgency about fundamental international monetary reform, and the onslaught of the oil crisis produced the spectacle—unfortunately temporary—of former fixed-rate diehards congratulating the world on its wisdom in having opted for a floating-rate system.

Since those halcyon days of winter 1973-74, however, the fixed-rate adherents and the commercial-banking and financial community have once again begun to find grave fault with the floating-rate system. They are not as yet anywhere near the point of recommending a return to fixed rates but only of recommending international cooperation and coordination in smoothing exchange-rate movements. Their criticisms are of two kinds, narrowly technical and broadly policy-evaluative.

The technical criticisms concern chiefly the magnitude of the exchange-rate movements that have occurred, which are judged to have been excessive and erratic in relation to the adjustment neces-

21

sary, and the failure of forward markets to develop for more than a few currencies, which greatly hampers the safe conduct of international financial business.

With respect to the first criticism, I must confess that I have always been astounded by the confidence with which "practical" men pass judgments about how much movement of a market price is sufficient to restore equilibrium. If the market consistently overshot, one would expect vast profits to be made by currency speculation. But a number of bankers who have tried it have had their fingers badly and most embarrassingly burnt. This is perfectly predictable from the results of economic research on forward exchange markets, using techniques taken over from analysis of the behavior of stock-market prices. These results show that, despite the frequent appearance of apparent purposive patterns, changes in the movement of foreign-exchange rates in a free market are a "random walk." One must also note that foreign-exchange markets, like stock markets, capitalize expected future price movements into current prices, so that prices may be expected to move more sharply in response to new information than consideration of current demands and supplies alone would lead one to expect. (To put this point another way, "practical" men are as guilty as the balance-of-payments theorists they criticize of assuming that international adjustment occurs only through the current account.) In addition, one suspects that there is an important degree of optical illusion in bankers' discussions of the magnitudes of exchange-rate changes. Financial attention tends to concentrate on the rates between currencies that are important in international finance, but these are not necessarily the exchange-rate relationships relevant to international trade and direct investment. Thus, for the United States the rates of the Canadian dollar and Japanese yen are far more important than the rate of the Swiss franc, and the Canadian dollar has remained within a few cents of parity with the American dollar since 1970, while the yen rate has been fairly stable since 1973.

The second technical criticism, concerning the failure of forward facilities to develop on the expected scale, also raises some questions about what one can reasonably expect. To be brief and colloquial, it would obviously be pleasant for me if someone were to operate an all-night bar on the corner of my street, so that I could get a drink if ever I needed one, but I cannot reasonably expect anyone to open such a bar solely on the expectation of having me for a customer. Forward facilities in foreign exchange will develop only if there is enough volume of business to yield a reasonably predictable profit. It would not surprise me, given all the other risks to which private trade

22

and investment are subject in the inflationary and oil-uncertain mid-1970s, if the establishment of futures markets in a broad range of currencies commanded a very low priority.

The policy-evaluative criticism of the floating-rate system is that the world has had more inflation, not less, since floating rates became the system of international monetary relationships. But floating-rate exponents have never argued that floating rates will guarantee more price stability than fixed rates. The original argument, which goes back to the 1930s depression, was that only with floating rates could a country pursue an independent employment and price-stabilization policy. Actually, in that period countries did not in fact employ floating rates at all boldly for that purpose. Recently, the argument for floating rates has been modified, quite logically, into the proposition that only with floating rates can a country pursue an independent price-stabilization policy *if it so desires*. Whether it chooses to do so or, on the contrary, chooses to permit more inflation than would have been consistent with fixed rates is a matter for its own political choice.

Given the head of steam that inflation was permitted to develop under the fixed-rate system, and particularly the unusual degree of synchronization of the up-phases of national business cycles and the consequent pressure of world demand on food prices in 1972-73, it is not surprising that countries should have opted to let inflation rip. There is, in fact, a dangerous parallelism of irrational interpretation building up in this connection: just as the defenders of fixed rates used to attribute the depression and the constriction of world trade of the 1930s to the 1930s floating-rate regime—which relieved the worst of the horror—instead of blaming the predecessor fixed-rate regime that produced them, so those currently hankering after a return to fixed rates are building toward laying the blame for world inflation on the 1970s floating-rate regime, rather than on the predecessor fixed-rate regime that made floating rates necessary.

In any case, it is far too early to conclude that the floating-rate regime is more inflationary than the predecessor fixed-rate regime. The present world recession may succeed in breaking inflationary expectations and restoring rough price stability if contractionary monetary policies are not reversed too sharply and expansively. And the U.S. determination to fight inflation probably owes something to the effects of devaluation and downward flotation of the dollar in bottling up American inflation within the U.S. economy.

Time alone will tell. Meanwhile, it is certain that the world will not return to the fixed-rate system for a long time ahead, if ever. This raises some interesting problems for national economies and their

23

policy-makers as to the best way of living with the floating-rate regime.

My own country, Canada, has chosen in my view the worst possible strategy—to stay virtually pegged to the American dollar while letting the energy-resources boom give the economy a still more inflationary impetus than pegging to the U.S. dollar and the U.S. inflation alone would have done.

In an earlier run of the Horowitz Lectures, my Chicago colleague Milton Friedman surprised some of his audience and, later, of his readers by recommending that Israel should abolish its central bank and instead peg irrevocably to the U.S. dollar. I agree entirely with Friedman's argument that a small country anxious to promote economic development should eliminate the temptations to inflation inherent in central banking by joining irrevocably to a larger currency area, but I do not find it self-evident that the U.S. dollar is the proper currency to peg to. For a country significantly involved in trade with a number of countries, pegging to the currency of the most rapidly inflating country or to the currency depreciating most rapidly in terms of other major currencies (which may not be the same thing) automatically guarantees the most rapid possible rate of domestic inflation. The only relevant argument for doing so—and it is a nonargument—is that a country heavily dependent on another for imported capital and unilateral transfers should peg to the currency of the investing country in order to encourage investor confidence and maintain the domestic-currency value of foreign gifts. The second objective makes no economic sense whatever—the real value of foreign gifts is what they can buy abroad—and the first makes no sense when the relevant risk is created by the irresponsibility of the investing country's financial management and not by that of the recipient country. To be concrete, if Israel as a small country finds it preferable to peg the Israeli pound to a larger foreign country's currency rather than to float independently, the question of whether to peg, say, to the German mark rather than the U.S. dollar should at least be looked at seriously.

I have said that, in my judgment, the regime of floating exchange rates is going to be with us for a long time. This raises the obvious question of whether eventually the world monetary system will return to a regime of fixed rates—or, more likely, "flexible" or "adjustably pegged" rates. My own hunch is that it will. One reason is historical: Britain after the Napoleonic War, the European countries after each of the two world wars, the United States after its period of floating (1860-79), not to speak of lesser countries practicing currency flotation for shorter or longer periods, all returned sooner or later to

24

fixed exchange rates. As an incidental point, worth meditating on and relevant in the event of another major war, note that, in contrast to a more distant past, the advent of World War II led countries to peg the exchange rates of previously floating currencies, presumably to fix the unit of calculation for the external transactions of the controlled war economy. The other reason for my expectation is theoretical. The case for floating exchange rates is always carefully framed to distinguish between exchange rates that are free to move in response to market forces and exchange rates that oscillate significantly over time. The point is that, under stable economic conditions and with stable national economic policies, rates free to move will actually change little and slowly over time. Such a situation of relative stability must come about, if only transiently, at some point in the future. With rates stable, it will seem a trivial step, well worth the additional benefits, to move from *de facto* to *de jure* fixity of exchange rates.

That possibility makes it more, not less, necessary to keep fresh the memory of the intensive debates that were proceeding up until early 1974, after the outbreak of the oil crisis, about the main lines that a reform of the international monetary system (conceived as a fixed-rate system) should take. There are, specifically, three problems that are likely to be more difficult to get to grips with in the light of floating-rate and oil-crisis experience than they already were in the last phase of the International Monetary Fund system: the conditions under which exchange-rate changes in both directions should be sanctioned and indeed internationally required (recall what I said earlier about the financial community's feeling that exchange-rate trends in the past two years have been uncomfortably severe and erratic); the future of the dollar as a nationally created and controlled international reserve currency substitutable for and against Special Drawing Rights under ill-defined and amorphous conditions; and the possible world central-banking role of the International Monetary Fund. The prime function of a world central bank should be to provide stable growth of the international money base of the world financial system. But the International Monetary Fund has two serious distractions from this primary objective. Because of its historical origins in 1930s depression thinking, it is psychologically dominated by the presumption that the main danger to be guarded against is a shortage of international liquidity. And owing to its character, being at least partially a democratically responsible world institution and obliged to develop and maintain popular support among the numerical majority of its constituents, it tends to concern itself with the lender-of-last-resort function of the ultimate source of international credit, to the neglect of the

25

money-supply-control function and of Bagehot's dictum that lending of last resort should be conducted *at a penalty rate*. It must obviously be demoralizing for a central-banking institution to be empowered, by virtue of low conventional lending rates in a period of rapid inflation, to lend at last resort at a negative real cost of borrowing to the borrower and to be under political pressure to ration credit on the basis of the borrower's need for real resources.

# References

Alexander, Sidney S., "Effects of a Devaluation on a Trade Balance," *IMF Staff Papers*, 2 (April 1952); reprinted as Chap. 22 in R. E. Caves and H. G. Johnson, eds., *Readings in International Economics*, Homewood, Ill., Irwin, 1969.

Archibald, G. C., and R. G. Lipsey, "Monetary and Value Theory: A Critique of Lange and Patinkin," *Review of Economic Studies*, 26 (October 1958), pp. 1-22.

Frenkel, Jacob A., and Harry G. Johnson, eds., *The Monetary Approach to the Balance of Payments*, London, Allen & Unwin, and Toronto, University of Toronto Press, 1976.

Gervaise, Isaac, *The System or Theory of the Trade of the World*, 1720; reprinted in Economic Tracts series, Baltimore, The Johns Hopkins Press, 1956.

Hume, David, *Political Discourses*, 1752; reprinted in E. Rotwein, ed., *David Hume: Writings on Economics*, London, Nelson, 1955.

Machlup, Fritz, *International Trade and the National Income Multiplier*, New York, Blakiston, 1943; reprinted, New York, Augustus M. Kelley, 1965.

Meade, James E., *The Theory of International Economic Policy*, Vol. I, *The Balance of Payments*, Oxford, Oxford University Press, 1951.

Metzler, L. A., "Underemployment Equilibrium in International Trade," *Econometrica*, 10 (April 1942), pp. 97-112.

Mundell, Robert A., "The Appropriate Use of Monetary and Fiscal Policy for External and Internal Stability," *IMF Staff Papers*, 9 (March 1962), pp. 70-79.

Ohlin, Bertil, "The Reparation Problem: A Discussion," *Economic Journal*, 39 (June 1929), pp. 172-178.

Patinkin, Don, *Money, Interest, and Prices*, 2nd ed., New York, Harper & Row, 1965.

Robinson, Joan, "The Foreign Exchanges," Chap. 4 in H. Ellis and L. A. Metzler, eds., *Readings in the Theory of International Trade*, London, Allen & Unwin, 1950.

Salter, W. E., "Internal and External Balance: The Role of Price and Expenditure Effects," *Economic Record*, 35 (August 1959), pp. 226-238.

Tinbergen, J., *On the Theory of Economic Policy*, Contributions to Economic Analysis series, Amsterdam, North-Holland, 1952.

26

# CURRENCY DEVALUATION
# IN DEVELOPING COUNTRIES

Richard N. Cooper

Currency devaluation is one of the most dramatic—even traumatic—measures of economic policy that a government may undertake. It almost always generates cries of outrage and calls for the responsible officials to resign. For these reasons alone, governments are reluctant to devalue their currencies. Yet under the present rules of the international monetary system, laid down in the Articles of Agreement of the International Monetary Fund, devaluation is encouraged whenever a country's international payments position is in "fundamental disequilibrium," whether that disequilibrium is brought about by factors outside the country or by indigenous developments. Because of the associated trauma, which arises because so many economic adjustments to a discrete change in the exchange rate are crowded into a relatively short period, currency devaluation has come to be regarded as a measure of last resort, with countless partial substitutes adopted before devaluation is finally undertaken. Despite this procrastination, over 200 devaluations in fact occurred between the inauguration of the IMF in 1947 and the end of 1970; to be sure, some were small and many took place in the years of postwar readjustment, especially 1949. In addition, there were five upvaluations, or revaluations, of currencies. Two more occurred in May 1971.

By convention, changes in the value of a currency are measured against the American dollar, so a devaluation means a reduction in the dollar price of a unit of foreign currency or, what is the same thing, an increase in the number of units of the foreign currency that can be purchased for a dollar. (The numerical measure of the extent of devaluation will always be higher with the latter measure than with the former; for example, the 1967 devaluation of the British pound from $2.80 to $2.40 was 14.3 per cent and 16.7 per cent on the two measures, respectively.) By law, changes in currency parities are against gold, but since the official dollar price of gold has been unchanged since 1934, these changes in practice come to the same thing. Except when many currencies are devalued at the same time—as they were in September 1949 and to a much less extent in November 1967 (when over a dozen countries devalued with the pound) and August 1969 (when fourteen French African countries devalued their currencies along with the French franc)—a currency devaluation against the dollar is also against

Reprinted from "Currency Devaluation in Developing Countries—Essays in International Finance No. 86, June 1971," pp. 3–31, with permission of Princeton University Press, Princeton, Copyright 1971.

the rest of the global payments system, that is, against all other currencies.

Only a baker's dozen of countries did not devalue their currencies at least once during the period 1947-70 (Japan, Switzerland, and the United States among developed countries, and ten less developed countries, mostly in Central America). Largely because they are so numerous, but partly also because they devalue on average somewhat more often than the developed countries do, less developed countries account for most currency devaluations. Yet the standard analysis of currency devaluation, which has advanced substantially during this period and is still being transformed and further refined, fails to take into account many of the features that are typical of developing countries today, and which influence substantially the impact of currency devaluation on their economies and on their payments positions.

This essay attempts to do three things. First, it sketches very briefly the analysis of currency devaluation as it stands at present. Second, it suggests how this analysis has to be modified to take into account the diverse purposes to which the foreign-exchange system is put in many less developed countries, and the extent to which these diverse purposes influence the nature of devaluation and its effects on the economy. Third, it draws on recent experience with about three dozen devaluations to see to what extent the anxieties of government officials, bankers, and traders, and even some economists, about devaluation and its effects are justified, and interprets some of this experience in light of the earlier theoretical discussion.

## I. A SUMMARY OF THE THEORY OF DEVALUATION

In analyzing devaluation, the exact nature of the initial disequilibrium is important, and much analysis misleads by its focus on economies that are assumed to be in equilibrium at the moment of devaluation. To set the stage precisely, suppose we have a country which for reasons past has money costs that are too high to permit it to balance its international payments at a level of domestic economic activity that is both desired and sustainable, and as a result it must finance a continuing payment deficit out of its reserves, a process that obviously cannot continue indefinitely. Thus by assumption we are not dealing with a case in which domestic demand is pressing against productive capacity to an extent that is regarded as undesirable ("inflationary"), although under the circumstances domestic expenditure does exceed domestic output, a necessity to maintain full employment. Correction of the payments imbalance by reducing aggregate demand (the rate of money spending) would lead to unwanted unemployment because of the rigidity of fac-

4

tor incomes in money terms, especially wages. Perhaps ultimately the pressure on costs and prices of a depression in activity would restore an equilibrium level of costs and prices that would lead to payments balance at full employment, but the transitional depression might have to be long and painful. The recommended alternative is devaluation of the currency, which at the stroke of a pen lowers the country's costs and prices when measured in *foreign* currency. Analysis of the effects of devaluation on the country's economy and of the mechanism whereby it eliminates the payments deficit has proceeded under three quite different and apparently contrasting approaches: the elasticities approach, the absorption approach, and the monetary approach.

## Three Approaches to Analysis

The *elasticities approach* focuses on the substitution among commodities, both in consumption and in production, induced by the relative price changes wrought by the devaluation. For an open economy such as the one we are considering here, the principal relative-price change is between goods, whether imported or exported, whose price is strongly influenced by conditions in the world market, and those home goods and services that are not readily traded. For a small country, we can assume that the prices in domestic currency of foreign-trade goods—exports, imports, and goods in close competition with imports—will rise by the amount of devaluation (the larger of the two percentages mentioned above is the relevant one here). This rise will divert purchases out of existing income to nontraded goods and services, thereby reducing domestic demand for imports and for export goods, releasing the latter for sale abroad. When the country is large enough to influence world prices, domestic prices may rise by less than the amount of the devaluation, since prices in foreign currency will fall somewhat in response to the reduction in our country's demand for imports or to the increase in its supply of exports. There is some presumption that most countries will have a greater influence on their export prices than on the prices at which they import, so the rise in local prices of exports will be less, and the terms of trade will deteriorate.

The shift in relative prices operates both on consumption and on production. Consumption will be diverted to lower-priced nontraded goods and services, releasing some existing output for export and cutting demand for imports. At the same time, increased profitability in the foreign-trade sector, arising from the fact that prices in domestic currency have risen more than domestic costs, will stimulate new production of export and import-competing goods, and will draw resources into these industries. If excess capacity happens to exist in these industries, the

5

resources drawn in will be variable ones—labor and materials. Otherwise new investment will be required; in agriculture, land may have to be recropped or herds rebuilt.

The elasticities approach gives rise to the celebrated Marshall-Lerner condition for an improvement in the trade balance following a devaluation: that the elasticity of demand for imports plus the foreign elasticity of demand for the country's exports must exceed unity, which is to say that the change in the quantity of imports and exports demanded together must be sufficiently great to offset the loss in foreign earnings consequent upon lowering the price of exports in foreign currency. This condition assumes initially balanced trade, finished goods, and elastic supply of exports both at home and abroad, but may be modified to allow for initial trade imbalance, for less than perfectly elastic supplies of export, and for intermediate products.

The *absorption approach* shifts attention from individual sectors to the overall economy. Its basic proposition is that any improvement in the balance on goods and services must, in logic, require some increase in the gap between total output and total domestic expenditure. It starts from the identity $E + X = Y + M$, where $E$ is total domestic expenditure on goods and services and $X$ is total foreign expenditure on our country's goods and services (exports), the sum of the two representing total "absorption" of the goods and services available to the country, which derive from its own aggregate output, $Y$, and imports from the rest of the world, $M$. Rearranging the terms yields $X - M = Y - E$, which shows that any trade surplus reflects an excess of output over domestic expenditure, and vice versa for a trade deficit. It follows that to reduce a deficit requires a corresponding reduction in the gap between output and expenditure. Excess capacity and unemployment will permit an increase in output; otherwise expenditure must be reduced. Without such a reduction, there can be no improvement in the balance, regardless of the elasticities. This analysis points to the policy prescription that devaluation must be accompanied by deflationary monetary and fiscal policy to "make room" for improvement in the balance, a prescription to which we shall return below.

The *monetary approach* to devaluation focuses on the demand for money balances and the fact that an excess demand for goods, services, and securities, resulting in a payments deficit, reflects an excess supply of money. It draws attention to the analytical parallel between a devaluation and a reduction in the supply of money that affects all holders in equal proportion. Devaluation is equivalent to a decline in the money supply and in the value of other financial assets denominated in local currency, when measured in *foreign* currency. Put another way, the

6

real value of the money supply will be reduced by devaluation, because the local prices of traded goods and services, and, secondarily, those of nontraded goods and services to which demand is diverted, will rise. The public will accordingly reduce its spending in order to restore the real value of its holdings of money and other financial assets, which reduction in expenditure will produce the required improvement in the balance of payments. For a country in initial deficit, the right devaluation will achieve just the right reduction in the real value of the money supply, and the deficit will cease. To restore lost reserves the country must devalue by more than that amount, in order to achieve a surplus. But once the public has reattained its desired financial holdings, expenditure will rise again and the new surplus will be eliminated. On this view, a devaluation beyond the equilibrium point has only a once-for-all effect. A key implication of this approach is that if the monetary authorities expand domestic credit following devaluation to satisfy the new demand for money, the effects of the devaluation on international payments will be undermined. (The money supply may of course increase in response to the inflow of reserves; indeed, if it does not, the surplus will continue until some other country takes steps to curtail it.)

These three approaches are complementary rather than competitive—they represent different ways of looking at the same phenomenon, and each has its strengths and weaknesses. The first has its roots in Marshallian, partial-equilibrium analysis, and is most suitable when the foreign-trade sector—like Marshall's strawberry market—is small relative to the total economy, or when there are ample unemployed resources—and even in the latter case it offers only a part of the story. The absorption approach is "Keynesian" in its focus on total output and expenditure, not differentiating among sectors and neglecting monetary effects. But it draws attention to the impact of changes in exchange rates on overall income and expenditure, which the elasticities approach fails to do. The third approach is the international counterpart of the recently revived monetary school of thought propagated by the Chicago-London School of Economics, but its intellectual roots go back to David Hume, where stock adjustments in the real value of money balances were all-important.

It is tempting to think of these three approaches in temporal sequence, with the first stage of the elasticity approach representing the short run, the absorption (income-expenditure) approach applying to the medium run, and the monetary approach applying to the long run, on the grounds that asset portfolios take a long time to adjust following a major dislocation. But this would oversimplify the matter. All factors are present to some degree even immediately following devaluation.

7

In the first instance relative prices normally do change, however, as assumed by the elasticities approach, and this in turn will alter the patterns of consumption and, in the right circumstances, of production, encouraging the necessary increase in net exports.

With initial excess capacity, these alterations will generate additional income, which by leading to additional expenditure will in turn damp down the improvement in the trade balance; without it, the switch in demand toward home goods will tend to bid up their prices. But unless the monetary authorities expand domestic credit the rise in prices will not be sufficient to eliminate the change in relative prices initially brought about by the devaluation, and some improvement in the trade balance will remain.

All this is consistent with the monetary approach. The initial disequilibrium reflects not only an excess supply of money but also a misalignment of relative prices between home and tradable goods, since the fixed-exchange-rate link with the world market diverts the impact of those excess holdings of money into demand for imports rather than higher prices in the foreign-trade sector. The appropriate devaluation simply corrects this disequilibrium set of relative prices and at the same time lowers the real value of money holdings and, hence, expenditures. It therefore has a durable effect. This contrasts with the case where the starting point is one of monetary equilibrium, as is usually assumed in the theoretical analysis despite the fact that devaluation seems superfluous in such circumstances, in that devaluation from equilibrium can have only a transitory effect, giving rise to the wholly misleading impression that devaluation cannot really "work."

Whether the second stage of the elasticities approach—the new investment in the foreign-trade sector—comes into play depends in large part on whether the structure of potential output was seriously affected during the disequilibrium period before devaluation. If the disequilibrium persisted for some time, or if investors were prompt to respond to profitable opportunities and failed to anticipate the eventual need for devaluation, then there would be excess capacity in the home-goods sector and deficient capacity in the foreign-trade sector from the viewpoint of long-run equilibrium, and the second stage would come into play. Otherwise, there would be sufficient capacity in the foreign-trade sector (not fully utilized before devaluation) and no change in the structure of potential output would be necessary.

The impact of growing cost-price disequilibrium on production in the export industries, and its subsequent reversal after devaluation, can be illustrated graphically by Finland's experience. Here a "zero line" marks the boundary north of which it is unprofitable to cut and

8

transport timber for export. As cost inflation proceeded in the 1950s, this zero line gradually moved southward, to the point in 1957 that it was only about 200 miles from the south coast. Following the 1957 devaluation, the line shifted markedly northward again.

## Distributional Effects

There is a distributional counterpart to these allocational changes which should be explicitly acknowledged, since distributional considerations are so important in less developed countries. A devaluation will raise the "rents" on all factors working in the foreign-trade sector, particularly, in the first instance, entrepreneurial returns in industries engaged in export and in competition with imports. At the same time, the real income of other groups (including the government) will decline because of the rise in prices of these goods. If the higher profits are expected to continue, managers in these industries will expand output and in so doing will bid up the prices of other factors of production used extensively in the foreign-trade sector, leaving a distributional effect in the end that may favor, say, labor, even though it favored certain profits initially. Since we started with a disequilibrium pattern of expenditure and a disequilibrium distribution of income (for a given tax regime), both produced by the misalignment of prices between traded and nontraded goods, the new position brought about by appropriate devaluation will persist unless it is disturbed by other factors.

But both the speed with which the initial distributional effect is transformed to the ultimate effect and the chance that the ultimate effect will not be disturbed will vary greatly. It is here that "money illusion" enters the picture, provided that term is interpreted broadly to cover cases where the decline in real income from a rise in prices is perceived (so there is no "illusion" in a literal sense) and accepted, even when a reduction in money wages would not be accepted. There are many reasons for such illusion to be present, not the least of which is the importance of contracts in business transactions. In the long run contracts can be renegotiated, but in the short run there are important costs to breaking and renegotiating them. Even when "contracts" are broken in any case, as when workers leave jobs in the home-goods sector to take up jobs in the foreign-trade sector, they may be willing to move at real wages lower than their pre-devaluation wages in the expectation of greater job security if they do so. Thus, while money illusion is not normally necessary for devaluation to be successful in improving the trade balance, the more widespread it is, and the longer it lasts, the greater will be the gain to reserves in the period following a devaluation of a given amount.

9

In another respect, however, money illusion is even more important. Some factors of production profited (at the expense of others, and of the national reserves) before the devaluation, when the domestic costs of foreign-trade goods were too high. This state of affairs was not sustainable in the long run, but those factors that did profit may be most reluctant to accept the reduction in real rewards that is in fact necessary, given the regime of taxes and other policies that affect the distribution of income. If through "bargaining power" (strong unions, administered prices) they succeed in raising their money incomes enough to restore their pre-devaluation level of real income, then the initial disequilibrium will also have been restored. The authorities will be forced to devalue again in the hope that it will work (or can be made to work) the second time. Or they may in the end have to reduce domestic demand, thereby creating unemployment and damaging all groups (although not equally) as the only way to resolve the incompatible objectives of payments equilibrium and level of real income (at full employment) acceptable to those who benefit from the disequilibrium. Money illusion will help to resolve the difficulty by permitting the groups in question to accept lower real incomes while still keeping up appearances with high and even somewhat enlarged money incomes.

## II. MODIFICATION IN THE ANALYSIS FOR DEVALUATION IN MOST DEVELOPING COUNTRIES

The foreign-exchange system of a country can be used to pursue many objectives other than clearance of the foreign-exchange market, and, faced with inadequate instruments of policy to achieve the many objectives expected of them, the governments of many less developed countries have called upon it to do so. These functions range from fostering industrialization, improving the terms of trade, and raising revenue to redistributing income among broad classes and even doling out favors to political supporters. A practice used frequently to accomplish all three of the first objectives, and also to redistribute income, is to give primary export products a rate of conversion into local currency lower than the rate that importers must pay to purchase foreign exchange (and that exporters of nontraditional products receive). Import-substituting investment is stimulated by the unfavorable rate on imports, foreign export prices are higher than they otherwise would be in the rare event that the country can influence world prices for its products, and the government gains revenue from the often substantial difference between the buying and selling prices of foreign exchange. Similarly, imported consumer goods are often charged a rate much higher than imported investment goods, in an effort to stimulate invest-

10

ment in manufacturing (and with the undesirable side-effect of encouraging modes of production that use relatively more capital and relatively more imported ingredients or components). Finally, and not least, the exchange system can be used to redistribute income between broad classes, as for example in Argentina when the exchange rate applied to traditional exports, meat and wheat, was deliberately kept low for a number of years with a view to keeping down the cost of living for urban workers.

All of these functions involve multiple exchange rates of some kind, either explicit or implicit, that is, charging different exchange rates according to the commodity or service, the origin or destination, or the persons involved in the transaction. As such, they inevitably invite arbitrage and require policing—but so of course do taxes, which they often replace in function.

Moreover, politicians have learned that an objective achieved indirectly is frequently socially acceptable when direct action would not be. This is not always because of an imperfect understanding of the indirect means in contrast to the direct means, although that plays an important role. It is much easier for an interest group to mobilize successfully against an export tax than it is to mobilize against an over-valued currency supplemented by high import tariffs and possibly accompanied by some export subsidies, even though the two systems might have precisely the same economic effects. As Fritz Machlup has said (in connection with Special Drawing Rights): "We have often seen how disagreements among scholars were resolved when ambiguous language was replaced by clear formulations not permitting different interpretations. The opposite is true in politics. Disagreements on political matters, national or international, can be resolved only if excessively clear language is avoided, so that each negotiating party can put its own interpretation on the provisions proposed and may claim victory in having its own point of view prevail in the final agreement." Machlup was speaking of language, but the same is true of action; a roundabout way of accomplishing a controversial objective will often succeed where direct action would fail, because it obscures, perhaps even from the policymakers themselves, who is really benefitting and who is being hurt.

The difficulty is that the pursuit of these diverse objectives too often leads to neglect of the function of the exchange rate in allocating the supply of foreign exchange. When balance-of-payments pressures develop (sometimes as a result of inflationary policies, which in the short run are often also a successfully ambiguous way to reconcile conflicting social objectives), officials then engage in a series of patchwork efforts and marginal adjustments to make the problem go away (raising tariffs

II

here, prohibiting payments there), which may disturb the original objectives as well as coping only inadequately with the payments difficulty. When devaluation finally occurs, in consequence, the occasion is also taken (sometimes under pressures from the IMF or from foreign-aid donors) to sweep away many of the ad hoc measures that have been instituted to avoid the necessity for devaluation.

This fact makes currency devaluation in many developing countries (and some developed ones) a good deal more complex than a simple adjustment of the exchange rate, and the analysis must be modified to take these other adjustments into account. Broadly speaking, one can distinguish four types of devaluation "packages": (1) straight devaluation (involving a discrete change in the principal exchange rate, as opposed to a freely depreciating rate or an administered "slide" in the rate, such as was adopted by Brazil, Chile, and Colombia in the late sixties, whereby the rate was depreciated by a small amount every two to eight weeks); (2) devaluation with a *stabilization* program of contractionary monetary and fiscal policy aimed at reducing the level of aggregate demand, or at least the rate of increase of demand; (3) devaluation accompanied by *liberalization*, whereby imports and other international payments that were previously prohibited or subject to quota are allowed to take place under much less restraint than before the devaluation; and (4) devaluation accompanied by partial or full *unification* of exchange rates, whereby a pre-existing diversity of exchange rates is collapsed into a single, unified rate, or at most two rates, the lower one applying to traditional exports of primary products and in effect amounting to a tax on these exports.

It is obvious that these categories are not mutually exclusive. Devaluation may involve simultaneously a stabilization program, liberalization, and exchange-rate unification, and in fact at least some elements of all are often present in devaluation in developing countries. For example, of 24 devaluations studied in some detail (and which will provide the basis for evidence cited below), ten involved a fairly substantial degree of trade liberalization, ten (partially overlapping) involved a major consolidation of rates, and virtually all were accompanied by at least token measures of stabilization. (It might be mentioned in passing that in most developing countries the distinction between monetary and fiscal policy does not have the same meaning it has in more advanced countries. Since capital markets are little developed and access to foreign capital markets is limited, budget deficits, after allowing for foreign assistance, must be financed by the banking system, which results directly or indirectly in monetary expansion. Thus, the usual focus on eliminating government deficits is merely an indirect way to limit the rate of

12

monetary expansion, provided, of course, that bank credit to the private sector is also kept under control.)

These various simultaneous adjustments must be taken into account in analyzing the economic effects of devaluation. In particular, it is necessary to distinguish between devaluation from a position of open payments deficit, such as we considered in the preceding section, and devaluation from a position in which a latent deficit is suppressed by import controls and related measures, which are removed upon devaluation. An additional complication is that less developed countries are more likely at the time of devaluation to be generating new money demand at a rate greater than can be accommodated by total domestic output plus foreign assistance and other long-term capital inflows from abroad; in short, they are pursuing inflationary policies, as opposed merely to having costs that have gotten out of line in the course of *past* inflation.

In fact, most devaluing countries have some combination of an open payments deficit and a suppressed one. But for clarity of exposition, and to bring out the contrast with the analysis above most clearly, we will consider devaluation from a position in which the payments deficit is fully suppressed by other measures, and where the devaluation is accompanied by liberalization and/or unification of the exchange system involving the removal of special taxes, subsidies, and prohibitions that have been installed earlier. In addition, we will suppose that the country is not pursuing inflationary policies at the time of devaluation.

*Elasticity Pessimism*

The first point to note is that the elasticity of demand for imports is likely to be low when imports are concentrated on raw materials, semifabricated products, and capital goods, a structure prevalent in less developed countries. With import substitution in an advanced stage, all the easy substitutions having already been made in the pursuit of industrialization; imports depend largely on output rather than income and are not very sensitive to relative price changes. There is more room for substituting home production for imports of foodstuffs, although it will usually take a season or longer to bring this about. Moreover, import liberalization and exchange-rate unification will actually result in a *reduction* of the prices of those imports most tightly restrained before the devaluation, so consumption of them will be encouraged.

There is greater diversity of experience with regard to exports. Some countries—producers of oil, copper, and cocoa, for instance—have virtually no domestic consumption of the export goods. In others, exports include the major wage good—beef in Argentina and fish in Iceland, for instance. In the former countries, increasing exports require enlarged output

13

and development of new export products, and neither of these courses may be easy in the short run, although tree crops can sometimes be more intensively harvested. In the latter countries, there is more room for immediate increases in exports permitted by reductions in domestic consumption of the export products, but this gain is brought about only by courting a wage-price spiral, on which more will be said below. In developed countries, by contrast, there are many domestically consumed goods that are actual or potential exports, and hence there is more room for short-term increases in export supply by diverting output from the home to the foreign market.

When it comes to incentives to enlarge output and expand capacity, the principal reallocation here is between import-competing goods and exports, rather than between home goods and all foreign-trade goods, as in the case of open economies. This is because by assumption imports have already been stringently limited by high tariffs, disadvantageous exchange rates, and quantitative restrictions, all of which create a strong price incentive for domestic production. Some exports may also have been subsidized and, where this is so, devaluation accompanied by removal of the subsidy may leave no new incentive to increase production for export. But, generally speaking, exports are heavily penalized under the regimes we are considering, and devaluation has the effect of reducing the premium for producing import-competing goods for the home market and increasing the premium for production for export, with the principal shift in incentives coming between these two sectors rather than with respect to the home-goods sector (although of course there will also be some incentive to shift resources into that sector from the import-competing sector and out of it to the export sector).

New investment in the capacity to export will require that investors expect the improvement in their position to last, that the devaluation and associated policies will establish a new regime that will not simply slide back into the old configuration of policies. Establishing these expectations is one of the most difficult tasks of those carrying out the reform. The same problem exists in principle in devaluation from open deficit too, but developing countries that have not relied on restriction of imports for payments reasons stand a better chance of success, because investors will expect any emerging disequilibrium to be corrected rather than suppressed by controls.

Furthermore, the required investment may differ in character from that in developed countries. Where manufactures can be competitively exported under the new regime, conversion from domestic manufacturing may be relatively easy; but opening up export markets for manufactured goods for the first time is a drawn-out process, requiring

14

the establishment of new marketing channels. The shift from domestic to export crops in agriculture—or the opening of new lands—is generally easier; but for livestock and for tree crops the required gestation period may be several years.

For all of these reasons, some pessimism with regard to price elasticities would be quite justified for many developing countries, at least in the short run, but as we will see below it does not usually go far enough to prevent devaluation from improving the trade balance.

*Effects on Aggregate Demand*

The absorption approach suggests that a devaluation that merely substitutes for other measures, leading to no net improvement in the balance on goods and services, requires no cut in aggregate expenditure or increase in total output. But it is still worth asking what pressure devaluation in these circumstances might put on aggregate expenditure and output, since this will give some guide to the possible need for compensatory macroeconomic policy. To provide a framework for discussion, rewrite the basic equation noted above as $Y = E + D$, where $D = X - M$, the balance on goods and services measured in domestic currency. In order to discover the impact effect on output, $Y$, we must ask what will be the effects of devaluation on its two components, the level of domestic expenditure and the external balance measured in domestic currency. The impact on output will in turn affect incomes, expenditure, imports, and output again in a multiplier process. But the impact effect will tell us the impetus to this multiplier process, and in particular whether it is expansionary or deflationary.

To take the external balance first, for the reasons given above this might actually worsen in the period immediately following devaluation, when measured in foreign currency, and this by itself would have a deflationary impact upon the economy. The worsening would occur if import liberalization takes effect immediately, giving rise to an increase in imports, while the stimulus to exports occurs only with a lag. In time, of course, the stimulus to exports will also stimulate the domestic economy; but the immediate impact would be a deflationary one. Furthermore, any discrepancy between the local-currency value of a dollar's worth of imports and a dollar's worth of exports, for example due to tariffs, means that even a parallel expansion of imports and exports will be deflationary, provided the government does not spend the additional revenue at once.

Thirdly, devaluation is deflationary to the extent that remaining quotas are replaced in their import-restricting effects by the depreciated exchange rate. Scarcity rents that went to privileged importers before the

15

devaluation would now accrue to the central bank as it sells foreign exchange. In effect, price rationing will have replaced quantitative rationing, with no ultimate effect on the *final* market price, but with a higher domestic-currency price to the importer or firm enjoying the license. (If the licenses are auctioned, of course, these scarcity rents accrue to the government even before devaluation; but auctioning of licenses is in fact rare.)

Finally, the inelasticity of demand for imports suggests that a sharp rise in their local-currency price will lead to an increase in *expenditure* upon them, even if the quantity and foreign-exchange value of imports fall. In this respect devaluation is like an efficient revenue-oriented excise tax, increasing the price far more than it reduces the quantity purchased. Since imports will substantially exceed exports, thanks to inflows of foreign grants and capital, exports will have to expand a great deal before the increased local-currency income from their sale exceeds the increased local-currency expenditure on imports.

For all these mutually reinforcing reasons, the initial impact of devaluation on the domestic economy of a developing country is likely to be deflationary in that it will reduce purchasing power available for expenditure on domestic output. This may be so, paradoxically, even when the trade balance improves in terms of foreign currency. Thus in 14 of 24 devaluations examined, the balance measured in domestic currency worsened following devaluation—without including increased tariff revenues on imports—and in seven of these this worsening occurred despite an improvement in the balance when measured in foreign currency.

The external sector, however, is only one component of demand. It is necessary also to ask how devaluation may affect the level of total domestic expenditure, $E$. Refined analysis is required to discuss the possible effects satisfactorily, but here it will be sufficient to identify six effects that are likely to be important in developing countries.

(1) There is first the *speculative effect*, which is also important in devaluations from open deficits. If devaluation has been anticipated and is expected to lead to a general increase in prices there will be anticipatory buying before the devaluation and the post-devaluation period will therefore commence with larger-than-usual holdings of goods. Total expenditure by the public may therefore drop in the period immediately following devaluation, until these inventories are worked off. (This effect would also lead to a rise in imports before and a drop after the devaluation, insofar as this is permitted by the system of licensing or other controls.) While the speculative effect will normally lead to a drop in expenditure, however, it may lead to an increase if the price

16

increases following devaluation are expected to lead to general inflation, or if another devaluation is in prospect, as it did immediately following Britain's devaluation in 1967.

(2) Devaluation will generally lead to a redistribution of income, and this *distributive effect*, while present for any devaluation, is likely to be especially important in developing countries with heavy reliance on primary products for export. Unless checked by special export taxes, a devaluation will lead to a sharp increase in rewards to those in the export industry, who are often landowners. Whether large or small, landowners are likely to have different saving and consumption patterns from urban dwellers, generally saving more out of marginal changes in income, at least in the short run. Thus, a redistribution of real income from workers to businessmen and from urban to rural dwellers is likely, in the first instance, to lead to a drop in total expenditure out of a given aggregate income, and this drop will be deflationary. But of course the redistributional effect could also go the other way, if as a result of devaluation the real income of those with a low marginal propensity to save is increased at the expense of others. The redistributional effect will also affect the level of imports out of a given total income, since consumption pattern of those who gain may differ from that of losers. But this effect is likely to be less marked than the total expenditure effect, partly because much of the import bill of developing countries represents inputs into domestically produced goods and services, so they are somewhat more widely diffused throughout the economy than would be the case for direct imports of manufactured consumer goods. Diaz-Alejandro has documented well the dominating importance of the redistributive effect following the Argentine devaluation of 1959, where the shift of income to the landowners led to a sharp drop in domestic spending and therefore to a secondary drop in imports.

(3) A devaluation will lead to a rise in the domestic costs of servicing *external debt* denominated in foreign currency. Where the liabilities are those of businessmen who do not benefit much from the devaluation, it may lead to bankruptcy and an attendant decline in business activity, even when businesses are otherwise sound. This factor allegedly figured in the decline in investment following the Argentine devaluation of 1962. Even where the debt is held officially, the problem of raising the local-currency counterpart of external servicing charges often poses a serious problem, and sometimes represents a serious inhibition to devaluation.

Indeed (to digress for a moment), these "accounting" relationships, usually ignored by economists, often preoccupy officials and bankers.

17

Local development banks that have borrowed abroad (for instance, from the World Bank or IDA) in foreign currencies and re-lent to local business in domestic currency have accepted an exchange risk that has occasionally provided the major barrier to devaluation: to allow its development bank to fail might psychologically undermine the government's development plans. But if the bank is to be saved, who is to absorb the devaluation loss, and how? (The obvious retrospective answer is that local borrowers should be charged interest rates sufficiently above what the development bank pays on its foreign debt to cover the exchange risk—with the added advantage that such rates will more closely approximate the true cost of capital in the developing country. But development banks have often failed to do this. Or, if they have done it, they have failed to set aside a sufficiently large reserve out of the difference in rates.) A similar problem arises for net *creditors* when the value of their foreign claims is reduced in terms of local currency by devaluation abroad or revaluation of the home currency. Thus, Hong Kong inadvisedly devalued its currency following the 1967 devaluation of sterling, apparently because the commercial banks in Hong Kong held large sterling assets against their local-currency deposits, and the banking system would have been threatened if the relationship between sterling and Hong Kong dollars had not been preserved. But the government thought better of this decision and revalued again four days later, in the meantime having worked out a way to indemnify the banks out of official reserves. By the same token, the German Bundesbank showed substantial paper losses (in marks) on its assets held in gold and dollars following the revaluations of 1961 and 1969. The 1961 revaluation was delayed until the German government would agree to indemnify the bank for its "losses" (which were entirely paper losses, arising from double-entry bookkeeping conventions) out of the budget over a period of seven years. Where private parties have incurred foreign debt, of course, the loss is real to the firm or bank, and that may have undesirable consequences for the economy as a whole. But a thorough discussion of this important issue is beyond the scope of the present essay.

(4) When the balance of goods and services has turned adverse in terms of domestic currency—as we have seen above may frequently be expected—then in the absence of countervailing monetary action a domestic *credit squeeze* may result, since importers and others will be paying more into the central bank for foreign exchange than exporters are receiving. This in turn may lead to a reduction in domestic expenditure.

(5) On the other hand, the improved earning opportunities in the export industries may (if they are expected to last) induce both *domestic*

18

*and foreign investment* in the country. Foreign investors bring their funds with them, as it were, and increase local credit by converting foreign exchange into domestic currency at the central bank. Domestic investors must either activate idle balances or find banks willing and able to lend, in the second instance leading to domestic credit expansion. Of course, the incentives to invest in import-competing industries will be reduced by the devaluation (in sharp contrast to the case of devaluation from a position of open deficit, where they will be stimulated by devaluation); but the stimulus to investment may on balance be positive, partly because there are limits to the rate at which disinvestment can take place. For reasons given earlier, however, the extent of new investment will depend on expectations about the durability of the new regime, and investors may wait awhile to see how things are going.

(6) In the monetary approach to devaluation from an open deficit, attention was drawn to the reduction in the real value of money holdings and reliance was placed on a desire to reconstitute these holdings to reduce expenditure. In the case of devaluation from a suppressed deficit, however, this *money-demand* effect is more complicated, and may not be present at all. If devaluation simply displaces other instruments of policy, with no effect on domestic prices, the real value of money balances will not be altered. If, as is more typically the case, devaluation displaces some other limits on imports but raises the local prices of exports, the effect on the real value of money holdings will depend upon the importance of export products in local expenditure. When export products are extensively purchased by residents, the monetary effect will tend to reduce domestic spending. Import liberalization, on the other hand, cuts the other way insofar as import prices actually fall. Moreover, in the long run another factor comes into play: to the extent that devaluation displaces measures that led to a less efficient use of resources, the devaluation package will lead (after the necessary reallocation of resources has taken place) to an increase in real income, and this in turn will require a supporting increase in money holdings. Unless it is supplied by the monetary authorities, this demand will depress expenditure relative to potential income.

The upshot of these various considerations is that devaluation in developing countries is likely to be deflationary in the first instance, and thus may "make room" for any improvement in the balance on goods and services, without active reinforcement from monetary and fiscal policy. Indeed, for reasons given below, it may sometimes be desirable to accompany devaluation with modestly *expansionary* policies. Frequently, however, the devaluation will take place against a background of excessively expansionary policies. In this case the devaluation-induced

19

deflation will be helpful in bringing the economy under control, but these effects must be taken into account if the government is to avoid overshooting the target with deliberately contractionary measures.

In short, unless the devaluation is very successful in stimulating exports or in stimulating investment, the absorption approach to devaluation is of less relevance to devaluation in developing countries except in manifestly inflationary situations—the real problem will often be getting adequate capacity in the export sector, not in releasing resources overall.

Before turning to the actual experience of devaluations in developing countries, it should be noted that a devaluation will have powerful short-run distributive effects (alluded to above in the discussion of the impact of devaluation on expenditure). When tariffs are reduced (unless they are offset by a reduction in subsidies), the government loses revenue; when quotas are eliminated, quota-holders lose the quasi-rents they enjoyed by getting a scarce resource (the right to import) at a price below its social value. When prices rise, all those on fixed money incomes suffer. Petty officials responsible for licensing or tariff collection may also lose the "fees" they can collect by virtue of their position of control. The gainers are those in the actual and potential export industries and, where a quota system is replaced by a dual exchange-rate system (the lower rate usually applying to traditional exports), the government. These prospective gains and losses influence sectional attitudes toward devaluation and their willingness to help make it succeed.

### III. SOME EVIDENCE ON THE IMPACT OF DEVALUATION

Having set out how the conventional analysis of devaluation may have to be adapted to devaluations in developing countries, we turn now to the actual experience of these countries with devaluation. As noted in the introduction, currency devaluations have occurred with some frequency in the last 25 years, averaging nearly ten a year, despite widespread reluctance to engage in them. Many of these were small, or were by countries with inadequate statistics, or were by developed countries, or were part of a larger movement of exchange rates of one block of countries as against another—the last kind of devaluation raising rather different issues for analysis than have been considered above. The evidence drawn on here derives from a study of 24 devaluations occurring over the period 1953-66 and including most of the major devaluations by developing countries in the early 1960s (a more complete description and analysis of these cases is found in Chapter 13 of G. Ranis, ed., *Government and Economic Development*, Yale University Press, 1971),

20

supplemented by some experience drawn from about a dozen devaluations in the late 1960s.

There are many questions that one can ask about the consequences of devaluation and its associated package of policies, which may have profound effects upon the allocation of resources, growth, and the distribution of income in developing economies. We are not concerned with these ultimate effects—although empirical work on them is all too rare—but, rather, with the immediate, impact effects of devaluation. These start the transition to the longer-term effects, if they are given a chance to work themselves out. The reason for focussing on impact effects is that they often determine whether the longer-term effects will be given a chance to work themselves out. Officials have notoriously short planning horizons, and their anxieties about the impact effects of devaluation often lead to a postponement of devaluation and the substitution in its place of numerous ad hoc measures, imposing substantial costs by impeding the efficient operation of the economy.

The reluctance of officials arises in large measure from the considerations adduced in the introduction: devaluation will disturb an implicit social contract among different segments of society—or at least will jar some groups out of their acquiescence in the existing state of affairs, with its numerous implicit compromises—and officials are understandably anxious about rocking an overloaded and delicately balanced boat. But sooner or later the decision may be forced upon them, when for external or internal reasons the external disequilibrium deepens and a suppressed deficit becomes an open deficit which can be corrected only by disturbing the social equilibrium anyway.

More specific anxieties are also expressed about the consequences of devaluation, however, and they can be grouped under four headings: (1) Devaluation, it is feared, will not achieve the desired improvement in the balance of payments, because neither imports nor exports are sufficiently sensitive to relative price changes within the acceptable range of such changes—in a phrase, elasticity pessimism. (2) Devaluation will worsen the terms of trade of the country and thus will impose real costs on it. (3) By raising domestic prices, devaluation will set in motion a wage-price spiral that will rapidly undercut the improved competitiveness that the devaluation is designed to achieve. (4) Whatever its economic effects, it is thought that devaluation will be politically disastrous for those officials responsible for it.

Let us see to what extent these fears are justified by experience, adopting the short-run (one year, say) perspective of the official.

21

## Impact on Trade and Payments

In nearly three-fourths of the three dozen devaluations examined the balance on goods and services, measured in foreign currency (as is appropriate for balance-of-payments analysis, although a number of countries record their payments positions in domestic currency), improved in the year following devaluation. In 90 per cent of the cases either this or the overall monetary balance (often both) improved in the year following devaluation. Of the four countries that showed a worsening on both counts, two involved important import liberalization resulting in a rise in imports, and one (Israel) was engaged in sporadic warfare and was running down reserves to build up its defense position.

Of course, these actual improvements could have taken place for reasons quite independent of the devaluation, for example an increase in world demand for the country's products or a drop in domestic expenditure due to a crop failure. Adjustment of the trade data to allow for movements in world demand and for changes in the level of domestic activity reveals a slight increase in the number of countries improving their trade balance following devaluation.

These improvements occurred despite good reasons for being an elasticity pessimist about developing countries, for the reasons given above. No doubt some part of the improvement both in trade and in overall payments can be explained by the speculative considerations already mentioned—a reversal of flows after the devaluation occurred. But not all of it can be explained in this way, for the second year following devaluation usually showed a preservation of, and sometimes a substantial increase in, the gains. The fact that supply elasticities are low in the short run helps in theory to assure that there is little or no loss in export receipts such as would arise if supply could be increased rapidly at unchanged *domestic* prices. A steadiness in export earnings, combined with some reduction in imports, will assure some improvement in the trade balance, but only a modest one. In only five of the cases examined did the improvement in the trade balance exceed the initial trade deficit, thereby swinging the country into trade surplus—a fact that should not be surprising for countries that normally import capital from the rest of the world.

Interestingly enough, most of the countries that liberalized imports experienced a *reduction* in the volume of imports in the year following devaluation—partly because of a decline in activity and a switching away from imports to domestic sources of supply, but even more because import liberalization was often delayed from three to nine months following the devaluation, apparently reflecting a wait-and-see attitude

22

on the part of the authorities toward the devaluation. In delaying, however, they increased the risk of a wage-price spiral.

*Impact on the Terms of Trade*

Many countries do not have even reasonably comprehensive data on the prices they pay for imports and receive for their exports, hence on their terms of trade. Among those that do, somewhat under one-half showed a deterioration in the terms of trade following devaluation. But some of these deteriorations were independent of the devaluation, and in any case all were small relative to the size of the devaluation—one or two per cent, compared with nominal devaluations ranging from ten to nearly 70 per cent.

The negligible deterioration observed in the terms of trade may of course have been due to preventive measures taken by the devaluing countries. Most of them imposed special taxes (or a disadvantageous exchange rate, lower than the new principal rate) on certain exports of primary products. But usually these taxes were imposed for distributive or revenue reasons, not to prevent a deterioration in the terms of trade through a fall in foreign-currency prices of exports. A standard pattern, for example, is to impose a tax roughly equivalent to the amount of devaluation on exports out of the current harvest, on the ground that the quantity of such exports can be increased only marginally (unless domestic consumption is substantial) and there is no reason to pass windfall gains on to the farmers. The new exchange rate is applied to subsequent harvests. In other instances the tax has been imposed to prevent an immediate rise in the domestic price of an export product important in local consumption, such as olive oil in Greece. In both cases it is a rise in domestic prices, not a fall in foreign ones, that the authorities are guarding against. Where only one or two foreign marketing organizations dominate a country's export sales, however, these buyers may retain their pre-devaluation buying price for domestic produce, which of course implies a decline in the price in terms of foreign currency. Thus, existing institutional arrangements may permit foreign buyers, in the short run, to improve their terms of trade at the expense of the devaluing country, and a tax will help to prevent this. In the long run, competition from potential foreign buyers will also prevent it, but by that time domestic supplies may also have increased. Finally, there are some commodities—such as hazel nuts in Turkey, jute in Pakistan, cocoa in Ghana—where one country does have a dominant position in the world market, and in these cases too the imposition of an export tax or its equivalent will prevent a deterioration in the terms of trade.

But preoccupation with the terms-of-trade effects of devaluation in

23

fact reflects a misunderstanding of the purposes of devaluation, or at best confuses devaluation theory with optimal-tariff theory. A country that dominates world markets in one or more of its export products can increase its welfare by imposing a tax on those exports up to the point at which the additional gains from further increases in the foreign-currency price (arising from the willingness of foreign buyers to pay part of the tax) just compensate for the additional welfare losses arising from the tax-induced reduction in trade. If the devaluing country has already imposed such optimizing export tariffs—import tariffs alone will not do here, because in equilibrium they also discourage manufactured exports, on which the optimal export tax is surely zero for developing countries—then devaluation will not require their alteration unless the causes of the payments imbalance also happen to have altered the optimum export tax. A pre-devaluation rise in domestic costs and prices, leading indeed to the need for devaluation, will have improved the country's terms of trade *beyond* the optimal point. The objective should be to maximize net returns on exporting, not merely to prevent a deterioration in the terms of trade, and in these circumstances some lowering of export prices in terms of foreign currency will be desirable to stimulate foreign purchases.

As a slight digression, it might be mentioned that at least one country, Jamaica, devalued because of a deterioration in its terms of trade caused by devaluation of *another* currency, the pound sterling. Britain buys Jamaica's sugar and bananas at prices fixed in sterling well above world-market prices. At the same time, Jamaica's imports are much more diversified as to source. When sterling was devalued in November 1967, the real value of Jamaica's export earnings therefore dropped and, more than that, the receipts of Jamaica's major export plantations would have dropped in terms of Jamaican dollars, while their expenditures (including wage bill) would not have dropped by nearly as much. To prevent bankruptcy and large-scale unemployment in these important agricultural industries, Jamaica therefore devalued its currency to maintain its parity with the pound. Similar considerations (as well as balance-sheet ones) may have led the French African countries to devalue their currencies with the French franc in 1969.

## Impact on Wages and Prices

Assessing the impact of devaluation on domestic prices and wages is exceptionally difficult, and only partly because price and wage data are sparse and of dubious quality for most developing countries. It is difficult also because exogenous events, expectational patterns based on the same history that led to the devaluation, and policies associated with

24

but sometimes also at variance with the devaluation all may have important influences on both wages and prices.

It is useful first of all to distinguish between demand-induced and cost-induced increases in prices and wages. By conventional analysis, both should be present following a successful devaluation, for the improved trade balance will increase the claims on domestic output, and the devaluation will lead directly to an increase in the local prices of imports and other foreign-trade goods. We have seen, however, that devaluation may lead to a decline rather than an increase in demand for domestic output, and this alone would tend to depress prices. The extent to which devaluing countries have taken the advice normally tendered to pursue deflationary monetary and fiscal policies will reinforce these devaluation-induced pressures. There is of course no contradiction between deflationary pressures and observed price increases; the devaluation here is very much like an excise tax, which reduces demand by withdrawing purchasing power from circulation, but also raises local prices. Where the devaluation merely substitutes for other measures to restrict imports, such as quotas or special tariffs, there need, of course, be no rise in these prices following devaluation, for under competitive conditions the local market prices will have already risen to reflect scarcity values.

In fact, some depression in economic activity is frequently found following devaluation in developing countries, sometimes lasting only a few months, not infrequently lasting more than a year. While it is impossible to disentangle the deflationary effects of devaluation from those of autonomous policy measures designed to facilitate success of the devaluation, there is much circumstantial evidence to suggest that the extent of depression is a surprise to the authorities in the devaluing countries, that they have not adequately taken into account the depressing effects of the devaluation itself, or that they have exaggerated its expansionary impetus. In too many cases, of course, the need to devalue arises from pre-devaluation inflation that has not been brought fully under control even after devaluation, and these cases reinforce the views of those who insist on strongly deflationary measures to accompany devaluation; in those cases further deflation is necessary to make the devaluation work. But in other cases further deflation is not necessary, and on the contrary may aggravate the difficulties of the authorities in keeping the situation under control just as exports are expanding most rapidly. We return to this possibility below.

Despite the theoretical argument that under some circumstances domestic prices need not rise following devaluation, in fact they in-

25

variably do. This is partly because there is normally some effective devaluation for imports and export products, even when export subsidies are removed and imports are liberalized, and partly because the instinctive reaction of importers is to pass along to their customers any increase in costs that they have incurred. If they are already charging what the market will bear, however, these higher prices are not sustainable in a given monetary environment, and in the course of time competition among importers will result in a subsequent drop in prices—not to below the pre-devaluation level, but toward it, to an extent governed by the degree to which devaluation substitutes for import quotas as a restraint on imports. Such a pattern can be observed for about half of the few countries for which adequate monthly data on local prices of imports are available: prices rise sharply following devaluation, reach a peak three or four months later, and then gradually drop back, sometimes substantially. In an inflationary monetary environment, of course, one does not observe a post-devaluation decline in prices, but the rate of increase is reduced temporarily.

Higher prices will raise costs directly (especially since many imports are intermediate products and capital goods) and they will also stimulate demands for higher money incomes by local factors of production, especially wage and salary employees. But the cycle of wage and price increases should be self-limiting, unless *all* parties (including the government) attempt to maintain their real incomes in the face of rising import prices, or unless the devaluation stimulates price increases that are quite unrelated to increases in costs. In addition, for either case the monetary authorities must support the increase in money incomes with domestic credit expansion if domestic prices and incomes are to rise by the full amount of the devaluation without generating unemployment.

As we saw in the first section, an open deficit will reflect both a level of expenditure and a distribution of income that is not sustainable at the existing level of output and with the existing structure of taxation and expenditures insofar as they affect distribution. Devaluation requires that some real incomes go down and that total expenditures go down, even though aggregate income need not drop. If, however, those who benefitted from the initial disequilibrium insist on retaining the same level of real income, and if they have the market power through administered prices or through wage bargaining to stake out that claim in monetary terms, then the devaluation cannot succeed without general deflation leading to unemployment—unless, of course, there is some unutilized capacity and the tax system can be so altered as to assure that enough of the increased output will go to the powerful factors in

26

the post-devaluation period. Even this will not work if these factors insist on maintaining their pre-devaluation *share* of income.

Second, the devaluation may stimulate price increases that were overdue in any case, but for reasons of law, custom, fear of public opprobrium, or simply inertia were not made earlier—the liquidation of unliquidated monopoly gains, to use Galbraith's term. This problem arises especially with public utilities subjected to an inflationary environment in the past. Being highly visible to the public, electric companies and bus companies do not readily raise their rates, and they are frequently under substantial government pressure not to do so. A currency devaluation, being little understood by the public, presents a natural occasion to raise such prices and lay responsibility on the devaluation. Several devaluations have led to rioting in the streets—as well as to larger wage claims—when an economically unrelated but psychologically related increase in urban bus fares occurred shortly afterward.

In either case the monetary authorities are confronted with a dilemma; it is here that management of a devaluation is trickiest. Economists have been too little interested in these matters of management, even though they affect the final result (that is, the path is important for determining the equilibrium, or indeed whether equilibrium is achieved). For, if the authorities do not allow some monetary expansion, unemployment and underutilization will result; and if they do allow it, the effects of the devaluation will be weakened and perhaps undermined. That various groups attempt to maintain their pre-devaluation incomes poses a more acute problem in the case of devaluation from open deficit than devaluation from suppressed deficit, since in the latter case much of the adjustment toward equilibrium income distribution will already have been made, except insofar as some firms and individuals are profiting from quantitative restrictions. Since developing countries generally do rely on quantitative restrictions before devaluation, and since they also generally have some open deficit in spite of their ad hoc adjustments, the problem remains a practical one.

In the event, price-wage spiraling does not generally get out of control, at least within the year or so following devaluation. Twelve months after devaluation, wholesale prices of imported goods will generally have risen, but by less than the devaluation (after having fallen from a peak reached three or four months after devaluation, as noted earlier), general wholesale prices will have risen less than this, consumer prices will have risen by about the same as wholesale prices, and, except where devaluations are small, manufacturing wages will have

27

risen by less than consumer prices, showing a decline in real wages following the devaluation. Thus nonwage incomes of employed factors—mostly profits and rents—show an increase in real terms a year later, and it is this increase that provides the incentive for the necessary reallocation of resources, which reallocation may ultimately restore and even raise real wages, depending on the relative factor intensities in the export industries as opposed to the protected industries.

Thus, to sum up briefly the experience following devaluations in less developed countries, it seems that official anxieties concerning the economic effects are exaggerated. The firmest generalization that can be made is that country experiences are highly diverse, which of course may be unsettling to cautious officials. But, for a hypothetical "representative" country, devaluation seems to improve both the trade balance and the payments position within the first year; it does not seem to lead to deterioration in the terms of trade of any consequence; it does lead to price increases, but not by amounts great enough to undermine the devaluation; price increases of imports are substantially less than the devaluation, suggesting that importer margins have been reduced; real wages fall; and there is a slump in economic activity following the devaluation.

## The Political Impact

The fourth apprehension concerns the political fate of those responsible for the decision to devalue, and here experience is not nearly so encouraging. A naive test is whether the government fell within a year of the devaluation. In nearly 30 per cent of the cases examined it did. Some of these changes in government were clearly unrelated to the devaluation—Costa Rica and Colombia each happened to have elections within the year, for example, and both countries have quite regularly voted out the incumbent government in recent history, devaluation or not. But in other cases the devaluation and associated policies for managing the economy were the main issue on which the government fell. And there were near misses in both Israel (1962) and India (1966), where the ruling government came under severe criticism for its decision to devalue, but survived the crisis for more than a year.

A check was provided by examining a random control group of similar countries that did not devalue; governments changed within the year in only 14 per cent of the control sample. Thus it appears that devaluation—or the policies that led to the need for devaluation or the policies that followed it—roughly doubles the chance that a ruling group will be removed from power. But the test will have to be refined considerably before it can be regarded as anything more than suggestive, in par-

28

ticular by selecting a control group from countries that seem to be in some balance-of-payments difficulty, either of an open or a suppressed type, rather than just from all developing countries.

Ministers of finance fared much worse. Nearly 60 per cent of them lost their jobs in the year following devaluation—half of them of course when their governments fell—compared with a turnover in a control group of only 18 per cent. So the chances of ouster for the official immediately responsible seems to increase by a factor of three as a result of devaluation. Again the test should be refined. And, in any case, losing one's job as finance minister does not necessarily end a political career; James Callaghan of Britain felt obliged to resign after devaluing sterling, but was immediately promoted to Home Secretary.

IV. CONCLUSIONS

Managing a devaluation through the transition phase to final success requires both judgment and delicacy in handling. Consider first the problem of aggregate demand. As we noted, this frequently falls following a devaluation, and unless the economy was badly overheated beforehand it may lead to a drop in profits and employment. If the slump is sufficiently severe and prolonged, it will evoke calls for expansionary action by the government, for few governments these days can escape responsibility for developments in their economies. If the government then yields to these pressures, the expansionary policies may come when devaluation-induced export expansion is also taking hold with a lag, and thereby increase demand pressures on the economy at just the wrong time. The better course of action, on these grounds, would be to mitigate the slump—that is, to take some modest *expansionary* action with or immediately following the devaluation, contrary to the usual advice—and then to draw back with monetary and fiscal policy when new export demand is becoming important. Properly timed, this would reduce the social and economic costs of the slump and would prevent belated expansionary action, in response to political pressure, from undermining the effects of the devaluations on the trade balance.

On the other hand, we have also seen that there is often a sharp increase in prices in the period immediately following devaluation, as importers attempt to pass on to their customers all or most of the increased cost of foreign goods. To the extent that these price increases, some of which are not otherwise sustainable, get built into wages and other local costs, they will undermine the devaluation. Timing here becomes crucial. The authorities should do what they can to reduce the temporary increase in prices (lest it become permanent), to make sure

29

that it comes quickly and is brief, and to delay any wage settlements or administered price increases until after the peak of import prices has been reached and they are falling.

The size of the temporary increase in prices can be influenced by the speed and extent of import liberalization, and this argues for liberalizing imports at once with devaluation (or even before, if that can be done without signalling the intention to devalue), instead of waiting several months as most countries have done. With respect to the promptness with which prices of imported goods begin to fall after their initial rise, the slump in total demand reinforces the desired outcome, and this factor cuts against the suggestion above that the slump should be mitigated. The timing of prospective wage settlements should if possible be taken into account in choosing the time to devalue, the aim being to allow a considerable lapse of time between devaluation and major wage settlements. Necessary increases in administered prices, such as those of public utilities or of industries in the public sector, should also be delayed until the temporary rise is past and some prices are falling. Finally, the seasonality of food prices should be taken into account; devaluations immediately after a good harvest are more likely to achieve prolonged success than are devaluations after a poor harvest or before the new harvest is in, when food stocks are low and food prices are rising. Bad harvests, in particular, have greatly weakened the impact of several devaluations, notably those of India in 1966 and Colombia in 1962.

New investment in the export sector will take place only if investors believe that the change in relative prices achieved through devaluation is a reasonably durable one. Thus, in terms of the timing of export response, *expectations* about the capacity and the will of the authorities to keep the economy under control are as important as their actual success in doing so. Here history lives in the present. A country with a poor record of monetary and fiscal management, and with a history of inflation, is likely to have greater difficulty in bringing about the required reallocation of resources than one with more favorable experience in these respects. A slump, deep if not prolonged, may (regrettably) be necessary in such a country in order to establish a new pattern of expectations.

Thus there is a dilemma with respect to macroeconomic management in the period immediately following a devaluation, and in the end the authorities must inevitably tailor their policies to the particular requirements of the country, to some extent even playing by ear. Short-term economic management of this type remains very much an art.

At the same time, the apparent political consequences of devaluation —an increased probability that governments will lose their positions

30

and ministers their jobs—is unsettling. For it means that there may be a sharp conflict between the personal interests of those in authority and the interests of the country, a conflict that has to be resolved by those same persons, and which too often may be resolved at the expense of the country. This conflict perhaps plays an even greater role than the "social contract" considerations outlined earlier in leading to procrastination over devaluation and an attempt to substitute ad hoc restrictions and subventions.

It would thus be desirable to depoliticize the whole question of devaluation, by making it less traumatic both for the officials and for the public. This suggests another reason, in addition to more strictly economic ones, for moving toward greater flexibility of exchange rates along the lines of the gliding parity, as Brazil and Colombia have done. Gradual changes in exchange rates would not only eliminate the political jolt and major economic dislocations following a large discrete devaluation, with its sharp alteration of relative prices and hence of factor incomes, but would also avoid the major misallocation of resources that takes place as a disequilibrium builds up under a fixed exchange rate. Taking exchange-rate changes in small, frequent steps would also help to resolve the dilemma posed above: a slump would not be necessary to redirect resources into export industries.

31

# Some Questions and Brief Answers About the Eurodollar Market

*John R. Karlik*

The Eurodollar market is perhaps well understood only by the practitioners, employed by banks and other financial institutions, who deal in it as their chosen way of making a living. Academic economists disagree about how the market functions and what is its real economic impact. To the uninitiated, the Eurodollar market seems to be a financial black box into which goes American money and from which comes credit for foreigners. Persons attempting to understand this phenomenon frequently pose a set of fundamental and important questions about what the market is and how it operates. Since these questions arise repeatedly, it seems appropriate to attempt to provide some brief answers for interested Members of Congress and other readers. The questions discussed are the following:

1. What is a Eurodollar deposit?
2. How did the Eurodollar market originate, what factors have been responsible for its growth, and what is its current size?
3. How does the Eurodollar market operate?
4. Does the Eurodollar market create money?

From John R. Karlik, *Some Questions and Brief Answers About the Eurodollar Market*, a Staff Study prepared for the use of the Joint Economic Committee, Congress of the United States (Washington, D.C.: U.S. Government Printing Office, February 7, 1977). Some tables omitted. John R. Karlik is a Senior Economist of the Joint Economic Committee, U.S. Congress

5. What is the impact of the Eurodollar market on the U.S. balance of payments?
6. What is the impact of the Eurodollar market on the foreign exchange value of the dollar?
7. Is the Eurodollar market an engine of inflation?
8. Is a cumulative credit collapse likely?
9. Does the operation of the Eurodollar market undermine the implementation of monetary policy in the United States?
10. Can the Eurodollar market be regulated? Is regulation desirable?

## 1. WHAT IS A EURODOLLAR DEPOSIT?

A Eurodollar deposit is a dollar deposit in a bank outside the United States. The depositors may be, for example, foreign manufacturers who have exported goods to the United States and obtained payment in dollars. Or they may be American residents who have withdrawn funds from their own accounts in the United States and placed them in a foreign bank, generally but not in ways to obtain a higher interest return than is available in the United States on savings account deposits, the purchase of certificates of deposit, Treasury bills, commercial paper, or the like.

Except for an insignificant amount, dollar deposits in foreign banks are not demand deposit liabilities of those banks. They are deposits for a specified time period and bearing a stated yield. The period of deposit may be for as short a time as overnight. But Eurodollars are typically not an immediate payments medium; one cannot generally write a check against a Eurodollar account. To be used to make payments a Eurodollar account must usually first be converted into a deposit with a bank located in the United States; it must become a normal dollar demand deposit. Investing in a Eurodollar account is therefore more like placing funds in a savings account or buying a certificate of deposit than like opening a checking account.

Occasionally reference is made to foreign currency deposits with European banks in currencies other than dollars. Such Eurocurrency deposits are placed with banks outside the nation issuing the currency. For example, an account in a German bank denominated in Swiss francs is a Eurocurrency deposit.

## 2. HOW DID THE EURODOLLAR MARKET ORIGINATE, WHAT FACTORS HAVE BEEN RESPONSIBLE FOR ITS GROWTH, AND WHAT IS ITS CURRENT SIZE?

The amount of credits extended through banks operating in the expanded Eurocurrency market, which now includes not only dollars but also sterling, German marks, Swiss francs, and other currencies and which encompasses Canada, Japan, Hong Kong, Singapore, and the Caribbean, as well as Europe, has grown from about $7 billion in 1963 to approximately $250 billion at the end of 1975. A deposit denominated in other than the domestic currency in a bank anywhere in the world is now loosely referred to as a Eurocurrency deposit.

The motivation underlying the inception of the Eurodollar market was the desire to avoid regulation, either regulations already in effect or additional restrictions that depositors feared might be imposed.

Among the first depositors of dollars in European banks were the Russians. Soviet enterprises were earning dollars both by selling gold and by exporting to the United States and to other countries. They feared that accounts opened in U.S. banks might be attached by Americans who had claims against the Soviet Government. The preferred alternative, therefore, was to place their dollar earnings in European banks. The 1958 abolition of most exchange controls in Europe permitted the growth of the Eurodollar market to accelerate. By the mid-1960's this market was a recognized force in European credit markets.

During the credit crunch of 1968 and 1969, U.S. commercial banks relied on the Eurodollar market to escape the effects of the interest rate ceiling imposed by the Federal Reserve under Regulation Q. The larger American banks directed their foreign branches to bid for dollars by offering yields above the level permitted in the United States. The head offices then borrowed heavily from their overseas branches. A portion of the new deposits attracted by overseas branches during this period apparently represented funds transferred out of the United States by Americans. Additional deposits were also attracted from foreigners, including foreign central banks. In October 1969 the Federal Reserve imposed a stiff reserve requirement on head office borrowings from abroad. The incentive for American banks to obtain funds overseas was further reduced in June 1973 when large denomination certificates of deposit were exempted from Regulation Q limitations on maximum interest yields.

Obviously a credit market does not grow to the present size of the Eurocurrency market purely on the basis of avoiding government regulations and reserve requirements. Indeed, the tacit approval and even the assistance of governments in the main Eurocurrency centers is required. Some central banks—both European and others—have deposited a portion of their dollar reserves in European commercial banks rather than investing in, say, U.S. Treasury bills. Banks in London and other financial centers have found accepting deposits in dollars and other foreign currencies and extending loans in these currencies to be profitable because no reserves are required against such deposit liabilities and because this business could be added to their normal functions at modest cost. The extra expense is small because these banks were already engaged in a large volume of international transactions and had well-established relationships with customers in a variety of countries. Most Eurocurrency transactions are for large amounts and can be handled at wholesale rates. European banks can for all these reasons offer somewhat higher deposit yields and lower loan charges than American banks and still make an acceptable profit. Depositors and borrowers appreciate this configuration, for obvious reasons.

The Eurocurrency markets, the largest of which is the Eurodollar market, have also had an important positive impact on economic activity in the countries where they have evolved. These markets constitute a highly efficient system for allocating credit among lenders and borrowers. They have facilitated higher levels of domestic and international commerce than would have been likely in their absence. The removal in 1958 of most European restrictions on the conversion of foreign exchange and the rapid growth of international trade in the 1960s, a large proportion of which was financed in dollars, created a need for dollar loan and deposit services in Europe during normal working hours. Banks understandably strove to satisfy this demand and finance additional commerce. The resulting gain in output and employment is the chief real economic benefit produced by the banks and other institutions that have jointly constructed the Eurocurrency financial network.

## 3. HOW DOES THE EURODOLLAR MARKET OPERATE?

The Eurodollar and other Eurocurrency markets are largely interbank markets. When a European bank accepts a dollar deposit, it naturally attempts to lend the funds at a higher interest rate than the yield it is paying to the depositor. In some cases the borrower will be the ultimate user of the funds, such as a European importer purchasing merchandise in the United States. In many instances, however, an individual or corporate borrower will not be immediately on tap as an acceptable investment opportunity for the bank. In this event, the bank will place the dollars, most likely for a short period, with another bank that is seeking funds. Similarly, if a worthy ultimate borrower appears when a European bank does not have surplus dollars to invest, it may temporarily borrow in the interbank market in preference to rejecting the customer.

Because of the volume of interbank transactions and the consequent double counting of available dollar credits that can easily result, estimates of the size of the Eurodollar market must be used cautiously. The chief source of data on Eurocurrency markets is the Bank for International Settlements (BIS) located in Basle, Switzerland. The BIS is the one surviving institutional remnant of the League of Nations. Eight European central banks, those of Belgium, France, Germany, Italy, The Netherlands, Sweden, Switzerland, and the United Kingdom, are the majority stockholders. In publishing Eurocurrency market data, the BIS attempts to eliminate double counting among the eight member countries of available Eurocurrency credit. The totals cited above have been deflated in an effort to eliminate the effects of redepositing within the eight BIS-reporting countries. However, since substantial Eurocurrency markets have now been established in Canada, the Caribbean, Japan, Hong Kong, Singapore, and the Middle East, the totals may still be inflated.

Of the approximately $250 billion of Eurocurrency credits granted during 1975, $205 billion were extended by banks in the eight BIS-reporting countries.[1] The bulk of these Eurocurrency loans in the eight countries were to banks; only $61 billion were to nonbank residents and foreigners. The great difference between total credits extended and the portion granted to nonbank users illustrates the extent to which the Eurocurrencey market is in fact an interbank market.

## 4. DOES THE EURODOLLAR MARKET CREATE MONEY?

Eurodollar deposits, as noted above, are not money in a strictly defined sense; they are time rather than demand deposits and cannot be drawn upon to make payments. However, if the definition of money is expanded from cash and demand deposits to include time deposits (i.e., from $M_1$ to $M_2$), should Eurodollar and other Eurocurrency accounts be included in this expanded measure of liquidity? Yes. Furthermore, if one adopts this expanded definition of the money supply, creation of a Eurodollar deposit will lead under certain circumstances to an equivalent increase in the global stock of liquidity. How does this consequence come about?

Suppose an American individual or corporation has a quantity of funds invested in a certificate of deposit or time deposit with a New York bank and decides to invest these funds in the Eurodollar market instead.[2] The certificate of deposit or time deposit must first be transformed into a demand deposit.[3] The individual or corporate treasurer then writes a check on his demand deposit in the New York bank and makes it payable to a European bank. At that point, the European bank has a demand deposit claim on a New York bank, and the individual or corporation has a time deposit with a European rather than an American bank.

The outcome of this series of transactions is that $M_2$ in the United States is unchanged but is increased in Europe by the amount of the Eurodollar deposit. The broadly defined global money supply has increased by this amount, since deposits by foreigners, including banks, are considered part of the U.S. money supply. But at this stage, the supply of credit to nonbanks has not changed.

If the European bank initially accepting the deposit relends it to another European bank, use of credit by the nonbanking sector is still not increased. This statement remains valid regardless of how many times the funds are redeposited among banks. Only when the funds are finally loaned to an ultimate nonbank user is the total quantity of credit available to support economic activity increased.

If the user is either a foreigner making payments to Americans or an American other than a bank, the story ends with the conclusion that the total amount of liquidity available globally is expanded by the amount of the Eurodollar deposit. If the user is an American bank, there is no increase in the total volume of credit available to the nonbanking sector of any other economy.

As another possibility, if the user is a foreigner who converts the dollars into his own currency, and if his central bank buys the dollars and redeposits them in a European commercial bank, another round of dollar credit expansion may occur. Similarly, if the foreign user pays the dollars to another foreigner and the recipient—depending on the yields available in New York and Europe—redeposits the dollars in a foreign bank, a second real economic transaction may then be financed.

Thus, an initial dollar deposit in a European bank can lead to a variety of outcomes. The amount of additional liquidity provided to nonbanks may be zero, equivalent to the deposit, or some multiple of the deposit.

This uncertainty about who may be the borrower of dollars from a European commercial bank and how these funds will be employed raises the question of the size of the "Eurocurrency multiplier." In other words, if a dollar sum is deposited in a European bank, will a multiple credit expansion occur? If so, what is the average amount of the multiple? Most importantly, what is the ultimate economic significance of the initial transfer?

Economists studying the Eurodollar market generally fall into either of two groups in responding to these questions. One group views the Eurodollar market as the product of a fractional reserve banking system that creates dollar credits. The reserves of Eurobanks, according to this conception, are checking account deposits in commercial banks located in the United States. Since Eurobanks are not required to hold minimal reserves as a fixed proportion of their dollar liabilities, one might expect that, by comparison with the ratio of reserves held in the United States against time deposits, Eurobanks would maintain a lower fraction of reserves. Most attempts to measure the ratio of "reserves" that Eurobanks hold voluntarily to liabilities indeed show a low proportion. The change in total Eurodollar balances implied by an initial change in "reserves"—if the fractional reserve banking analogy is accepted—is therefore quite high.

The other school of economists views the Eurobanks as financial intermediaries that do not create money but shunt available credit from lenders with excess liquidity to borrowers short of funds. These analysts emphasize the "leakages" to which Eurobanks are exposed. There is little reason, they say, to expect that the dollars a borrower has obtained from a European bank and subsequently paid to a third party will necessarily be put back into the Eurodollar market. Therefore, each Eurobank must, according to this view, maintain a more-or-less balanced term structure of dollar claims and liabilities. Furthermore, they maintain, Eurobanks may prefer to safeguard their ability to meet withdrawals by arranging standby lines of credit with U.S. banks, rather than by maintaining checking account balances, which earn no interest. Thus, according to this second school of thought, the low apparent reserve ratios of Eurobanks do not necessarily indicate that the market is a powerful machine for generating additional liquidity.

An indication of the extent to which the Eurodollar market creates money is the size of loans to nonbank borrowers as compared with all loans. At the end of 1975, banks in the eight countries reporting to the BIS indicated that out of dollar loans totaling $190 billion, only $41 billion were to nonbank borrowers. Of loans denominated in other currencies totaling the equivalent of $68 billion, $20 billion were to nonbanks. These totals do not include the activities of the Eurocurrency markets located in the Caribbean and the Far East, but they do encompass the bulk of Eurocurrency credit creation.

## 5. WHAT IS THE IMPACT OF THE EURODOLLAR MARKET ON THE U.S. BALANCE OF PAYMENTS?

It is sometimes maintained that growth of the Eurodollar market is dependent upon net capital outflows from the United States or upon U.S. payments deficits. In the example discussed above, an American resident transferred a sum of dollars from an account with a New York bank to an account with a European bank. This action produces a capital outflow from the United States. But as is also evident in the above example, the full story extends far beyond the initial transaction. Particularly if the European bank receiving the funds is the branch of an American institution, the head office may borrow the dollars back from its branch. In this case, a subsequent capital inflow offsets the initial outflow, and there is no net transfer of funds internationally.

On the other hand, if the foreign bank receiving the dollars sells them to the central bank in exchange for the domestic currency, U.S. liabilities to foreign official institutions increase, and a U.S. official settlements deficit will be expanded (or a surplus diminished) by the amount of the transaction. If the foreign central bank then invests the dollars in U.S. Treasury bills, the impact on the U.S. official settlements balance is not further changed. Thus, while an initial transfer of dollars out of the United States arising from a trade, services or capital transaction, or the purchase of dollars in the exchange market with foreign currency is a requisite for the establishment or enlargement of a Eurodollar balance, the ultimate consequences of this action on the U.S. balance of payments are by no means clear. . . .

Financial markets in New York and Europe are competitors. The rate of growth of the Eurodollar market has been determined, more than anything else, by the relative attractiveness of investing short-term either in New York or in Europe and by the relative availability and cost of funds in the two areas. The particularly rapid growth of the Eurodollar market in 1968 and 1969 resulted from a credit crunch in the United States and Regulation Q ceilings on the interest rates that American banks could offer. The reaction of American banks to this combination of factors was to bid for deposits through their European branches. Although the widespread adoption of floating exchange rates brought about a sharp decrease in the U.S. payments deficit during 1973, this development did not inhibit the market's growth. In 1974, when the market expanded by one-third, the chief motivating factors were apparently the desires of oil producing countries to invest their expanded earnings in highly liquid bank deposits and the needs of both industrial and developing countries to finance high-cost oil imports. The same factors remained important in 1975.

To conclude, a transfer of dollars from the United States and into a European or other foreign bank cannot be presumed to produce a U.S. payments deficit of even approximately the same magnitude. The growth of the Eurodollar market is not linked with U.S. payments deficits in any readily identifiable way.

## 6. WHAT IS THE IMPACT OF THE EURODOLLAR MARKET ON THE FOREIGN EXCHANGE VALUE OF THE DOLLAR?

The foreign exchange value of the dollar tends to fall when Americans need to make increased payments to foreigners or when individuals desire to hold additional assets valued in other currencies. Conversely, the external value of the dollar tends to rise when foreigners' payments to Americans increase or when individuals desire to hold more dollar assets. The Eurodollar market has established convenient mechanisms for the temporary investment of excess dollar balances. It also offers another source of dollar loans for periods ranging from overnight to several years. In general, the market has made the dollar more useful and desirable relative to other currencies. Therefore, its net effect has probably been to increase the value of the dollar somewhat in comparison with other currencies.

From time to time, however, transactions have occurred in the Eurodollar market that have had a depressing impact on the external value of the dollar. Speculators believing that a particular foreign currency was likely to increase in value have occasionally drawn on Eurodollar credit lines and sold borrowed dollars to buy another currency. They hoped to realize profits by repaying the dollar loans after the expected upward revaluation of other currency.

If the anticipated exchange rate change indeed occurred, repaying the loan with interest consumed most but not all of the dollars obtained from converting the foreign currency balance at its new higher value. A margin constituting the profit remained. If the expected exchange rate change did not occur, speculators losses were limited to the cost of interest on the loan and the cost of two currency conversions. The shift in 1973 from fixed to flexible exchange rates eliminated many of the opportunities for speculative gain that had previously existed. The large international transfers of liquid capital that had resulted from this incentive have also largely disappeared.

Because the Eurodollar market has grown to constitute a major international financial market, the transfer of a dollar balance out of the United States can no longer be presumed to have an impact on the exchange value of the dollar. An international capital flow will produce exchange rate repercussions only if there is an exchange market transaction. But because of the Eurodollar market, dollars can be transferred out of the United States and easily be invested abroad as dollars; they need not be converted into any other currency. Similarly, dollars moved into the United States need not have been acquired through a previous sale of foreign currencies. The growth of the Eurodollar market in the last decade has considerably enhanced the usefulness of the dollar as an international transaction or vehicle currency and has therefore probably increased foreigners' desired dollar holdings. The foreign exchange value of the dollar is most likely a bit higher than it would have been in the absence of a Eurodollar market.

## 7. IS THE EURODOLLAR MARKET AN ENGINE OF INFLATION?

Would inflation rates experienced during recent years have been substantially lower if there had been no Eurodollar market? Of course, some inflation would have occurred anyway as a result of (a) increased prices for oil, food, and raw materials, as a consequence of (b) generally overstimulative monetary and fiscal policies in 1973 and 1974, and as an aftereffect of (c) dollar purchases by foreign central banks during the last throes of the fixed exchange rate system in 1971 and 1972. Central bank dollar purchases had the effect of increasing commercial bank reserves and money supplies in some countries.

If all these other factors are taken into consideration, has there been an additional increment of inflation that can be attributed to the operation of the Eurodollar and other Eurocurrency markets? (When considering the impact of these financial markets on prices and total economic activity, focusing on only the dollar component would omit an important segment). An answer to this question can be inferred from the data presented in Table 1. The first column in this table gives the level of $M_2$ in the eight European countries reporting to the Bank for International Settlements at the end of each calendar year from 1970 through 1975. It is appropriate to use $M_2$, the domestic money supply in these countries defined to include not only currency and demand deposits but also time deposits and certificates of deposit, as a basis for comparison because Eurocurrency deposits are also made for a specified time period. The second column gives for the same years the amount of Eurocurrency claims against nonbank residents of the eight BIS-reporting countries. . . .

If the Eurocurrency market is an engine of inflation, it must create money in excessive amounts in addition to the volume of credit created by domestic banking systems. But examination of the data presented in Table 1 shows that in most recent years the Eurocurrency market has usually contributed only marginally, and at most modestly, to the supply of credit available in Europe.

In 1971, $M_2$ grew in the eight BIS-reporting countries by nearly $80 billion, and in 1972 by a slightly greater amount. Yet in 1972 Eurocurrency claims against nonbank residents of the eight countries grew merely $1 billion, and on December 31, 1972, the total amount of these claims was

TABLE 1

Data Provided by Banks in the 8 BIS-Reporting Countries (In billions of dollars)

| | $M_2$ | Eurocurrency claims against nonbank residents |
|---|---|---|
| Level at the end of: | | |
| 1970 | 348.7 | (¹) |
| 1971 | 428.1 | 7.6 |
| 1972 | 511.2 | 8.6 |
| 1973 | 640.3 | 14.0 |
| 1974 | 773.5 | 23.7 |
| 1975 | 831.8 | 24.0 |
| Change during: | | |
| 1971 | 79.4 | (¹) |
| 1972 | 83.1 | 1.0 |
| 1973 | 129.1 | 5.4 |
| 1974 | 133.2 | 9.7 |
| 1975 | 58.3 | .3 |

¹Not available.

*Sources:* "International Financial Statistics" and the "Forty-Sixth Annual Report of the Bank for International Settlements."

only $8.6 billion, as contrasted with an $M_2$ of over $500 billion. $M_2$ expanded during the next 3 years by $129 billion, by $133 billion, and by $58 billion respectively. In comparison, from 1973 through 1975 Eurocurrency claims against residents expanded by $5 billion, by $10 billion, and by less than $1 billion respectively. In 1974 the Eurocurrency market made its largest percentage contribution to the supply of credit in Europe; that year the expansion of Eurocurrency claims against nonbank residents was the equivalent of 7.3 percent of the growth of $M_2$. . . . Therefore, even if one were to accept the thesis that excessive monetary expansion were an important cause of inflation, Eurocurrency markets hardly appear to be a major source of that expansion.

When appraising the inflationary impact of Eurocurrency markets, one should keep two additional considerations in mind.

First, not all credit generated by the Eurocurrency market is necessarily additional credit. In at least some years, central banks in Europe would probably have induced commercial banks to create more liquidity than they actually did had the Eurocurrency market never come into being.

Second, on the other side of the ledger, the Eurocurrency market should be recognized as having helped combat recession during periods when demands for credit were unusually strong. Such a period was 1974, the year following the quadrupling of oil prices.

The Eurocurrency market provided a vital service in accepting large deposits from oil producing countries and lending the funds to hard pressed oil importing nations. Developing countries contending with increased energy and food costs and, subsequently, with a drop in earnings for their own commodity exports, have been especially aided by credits obtained in the Eurodollar market. Although the problems of these nations are by no means solved and may become more serious, their transitional pains following the abrupt international price changes of recent years would have been far more severe without the financial cushion provided through the Eurocurrency market.

No authoritative summary measure can be offered of the inflationary costs versus the real benefits of credit creation in the Eurocurrency market. Part of the reason that costs and benefits cannot be simply set off against one another is that they have been experienced by different individuals in widely separated countries and with vastly divergent incomes. But in the record of the Eurocurrency market over the past five years, there is scant evidence to support an assertion that it has served as an engine of inflation.

Indeed, if there is a monetary engine of inflation in Europe, it is more likely to be discovered in the operation of domestic banking systems than in the Eurocurrency markets. From 1971 through 1974, $M_2$ in the eight BIS-reporting countries grew each year by nearly 20 percent or more.

## 8. IS A CUMULATIVE CREDIT COLLAPSE LIKELY?

The Eurodollar market, as explained above, operates efficiently because the banks and other financial institutions participating in it can invest or obtain funds easily via the market for periods of from one day to over a year. Interbank transactions constitute the bulk of the volume in the market, although it is the initial depositors and final borrowers who experience its real economic consequences. Because this market, like the foreign exchange market, operates on verbal commitments backed by mutual trust, and because fluid interbank operations are essential to efficient operation, the Eurodollar market would appear to be particularly vulnerable to the failure of even a modest-sized institution.

During the 6 or 9 months following the quadrupling of oil prices in the fall of 1973, many observers feared that the Eurodollar market would not be able to invest profitably the volume of liquid assets that would most likely be deposited by oil producing countries. The worriers went on to speculate that even if the institutions in the market somehow managed to accept and disburse the funds, an economic collapse in a European country or a major default by a developing nation that had borrowed heavily would provoke a financial crisis that gathered strength like an avalanche.

These worst fears have not been borne out for at least two reasons. First, when banks operating in the Eurodollar market began to run out of profitable opportunities for short-term investment of deposits subject to quick withdrawal, they lowered their deposit rates and announced their reluctance to accept additional large deposits. Second, as a consequence of both self-discipline and chiding by various central banks—notably the Bank of England—Eurodollar banks have tightened their lending requirements. At present there seems to be no imminent danger of a crisis, but numerous substantial loans to developing country borrowers remain to be repaid or refinanced.

Officials have taken two steps to help bolster the stability of the Eurodollar market and to curb excessive credit creation and the risk of a crisis.

First, the central banks of the major industrial countries agreed in the Spring of 1971 to limit the extent to which additions to their own dollar reserves are redeposited in the market. If redepositing became standard procedure, the increase in the money supplies of the nations encompassing the market could theoretically be limitless. Therefore, controlling the extent of redepositing is a step toward governing the credit-creating impact of the market.

Second, the central banks of the major industrial countries agreed in 1974 that in the event of a crisis, each will stand behind its own banks and the overseas branches of domestic banks to keep the crisis from spreading.

The precise terms of this mutual acceptance of responsibilities have not been spelled out, but the principle seems clear. For example, Federal Reserve Board member Henry C. Wallich said in testimony before the Senate Permanent Investigations Sub-committee in October 1974:

> The Federal Reserve is prepared, as a lender of last resort, to advance sufficient funds, suitably collateralized, to assure the continued operation of any solvent and soundly managed member bank which may be experiencing temporary liquidity difficulties associated with the abrupt withdrawal of petrodollar—or any other—deposits.

This commitment to back "any solvent and soundly managed member bank" extends to overseas branches as well.

Central bankers of the major industrial nations meeting in Basle, Switzerland, issued the following statement on September 9, 1974:

> The Governors also had an exchange of views on the problem of the lender of last resort in the Euromarkets. They recognized that it would not be practical to lay down in advance detailed rules and procedures for the provision of temporary liquidity. But they were satisfied that means are available for that purpose and will be used if and when necessary.

## 9. DOES THE OPERATION OF THE EURODOLLAR MARKET UNDERMINE THE IMPLEMENTATION OF MONETARY POLICY IN THE UNITED STATES?

In considering the impact of the Eurodollar market on the implementation of monetary policy in the United States, one must distinguish between recent developments in international financial institutions that merely make life more complicated for Federal Reserve authorities and other changes that could prevent or counteract the working of monetary policy in the United States. Some observers might conclude that the Eurodollar market has made life only a little more complex for American money managers, while others, at the opposite end of the spectrum, would argue that the existence of the Eurodollar market as an alternative source of credit can at critical times totally vitiate the intent of Federal Reserve policy. . . .

The record of policy actions by U.S. monetary authorities is an important indication of whether they perceive that international capital flows frustrate the implementation of domestic monetary policy. Such flows are certainly a complicating factor. But the progressive [1970s] reduction of reserve requirements [on the borrowings of domestic offices of U.S. banks from their foreign branches and from foreign banks] and the elimination of constraints on capital outflows suggests that in the minds of the authorities, the benefits of open money and exchange markets outweigh the disadvantages of the resulting complications. In any event, if serious problems did arise at some time, nothing prevents the authorities from introducing controls over international capital flows and exchange transactions.

## 10. CAN THE EURODOLLAR MARKET BE REGULATED? IS REGULATION DESIRABLE?

Numerous individuals have from time to time urged that the Eurodollar market be regulated to limit credit creation or to reduce the risk of a credit collapse. Regulation can be discussed from two perspectives—feasibility and desirability. While somewhat greater regulation might be desirable, to date the inflationary consequences of excess credit creation have not been sufficiently demonstrable and the risk of an avalanching credit collapse has not been sufficiently evident to prompt monetary authorities to achieve the high degree of cooperation that would be necessary to regulate the Eurodollar market effectively. Even the eight central bank members of the Bank for International Settlements have not been able to agree on mechanisms for controlling the growth of the Eurodollar market or on standards of credit worthiness to be applied to lenders. At the present time, therefore, only the most modest degree of regulation seems possible.

Another factor severely limits the feasibility of any efforts that might be undertaken to regulate the Eurodollar market. In recent years the market has spread rapidly from its origins in the City of London and the financial centers of continental Europe to the Caribbean, the Mideast, Singapore, Hong Kong, and Tokyo. If burdensome regulations were imposed in the existing centers of Eurocurrency activity, most of the market's functions might well be transferred to some other area, particularly to a bastion of free enterprise. In the event of such a relocation, the profits and the jobs derived from the market's activities would move also. Reluctance to forgo these benefits, particularly in London, have deterred authorities from imposing as comprehensive regulations as they otherwise might have.

Should the evident difficulty of regulating the Eurodollar market be a source of concern? How much concern, since the possibility of a serious crisis can never be entirely excluded? Following the 1973 increase in oil prices, the Eurodollar market has gone through at least two distinct periods of stress. First, there was the danger—discussed above—that Eurocurrency banks would not be able to accommodate the huge volume of deposits from oil producing countries and lend these funds out at profitable rates of return. The banks did accept a major increase in deposit liabilities. But they eventually lowered their interest rates to discourage further acquisitions of massive short-term deposits and gradually tightened their lending criteria. Second, a few banks—most notably Franklin National and Herstatt—speculating in the foreign exchange market, not in the Eurocurrency markets, suffered severe losses. These events brought into question the quality of bank management and their ability to control the exposure of their institutions. For a subsequent period all new deposits were placed only with the largest and most respected institutions, and some funds were withdrawn from smaller banks. The announcement of central banks' commitment to stand behind their own national banks and these banks' overseas branches helped reassure depositors.

A third time of stress is presently foreseen. Developing countries have borrowed heavily in the Eurodollar market to finance oil imports and to compensate for the loss of earnings resulting from the subsequent drop in export prices for many of their commodities. Some of these nations are approaching the limits of their borrowing and loan servicing capabilities. How well Eurocurrency banks would be able to withstand defaults on outstanding loans to some developing countries or the rescheduling under duress of loan repayments is the subject of present concerns.

The real economic adjustments to the increased prices of oil, bauxite, and perhaps other commodities will continue. Some industrial and developing countries will be able to continue borrowing in the Eurocurrency markets to help lengthen the period during which real adjustment will occur and so mitigate the pain of that transition. In others, the bite has begun to take hold, and the need to curtail some incomes and transfer resources is imperative. However, the adjustments need not and will not occur everywhere simultaneously. Exporters in industrial countries are benefiting from growing sales to oil producing nations. Some of these industrial exporting countries will have excess funds to deposit in U.S. and Euro banks. Oil producers will also continue to make deposits. Banks operating in the Eurocurrency markets will most likely be able to adjust to strains of future demands as flexibly and as successfully as they have in the recent past.

All participating financial institutions recognize that the maintenance of stability in the Eurocurrency markets is in their own best interest. The issue is whether competitive instincts among institutions can be sufficiently curbed through self-discipline to preserve the soundness of the entire structure.

## CONCLUSION

The Eurodollar market, like virtually all modern economic institutions, is a mixed blessing. It has produced important benefits in terms of helping to expand international trade, to stimulate economic growth, and most recently to distribute the excess earnings of oil producers among consumers needing credit to finance their imports. On the other hand, it may have raised rates of inflation somewhat. It has generated substantial business for the countries in which it is located—most of all for banks in the City of London. The financial institutions and individuals operating in the market can and will elude extensive regulation, if attempted. Unwillingness to forgo profits generated by the market and inability among central banks to agree on appropriate operating guidelines and on joint monetary policies have enabled the Eurodollar market to continue enjoying virtually no formal regulation. At the same time, the banks operating in the market know that the maintenance of stability is in their own best interest. Given the record of what is now a tested and mature market, there is reason to hope that—under the surveillance of concerned officials—the sometimes uneasy balance in the Eurocurrency markets is maintained so that lenders and borrowers can continue to enjoy its benefits.

## BIBLIOGRAPHY

The literature on the Eurodollar market is extensive and difficult. Much of the difficulty results from conceptual muddiness and confusion about how the market operates. The following is a selected list, with brief comments, of readings that may be useful to individuals desiring to pursue further the issues raised in the preceding discussion.

### Books

Stem, Carl H., Makin, John H., and Logue, Dennis E., editors. *Eurocurrencies and the International Monetary System.* Washington, D.C.: American Enterprise Institute for Public Policy Research, 1976.
    Contains selections by Thomas D. Willett on the inflationary impact of Eurocurrency growth (pp. 214–221), by Carl H. Stem on Eurocurrency credit expansion and regulation (pp. 283–332), and by John H. Makin on the "multiplier" versus the "new-view" analysis of how the Eurocurrency market operates.

### Articles

Bell, Goeffrey L. "Credit Creation Through Eurodollars?" *The Banker*, vol. 114 (August 1964), pp. 494–502.
    A good discussion of whether the Eurodollar market creates money or merely redistributes existing liquidity.
Crockett, Andrew D. "The Eurocurrency Market: An Attempt to Clarify Some Basic Issues." *International Monetary Fund Staff Papers*, vol. 23, no. 2 (July 1976), pp. 375–386.
    Argues that the fractional reserve banking model is inappropriate for analysis of the Eurocurrency market.
Friedman, Milton. "The Eurodollar Market: Some First Principles." *Federal Reserve Bank of St. Louis Review*, vol. 53, no. 7 (July 1971), pp. 16–24.
    Basic presentation of the viewpoint that the Eurodollar market creates money in a way similar to a fractional reserve banking system.
de Grauwe, Paul. "The Development of the Eurocurrency Market." *Finance and Development*, vol. 12, no. 3 (September 1975), pp. 14–16.
    Historical review for readers unfamiliar with the Eurocurrency market.
Hewson, John and Sakakibara, Eisuke. "The Eurodollar Deposit Multiplier: A Portfolio Approach." *International Monetary Fund Staff Papers*, vol. 21, no. 2 (July 1974), pp. 307–328.
———. "The Eurodollar Deposit Multiplier: A Note." *International Monetary Fund Staff Papers*, vol. 22, no. 2 (July 1975), pp. 565–568.
    Contains estimates of the Eurodollar multiplier based on a portfolio rather than a fractional reserve model.

Klopstock, Fred H. "The International Money Market: Structure, Scope and Instruments." *Journal of Finance*, vol. 20, no. 2 (May 1965), pp. 182–208.
      A comprehensive discussion of international money markets, including how the Eurodollar market links national money markets.
——. "Money Creation in the Eurodollar Market—A Note on Professor Friedman's Views." *Federal Reserve Bank of New York Monthly Review*, vol. 52, no. 1 (January 1970), pp. 12–15.
      Disputes the Friedman view that the Eurodollar market is a source of multiple credit creation.
Lutz, F. A. "The Eurocurrency System." *Banca Nazionale del Lavoro Quarterly Review*, vol. 27, no. 110 (September 1974), pp. 183–200.
      Discussion of alternative conceptions of how Eurocurrency markets work and of whether they redistribute available credit or create money.
Mayer, Helmut, W. "Multiplier Effects and Credit Creation in the Eurodollar Market." *Banca Nazionale del Lavoro Quarterly Review*, vol. 24, no. 98 (September 1971), pp. 233–262.
      Argues that the credit creation mechanism of a self-contained commercial banking system cannot be applied to credit creation in the Eurodollar market.
——. "Some Analytical Aspects of the Intermediation of Oil Surpluses by the Eurocurrency Market." *Banca Nazionale del Lavoro Quarterly Review*, vol. 27, n. 110 (September 1974), pp. 201–226.
      Discusses the capability of the Eurocurrency market to help finance the payments deficit resulting from the increase in oil prices.
McClam, Warren D. "Credit Substitution and the Eurocurrency Market." *Banca Nazionale del Lavoro Quarterly Review*, vol. 25, no. 103 (December 1972), pp. 323–363.
      This discussion argues that in the absence of the Eurocurrency market, national central banks would have had to expand domestic money supplies more.
Meulendyke, Ann-Marie. *Causes and Consequences of the Eurodollar Expansion.* Research Paper No. 7503. Federal Reserve Bank of New York, March 1975.
      Analyzes the Eurodollar market as a fractional reserve banking system a la Friedman and measures the money creating potential of the market.
Sakakibara, Eisuke. "The Eurocurrency Market in Perspective." *Finance and Development*, vol. 12, no. 3 (September 1975), pp. 11–13, 41.
      Basic explanation of the Eurocurrency market not as a fractional reserve banking system, but as a mechanism for redistributing available credit.
Sweeney, Richard J. and Willett, Thomas D. "Eurodollars, Petrodollars, and World Liquidity and Inflation." *Journal of Monetary Economics*, forthcoming.
      Good evaluation of the extent to which money creation in the Eurocurrency market contributed to inflation.
Willms, Manfred. "Money Creation in the Eurocurrency Market." Unpublished paper DM/75/112, International Monetary Fund Research Department, December 17, 1975.
      Compares and evaluates the operation of the Eurocurrency market as a fractional reserve banking system using different definitions of reserves.

## Notes

1. The data cited here are from the *"Forty-Sixth Annual Report of the Bank for International Settlements,"* published June 14, 1976.
2. Since most Eurocurrency transactions are interbank transactions, the series of events recounted in the following paragraphs is not intended to be typical. A more typical Eurocurrency transaction might be between two commercial banks, or among a central bank and several commercial banks. Such alternative transactions would have effects on the supply of available credit in various countries different from the sequence discussed in the text. The example presented in

the text illustrates in a simple way the diverse impacts on the global availability of credit that may result from the transfer of a dollar balance from a U.S. bank to a European bank. Variations on this theme would include, for example, a decision by a foreign exporter to place his dollar earnings in a bank in London rather than in New York, or a decision by the central bank of, say, a Latin American country to deposit dollars in Frankfurt rather than buy U.S. Treasury bills.

3. This action will increase the total amount of reserves the U.S. banking system is required to hold, since demand deposits carry a higher reserve requirement than time deposits. But suppose the Federal Reserve through open market operations increases the total stock of reserves by the incremental amount required to permit this marginal increase in the U.S. money supply (narrowly defined).

*Jacques van Ypersele de Strihou*

# Operating Principles and Procedures of the European Monetary System

Before analyzing the operating principles and procedures of the European Monetary System, I believe that it is useful to describe some of the motivations behind this effort. I will then discuss the basic principles of the EMS and its conditions for success. Finally, I will try to answer some of the criticisms of the EMS.

## Motivations

A principal economic motivation for the creation of the EMS has been dissatisfaction with floating exchange rates during the past few years and the conviction that this monetary situation was having adverse effects on economic integration in Europe and, in general, on growth and employment in the European Community (EC). Expressed in a positive way, the objective of the EMS is to contribute to a lasting improvement of the present economic growth and employment situation of the Community and to its economic integration through greater exchange rate flexibility. This objective will be met only if the system is conceived in such a way that it will be durable and credible and contains neither a deflationary nor an inflationary bias.

Before explaining how the EMS should contribute to growth and employment, let me first talk about greater exchange rate stability. The EMS can help in two ways:

—in a short-term sense, through ironing out excessive fluctuations;
—in the longer term, through fostering greater convergence of the Community economies.

First, the European system, with its intervention rules and credit mechanism, should be able to effectively fight the phenomenon of "overshooting."

Reprinted from "The European Monetary System" (P.H. Trezise, ed.), pp. 5–24, with permission of The Brookings Institution, Washington, D.C., Copyright 1979.

By this I mean movements of the exchange rate in excess of what would be warranted by differences in inflation rates between countries.

Overshooting has often occurred in the past. It can be initiated by strictly national causes. It can also be initiated—and it often has been—by movements of third currencies, particularly the dollar. When people move out of the dollar because of a lack of confidence in that currency, they do not move equally into all the European currencies. They often move specifically to one EC currency, the deutsche mark. This pushes the D-mark up and it widens the relationships between the D-mark and the French franc or the pound sterling. So one can say that sharp fluctuations of the dollar have also contributed to excessive swings, or overshooting, in European currencies.

Expressing the same idea in the economist's jargon, one can say exchange rates between major currencies have frequently been determined by portfolio adjustments. Such changes have often overshot the purchasing power parity level between these currencies themselves and also between the major currencies and others that are less used as instruments for investments in financial assets. These excessive movements are usually accommodated ex post by price movements, especially in the more open economies, and this tends to exacerbate inflation differentials. The new European exchange rate system, with its provisions for intervention and the available resources to carry out this intervention, should help prevent overshooting.

Second, there is a more fundamental way in which the EMS should contribute to greater exchange rate stability. Participation in this system assumes that in the adjustment process countries will have to give a high priority to internal policy measures rather than rely on exchange rate changes. Otherwise the effectiveness of the system itself would be jeopardized. Participating countries therefore have to realize that, by adhering to this scheme, they compel themselves to aim at greater convergence, through domestic measures, of the fundamentals of their economies. This factor is sometimes called the disciplinary element in the system. But the term should not be misinterpreted. It should not be taken to mean that adjustment is wholly a matter of restrictive policies by deficit countries. Rather the clear intention is that adjustment should take place in a symmetrical way through actions by surplus as well as by deficit countries.

How will greater exchange rate stability contribute to higher growth and employment? There are several ways.

In the first place, it should allow a higher level of both foreign and domestic demand to develop. Monetary instability in Europe has had a deflationary impact in both surplus and deficit countries. In countries with strong currencies, excessive appreciation has contributed to deflationary pressures by reducing profits in export industries and by reducing prospects for sales. This was one of the causes of the downward revisions of growth in Germany in 1977 and 1978.

On the other hand, in countries whose currencies have depreciated too much in relation to stronger currencies, downward overshooting has led to inflationary pressures through increased import prices and wage indexation. These inflationary implications have acted as a brake on economic revival. Governments have been afraid to allow their economies to grow faster lest the expansion increase the pressures on balance of payments and cause further currency depreciation and more inflation.

Thus greater monetary stability should have a positive impact on economic revival by making measures to stimulate higher levels of demand more feasible. This should have important multiplier effects, in view of the openness of EC economies and the high proportion of intra-Community trade in total trade. Trade with other EC partners represents 69 percent of total Belgian trade, between 45 and 50 percent of the trade of France, Germany, Denmark, and Italy, and 38 percent of British trade.

Greater monetary stability would also encourage business confidence and investment. In talks with European business executives, one often hears complaints that they are unable to give their companies a full European dimension because of the ever-present exchange risks and uncertainty about inflation rates. It has been difficult to forecast correctly the cost in national currency of inputs from abroad or the revenue in national currency from exports. These uncertainties contribute to the fact that businesses are not harvesting the potential benefits of a market as large as Europe. Furthermore, they reinforce protectionist pressures and paralyze investment.

In fact, one can safely say that exchange rate fluctuations have in part replaced the old customs barriers in their negative effects on growth and on the development both of a large European market and of enterprises with such a dimension. The dismantling of customs barriers and the progress toward integration contributed to the fast growth in Europe in the 1960s, but the instability of exchange rates between European currencies in the 1970s has been a brake on integration and on growth.

In short, I argue that monetary instability in Europe has had the deflationary bias that some people have wrongly attributed to the EMS design. I will come back to this point later.

## Operating Principles and Requirements for Successful Functioning of the EMS

Having analyzed the economic motivation for the creation of the EMS, the main elements of which are described in the appendix to this paper, I will now discuss the operating principles of the EMS and the conditions for its successful functioning. I stress three factors:
—the convergence of underlying economic conditions in the EMS countries;
—flexibility in the operation of the system;
—greater stability between EMS currencies and other currencies.

### *Convergence of the Community Economies*

To be successful, the EMS, first of all, will have to be accompanied by policies designed to achieve a greater convergence of the economies of member states. The EMS cannot be durable and effective unless it is backed by complementary policies. As there are still important divergences in the situations of member states, great effort on the part of all countries and in all areas of policy will be needed if the system is to last.

Unless central rate changes are going to be very frequent, which would in itself limit the usefulness of the EMS, countries must, as noted, in principle give a higher priority to adjustment through internal policy measures than to changes in exchange rates.

Among these efforts, coordination of monetary policies deserves a special role. This is meant to assure a compatibility of the internal monetary objectives of member states with exchange rate objectives and with larger economic objectives. In this framework I believe that attention should be focused more on the coordination of domestic components of money creation, that is, on domestic credit expansion, than on one or more measures of the money stock. It would facilitate the monitoring of the EMS if members would broadly follow the principle of nonsterilization, through open market operations or other means, of exchange-market interventions. This would mean that countries losing reserves would allow tighter money and higher interest rates to reflect the liquidity effects of these losses, as has been the practice of the smaller countries in the "snake." A country facing a temporary accumulation of reserves will also have to remain calm and not try to offset quickly and completely the liquidity effect of sudden inflows of reserves.

Another approach to convergence is through coordination of global demand management policies. The concerted economic action decided upon in Bremen in July 1978, which modulated the extent of expansion of countries according to both balance-of-payments and inflationary problems, was an important approach to convergence. In the Bremen framework the strongest economy (Germany) took expansionary measures. This helped other countries to make necessary adjustments and makes it more probable that the right sort of adjustment and convergence policies will be followed under the EMS. This is one factor that is favorable to the initial functioning of the EMS.

Other elements of domestic policy also have an important role to play. In fact, the immediate issues affecting convergence these days are in the area of incomes policy, particularly in Ireland, Denmark, and Italy, where important wage negotiations are being discussed. The outcome of these negotiations will certainly affect the degree to which convergence can be achieved.

Let me also make a short comment here on the role of the divergence indicator, which is the main novelty in the exchange rate system and which is described more fully in the appendix. When we proposed it, our purpose quite clearly was not only to find a compromise between the two views about what to use as a trigger for mandatory intervention—the parity grid or the European currency unit (ECU)—but also to find an objective indicator as a trigger for policy coordination. This the snake did not have, for it did not indicate who should take measures. Thus the divergence indicator should become one element of a more equilibrated adjustment process and should help induce convergence. It will be very important that all countries make this new element function effectively, as it could be a means of fostering real convergence. Its role is to signal early in the game where divergences are appearing and to induce countries to take corrective actions.

A second condition (in addition to monetary coordination) for successful operation is flexibility in the system. While the EMS by itself should help to reduce differences in inflation rates among countries, it should not prevent remaining significant differences from being reflected in exchange rates. It is necessary to avoid the rigidity of the Bretton Woods system and to "de-dramatize" exchange rate adjustments. Experience in the snake during the last three years has been positive on that score. A number of adjustments have been made, with exchange markets remaining calm. Several of these adjustments involved a general realignment. This was, for instance, the case in the October 1976 snake realignment, which gave new life to the snake when many outside observers were forecasting its imminent death. In a sense, the realignment two years later was also a very successful one. It was a kind of preemptive strike, which anticipated market tensions as the January 1 deadline for the EMS approached. This operation permitted the system to start in a quiet way, first unofficially in January, then officially in March.

If changes in exchange rates can be kept small, it will be an important element in deterring speculation. Often one hears that speculation cannot but gain from a system of stable and adjustable exchange rates. That is not right. To the extent that changes in central rates are smaller than twice the width of the margin of fluctuation it is not at all sure that speculation will gain. If before the change of the central rate a currency is at the floor rate, and after the change is at the ceiling, speculation will not have gained, provided the change in the central rate is smaller than twice the margin.

Experience with the snake has shown that central rates may be adjusted by as much as 4 percent without having much effect on market rates if a depreciating currency manages to shift position with the past strongest currency inside the regular EMS band.

Some commentators on the EMS have criticized it on the ground that it does not provide clear criteria for adjustments of central rates. I disagree. If you set specific criteria for what is to be allowed, you will activate market forces that push you to make those changes. As soon as a country moves toward the indicator, speculation will be triggered. There are many cases in which you might justifiably want to resist a move, even if a sophisticated indicator tells you otherwise. I have often mentioned the case of Belgium in this respect. If Belgium had slavishly followed indicators, it would have been led to adjust in a more significant way vis-à-vis the D-mark in the last few years. Its policy of staying close to the D-mark has allowed it, on the contrary, to rapidly decrease its inflation differential with Germany, from 7 percent in 1975 to less than 0.5 percent in the spring of 1979.

This, of course, is not to say that Belgium does not accept the role of an objective indicator, as is evidenced by its initiative in proposing the divergence indicator. However, this latter indicator can set off different kinds of action, among which I would especially emphasize domestic policy actions. It is true that adjustment in rates is one of the possible actions to be taken, but it is by no means the only one.

Let me now move to a third factor for success of the EMS. A stable relation between the dollar and European currencies is not an absolute condition of success but would greatly contribute to it. Obviously this is an element that is to a large extent outside the direct control of Europeans.

Erratic movements of the dollar have often contributed to the phenomenon of overshooting between European currencies. From this point of view, the smooth start of the EMS has been helped by the relative stability of the dollar. This reflects largely the new and effective concern of the U.S. authorities about the dollar, which has been manifested in monetary and budgetary policies since November 1, 1978. Continuation of such policies by the United States will be helpful to the EMS.

I wonder whether in the future we should not try to formalize somewhat the effort on both sides of the ocean to achieve greater stability. I wonder also whether it would not be feasible to devise a more comprehensive kind of divergence indicator, which would induce a divergent country or regional grouping to take action. Such a divergence indicator could be based on the IMF's special drawing rights. If the dollar, the ECU, or the Japanese yen diverged by a certain percentage against the SDR, this would trigger consultations in which possible action by the divergent country or group of countries would be considered.

## Objections

Many criticisms of the EMS have been heard in the Community as well as outside it. I will deal with some of them, realizing fully, however, that the best answers will not come from reasoning in the abstract but rather from the behavior of the new system.

The first and most important objection is that the economic situation in Europe is too divergent to allow a system of stable exchange rates among the European currencies. Those who raise this objection point out that the inflation rates of the nine members of the European Community vary at present between some 3 percent in Germany and 13 percent in Italy. This objection should be examined seriously.

Although comparisons of consumer price indexes are not, in my view, the best criterion for measuring existing inflation differentials, I think the answer to this objection is threefold.

First, it must be recognized that the exchange rate mechanism, if it is the only instrument of coordination, is of limited use. It seems essential that a system intended to stabilize exchange rates must go hand in hand with effective coordination of economic policies, in particular of internal monetary and budgetary policies, but also of incomes policies. It is not so much a question of imposing this convergence from outside. I believe countries have come to realize better in the last few years that it was in their own interests to take domestic measures toward convergence and that floating rates did not in fact grant independence to domestic policy. In other words, one can rightly say that those who adhere to the exchange rate system should be ready to adjust their internal monetary and economic policies accordingly.

Second, agreement on this point, however, does not imply that introduction of the system must wait for a complete disappearance of differences in inflation rates. Action should be taken to reduce them, but they need not be eliminated before the system can become operative. This EMS has sufficient flexibility to allow remaining real disparities to be reflected in exchange rates. In the snake mechanism, it will be noted, divergences have been reflected in changes in exchange rates, changes that have been carried out efficiently during the past few years.

Finally, the EMS includes an element of supplementary flexibility for the member states that did not participate in the snake in 1978. These countries may opt for wider margins (6 percent) around central rates, as Italy has done.

Another objection, partly linked with the first one, says that the system will necessarily be deflationary and will adversely affect employment and economic activity in the Community. The reasoning leading to this conclusion is as follows: those countries with higher inflation rates will be forced to adopt more restrictive monetary, budgetary, and incomes policies, which are detrimental to growth, in order to meet EMS exchange rate objectives.

I cannot agree with this objection, based as it is on what economists call "the Phillips curve," or on an assumption that there is a positive correlation between growth and increases in the price level. This comes down to saying that more inflation is necessary to growth. It is not a proven case. On the contrary, in many cases countries with a low inflation rate but greater confidence have had good rates of growth. The British in recent years have been compelled to recognize the error of this reasoning. Only after the introduction of anti-inflationary monetary, budgetary, and incomes policies did the performance of the British economy improve.

I do not intend to say that there may not be transitional problems for the poorer EC countries in the EMS. Demand management policies may be more difficult to apply in these countries. It is to meet this kind of difficulty that the issue of resource transfers to the poorer countries—Ireland and Italy in particular—has been raised.

The fear that the EMS will have a deflationary bias is also based on the proposition that the D-mark will pull up the other Community currencies above their purchasing power parity and that this will necessarily have deflationary effects through decreased competitiveness. This is an objection that cannot be met in the abstract. It would only be valid if one assumed that the country with a strong currency would refuse to take internal measures to prevent an excessive increase in the value of its currency and would also refuse to have its currency revalued in relation to other currencies. One answer is that the ECU divergence indicator is designed to induce countries whose currency is diverging to take the domestic measures necessary to prevent persistence of the divergence. Furthermore, experience with the snake has already shown that needed changes of the central rates can be carried out efficiently and flexibly.

A third objection is the opposite of the second: that the EMS will have an inflationary bias. Simplified, the reasoning is as follows: differences between inflation rates will continue, and speculation will take place on a large scale. Germany then will be obliged to grant important credits in order to support the weaker currencies. These credits in turn will raise the German money supply and lead to inflationary pressures in that country.

Here again, an answer cannot be given in the abstract. The objection assumes that inflation differentials will remain high and that adequate changes in central rates will be resisted. Let me repeat that a major factor in the efficient functioning of the system will have to be greater effective coordination of economic policies so as to reduce differences in inflation rates. Let me also repeat that, while the EMS itself ought to contribute to reducing divergences in economic performance, it should not prevent remaining real disparities from being reflected in exchange rates. The experience with the snake in the last few years shows that the normal adjustments have not been resisted. There have been periods of heavy intervention to fight speculation, of course, but most of these movements have been reversed within a short time.

## Conclusion

An important initiative has been taken in Europe toward greater exchange rate stability. To function successfully, the EMS will have to foster convergence of the economic situations of member countries and be operated in a flexible way.

This initiative should also be seen as an element that can bring greater worldwide monetary stability. In this context the continuance of the recent American efforts to increase the stability of the dollar is an important consideration.

## Appendix: Contents of the EMS Agreement

The EMS agreement contains three parts: an exchange rate system; the creation of a European currency, the ECU; and the first steps toward a European Monetary Fund.

## The Exchange Rate System

*Central rates and intervention rules.* Each currency has a central rate related to the ECU. These central rates have been used to establish a grid of bilateral exchange rates around which fluctuation margins of ±2.25 percent are established. EC countries whose currencies did not belong to the snake in December 1978 could opt for wider margins of up to ±6 percent at the outset of the EMS. Italy has availed itself of this opportunity. This wider margin should be gradually reduced as soon as economic conditions permit.

A member state that does not participate in the exchange rate mechanism at the outset—this is the case for the United Kingdom—may participate at a later stage.

Adjustments of central rates will be subject to a common procedure through mutual agreement of all countries participating in the exchange rate mechanism and the Commission of the European Community. When the intervention points defined by the fluctuation margins are reached, intervention in participating currencies is compulsory.

Intervention is also allowed before the margins are reached. In principle, such intervention will also be made in participating currencies, but intervention in third currencies is not excluded. The EMS agreement provides also for "coordination of exchange rate policies vis-à-vis third countries and, as far as possible, a concertation with the money authorities of these countries."

*Indicator of divergence.* One of the new elements of the EMS that makes it different from the snake, which involves only the parity grid system, is the indicator of divergence. It is a kind of warning system and will signal whether a currency is experiencing a movement differing from the average. The indicator is based on the spread observed between the variable value of the ECU and the ECU numéraire. It flashes when a currency crosses its "threshold of divergence." The formula chosen to calculate this threshold is: 75% × (2.25% or 6%) × (1 less the weight of the currency in the ECU basket). This means that the threshold is set at 75 percent of the maximum spread of divergence allowed for each currency.

The divergence indicator also is calculated so as to eliminate the influence of the weight of each currency on the probability of reaching the threshold. If this had not been done, currencies that have a large weight in the ECU would reach the divergence indicator later than other currencies since they affect the ECU more than the currencies with smaller weights.

Before measuring the effective divergence compared to the threshold, the effective divergence must be adjusted to eliminate the effect of movements of some currencies—the lira and the pound sterling—in excess of 2.25 percent.

Indeed, the lira has a margin of 6 percent and the pound sterling is subject to no margin. This is done so that, for instance, a wide movement of the pound would not by itself lead a currency across its divergence threshold.

When a currency crosses its threshold of divergence, the presumption is that the authorities concerned will correct the situation by adequate measures, such as the following:

—Diversified intervention. This means intervention in various currencies rather than in only the currency that is furthest away from the currency of the intervening country. Diversified interventions allow a better spread of the burden of intervention among currencies of the EMS.

—Measures of domestic monetary policy. This includes, among others, measures affecting the interest rate that have a direct effect on the flow of capital. In the snake system interest rate movements were an important instrument to alleviate tension.

—Changes in central rates. While the EMS itself ought to contribute to reducing divergences in economic performance, it should not prevent remaining real disparities from being reflected in exchange rates.

—Other measures of economic policy. These could include, for instance, changes in budgetary policy or incomes policy.

In case such measures, because of special circumstances, are not taken, the reasons for this shall be given to the other authorities, especially in the "concertation between Central Banks." Consultation will, if necessary, then take place in the appropriate Community bodies, including the Council of Ministers.

After six months, these provisions shall be reviewed in the light of experience.

To summarize this first part, the present EMS differs from the snake, especially in the following ways: (1) membership has been increased by the inclusion of the French franc, the lira, and the Irish pound; (2) one currency has a larger margin, 6 percent, than the standard 2.25 percent; and (3) the system is not only based on a parity grid but also has a new element, the divergence indicator.

### The ECU and Its Functions

A European currency unit is at the center of the EMS. The value and the composition of the ECU are identical with the definition of the European unit of account.

The relative weights of the currencies in the ECU were as follows in early March 1979:

| Deutsche mark | 33.02 | Belgian franc | 9.23 |
|---|---|---|---|
| French franc | 19.89 | Danish krone | 3.10 |
| Pound sterling | 13.25 | Irish pound | 1.11 |
| Dutch guilder | 10.56 | Luxembourg franc | 0.35 |
| Italian lira | 9.58 | | |

The ECU will be used (1) as the denominator (numéraire) for the exchange rate mechanism; (2) as the basis for the divergence indicator; (3) as the denominator for operations in both the intervention and the credit mechanisms; (4) as a means of settlement between monetary authorities of the EC.

The weights of currencies in the ECU will be reexamined and if necessary revised within six months of the entry into force of the system and thereafter every five years or, on request, if the weight of any currency has changed by 25 percent. Revisions have to be mutually accepted; they will, by themselves, not modify the external value of the ECU on the day of the change. They will be made in line with underlying economic criteria.

To serve as a means of settlement, an initial supply of ECUs will be provided by the European Monetary Cooperation Fund (EMCF) against the deposit of 20 percent of gold and 20 percent of dollar reserves currently held by central banks. This operation will take the form of specified, revolving swap arrangements. The deposits will be valued in the following ways:

—for gold, whichever of these two prices is lower: the average of the price, converted into ECUs, noted each day at the two fixings in London during the previous six months, or the average of the two fixings noted the day before the last one of the period (so as to avoid a price above the current market value);

—for the dollar, the market rate two days before the date of the deposit.

Every three months, when they renew the swap arrangements, central banks will make the adjustments necessary to maintain deposits with the EMCF corresponding to at least 20 percent of their reserves. This will be done to the extent that their reserves in gold and dollars have changed. The amounts of ECUs issued will also be adjusted according to changes in the market price of gold or in the exchange rate of the dollar.

A member state not participating in the exchange rate mechanism (the United Kingdom) may participate in this initial operation on the basis described.

## The European Monetary Fund and Present Credit Mechanisms

The agreement of December 1978 stated:

We remain firmly resolved to consolidate, not later than two years after the start of the scheme, into a final system the provisions and procedures thus created. This system will entail the creation of the European Monetary Fund as announced in the conclusions of the European Council meeting at Bremen on 6/7 July, 1978, as well as the full utilization of the ECU as a reserve asset and a means of settlement. It will be based on adequate legislation at the Community as well as the national level.

In the meantime, existing financing and credit mechanisms will continue, adjusted in the following ways. The very short-term financing facility of an unlimited amount will be continued. Settlements will be made forty-five days after the end of the month of intervention with the possibility of prolongation for another three months for amounts limited to the size of debtor quotas in the short-term monetary support. Under the snake system, settlements had to be made thirty days after the end of the month of intervention. Debtor quotas in the short-term monetary support (which serve as ceiling for the three-month extension privilege) have been multiplied by about 2.5.

The credit mechanisms will be extended to an amount of 25 billion ECUs of effectively available credit. This is about 2.5 times the previous amount. Its distribution will be: for short-term monetary support, 14 billion ECUs; for medium-term financial assistance, 11 billion ECUs.

The substantial increase in the amounts of credit available and the lengthened duration of some credit mechanisms are important elements for strengthening the credibility of the system; they guarantee that, in case of need, large means can be made available to countries to fight speculative movements.

## Comments by Ralph C. Bryant

Jacques van Ypersele's paper provides a clear description of the EMS and also a subtle statement covering its rationale, some of the problems it will encounter, and some of the doubts it evokes.

It seems to me I can most usefully focus my comments on the objectives and principles that underlie the EMS initiative. I hope to avoid offering merely an "American" view, and I will not attempt to comment on the possible implications of the EMS for the interests of the United States. Rather I intend to keep mainly to analytical observations about exchange rate variability—to sketch an analytical framework that may help in thinking about the issues raised by the EMS from the perspective of either side of the Atlantic.

As a point of departure, let me summarize the two dominant intellectual approaches to exchange rate variability.

The first may be termed the *untrammeled market position.* It is popular among North American economists and officials but is encountered elsewhere as well. It holds that every nation should pursue appropriate domestic macroeconomic policies and then permit currency values to be determined in the foreign exchange market without intervention by central banks or governments.

The second view, most frequently met in Western Europe and Japan, may be labeled the *minimum variance position*. Those holding this view emphasize the uncertainties and disruption that may be associated with exchange rate movements. They argue that governments should act to "maintain as much exchange rate stability as possible." EMS advocates who identify a "zone of monetary stability" in Europe as the prime objective of the experiment appear to take this position, at least for fluctuations of exchange rates between Community currencies.

To provide an analytical framework for evaluating these views about exchange rate variability, imagine an open economy in which external trade is important, banks and nonbanks have significant amounts of assets and liabilities vis-à-vis foreigners, and restrictions on capital flows are not sufficiently comprehensive to prevent shifts in assets when changes occur in expectations about exchange rates, interest rates, or asset prices. Policy actions or nonpolicy disturbances originating abroad have more than trivial effects on the home economy, and vice versa; but typically the effects are greatest in the originating country. In short, suppose the situation of *intermediate interdependence* that actually characterizes Western Europe and North America today.

Make the further, realistic assumption that the primary concern of a nation's policymakers is with national objectives—with levels of employment and prices at home. Little or no welfare value is attached to developments in employment or prices abroad, except as feedbacks are perceived to affect the home economy. Policymakers consider themselves free to adjust the instruments of macroeconomic policy as they deem best. As is the case under the current Articles of Agreement of the IMF, assume that there are no binding supranational constraints on movements of exchange rates and external reserve assets.

How should the home nation's policymakers respond—in particular, should they or should they not allow the exchange rate to move—when various types of nonpolicy disturbances occur?

Consider first a recession that originates abroad, caused by an unexpected increase in the foreign savings rate and an unexpected slump in foreign business investment. That disturbance brings incipient pressure on both the foreign and the domestic economies; market interest rates abroad tend to fall, the domestic currency tends to depreciate, and domestic interest rates tend to fall. Actual adjustments in financial variables depend on how domestic monetary policy is implemented at home and abroad and on the extent, if any, of exchange market intervention.

If the country's central bank chooses to resist its currency's tendency toward depreciation, its external reserve position deteriorates. Exports decline, domestic real output falls, and some downward pressure is exerted on the domestic price level. When the exchange rate is held fixed, the foreign recession is transmitted to the maximum possible extent to the home economy.

If, on the other hand, the domestic currency is allowed to depreciate, the adverse effects of the foreign recession on the home economy are lessened. Domestic output and prices are likely to be lower than they would be if there were no foreign recession, but by smaller amounts than if the exchange rate was held fixed.

Policymakers of the country would thus have reason to reject exchange rate "stability" in these circumstances. Depreciation of the domestic currency, while not desired for its own sake, would help to buffer domestic variables from a disturbance originating outside the home economy.

Consider next a recession that begins at home in response to an unexpected decline in consumption and investment spending by the country's residents. The incipient pressures on financial variables now include tendencies for the home currency to appreciate and for market interest rates to fall at home and abroad.

If the central bank holds the exchange rate stable, external reserves rise. Domestic output and employment fall and prices are under downward pressure. But as imports decline, some of the contractional impetus of the recession is transmitted abroad through the current account of the balance of payments.

If the exchange rate is permitted to appreciate, however, less of the contractional impetus is transferred abroad. Appreciation of the domestic currency tends to reinforce the excess supply created by the initial disturbance in goods markets at home: for example, it causes a fall in import prices relative to those for domestic goods and makes export goods less competitive, thereby leading both indigenous and foreign residents to switch expenditures away from domestic to foreign goods. The disruptive effects of the disturbance are therefore confined more to the home economy because of the appreciation of the domestic currency. From the point of view of the nation's policymakers, exchange rate variability in these circumstances is an inferior alternative to exchange rate stability.

As a third illustrative type of disturbance, consider an unexpected change in asset preferences where private investors, foreign or domestic, decide to increase the proportion of domestic-currency assets in their portfolios. The incipient pressure on financial variables in this case includes tendencies for the home currency to appreciate, for home market interest rates to fall, and for foreign interest rates to rise.

Assuming no central bank intervention in exchange markets causing the domestic currency to appreciate, native and foreign residents switch expenditures away from domestic to foreign goods; imports rise and exports tend to decline. The net result is an unanticipated and unwanted tendency toward contraction in domestic output and prices. (The contractional result is unambiguous if the central bank keeps domestic interest rates from falling while it allows the domestic currency to appreciate.)

If, however, exchange market intervention is carried out to prevent the exchange rate from changing, the external reserves of the country increase but there is no unwanted impact on output and prices (at home or abroad). In effect, the exchange market intervention accommodates the private sector's shift in asset preferences at the existing interest rates and exchange rate by means of offsetting changes in central bank balance sheets.

For this case of an autonomous change in private asset preference, therefore, both the home economy and the foreign economy are better off with exchange market intervention that keeps the exchange rate unchanged.

Analysis of the type sketched out above leads to an eclectic view about exchange rate variability. It shows that the variability of a nation's currency facilitates adjustment to some types of nonpolicy disturbances but aggravates the adverse consequences of others. (An analogous conclusion applies to policy "mistakes"—macroeconomic policy actions that turn out in retrospect to have been undesirable.) Hence it is sometimes desirable for exchange rates to fluctuate. At other times stability is advantageous. As an analytical matter, neither the untrammeled market nor the minimum variance position is sound. It is inappropriate to set either the presence or the absence of exchange rate stability as a goal of national macroeconomic policy. The traditional choice between fixed rates and flexible rates is an artificial one that policymakers do not have to, and should not, make. The proper approach is a discretionary one—managed fixing or managed floating, with "managed" the key word.[1]

With appropriate modifications for the specific circumstances, the framework for analysis that I have outlined can be applied in principle to individual European countries, to the Community or Western Europe as a whole, and to individual non-European economies. It provides a useful background for an

---

1. The argument summarized here draws on the discussion in part 5 of my forthcoming book, *Money and Monetary Policy in Interdependent Nations.*

inquiry into the prospects for the success or failure of the EMS, especially its planned treatment of variability among the currencies of the Community countries.

In his paper, van Ypersele mentions several motives or objectives for the EMS initiative. He does not differentiate them clearly, nor does he, perhaps understandably, weight them according to their importance. It is illuminating, however, to examine and evaluate them separately.

One objective is to foster exchange rate "stability." Apparently a "zone of monetary stability in Europe" is for some people virtually an end in itself. In any case, avoidance of the discomforts and costs of floating is an important aim, broadly shared by the Community member states.

A second motivation is to facilitate the achievement of national economic goals in the individual EC countries. For example, it is argued that the EMS will make it easier to keep national price levels under control or to promote higher and more stable levels of national employment. One can readily imagine individuals and governments being drawn to the EMS for these national motives without being attracted by the idea of greater European economic integration or political unity.

Finally, a third objective motivates some of the proponents of the EMS: to hasten the economic and political integration of Europe. You may recall that scientific measurement of beauty—the millihelen. One millihelen is the amount of pulchritude needed to launch one ship. It may be that politicians and economists require an analogous scientific measure of progress—genuine progress— toward European integration. Let me offer the "gaullicycle," that is, the amount of progress toward European unity calculated to make Charles de Gaulle turn one full revolution in his grave. One question to ask when evaluating the motives of those Europeans who are ardent supporters of the EMS is, how many gaullicycles do they intend and expect beyond the purely national benefits that are sought?

Conclusions about the likely viability and longevity of the EMS should depend heavily on the balance between these differing motivations and, hence, on the supporting actions that will or will not be taken to advance the several objectives. I will discuss the possibilities by suggesting several hypotheses.

Hypothesis A says that most of the support for the EMS may be ascribed to dissatisfaction with floating exchange rates per se, particularly with the fluctuations between the snake currencies and the dollar. The 1977-78 depreciation of the dollar might then be seen as the mother of the EMS, and the stability of the dollar since November 1978 as the good midwife who assured a smooth birth.

If this hypothesis is correct, one's verdict about the longevity of the EMS would have to be fairly pessimistic. The primary objective of the system is then close to what I have termed the minimum variance position about exchange rate variability. That position does not rest on solid analytical foundations. If fluctuations of the snake currencies against the dollar are viewed as the main difficulty, moreover, the EMS remedy is not suited to the disease it is intended to cure. Stability of exchange rates within Europe can protect against some kinds of disturbances—for example, those having their source in asset markets. But stability, even if it could be maintained, will not protect against other types of disturbance.

Hypothesis B is that support for the EMS is based mainly on the conviction that exchange rate stability—a zone of monetary stability in Europe—will help advance national macroeconomic objectives (for example, anti-inflationary policies). The emphasis here is less on the dollar's fluctuations against snake currencies than on the alleged positive gains to European economies from stability among European currencies themselves.

Despite the difference in emphasis, this view is again close to the minimum variance position about exchange rates. It is similarly liable to periodic frustration. The situations and interests of the European nations are so divergent that economic disturbances may often create circumstances in which it is preferable for some European nations to hold exchange rates steady and for others to let them vary. Could one think, for instance, that it would be in the interests of both Germany and France to deal with a wage explosion in France by intervention to stabilize the deutsche mark-franc exchange rate?

Under the first two hypotheses, it would be appropriate to interpret the initiation of the EMS as an experiment little different from earlier efforts at European monetary unification. The EMS would be based on views that are no more durable than those incorporated in the earlier Barre and Werner reports or in the agreement establishing the snake in the tunnel. A temporary convergence of short-run national interests in promoting exchange rate stability under "fair weather" conditions would disguise fundamental differences in national interests. With such shaky foundations, the EMS would be likely to run into trouble in future years when the weather was stormy.

Hypothesis C asserts that the primary aim of the EMS is to facilitate better achievement of national macroeconomic objectives, but to do so through more "ordered management" of exchange rates, including exchange rate changes among Community currencies as may be desirable from time to time. If this hypothesis is correct, the EMS still represents a negligible number of gaullicycles toward greater European integration. Unlike hypotheses A and B, however, this approach is not based on the minimum variance position about exchange rate variability. If orderly, prompt changes in the EMS parity grid are indeed going to take place when needed, one's judgment about the likely durability and longevity of the EMS can be more favorable. The system will emphasize not a particular set of exchange rates, but rather the pragmatic management of intra-European exchange rate variability. Note that under this hypothesis the zone of monetary stability becomes a much subordinated objective of the Community.

Now assume, contrary to the preceding hypotheses, that the hope for greater progress toward European integration is a major driving force behind the EMS experiment. If that assumption is true, one can presume that when the economic or political interests of an individual EC nation conflict with the interests of others or the Community as a whole (as will happen) that nation will be willing to sacrifice its national goals to a greater extent than in the past. Not only in fair weather but also in foul, the interests of the Community as a whole are to move up the scale of priorities relative to national interests.

Hypothesis D is one variant of the view that "integration" should get a higher priority than in the past. It asserts that the "Europeanist" strategy, often attributed to Jean Monnet, lies behind the creation of the EMS. That strategy, you recall, relies on an apparently technical initiative in the monetary field to catalyze greater integration. But it does not simultaneously involve direct steps to foster greater convergence in nations' domestic macroeconomic policies.

A zone of monetary stability within Europe—if that indeed meant unchanging exchange rates regardless of national objectives—would of course force more integration by requiring a convergence of domestic policies. But substantial differences still exist among European nations—in the ultimate goals of their economic policies, in their experiences with internal nonpolicy disturbances, in the behavior patterns of their residents, and in their institutional, social, and legal environments. If European policymakers seek to bring about greater integration among their heterogeneous economies solely by trying to stabilize their currencies vis-à-vis each other, the attempt seems certain to be abortive. So long as the more basic differences in national objectives, behavior patterns, and institutions remain, the means (exchange rate stability) will have to be abandoned long before the end (greater integration) has been accomplished.

That brings me to a final hypothesis E, which supposes not only that the Community nations now intend to give European integration a high political priority, but also that they plan to institute a number of complementary direct measures to bring about a convergence of domestic economic policies. This will call for a large number of gaullicycles of progress, both in the intent of policymakers and in their success in carrying out their intentions. If this ambitious agenda is attempted, the prognosis for the EMS should be very different from the prognoses under the other hypotheses. If the political will were to be strong enough to forcefully promote a much greater convergence of national policies, one could reasonably foresee progress toward genuine integration and, following from it, good prospects for exchange rate stability.

For purposes of this discussion I have sharply differentiated the possible objectives for and motivations behind the EMS. In practice, of course, they are intermingled. I, for one, am not clear about how to weight them. I suspect that the several founders of the EMS give them substantially different weights. I am therefore led to conclude with an agnostic judgment, which may be encapsulated in two contradictory aphorisms attributed to English scholars.

On the one hand, the EMS seems to warrant the verdict of Samuel Johnson on second marriages. "A second marriage," said Johnson, "is the triumph of hope over experience." For the lira, the EMS represents a second marriage; for the French franc, it is a third.

On the other hand, it is conceivable that the time has arrived for new political moves toward European unity. It may even be the case that new efforts will be made, in association with the operation of the EMS, to promote a genuine convergence of domestic macroeconomic policies. In that case, the appropriate aphorism is the remark made to a young woman by the classical scholar and master of Balliol College, Oxford, Benjamin Jowett. "You must believe in God, my dear," Jowett said, "despite what the clergymen say."

It thus seems prudent to leave the issue open. Skepticism about the EMS is easy to justify. But, just possibly, one should believe in its viability despite what the economists say.

RUDIGER DORNBUSCH
*Massachusetts Institute of Technology*

# Exchange Rate Economics: Where Do We Stand?

THE BROOKINGS PANEL on Economic Activity for the past ten years has mirrored much of the exciting theory and empirical work in open-economy macroeconomics. In the spirit of Brookings, the papers have explored what issues openness raises for macroeconomic management. The range of interests has been quite broad, beginning with William Branson's "new view of international capital movements" and including Marina Whitman's dismissal of "global monetarism" and many of the topics of the day from trade equations and oil to commodity booms, debt, and portfolio selection.[1] The questions have been similar—how much independence there is for macroeconomic policy in an interdependent world; how important monetary factors are; or how can the interest rate be kept lower than the market will bear. The papers have emphasized the evolution of open-economy macroeconomics from the structure of the 1960s—the Mundell-Fleming model—to a framework better suited to the analysis of inflation, expectations, and portfolio substitution.

This paper maintains the tradition of asking how international interdependence has impinged on macroeconomic variables and policy options. The paper takes as its frame of reference the experience with floating ex-

I am grateful for comments from members of the Brookings panel and from Stanley Fischer. Robert E. Cumby made many suggestions and provided generous research assistance. Financial support was provided by a grant from the National Science Foundation.

1. See William H. Branson, "Monetary Policy and the New View of International Capital Movements," *BPEA*, 2:1970, pp. 235–62; and Marina v. N. Whitman, "Global Monetarism and the Monetary Approach to the Balance of Payments," *BPEA*, 3:1975, pp. 491–536.

change rates and seeks to explain, in the light of today's theories, the pattern of exchange rate movements and policy responses.

The main lessons that emerge from the analysis concern the inadequacy of the monetary approach as a complete theory of exchange rate determination, the central role of the current account in influencing exchange rates, the suggestion that there is a deutsche mark shortage and, finally, the conclusion that an interest rate policy not oriented toward the external balance has aggravated exchange rate instability.

The paper is divided into two parts. In the first part, developments in exchange rates are analyzed using a variety of models, starting with the monetary approach, and leading from there to models of exchange rate dynamics and the current account. I show that unanticipated disturbances to the current account have been an important source of unanticipated movements in exchange rates. In addition, the structure of real returns on securities denominated in different currencies suggests that the deutsche mark should be occupying an important share in internationally diversified portfolios, and that substitution in that direction may well explain the persisting tendency for that currency to appreciate in real terms.

In the last part of the paper I address the important question of how the system of flexible exchange rates has been operated. A review of intervention and interest rate policies in key countries suggests that external constraints have not been predominant. On the contrary, interest rate policies have been pursued quite independently of a desire to finance imbalances in current accounts through capital flows; and that independence has led to growing requirements for intervention. The proposal by James Tobin for a tax on foreign exchange transactions is considered in this context.

The paper concludes with the demonstration that much of the observed instability in exchange rates has been due to unanticipated disturbances, with the forecasting errors broadly shared by governments and the public alike. The instability has been aggravated, however, by a failure to use monetary policy with a view to the external balance and by a failure to recognize portfolio shifts toward marks as part of the adjustment process to the regime of flexible exchange rates.

## Exchange Rate Theories and Empirical Evidence

There are basically three views of the exchange rate. The first takes the exchange rate as the relative price of monies; the second, as the relative price of goods; and the third, the relative price of bonds. I regard any one of these views as a partial picture of exchange rate determination, although each may be especially important in explaining a particular historical episode. Still, it is useful to approach exchange rate theory not from the complex perspective of an all-encompassing model, but rather from the vantage point of a sharply articulated, partial model. The monetary approach is a good place to start. Although in the opinions of some of its proponents it represents a quite complete theory of the exchange rate, I will expand it to a more general theory by relaxing some of the special assumptions that are required if it is to stand on its own.

## THE MONETARY APPROACH

At the outset of flexible rates in the 1970s, the literature emphasized a monetary interpretation of exchange rate determination.[2] Most versions of the monetary approach assume strict purchasing power parity (PPP). Exchange rates move promptly in order to maintain the international linkage of prices. Thus there is no room for changes in the terms of trade. With $e$ denoting the logarithm of the home currency price of foreign exchange, and $p$ and $p^*$ denoting the logarithm of home and foreign prices, respectively, PPP implies[3]

$$(1) \qquad\qquad e = p - p^*,$$

where here and throughout the paper variables in lowercase (except interest rates) represent logarithms.

The next step in the monetary approach is to take prices as determined by domestic nominal money supply and real money demand. With real money demand depending on real income and the nominal interest rate, the expression becomes

$$(2) \qquad\qquad p = m - ky + hi$$

$$p^* = m^* - ky^* + hi^*,$$

2. See the collection of papers in *Scandinavian Journal of Economics*, vol. 78, no. 2 (1976), pp. 133–412; the papers collected in Jacob A. Frenkel and Harry G. Johnson, eds., *The Economics of Exchange Rates: Selected Studies* (Addison-Wesley, 1978); John F. O. Bilson, "The Monetary Approach to the Exchange Rate: Some Empirical Evidence," *IMF Staff Papers*, vol. 25 (March 1978), pp. 48–75; and Jacob A. Frenkel, "Exchange Rates, Prices, and Money: Lessons from the 1920's," *American Economic Review*, vol. 70 (May 1980, *Papers and Proceedings, 1979*), pp. 235–42.

3. Throughout the paper an asterisk denotes a foreign variable.

where

$m$ = logarithm of nominal money
$k$ = income elasticity of real money demand
$y$ = logarithm of real income
$h$ = semilogarithmic interest response of real balances
$i$ = nominal interest rate.

Combining equations 1 and 2 yields the exchange rate equation of the monetary approach:

$$(3) \qquad e = m - m^* + h(i - i^*) - k(y - y^*),$$

where coefficients are assumed to be equal for all countries.

The model establishes that relative changes in money supply, interest rate, and real income affect the exchange rate. An increase in the money supply at home leads to an equiproportionate depreciation. Because an increase in domestic real income raises the demand for real balances and thus leads to a fall in domestic prices, it induces an offsetting exchange appreciation. Relatively higher domestic interest rates, by contrast, reduce the demand for real balances, raise prices, and therefore bring about an exchange depreciation.

There are two ways to test the monetary approach. One recognizes that instantaneous PPP is an essential part of the monetary approach and directly tests whether PPP prevails. The second examines the explanatory power of econometric equations specified like equation 3.

There is ample evidence accumulating that this assumption is *not* warranted. Not only does the short-term exchange rate deviate from a PPP path, but there are also cumulative deviations from that path that show substantial persistence. This is clearly brought out by table 1, which shows annual inflation rates for consumer prices in the United States, five other major industrial countries, and a trade-weighted index of those countries. The table also shows the average annual appreciation of the foreign currencies relative to the dollar, bilaterally and as a group. Contrary to PPP theory, *real* exchange rates have not remained constant. The striking fact is that during the period from 1973 to 1979, the annual rate of inflation in the United States averaged about 1 percentage point less than in the group of foreign countries, yet the dollar has depreciated at an average rate of over 1 percent a year.[4] There has thus been an average annual

4. The comparison here is based on consumer prices; it holds, in general, for other price indexes also.

**Table 1. Inflation and Currency Appreciation in Major Industrial Countries, 1973–79**
Annual average, in percent

| | Measure | |
|---|---|---|
| Country | Consumer price inflation | Appreciation on the dollar |
| United States | 8.5 | ... |
| Other major industrial countries[a] | 9.4 | 1.4 |
| Canada | 9.2 | −2.6 |
| France | 10.7 | 0.7 |
| Germany | 4.6 | 6.4 |
| Japan | 9.9 | 3.6 |
| United Kingdom | 15.6 | −2.4 |

Source: International Monetary Fund, *International Financial Statistics*, vol. 33 (March 1980), series ahx for exchange rates and series 64 for prices.
a. These series are weighted averages of the respective individual series for the five foreign countries. The relative weights are derived from the International Monetary Fund's multilateral trade model. They are: Canada—0.2405, France—0.1640, Germany—0.2340, Japan—0.2160, and the United Kingdom—0.1440.

change in relative price levels adjusted for exchange rate movements (or a *real* depreciation of the dollar) of more than 2 percentage points. This substantial rate of real depreciation should attract attention and study rather than being confined to the error term. The evidence of table 1 is also reflected in figure 1, which shows that the International Monetary Fund's multilateral nominal and real effective exchange rates of the dollar have moved together. Figure 2 illustrates how the nominal effective exchange rate has departed from, rather than simply offset, inflation differentials.[5]

The alternative approach to testing the monetary theory relies on evidence from regression equations. The empirical evidence reported in table 2 tests the explanatory power of the theory as specified by equation 3, using the dollar-mark exchange rate. The explanatory variables are relative nominal money supplies, relative real income levels, and nominal long-term and short-term interest differentials.

The long-term interest differential appears in the exchange rate equation either because, in addition to short-term interest rates, long-term rates measure one of the alternative costs of holding money or because they are taken as a proxy for anticipated inflation differentials. In either

5. Throughout the remainder of this paper the nominal effective exchange rate is this trade-weighted index of the five foreign countries of table 1, rather than the International Monetary Fund's published multilateral trade-weighted index.

**Figure 1. International Monetary Fund's Effective Exchange Rate of the Dollar, Nominal and Real, 1973:1–1979:4**

*Index, 1975 = 100*

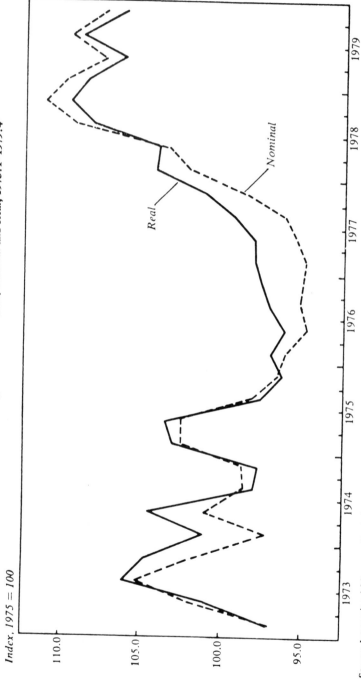

Source: International Monetary Fund, *International Financial Statistics*, various issues. The real and nominal rates are the inverses of, respectively, the index of relative wholesale prices (series 63 110) and the nominal effective exchange rate index (series amx).
a. The data are quarterly averages.

**Figure 2. Relative Price Level and Nominal Effective Exchange Rate in the United States, January 1973–September 1979[a]**

*Index, 1975 = 100*

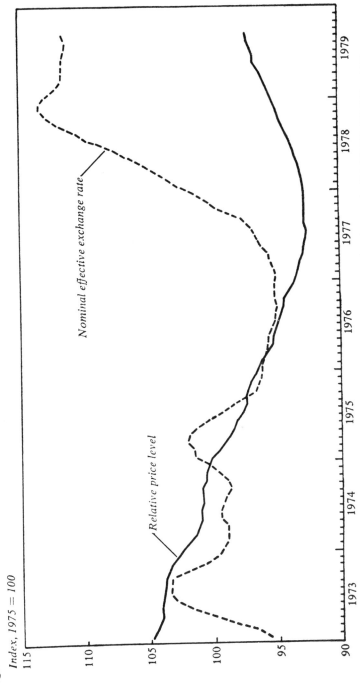

Nominal effective exchange rate

Relative price level

Sources: Nominal effective exchange rate—same as table 1; and prices—*Business Conditions Digest*, various issues, series 320, 732, 733, 735, 736, and 738.
a. The data are seven-month centered moving averages. The nominal effective exchange rate is calculated as a weighted average of the index of five foreign countries, as described in table 1, note a. The relative price level is calculated as the ratio of the U.S. index of consumer prices to a weighted index of consumer prices in the five countries, with weights equal to those used in calculating the exchange rate.

**Table 2. Equations Explaining the Monetary Approach to Exchange Rate Determination, Using the Dollar-Mark Exchange Rate, 1973:2–1979:4 and Subperiod[a]**

| Equation and sample period | Independent variable | | | | | | Summary statistic | | | |
|---|---|---|---|---|---|---|---|---|---|---|
| | Constant | $(e + m^* - m)_{-1}$ | $m - m^*$ | $y - y^*$ | $(i - i^*)_S$ | $(i - i^*)_L$ | $R^2$ | Durbin-Watson | Standard error of estimate | Rho |
| 2-1 1973:2–1979:4 | 5.76 (2.81) | … | −0.03 (−0.07) | −1.05 (−0.97) | 0.01 (1.90) | 0.04 (2.07) | 0.33 | 1.83 | 0.05 | 0.88 |
| 2-2 1973:2–1978:1 | 4.82 (2.51) | … | 1.00 | −0.93 (−0.90) | −0.00 (−0.29) | 0.07 (5.94) | 0.66 | 1.69 | 0.04 | 0.06 |
| 2-3 1973:2–1979:4 | 4.63 (2.12) | … | 1.00 | −0.76 (−0.66) | 0.01 (1.62) | 0.04 (1.82) | 0.08 | 1.80 | 0.05 | 0.99 |
| 2-4 1973:2–1979:4 | 0.23 (0.12) | 0.83 (8.26) | 1.00 | 0.16 (0.17) | 0.01 (1.36) | 0.01 (0.67) | 0.88 | 1.85 | 0.05 | … |

Sources: Exchange rate—Board of Governors of the Federal Reserve System, *Federal Reserve Statistical Release*, G.5, "Foreign Exchange Rates," various issues; U.S. money supply—Board of Governors of the Federal Reserve System; German money supply—International Monetary Fund, *International Financial Statistics*, various issues and Deutsche Bundesbank; real income—Organisation for Economic Co-operation and Development, *Main Economic Indicators*, various issues; and interest rates—Morgan Guaranty Trust Company of New York, *World Financial Markets*, various issues.

a. The equations were estimated using quarterly data. The independent variables are: $e$—logarithm of the dollar-mark exchange rate; $m$—logarithm of the money supply ($M_1$), seasonally adjusted; $y$—logarithm of gross national product at 1975 prices, seasonally adjusted at annual rates; $is$—yield on representative money-market instruments; and $i_L$—yield on domestic government bonds. An asterisk denotes a variable for Germany. The numbers in parentheses are $t$-statistics.

view, a rise in the domestic long-term interest rate differential leads to a reduction in real money demand and thus to higher prices and depreciation.[6]

The theory suggests that a rise in domestic relative income induces appreciation and that an increase in domestic interest rates induces depreciation. Equation 2-1 in table 2 tests this theory with quarterly data, with coefficients constrained to be equal for all countries. It offers little support for the monetary approach. Only a small fraction of the variance in the exchange rate is explained, and there is a high (0.88) estimated coefficient of serial correlation. Although interest rates have the expected sign and are significantly different from zero, the coefficient of relative monies is actually negative, but it is insignificant.

The coefficient of relative monies in the remaining equations is constrained to unity. Equations 2-2 and 2-3 differ in sample period and demonstrate the instability of equation 3. For the complete sample period the equation has negligible explanatory power. Equation 2-4 allows for lagged adjustment in real balances by introducing the lagged dependent variable as an explanatory variable.[7] Only the lagged adjustment term appears significant in this formulation.

The evidence on PPP and the econometric evidence reported here leave little doubt that the monetary approach in the form of equation 3 is an unsatisfactory theory of exchange rate determination. The key link between the exchange rate and PPP fails to hold, and any reasonable model must include a theory of real exchange rate determination.

The monetary approach was an important stepping stone of empirical research in international monetary economics and a plausible, if bold, hypothesis. Together with the asset market approach, it reflected a reaction to elasticity models of the exchange rate and, in that respect, was a substantial contribution. Both approaches share the partial equilibrium

6. For further discussion of the roles of long-term and short-term interest differentials, see Jeffrey A. Frankel, "On the Mark: A Theory of Floating Exchange Rates Based on Real Interest Differentials," *American Economic Review*, vol. 69 (September 1979), pp. 610–22.

7. For further discussion see Rudiger Dornbusch, "Monetary Policy under Exchange-Rate Flexibility," in Federal Reserve Bank of Boston, *Managed Exchange Rate Flexibility: The Recent Experience*, Conference Series, 20 (FRBB, 1979), pp. 90–122; Frankel, "On the Mark"; and P. Hooper and J. Morton, "Fluctuations in the Dollar: A Model of Nominal and Real Exchange Rate Determination" (Board of Governors of the Federal Reserve System, October 1979).

view that exchange rates are determined by the conditions of stock equilibrium in the asset markets. They ignore other factors important to a general equilibrium analysis. I turn next to a broader model that reintroduces the more traditional aspects of exchange rate determination—the current account, wealth effects, expectations, and relative prices.

## A GENERAL MODEL OF EXCHANGE RATES

If strict PPP is abandoned, the way is clear for a broad approach to modeling exchange rate determination. A first step here is the traditional Mundell-Fleming model that remains, with some adaptations, the backbone of macroeconomic models of the exchange rate.[8] This model assumes that domestic prices are fixed in each home currency so that the exchange rate sets the terms of trade or the price of domestic goods relative to imports. Capital is fully mobile internationally and, with perfect substitutability between home and foreign securities (ignoring exchange rate expectations), interest rates are equalized internationally. Output is demand-determined.

Suppose, in this setting, that monetary expansion occurs at home. The resulting decline in interest rates leads to an international differential that brings about an incipient capital outflow. The exchange rate depreciates and, with elasticity conditions satisfied, demand shifts toward domestic goods. The induced increase in output leads to a rise in income and money demand until equality among international interest rates is restored at a higher level of output with a lower real exchange rate.

An expansion in demand for home output arising from fiscal policy or an exogenous shift in demand leads to an increase in income and money demand, and hence a tendency for interest rates to increase. The induced capital inflows bring about exchange rate appreciation, a loss in competitiveness, and hence a deterioration in the current balance that dampens or offsets the expansion. This result is clearly a curiosity, and I return to it below.

An extended Mundell-Fleming model can be derived by relaxing five key restrictive assumptions: fixed prices, the fully demand-determined level of output, the absence of exchange rate expectations, the absence of a role for the current account in exchange rate determination, and the

8. For an exposition and further references, see Rudiger Dornbusch, *Open Economy Macroeconomics* (Basic Books, forthcoming in 1980).

perfect substitutability of domestic and foreign securities. The first three assumptions are readily relaxed.[9]

Rational expectations and long-run neoclassical features such as full employment are included in the extended Mundell-Fleming model. The increase in demand again brings an immediate nominal and real appreciation that restores demand to the full employment level through an offsetting deterioration in the current account, and monetary expansion leads to an immediate depreciation of the nominal and real exchange rate. Moreover, the exchange rate must overshoot, depreciating proportionately more than the expansion in money, if asset markets adjust more rapidly than goods markets. The domestic interest rate falls relative to those abroad, and asset markets will be in balance only if the exchange rate initially overshoots, so that there are corresponding expectations of currency appreciation.[10]

The extended Mundell-Fleming model is a first approach to expanding exchange rate theory in the absence of PPP that allows for short-run real effects of monetary disturbances and that permits the possibility of permanent changes in relative prices in response to changes in the pattern of world demand. By introducing rational expectations, the model focuses on "news" as the determinant of unanticipated changes in the exchange rate. Over time the exchange rate follows a path delineated by interest differentials. News about monetary developments or the state of demand bring about immediate changes in the level and path of the exchange rate. These ideas can be incorporated by distinguishing between actual and antici-

9. See Rudiger Dornbusch, "Expectations and Exchange Rate Dynamics," *Journal of Political Economy*, vol. 84 (December 1976), pp. 1161–76, and *Open Economy Macroeconomics*.

10. The model is made up of the condition of monetary equilibrium,

$$m - p = ky - hi;$$

the condition of equalization of interest rates, adjusted for anticipated depreciation, $\dot{e}$,

$$i = i^* + \dot{e};$$

and the condition of equilibrium in the goods market,

$$y = a(e - p) + u,$$

where it is assumed, for expository simplicity, that there is no direct effect of interest rates on aggregate demand. The rate of inflation (relative to trend) is determined by the output gap, $y - \bar{y}$; that is, $\dot{p} = b(y - \bar{y})$. The model determines at a point in time the level of output and the exchange rate, as well as the rate of inflation and depreciation, as a function of prices. Shifts in demand, shown by shifts in $u$, lead to immediate offsetting changes in the real exchange rate.

pated depreciation, $\dot{e}'$ and $\dot{e}$, respectively. With perfect asset substitutability, the actual rate of depreciation is the sum of anticipated depreciation, which equals the nominal interest differential, $i - i^*$, and the effect of news, which is given by the difference between actual and anticipated depreciation,

$$(4) \qquad\qquad \dot{e}' = (i - i^*) + (\dot{e}' - \dot{e}).$$

The relevant news in this model is changes in monetary conditions and in the demand for domestic output.

The model retains the uncomfortable property that *any* increase in demand for home output, whether through fiscal expansion or increased net exports, leads to nominal and real appreciation because the only role of the current account is as a component of demand. Imbalances in the current account have no medium-term feedback on the economy, either in goods markets or in asset markets. The analysis can now be expanded to introduce the role of the current account.

Suppose that in the goods market demand for home output depends not only on income and the terms of trade but also on real wealth. A rise in real wealth would be expected to increase real spending and demand for domestic goods. A rise in wealth thus creates an excess demand, which, to maintain output at full employment, would have to be offset by the expenditure-shifting effect of a real exchange rate appreciation. In the diagram below the *y* schedule is seen as the combination of real exchange rates—defined as the ratio of the price of imports to domestic goods imports—and the level of real wealth, *w*, which is consistent with output at full employment.[11]

The current account is balanced along the schedule $\dot{w} = 0$. With more wealth there is increased spending and thus a tendency for an external deficit. To restore external balance, the real exchange rate must depreciate, thus shifting demand from foreign goods toward home output, and

---

11. In terms of note 10, the equilibrium condition in the goods market now becomes $y = J(e - p, w, u)$, where $w$ denotes the level of real wealth and a rise in real wealth increases demand for home output. Real balances are excluded from the definition of real wealth. The current account is equal to the rate of change of real wealth, $\dot{w}$; that is, $\dot{w} = H(e - p, w, y, v)$, where $v$ is a shift parameter. The current account improves with real depreciation but deteriorates with an increase in income or wealth as both induce increased spending. For a more complete model along these lines see Rudiger Dornbusch and Stanley Fischer, "Exchange Rates and the Current Account," *American Economic Review* (forthcoming in December 1980).

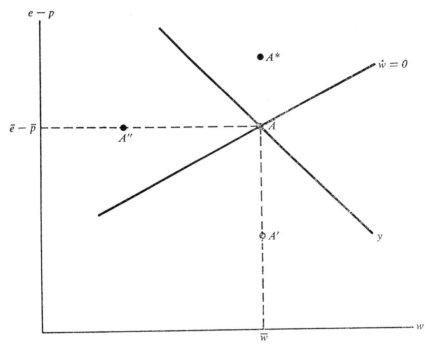

thereby restoring external balance. Accordingly, the external balance schedule is positively sloped; points above the schedule correspond to a surplus and points below to a deficit. Furthermore, a surplus implies net acquisition of claims on the rest of the world and hence growing real wealth; the converse is true for a deficit.

The extended framework is helpful in identifying the long-run equilibrium of the economy, its determinants, and some of the factors that affect the dynamics. The diagram shows that long-run equilibrium occurs for real variables—real wealth and the terms of trade or real exchange rate. At point $A$, demand for domestic output is at full employment and the current account is in equilibrium or, equivalently, income equals expenditure. In the background is the monetary sector that specifies the price level and the nominal exchange rate.

The expanded model makes possible the immediate interpretation of a demand shift or increase in net exports. With a permanent increase in net exports there is an excess demand for domestic goods and an equal surplus. To restore internal and external balance simultaneously, all that is required is nominal and real appreciation. A demand shift thus leads to an instantaneous real and nominal appreciation to a point like $A'$, with no further adjustments needed. By contrast, a rise in spending on both home and traded goods in the pattern of average expenditures will leave the equilibrium composition of spending unchanged, and thus only leads to a change in long-run wealth at point $A''$. Over time the economy will reduce its stock of assets until spending has declined sufficiently for the initial real exchange rate to be reestablished. The adjustment process depends, of course, on the interaction between goods markets and the monetary sector.

The uncomfortable fact remains that even in this model there is a short-run tendency for an expenditure increase to induce appreciation. The reason is, once again, that the increase in demand leads to a rise in income and thus to higher money demand and increased interest rates. Because the long-run real *and* nominal exchange rate are unchanged, higher interest rates are only compatible with equilibrium in the international capital market if there is the expectation of depreciating currency. That expectation will arise through an initial real and nominal appreciation. Thus in the diagram above the real exchange rate would appreciate in the short run to a point like $A'$. Over time, as the stock of assets is reduced through the current account deficit and demand falls, the real exchange rate depreciates until point $A''$ is reached. An immediate appreciation is again implied when the dynamics are governed by short-run price stickiness and rational expectations in asset markets.

Expansionary fiscal policy will only lead to an initial depreciation of the nominal and real exchange rate if, in addition to the expectation of an unchanged long-run *real* exchange rate, the expectation of a nominal depreciation is introduced. There is good reason for such an assumption if one considers a fiscal expansion as one that is accommodated by an expansion in nominal money so that the nominal interest rate is unchanged. And it is the only way to generate this result in the model. With an accommodating nominal money expansion, the expectation of a higher long-run level of prices with unchanged terms of trade leads to an immediate depreciation of the real exchange rate to a point like $A*$ in the diagram. At $A*$, assuming smooth adjustment, there is a current account deficit (the $\dot{w} = 0$ locus shifts leftward, as does the $y$ schedule) combined with an output expansion. From $A*$ the economy moves toward $A''$; wealth declines, and the real exchange rate appreciates to restore the initial terms of trade.

The final exercise to be considered is a sustained increase in the rate of money creation. The expectation of higher long-run inflation, and of the induced increase in velocity, implies a one-time rise in the cost of foreign currency. With rational expectations, the currency immediately depreciates before prices rise and the economy moves to a point like $A^*$ in the diagram. But because in the long run the real exchange rate and real wealth are unchanged, and because the real depreciation induces a current account surplus at $A^*$ (this time the schedules remain the same), a clockwise adjustment occurs until the economy returns to point $A$. Output is initially above full employment in the adjustment process as a consequence of the overdepreciation; assets are accumulated through the current account; and the real exchange rate appreciates. The current account surplus and the income expansion are, of course, only transitory, as is the real depreciation.

I have described a fairly eclectic general equilibrium model of goods markets and asset markets, expectations, and current account adjustment. The model is capable of accounting for some of the exchange rate experience in the United States, in particular the transitory deviations from PPP, permanent changes in the real exchange rate, and jumps of exchange rates in response to new information. This latter phenomenon is a key feature of the model and implies that, because of the differential speed of adjustment in goods markets and asset markets, even purely monetary disturbances have transitory real effects.

TESTING THE NEWS

In this section I offer some tests of the exchange rate model developed above. I showed there that unanticipated changes in aggregate demand or in net exports affect the equilibrium exchange rate. In particular, an accommodated increase in demand leads to depreciation and a current account deficit; an unanticipated increase in net exports leads to an appreciation. A monetary expansion induces depreciation, income growth, and a transitory current account surplus.

Perhaps the central implication of the rational expectations model is that it must be tested in "news form." With the assumption that asset markets are efficient, all available information is immediately embodied in asset prices and exchange rates. If one disregards for now the possibility of a risk premium, deviations of exchange rates from the path implied by interest differentials are thus entirely due to news.[12]

The extended model first distinguishes news of three kinds as important determinants of unanticipated changes in exchange rates: news about the current account, cyclical or demand factors, and interest rates. To test this model empirically, I use the definition of unanticipated depreciation as the difference between the actual depreciation and interest differentials, $\dot{e}' - (i - i^*)$. The theory suggests that an unanticipated surplus in the current account leads to appreciation, while an unanticipated increase in demand that is accommodated will lead to depreciation. Denoting news about the current account, cyclical movements, and interest rates as *CAE, CYC,* and *INN,* respectively, the equation becomes

(5)     $\dot{e}' - (i - i^*) = \alpha_0 - \alpha_1 CAE + \alpha_2 CYC - \alpha_3 CYC^* + \alpha_4 INN,$

where in the absence of a risk premium, $\alpha_0$ is expected to be zero.

As measures of the current account and cyclical news I use the official forecast errors of the Organisation for Economic Co-operation and Development, which publishes biannual six-month forecasts for current account balances and real growth of major industrial countries.[13] Combined with the subsequently realized current account balances and growth rates, these forecasts yield time series data for the news shown in the explanatory variables. Because these forecasts are prepared through multilateral intergovernmental consultation, they are broadly representative of informed opinion about growth and current account balances.

Consider next the unanticipated depreciation, $\dot{e}' - (i - i^*)$, for the nominal effective exchange rate of the dollar (defined in table 1). The

12. The idea of testing rational expectations models in news form is familiar from the work of Robert J. Barro in macroeconomics. In the context of exchange rate problems the idea is rapidly becoming accepted. See in particular Dornbusch, "Monetary Policy"; Peter Isard, "Expected and Unexpected Changes in Interest Rates," International Finance Discussion Paper 145 (Board of Governors of the Federal Reserve System, June 1979); Michael P. Dooley and Peter Isard, "The Portfolio-Balance Model of Exchange Rates," International Finance Discussion Paper 141 (Board of Governors of the Federal Reserve System, May 1979); extensive work by Michael Mussa, in particular his "Empirical Regularities in the Behavior of Exchange Rates and Theories of the Foreign Exchange Market," in Karl Brunner and Allan H. Meltzer, *Policies for Employment, Prices, and Exchange Rates,* Carnegie-Rochester Conference Series on Public Policy, vol. 11 (Amsterdam: North-Holland, 1979), pp. 9–57, as well as references given there.

13. See Organisation for Economic Co-operation and Development, *OECD Economic Outlook,* various issues.

monthly series is shown in figure 3, together with the series for anticipated depreciation given by $i - i^*$ (both expressed as annual percentage
rates). As the figure clearly illustrates, unanticipated changes constitute
nearly all the actual variation in exchange rates.

Regression equations explaining unanticipated depreciation of the dollar against a trade-weighted mixture of other currencies are shown in
table 3. Equation 3-1 explains the unanticipated depreciation of the dollar by the current account and cyclical errors. The cyclical errors for the
United States and five foreign countries are constrained to be of equal
and opposite sign in this equation. The equation accounts for much of
the unanticipated depreciaticn, and evidence of serial correlation does
not appear in the errors. The coefficients do have the expected signs. The
coefficient on the current account news is significant. An unanticipated
current account surplus in the United States of $1 billion is worth half a
percent of appreciation. The coefficient on the cyclical forecast error
indicates that unanticipated growth leads to depreciation. But it is not
significantly different from zero. Perhaps this reflects the fiscal expansion
phenomenon discussed above.

Equations 3-2 and 3-3 include unanticipated changes in interest rates.
Ideally the term structure of interest rates should be used to measure innovations; but here, because of the complexity of deriving such series,
residuals from an autoregression of the short-term interest differential
have been used. The equations show that unanticipated increases in short-
term interest differentials appear with a positive coefficient that is significant. The interest differential may reflect a causal role for unanticipated
changes in the term structure, inflation news, or cyclical effects as suggested by a comparison of equations 3-1 and 3-2 in the table.[14]

Table 4 presents similar equations for the dollar-mark and dollar-yen
exchange rates. Consider first the case of Japan. Equation 4-1 shows quite
strikingly the role of current account errors and cyclical errors. An unanticipated surplus in the Japanese current account leads to dollar depreciation or yen appreciation. A cyclical expansion in Japan induces a yen
depreciation. Both the coefficients of $CAE$ and $CYC$ are significantly dif-

14. Frenkel reports regressions of the level of the exchange rate on lagged forward rates, interest differentials, and interest innovations, the last appearing with a
positive coefficient. He attributes the positive coefficient to inflation news. See Jacob
A. Frcnkel, "Flexible Exchange Rates in the 1970's," Working Paper 450 (National
Bureau of Economic Research, 1980), pp. 34–37. In my equations the introduction
of inflation news yields a negative, insignificant coefficient.

**Figure 3. Anticipated and Unanticipated Depreciation of the Dollar, February 1973–January 1980**ᵃ

*Percentage points*

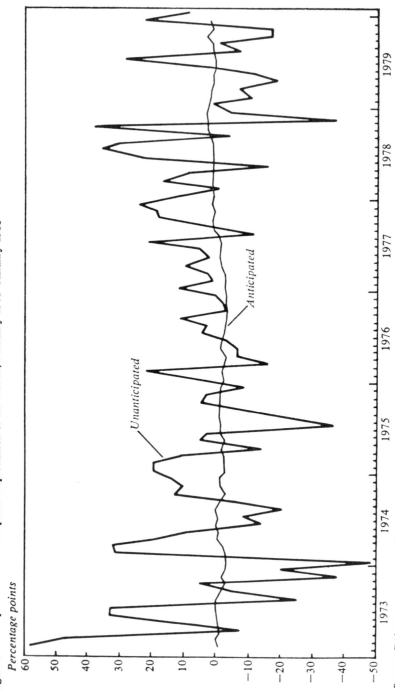

Sources: Exchange rate—same as table 1; and interest rates—same as table 2.

a. The data are monthly, expressed at annual percentage rates. The unanticipated depreciation of the dollar is the difference between actual depreciation of the dollar and anticipated depreciation. The exchange rate is the nominal effective exchange rate described in table 1, note a. Anticipated depreciation is measured by short-term interest differentials between the United States and a trade-weighted average of the interest rates of five foreign countries. The weights equal those used in calculating the exchange rate.

**Table 3. Equations Explaining Unanticipated Depreciation of the Nominal Effective Exchange Rate of the Dollar, Second Half of 1973 through Second Half of 1979**[a]

| Equation | Independent variable | | | | | Summary statistic | | | |
| | Constant | CAE | CYC | CYC* | INN | $R^2$ | Durbin-Watson | Standard error of estimate | Rho |
|---|---|---|---|---|---|---|---|---|---|
| 3-1 | 3.5 (1.88) | −0.49 (−2.62) | 1.86 (1.35) | −1.86 | ··· | 0.41 | 2.13 | 6.2 | ··· |
| 3-2 | 2.7 (1.69) | −0.31 (−1.82) | 0.47 (0.33) | −0.78 (0.57) | 13.33 (1.99) | 0.63 | 2.03 | 5.5 | −0.24 |
| 3-3 | 3.1 (2.47) | −0.27 (−2.19) | ··· | ··· | 13.79 (2.53) | 0.61 | 2.13 | 5.1 | −0.28 |

Sources: Forecast and actual current account balances and real output growth—Organisation for Economic Co-operation and Development, *OECD Economic Outlook*, various issues; exchange rate, unanticipated depreciation of the dollar, is described in figure 3, note a. *CYC* and *CYC\** are unanticipated growth in real output of, respectively, the United States and a trade-weighted average of the five foreign countries in table 1. Unanticipated growth is the difference between the OECD's six-month forecast and realized growth. The data are seasonally adjusted annual rates of growth. *CAE* is the forecast error for U.S. current account balances, using the OECD's forecasts. The data are measured in billions of dollars, seasonally adjusted. *INN* denotes the residuals from an autoregression of short-term interest differentials between the United States and a trade-weighted average of five foreign countries. Trade-weighted variables use the weights in table 1, note a. The numbers in parentheses are *t*-statistics.

**Table 4. Equations Explaining Unanticipated Depreciation of the Dollar-Yen and Dollar-Mark Exchange Rates, Second Half of 1973 through Second Half of 1979[a]**

| Country and equation | Independent variable | | | | | | Summary statistic | | | |
|---|---|---|---|---|---|---|---|---|---|---|
| | Constant | CAE | CYC | CYC* | INN | RES | $R^2$ | Durbin-Watson | Standard error of estimate | Rho |
| **Japan** | | | | | | | | | | |
| 4-1 | -3.81 (-0.45) | 1.35 (4.21) | -2.63 (-4.03) | 1.27 (1.13) | ... | ... | 0.73 | 2.11 | 8.7 | 0.75 |
| 4-2 | 0.60 (0.27) | 1.40 (5.70) | -1.71 (-2.12) | 1.71 | 13.0 (2.03) | ... | 0.82 | 2.09 | 8.0 | ... |
| 4-3 | -0.29 (-0.08) | 1.39 (4.43) | -2.45 (-2.81) | 2.45 | ... | -2.14 (-0.73) | 0.71 | 1.85 | 9.1 | 0.32 |
| **Germany** | | | | | | | | | | |
| 4-4 | 2.29 (1.22) | 1.38 (1.93) | -0.53 (-0.76) | ... | 26.1 (3.51) | ... | 0.62 | 2.22 | 7.6 | -0.40 |
| 4-5 | 1.37 (0.42) | 1.01 (1.50) | ... | ... | 13.0 (1.36) | ... | 0.32 | 2.09 | 10.2 | ... |
| 4-6 | 1.80 (0.51) | 1.10 (1.48) | ... | ... | ... | -0.52 (-0.13) | 0.19 | 2.01 | 11.1 | ... |

Sources: Exchange rates and interest rates—same as table 2; and forecast and actual current account balances and real output growth—same as table 3.

a. CAE and CYC represent forecast errors for current account balances and real output growth in Japan or Germany. In the equations for both countries CYC* represents forecast errors for real output growth in the OECD countries. INN denotes the residuals from an autoregression of the U.S.-Japanese or U.S.-German short-term interest differential. RES denotes the residuals from a regression of U.S.-Japanese or U.S.-German short-term interest differentials on differentials between the United States and the respective country on long-term and short-term interest rates, unemployment rates, and inflation rates, with all explanatory variables lagged one period. CYC and CYC* are seasonally adjusted annual rates of growth; and CAE is seasonally adjusted, in billions of dollars. In equations 4-2 and 4-3 the coefficients of cyclical errors are constrained to be equal and opposite in sign. The numbers in parentheses are t-statistics.

ferent from zero. The equation explains a large portion of the unanticipated depreciation. Unlike equation 3-3 for the United States in table 3,
the constant terms are not significantly different from zero.

Equations 4-2 and 4-3 include interest rate news. The innovations in
equation 4-2 are from an autoregression of the interest differential. In
equation 4-3 the interest variable is residuals from an interest differential
equation. The roles of the two interest rate innovations are quite different.
The former have a significant positive coefficient reducing the magnitude
and significance of the cyclical effects; the latter, which are more nearly
orthogonal to cyclical effects, appear with a negative and insignificant coefficient. The same pattern is observed in the equations for Germany.

Unlike the dollar-yen exchange rate, unanticipated movements in the
dollar-mark rate are not dominated by news about cyclical or current account events. Unanticipated improvements in the current account of
Germany lead to a dollar depreciation, but the coefficient on the current
account and cyclical innovations variables are not significantly different
from zero. Innovations from an autoregression of interest differentials do
play a part in explaining exchange rate movements in equation 4-4. But in
4-6 the residuals from a reaction function for the interest differential,
which is discussed below, turn out to be insignificant. I argue there that
portfolio shifts may well be the explanation for these results.

The empirical analysis confirms that unanticipated real and financial
disturbances bring about unexpected movements in the exchange rate. To
that extent, the preceding theory is confirmed. Whether the size of exchange rate movements stands in reasonable relation to the disturbance
remains an open question. Clearly the answer depends not only on the
structural parameters, including trade elasticities, but also on the expected
persistence of the disturbance. The more persistent the disturbance, other
things being equal, the larger the required change in the real exchange
rate.

## PORTFOLIO DIVERSIFICATION AND THE
## DEUTSCHE MARK SHORTAGE

The analysis so far has largely excluded portfolio balance and its implications for exchange rates. The models considered share the assumption
of perfect substitutability of home and foreign securities on a depreciation-
adjusted basis, thus leaving no room for shifts in wealth or relative asset
supplies to affect the balance in asset markets. I now depart from this as-

sumption to see what insights a broader treatment of portfolio choice will yield.

A starting point is the hypothesis that money demand depends not only on income, the conventional transactions variable, but also on wealth. Shifts in wealth induced by current account imbalances create monetary imbalances leading to adjustments in long-run price level expectations and thus to exchange rate movements. This effect does not presuppose imperfect asset substitutability, although it is entirely compatible with it. With perfect mobility of capital, this specification of money demand implies that the real money demand of a country with a surplus rises while it falls abroad. The relative price level of the country with a surplus declines and, therefore, exchange rates for given terms of trade tend to appreciate.[15]

The results, of course, follow from a strong assumption about distribution effects. Monies are treated as nontraded assets, the demand for which is affected by an international redistribution of wealth. In the absence of an empirically significant wealth effect on money demand, this theory probably does not go very far in explaining exchange rates.

An alternative and more persuasive role for portfolio effects arises in the context of imperfect asset substitutability. With uncertain real returns, portfolio diversification makes assets imperfect substitutes and gives rise to determinate demands for the respective securities and to real yield differentials or a risk premium.[16]

15. This variant of the current account theory of exchange rates is emphasized in Rudiger Dornbusch, "Capital Mobility, Flexible Exchange Rates and Macroeconomic Equilibrium," in E. Claassen and P. Salin, eds., *Recent Issues in International Monetary Economics,* Studies in Monetary Economics, vol. 2 (Amsterdam: North-Holland, 1976), pp. 261–78; and Pentti J. K. Kouri, "The Exchange Rate and the Balance of Payments in the Short Run and in the Long Run: A Monetary Approach," *Scandinavian Journal of Economics,* vol. 78, no. 2 (1976), pp. 280–304.

16. This line of research has been particularly pursued in W. H. Branson, "Asset Markets and Relative Prices in Exchange Rate Determination," *Sozialwissenschaftliche Annalen,* vol. 1 (1977), pp. 69–89; and William H. Branson, Hannu Halttunen, and Paul Masson, "Exchange Rates in the Short Run: The Dollar-Deutschemark Rate," *European Economic Review,* vol. 10 (December 1977), pp. 303–24. See also Michael G. Porter, "Exchange Rates, Current Accounts and Economic Activity—A Survey of Some Theoretical and Empirical Issues" (Board of Governors of the Federal Reserve System, June 1979); Dooley and Isard, "Portfolio-Balance Model"; Maurice Obstfeld, "Capital Mobility and Monetary Policy under Fixed and Flexible Exchange Rates" (Ph.D. dissertation, Massachusetts Institute of Technology, 1979); Pentti J. K. Kouri and Jorge Braga de Macedo, "Exchange Rates and the Inter-

The portfolio model provides an explanation of the unanticipated mark appreciation that is only poorly accounted for by the current account and cyclical innovations. I argue that the system of flexible exchange rates and the macroeconomic policies and disturbances have created an incentive for portfolio diversification, that the mark would occupy a large share in an efficiently diversified portfolio, and that the resulting portfolio shifts or capital flows account for some of the unanticipated appreciation.

Table 5 shows the realized means and variances of the *real* returns on assets denominated in different currencies. The real yield in each instance is the nominal short-term interest rate plus the depreciation of the dollar relative to the particular currency, thus creating dollar returns, less the rate of inflation of the dollar price index of manufactures in world trade. The real return data thus are comparable and appropriate for an investor that does not have a particular local habitat.

Concentrating on the 1976–79 period, note that both the mark and the dollar are relatively stable (low-variance) assets and that their returns are negatively correlated. The dollar has a negative mean return, while the mark has a positive one.

In principle, an efficiently diversified portfolio is a wide-ranging one, including bonds, amusement parks, old-age homes, and so on. In practice, investors develop a narrow portfolio, highly concentrated in home securities with a small range of international claims. Suppose, to make a point, that only dollars and marks are part of the portfolio of international assets. What would be their respective shares? The relevant model of utility-maximizing portfolio diversification shows that the share of mark assets, using the distribution of returns of table 5, is 56 percent. This corresponds to a 50 percent share of bonds denominated in marks in the minimum-variance portfolio plus a 6 percent share in a speculative mark position.[17] The speculative position in marks, motivated by the differen-

---

national Adjustment Process," *BPEA*, 1:1978, pp. 111–50; and Rudiger Dornbusch, "A Portfolio Balance Model of the Open Economy," *Journal of Monetary Economics*, vol. 1 (January 1975), pp. 3–20.

17. Let $w$ be the initial level of real wealth; $r$ and $r^*$, the random real returns on home and foreign securities; and $x$, the portfolio share of foreign securities. End-of-period wealth then is random and equal to $\bar{w} = w(1 + r) + xw(r^* - r)$. Utility is a function of the mean and variance of end-of-period wealth:

$$U = U(\bar{w}, s_{\bar{w}}^2).$$

The mean and variance of wealth are defined as

Table 5. Means, Variances, and Covariances of Real Returns in Four Major Industrial Countries, 1973:3–1979:2 and Subperiod.
Percentage points

| Statistic | 1973:3–1979:2 | | | | 1976:1–1979:2 | | | |
|---|---|---|---|---|---|---|---|---|
| | United States | Germany | Japan | United Kingdom | United States | Germany | Japan | United Kingdom |
| Mean | -2.48 | 0.55 | 0.94 | -0.98 | -3.54 | 3.74 | 4.31 | 2.49 |
| Variance or covariance | | | | | | | | |
| United States | 139.7 | ... | ... | ... | 49.7 | ... | ... | ... |
| Germany | -26.9 | 98.3 | ... | ... | -9.9 | 48.8 | ... | ... |
| Japan | 34.6 | 36.5 | 320.3 | ... | -55.8 | -32.6 | 350.1 | ... |
| United Kingdom | 2.7 | 22.1 | -16.9 | 214.0 | -1.7 | -11.4 | -97.2 | 336.3 |

Sources: Interest rates and exchange rates—same as table 2; and price index for manufactures in world trade—United Nations, *Monthly Bulletin of Statistics*, various issues.
a. Real returns are calculated as the nominal short-term interest rate, plus the depreciation of the dollar relative to the country's currency, minus the rate of inflation of the price index of manufactures in world trade. The data are quarterly averages.

tial in mean real yields, is quite small because of the large variance of the nominal rate of depreciation that makes speculation risky. The share in the minimum variance portfolio is substantial, though, because the mark is an attractive asset—it has a relatively low variance of the real yield and a negative covariance with the dollar. The exercise, while merely an illustration, does suggest that the mark has characteristics that should make it play a large role in portfolios, and indeed, an even greater role as an international asset than was the case in the 1960s or early 1970s.

The argument may overstate the case in a number of ways. First, the realized returns may not equal the return distribution that investors anticipate. This is even more true if much of the differential in mean real returns reflects unanticipated mark appreciation.[18] Second, other currencies may enter the portfolio, some with features more attractive than those of the mark. Third, international differences in consumption patterns may bias the portfolio shares away from those implied by the return distribution of table 5. Each of these arguments has some force, although none

---

$$\bar{w} = w(1 + \bar{r}) + xw(\bar{r}^* - \bar{r}); \qquad s_{\bar{w}}^2 = w^2[(1 - x)^2 s_r^2 + x^2 s_{r^*}^2 + 2x(1 - x)s_{rr^*}],$$

where a bar denotes a mean. Maximizing utility with respect to $x$ yields the optimal portfolio share,

$$x = \frac{(\bar{r}^* - \bar{r}) + \theta(s_r^2 - s_{rr^*})}{\theta s_n^2},$$

where $\theta \equiv -U_2 w/U_1$ is the coefficient of relative risk aversion, $s_{rr^*}$ is the covariance of real returns, and

$$s_n^2 \equiv s_r^2 + s_{r^*}^2 - 2s_{rr^*}$$

is the variance of the nominal rate of depreciation. The first term, $(\bar{r} - \bar{r}^*)/\theta s_n^2$, corresponds to the speculative portfolio share in marks and depends on the mean real yield differential and the variance of the nominal rate of depreciation. The second term represents the hedging, or minimum-variance, portfolio that depends only on variances. For further discussion, see Rudiger Dornbusch, "Exchange Risk and the Macroeconomics of Exchange Rate Determination" (Massachusetts Institute of Technology, April 1980), and the references cited there.

18. Table 5 cannot strictly be used to establish the case for diversification since the data reflect both the "fundamentals" *and* the effect of the alleged portfolio diversification. To the extent that the incidence of the latter was unanticipated, the reported means and variance are not those the asset holders had in mind and accordingly cannot be used to establish the case for portfolio diversification. In a short time-series for the flexible exchange rate system there is no apparent way of extracting the fundamentals, nor is it possible to tell how serious the discrepancy has been between previous beliefs and ex post returns.

of them necessarily suggests a lower mark share in an international portfolio.[19]

The main point is simply that the transition to flexible rates has quite decisively changed the structure of real returns confronting international investors—central banks, firms, or households. With the new return structure, and by virtue of size, the mark should occupy a large share of portfolios, much larger than would have been expected in 1970–73, before the period of flexible exchange rates. Investors can be expected to make a gradual transition to the new diversification pattern. But, as the poorly understood process of substitution from $M_1$ to negotiable-orders-of-withdrawal (NOW) accounts and money-market funds in the United States suggests, little is known about the dynamics of portfolio adjustment.

As the substitution process takes place, the mark will tend to appreciate unless there is an offsetting increase in the relative supply of assets denominated in marks. Such an increase could be created through deficit finance, arising from sterilized exchange rate intervention, or take the form of Carter bonds (bonds issued by the U.S. government denominated in marks). In fact, as I show below, there has been a large increase in the relative supply of these assets because of larger German deficits. Sterilized intervention has made up a further part of the increased demand. The remainder has been met by appreciation of the mark, revaluing the share of marks already existing in international portfolios.

A first implication of the portfolio model then is to help identify a shortage of marks. The adjustment process to the new role of the mark as an international asset has brought about a curious reversal of the old intermediation view of the U.S. balance of payments. Germany has been showing a sustained short-term capital account surplus with a direct investment and portfolio investment deficit. Germany displays the pattern typical of the United States when the dollar took an increasing role in international portfolios after the restoration of currency convertibility in the late 1950s.

19. Kouri and Macedo found an optimal mark share of 37 percent in a multiple-currency portfolio with local habitats. See their "Exchange Rates and the International Adjustment Process," p. 129. See, too, the analysis in William Fellner, "The Bearing of Risk Aversion on Movements of Spot and Forward Exchange Relative to the Dollar," in John S. Chipman and Charles P. Kindleberger, eds., *Flexible Exchange Rates and the Balance of Payments: Essays in Memory of Egon Sohmen* (Amsterdam: North-Holland, forthcoming).

## RELATIVE ASSET SUPPLIES, WEALTH, AND EXCHANGE RATES

I now explore the portfolio model further to see whether there are implications that reinforce or put in question the conjecture discussed above: that for given asset supplies and wealth the structure of real returns implies a shift in portfolios toward assets denominated in marks, thus explaining the persistent appreciation of the mark.

The portfolio-diversification model implies a relationship between the nominal interest differential, the expected rate of depreciation, and the risk premium, $R$:

$$(6) \qquad \dot{e} = i - i^* + R\left(\frac{EV^*}{V + EV^*}, \frac{W^*}{W + W^*}\right),$$

where

$E$ = level of domestic currency price of foreign exchange

$W$ = level of wealth

$V$ = supply of nominal debt.

The risk premium in equation 6 is an increasing function of the relative supply of assets denominated in foreign currency, $EV^*/(V + EV^*)$, and a decreasing function of foreign relative wealth.[20] What matters for the risk premium are the relative supplies of *outside* bonds (net assets of the private sector) denominated in the two currencies, independently of the issuing source.[21] Risk is here a question of the variability of real returns due to uncertain inflation and exchange rate depreciation, not a

20. The risk premium can be written as

$$R = \theta\left[s_n^2\left(\frac{EV^*}{V + EV^*} - \beta\right) - s_q^2(\phi - \phi^*)\frac{W^*}{W + W^*}\right],$$

where $V$ is domestic currency outside bonds, and $W$ is domestic nominal wealth; $s_n^2$ and $s_q^2$ are the variances of the rates of nominal and real depreciation; $\theta$ is the coefficient of relative risk aversion; $\phi - \phi^* \geq 0$ equals the difference between domestic and foreign expenditure shares of domestic goods; and $\beta$ is the minimum-variance portfolio share defined in note 17. For a derivation, see Dornbusch, "Exchange Risk."

21. Frankel and Kouri emphasized that the risk premium involves outside assets independent of the issuer. See Jeffrey A. Frankel, "The Diversifiability of Exchange Risk," *Journal of International Economics*, vol. 9 (August 1979), pp. 379–93; and Pentti J. K. Kouri, "The Determinants of the Forward Premium," Seminar Paper 62 (University of Stockholm, Institute for International Economic Studies, August 1976).

question of default. Note also that the relative wealth term will give rise to a risk premium only to the extent that there are differences in consumption patterns *and* that there is variability in the *real* exchange rate.

Suppose now that interest rates and anticipated rates of depreciation are given, perhaps determined by, the monetary sector of the more general model. The risk-premium model has implications for the relationships among wealth, asset supplies, and the exchange rate. In particular, the model implies that an increase in foreign relative wealth, say arising through a cumulative foreign current account surplus, will bring a relative increase in the demand for securities denominated in foreign currency. The resulting disequilibrium in the asset market is resolved by an appreciation of the foreign currency that revalues existing stocks of securities denominated in foreign currency. This must be an unanticipated wealth redistribution; otherwise, speculators would have anticipated the jump in the exchange rate.

Unanticipated changes in the relative supplies of securities likewise affect the exchange rate. For example, an unanticipated fiscal deficit that expands the supply of bonds denominated in foreign currency leads to a depreciation of the foreign currency, which restores portfolio balance at unchanged yields. (In general, exchange rates and asset yields are jointly determined.)

The risk-premium model has served as the basis for extensive research attempting to explain exchange rate movements by changes in relative wealth (using changes in net foreign assets as a proxy) and in relative asset supplies.[22]

The model has had mixed results in empirical tests, largely because of the difficulty in developing measures of relative nominal outside assets and relative nominal wealth. Part of the problem may also have been the use of actual versus unanticipated variables. Given these difficulties, the existing results must be considered very tentative. Even so, the risk-premium model is of interest because it offers, through the wealth channel, a role for the current account to affect exchange rates. At the same time, this

22. Early work, in particular Branson, Halttunen, and Masson, "Exchange Rates in the Short Run," gave particular emphasis to the current account, taking wealth to be represented by the cumulative current account. A more balanced treatment that recognizes the central role of asset supplies, as opposed to the distribution effects induced by current account imbalances, is found in Obstfeld, "Capital Mobility," and John P. Martin and Paul R. Masson, "Exchange Rates and Portfolio Balance," Working Paper 377 (National Bureau of Economic Research, August 1979).

**Table 6. Current Account Balances and Net Borrowing in Germany, and Ratios of German to U.S. Debt, 1973-79**

Billions of deutsche marks, except as noted

| Item | 1973 | 1974 | 1975 | 1976 | 1977 | 1978 | 1979 |
|---|---|---|---|---|---|---|---|
| Current account balance | 12.3 | 25.5 | 8.5 | 8.6 | 9.8 | 17.6 | −9.0 |
| Net government borrowing | 6.1 | 10.8 | 36.4 | 20.0 | 21.7 | 27.4 | 25.1 |
| Ratio of German to U.S. government debt (percent) | | | | | | | |
| Measured in dollars | 6.7 | 8.5 | 9.5 | 10.7 | 12.7 | 15.7 | 17.8 |
| Measured in respective currencies | 18.0 | 20.5 | 24.8 | 25.4 | 26.6 | 28.7 | 30.8 |

Sources: Government debt and borrowing—International Monetary Fund, *International Financial Statistics*, vol. 33 (May 1980), series ae, series 84 and 88, and series 88, pp. 164, 166, and 404, respectively; and current account balances—Deutsche Bundesbank, *Monthly Report of the Deutsche Bundesbank*, vol. 32 (March 1980), p. 70.

model introduces a potential link between deficit finance and exchange rates through the relative supply of assets. It thus supplements the extended Mundell-Fleming model and offers alternative channels through which current account and fiscal innovations can affect the exchange rate. Indeed, the equations reported in tables 3 and 4 may well reflect in part the effects of the risk-premium model.

MARK APPRECIATION

The risk-premium model may help explain the mark appreciation of recent years. In table 6, I report the German current account balance, net public sector borrowing, and the ratio of German to U.S. debt (valued both in dollars and in the respective currencies). The first point to note is that since 1975 the current account has been entirely dominated by the fiscal deficit. The demand for mark assets created by the redistribution of wealth toward Germany through the current account must have been met quite amply by the deficit finance. The German debt has increased much more rapidly than that of the United States. Thus if a risk-premium view were taken, one would expect the mark to show a cumulative depreciation, not an appreciation.

The risk-premium model suggests that a demand shift toward assets denominated in marks has dominated the downward pressure on the exchange rate arising from the combination of changes in relative wealth and the relative supplies of mark assets. Given the attempt to attain optimal diversification, the mark was appreciating because of an insufficient creation of mark assets.[23]

The risk-premium model has one further implication that has relevance for the equations in tables 3 and 4. The existence of a risk premium implies that not all the difference between interest differentials and actual depreciation is unanticipated; part corresponds to the risk premium and only the residual represents news. Thus equation 4 becomes

(7)                    $\dot{e}' - (i - i^*) = (\dot{e}' - \dot{e}) + R,$

$$= news + risk\ premium.$$

The risk premium accordingly can account for a significant constant or for serial correlation in the equations above.[24]

## The Flexible Exchange Rate System

I now examine some key features of the system of flexible exchange rates to form a judgment about its shortcomings and the possibilities for reform. Has the system been critically defective? In this section I investigate some firmly established working characteristics of the system, including intervention, interest rate policies, current account adjustment, and current account financing. The issues are whether intervention policies have been designed to frustrate real exchange rate adjustment; whether interest rate policies were significantly restricted by actual or potential exchange rate developments; and finally, whether current account imbalances have been sustained and officially funded rather than adjusted and financed through capital flows.[25]

---

23. The data in table 6 understate the increase in these assets because they omit items such as Carter bonds or debt created through sterilized intervention.

24. Cumby and Obstfeld do find evidence of a risk premium in weekly data for all major currencies. See Robert E. Cumby and Maurice Obstfeld, "Exchange-Rate Expectations and Nominal Interest Differentials: A Test of the Fisher Hypothesis," Discussion Paper 34 (Columbia University, Department of Economics, July 1979).

25. For an extensive discussion see the papers by Jacques R. Artus and John H. Young, "Fixed and Flexible Exchange Rates: A Renewal of the Debate," *IMF Staff Papers*, vol. 26 (December 1979), pp. 654–98; Morris Goldstein, *Have Flexible Exchange Rates Handicapped Macroeconomic Policy?* Special Papers in International Economics, 14 (Princeton University, International Finance Section, June 1980); and Steven W. Kohlhagen, "The Experience with Floating: The 1973–1979 Dollar" (University of California at Berkeley, n.d.).

I show that the capital mobility problem is summarized by the observation that when the current account gets bad the capital account gets worse. The reason is that interest rate policies are oriented toward internal balance, which aggravates the exchange rate consequences of cyclically unsynchronized movements in economic activity in the world economy.

OFFICIAL INTERVENTION

The reported changes in official reserve holdings have increased sharply during the 1970s. Have intervention policies had systematically stabilizing characteristics?

Figure 4 shows an adjusted series for changes in U.S. net liabilities to foreign official reserve agencies. The figure indicates sizable swings in intervention, which were larger than the swings in the U.S. current account. I present equations on the determinants of intervention in table 7. Given the size of reserve holdings and the level of nominal interest rates, much of the reported increase reflects the accrual of interest earnings rather than active market intervention. I thus use as a dependent variable an adjusted series that subtracts from changes in reserves an amount equal to the U.S. Treasury bill rate times the lagged stock of reserves. This series is measured as a fraction of lagged reserves. Equations 7-1 and 7-2 use unanticipated depreciation rates to explain U.S. net liabilities to foreign official holders. With a policy of "leaning against the wind," foreign central banks would acquire dollars through intervention whenever the dollar showed unanticipated depreciation. The equations strongly support that view, although only a small fraction of the variance is explained.

Equation 7-2 suggests that unanticipated depreciation of 1.0 percentage point (at an annual rate) leads to a cumulative intervention of 0.4 percent of foreign net claims on the United States, which at current levels of foreign net reserve holdings is about $600 million. The constant term of 1.0 suggests that the absolute size of intervention is growing along with nominal reserve holdings.

Equation 7-3 considers German intervention policy. There is more evidence of leaning against the wind. Unanticipated depreciation of 1 percentage point, at an annual rate, leads to an intervention at 1979 reserve levels of about $140 million. Interestingly, macroeconomic conditions affect the level of German intervention. A high rate of unemploy-

**Figure 4. Current Account Balances and Official Intervention in the United States, 1972:1–1979:4ᵃ**

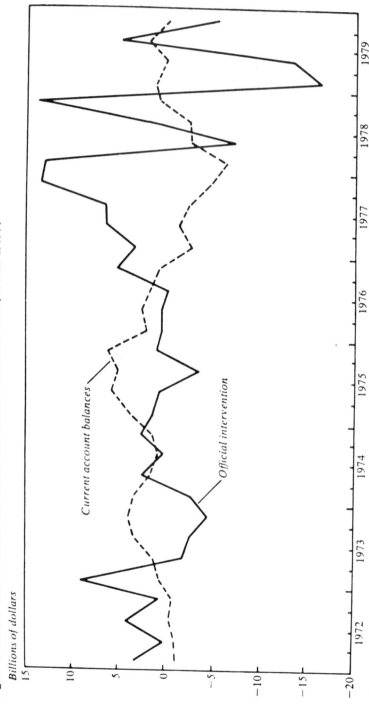

Sources: Current account balances and changes in official reserve holdings—*Survey of Current Business*, vol. 59 (June 1979), pp. 38–41, lines 38, 57, 61, and 79; and U.S. Treasury bill rate—Board of Governors of the Federal Reserve System.

a. The official intervention series is the change in official reserve holdings, adjusted to exclude the accrual of interest earnings. Interest earnings were estimated as the market yield on U.S. three-month Treasury bills times the lagged stock of official reserve holdings. The stock of reserve holdings was cumulated from its level of $63.2 billion in 1972:4, using the series on changes in official reserve holdings.

**Table 7. Estimates of the Determinants of Intervention in Germany, Japan, and Major Industrial Countries, 1973:3–1979:4[a]**

| Country and equation | Independent variable | | | | | Summary statistic | | | |
|---|---|---|---|---|---|---|---|---|---|
| | Constant | $\dot{e}_u$ | $\dot{e}_d(-1)$ | $U$ | $\dot{p}$ | $R^2$ | Durbin-Watson | Standard error of estimate | Rho |
| **Major industrial countries** | | | | | | | | | |
| 7-1 | 1.01 (103.8) | 0.003 (3.25) | ... | ... | ... | 0.31 | 2.01 | 0.05 | ... |
| 7-2 | 1.00 (104.8) | 0.003 (3.22) | 0.001 (1.68) | ... | ... | 0.38 | 1.81 | 0.05 | ... |
| **Germany** | | | | | | | | | |
| 7-3 | 0.96 (22.4) | 0.003 (3.99) | ... | 0.018 (1.88) | −0.007 (−1.89) | 0.54 | 2.00 | 0.06 | −0.15 |
| **Japan** | | | | | | | | | |
| 7-4 | 0.97 (50.5) | 0.004 (4.41) | ... | ... | ... | 0.44 | 1.94 | 0.10 | ... |

Sources: U.S. official intervention—same as figure 4; German and Japanese official intervention—International Monetary Fund, *International Financial Statistics*, various issues, series 1d.d; U.S. Treasury bill rate—same as figure 4; unanticipated depreciation of the dollar—same as figure 3 for the United States, and table 4 for Germany and Japan; prices—same as figure 2; and German unemployment—Organisation for Economic Co-operation and Development, *Main Economic Indicators*, various issues.

a. The dependent variable is the change in reserves net of interest earnings as a fraction of the lagged stock of reserves. Interest earnings are estimated as the market yield on U.S. three-month Treasury bills times the lagged stock of reserves. In equations 7-1 and 7-2, $\dot{e}_u$ is the unanticipated depreciation of the nominal effective exchange rate of the dollar described in table 1, note a. In equations 7-3 and 7-4 it is the unanticipated depreciation of the dollar-mark and the dollar-yen exchange rates, respectively. The symbols $U$ and $\dot{p}$ denote unemployment rates and inflation rates. The major industrial countries are Canada, France, Germany, Japan, and the United Kingdom. The numbers in parentheses are *t*-statistics.

ment increases the rate of intervention, while high inflation reduces intervention. With more unemployment, authorities use intervention to slow down real dollar depreciation to achieve a "beggar-my-neighbor" effect. With faster inflation, unanticipated dollar depreciation is opposed less strongly in order to achieve a reduction in inflationary pressure or to avoid imported inflation. The coefficients on the cyclical variables suggest a policy that goes significantly beyond leaning against the wind.[26] I found no evidence of real exchange rate targets.

Equations of the form reported in table 7, which use unanticipated depreciation to explain reserves adjusted for interest earnings, are more successful than actual reserve changes and actual depreciation. This can be interpreted to mean that nominal interest payments roughly maintain the stock of real reserves in the face of dollar depreciation. Unanticipated depreciation as the explanatory variable is compatible with a PPP evolution of nominal exchange rates and with an adjustment of real exchange rates that is dampened, but not offset, by intervention.

There also is strong evidence of leaning against the wind in the equations for Japan. Unanticipated dollar depreciation again appears as the relevant determinant. The size of the reaction coefficient is similar to those reported for Germany and for the rest of the world. For Japan, however, there is no evidence of cyclical influences on intervention policy.

The intervention equations support the view that monetary authorities largely aimed their operations at smoothing unanticipated movements in the exchange rate. For Germany, the presence of cyclical variables also suggests an element of beggar-my-neighbor policy in exchange intervention.

26. On intervention policy and specifically "leaning against the wind" see Paul Wonnacott, "Exchange Stabilization in Canada, 1950–4: A Comment," *Canadian Journal of Economics and Political Science,* vol. 24 (May 1958), pp. 262–65; Paula A. Tosini, *Leaning against the Wind: A Standard for Managed Floating,* Princeton Essays in International Finance, 126 (Princeton University, International Finance Section, December 1977); Jacques R. Artus, "Exchange Rate Stability and Managed Floating: The Experience of the Federal Republic of Germany," *IMF Staff Papers,* vol. 23 (July 1976), pp. 312–33; Peter J. Quirk, "Exchange Rate Policy in Japan: Leaning Against the Wind," *IMF Staff Papers,* vol. 24 (November 1977), pp. 642–64; David John Longworth, "Floating Exchange Rates: The Canadian Experience" (Ph.D. dissertation, Massachusetts Institute of Technology, 1979); and Stanley W. Black, "Central Bank Intervention and the Stability of Exchange Rates," Seminar Paper 136 (University of Stockholm, Institute for International Economic Studies, February 1980).

## INTEREST RATE POLICIES

The sensitivity of exchange rates to monetary policy interferes with the ability of monetary policy to achieve a noninflationary real expansion. Lowering interest rates leads to exchange rate depreciation and faster inflation through rising import prices. Exchange rate sensitivity thus steepens the Phillips curve when monetary policy is used to affect real output. It is not possible to determine whether the worsened trade-off has significantly reduced the use of monetary policy as an instrument. What can be investigated is whether interest rates have shown the cyclical pattern associated with domestic stabilization, declining during a recession and increasing with inflation. One can also ask whether exchange rate depreciation exerted a significant effect on interest rate policy.

Table 8 reports regression equations for the German-U.S. and Japanese-U.S. differential in short-term interest rates. The differential is used on the assumption that international cyclical movements have not been closely synchronized. The German-U.S. differential in nominal interest rates is explained by the current inflation differential, unemployment in the respective countries, and the lagged nominal interest rate differential. Higher inflation differentials are reflected in a higher nominal interest differential. An increase of 1 percentage point in the German unemployment rate leads to a decline of about 2 percentage points in the nominal interest differential. It cannot be established that the flexible rate system did not weaken the use of countercyclical monetary policy. But the evidence is that relative interest rates continued to have a clearly cyclical pattern.

In the German-U.S. case, I found no evidence for either monetary growth targets, intervention, or exchange depreciation as a significant influence on interest differentials.[27]

Equations 8-2 to 8-4, explaining the Japanese-U.S. interest rate differential, provide more evidence of a cyclically stabilizing pattern of nominal interest rates. Higher inflation differentials lead to higher nominal yield differentials. Higher unemployment in Japan reduces the relative Japanese interest rate, while higher unemployment in the United States raises it.

27. For further evidence see Jean Tirole, "Exchange Rate Expectations and Monetary Policy: A Structural Approach for France, Germany, U.K." (Massachusetts Institute of Technology, n.d.).

**Table 8. Estimates of the Determinants of the German-U.S. and the Japanese-U.S. Interest Rate Differentials, 1973:2–1979:4[a]**

| Country and equation | Independent variable | | | | | | Summary statistic | | | Rho[b] | |
|---|---|---|---|---|---|---|---|---|---|---|---|
| | Constant | $\dot{p}^* - \dot{p}$ | $U^*$ | $U$ | $(i^* - i)_{-1}$ | $\dot{e}_u$ | $R^2$ | Durbin-Watson | Standard error of estimate | (1) | (2) |
| **Germany-U.S.** | | | | | | | | | | | |
| 8-1 | 1.86 (1.4) | 0.20 (2.29) | -0.97 (-2.24) | 0.45 (1.2) | 0.57 (3.91) | ... | 0.82 | 1.89 | 1.3 | ... | ... |
| **Japan-U.S.** | | | | | | | | | | | |
| 8-2 | -0.58 (-0.2) | 0.14 (3.97) | -5.15 (-3.84) | 2.45 (6.4) | ... | ... | 0.73 | 1.71 | 1.1 | 0.57 | ... |
| 8-3 | 1.72 (1.0) | 0.12 (3.42) | -6.42 (-6.71) | 2.50 (10.4) | ... | 0.03 (3.23) | 0.90 | 1.80 | 0.9 | 0.27 | ... |
| 8-4 | 1.84 (1.5) | 0.09 (2.54) | -3.51 (-3.39) | 1.14 (2.9) | 0.53 (4.01) | ... | 0.96 | 2.03 | 1.0 | 0.07 | 0.53 |

Sources: Short-term interest rates—same as table 2; prices—same as figure 2; unanticipated depreciation—same as table 4; and unemployment rates—same as table 7.
a. The dependent variable is $i^* - i$, the short-term interest rate differential between the two countries; $\dot{p}^* - \dot{p}$ is the inflation differential; $U$ is the unemployment rate; and $\dot{e}_u$ is unanticipated depreciation of the dollar. An asterisk denotes a variable for a foreign country. The numbers in parentheses are $t$-statistics.
b. First- and second-order autocorrelation corrections, respectively.

Unlike the German-U.S. case, the equations for Japan show high serial correlation of errors and are reported with rho corrections. Unanticipated depreciation is introduced in equation 8-3 and shows a significant coefficient but with the wrong sign—higher dollar depreciation leads to an increased spread in favor of Japan. The variable may represent joint errors in the interest and exchange rate equations; or it may merely pick up lagged adjustment effects, as equation 8-4 suggests.

From the interest rate evidence it seems apparent that, whatever limitations on monetary policy may exist, interest spreads internationally have had the cyclical pattern called for by stabilization objectives. To that extent, at least, there is no clear demonstration that the flexible exchange rate system has limited the use of instruments. Furthermore, there is no evidence that interest rates have been systematically affected by intervention or exchange rate targets.

### CURRENT ACCOUNT ADJUSTMENT AND CAPITAL FLOWS

The next question is whether the flexible exchange rate period has been one of persistent and large current account imbalances with exchange rate movements exerting relatively little impact to restore balance. Table 9 shows means, standard deviations, and serial correlation of current accounts for four major industrial countries. The 1960–73 period of fixed exchange rates is compared with that of flexible exchange rates, 1973–79. No substantial change in current account behavior is apparent. Imbalances did not become more persistent, and, in particular, the United States did not have a persistent deficit.

The surprise of the last few years, if anything, is the fact that current account imbalances are not at all the "sticky mass" that Keynes thought they were. Instead, the large effect of variations on current accounts and the responsiveness of trade flows and direct investment to real exchange rates lead to a view of great flexibility in all important dimensions of the balance of payments.

How have current account imbalances been financed? In particular, to what extent have the large swings in current accounts been financed by stabilizing private capital flows? As figure 4 shows, exchange market intervention in the dollar, both transitory and cumulative, has been substantial compared to current account imbalances, frequently exceeding the latter by a large margin. In fact, rather than financing those im-

**Table 9. Current Account Balances as a Percent of GNP for Four Industrial Countries, 1960–73 and 1973–79**

Percentage points or correlation

| Period and country | Statistic | | |
|---|---|---|---|
| | Mean | Standard deviation | Serial correlation |
| **1960–73** | | | |
| United States | 0.4 | 0.4 | 0.58 |
| Germany | 0.7 | 0.9 | 0.41 |
| Japan | 0.5 | 1.2 | 0.39 |
| United Kingdom* | 0.1 | 1.1 | 0.37 |
| **1973–79** | | | |
| United States | 0.1 | 0.6 | 0.28 |
| Germany | 1.0 | 0.9 | −0.11 |
| Japan | 0.5 | 0.9 | 0.61 |
| United Kingdom* | −0.9 | 1.8 | 0.62 |

Source: Organisation for Economic Co-operation and Development.
a. The output measure is gross domestic product.

balances, net capital flows add to them. Deficits are accompanied by net capital outflows and surpluses by net inflows. In 1977 and 1978, for example, the United States ran current account deficits of about $14 billion, while the holdings of foreign official reserve agencies increased by $35 billion and $32 billion, respectively. In net terms the foreign private sector's claims on the United States were reduced at a rate of more than twice as great as the U.S. deficit. In 1979, in turn, the U.S. current account was nearly balanced; central bank intervention, this time in support of foreign currencies, amounted to nearly $16 billion.

It appears that interest rate policy, adjusted for depreciation, was not at all geared toward financing current account imbalances and stabilizing exchange rates. On the contrary, the independent pursuit of interest rate policy, together with current account surprises, has given rise to exchange rate instability, capital flows, and intervention. This has led to a clear positive relation between the U.S. current account and the return on U.S. assets, which is illustrated in figure 5. When the United States was in deficit, the return on dollar assets, adjusted for depreciation, was negative. Conversely, when the United States showed a surplus, the return differential, adjusted for depreciation, was positive.

A coherent story emerges from combining the evidence in figure 5 with that for intervention, exchange rate determination, and portfolio selection.

**Figure 5. Current Account Balances and Real Interest Differentials for the United States, 1973:1 to 1979:4ᵃ**

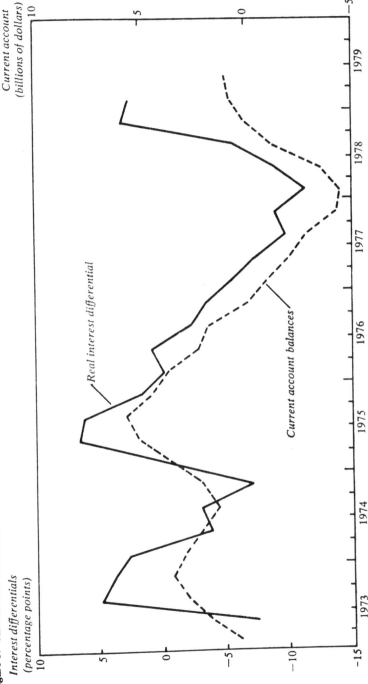

*Interest differentials (percentage points)*

*Current account (billions of dollars)*

Real interest differential

Current account balances

Sources: Current account balances—same as figure 4; real interest differentials—same as table 5.

a. The data are three-quarter centered moving averages. The real interest differential is the difference between the real interest rate for the United States and a trade-weighted average of five foreign countries, using the weights in table 1, note a. Real interest rates are calculated as described in table 5, note a.

Current account surprises give rise to unanticipated fluctuations in the exchange rate. There is no offset through interest rate policy and, accordingly, real interest differentials worsen for the deficit country. The unanticipated depreciation leads central banks to intervene in support of the depreciating currency, and the adverse depreciation-adjusted interest differential leads portfolio holders to shift from the depreciating currency.

Central bank intervention provides the umbrella for portfolio holders to shift their portfolios in response to anticipated interest differentials. Sterilization of the intervention implies that central banks can largely pursue their interest rate policy, albeit at the cost of larger and more dramatic intervention operations.

## Exchange Rate Flexibility and the Capital Mobility Problem

The preceding review of theory and empirical evidence indicates the fundamental problems that confront the design of an exchange rate and payments system. The system must meet conflicting needs. On the one hand, it should have flexible real exchange rates to provide for adjustment of current account imbalances through channels besides deflation or protection. On the other hand, short-term disturbances in the real sector should be largely accommodated at unchanging real exchange rates so that unnecessary variability will not be introduced in the allocation of resources. This accommodation requires a mechanism that ensures the financing of current account imbalances, cyclical or otherwise, through capital flows. Furthermore, financial disturbances should be substantially accommodated through asset management—trading one debt for another—and should not affect real activity or the real exchange rate. This requires institutional arrangements that make possible large-scale sterilized intervention or the issuance of debt denominated in foreign currency.

In the 1960s governments opted for an exchange rate regime with fixed nominal exchange rates, full accommodation of financial disturbances through pegging of exchange rates, and a lack of effective medium-term adjustment in the real exchange rate. When the dollar became overvalued under this regime, it led to the collapse of the system of fixed exchange rates and has left observers with the impression that a flexible real exchange rate is an essential part of a viable exchange and payments system. The large disparity of current inflation rates among countries and the imprecision in estimating their respective underlying trend rates of inflation make it difficult to formulate viable rules for pegging nominal rates, even if there could be agreement on the appropriate real exchange rate.

Once it is accepted that the medium-term real exchange rate should be flexible and that tight pegging of nominal rates is infeasible, the range of options is reduced to a form of floating rates. There does remain, however, a dimension of choice that may add to the stability of the macroeconomy and that concerns the treatment of capital flows. Should capital be free to move in response to expected yields and risks, or should it be immobilized? James Tobin has summarized one main concern about complete freedom of capital movements:

Under either exchange rate regime the currency exchanges transmit disturbances originating in international financial markets. National economies and national governments are not capable of adjusting to massive movements of funds across the foreign exchanges, without real hardship and without significant sacrifice of the objectives of national economic policy with respect to employment, output, and inflation.[28]

Tobin proposes "to throw some sand in the wheels of our excessively efficient international money markets."[29] Specifically, he advocates placing an internationally agreed, uniform, proportional tax on all spot conversions of one currency into another. The tax would reduce the round trip return on international portfolio shifts, and thereby open up an interest spread that would leave monetary authorities more freedom. The proposal would virtually eliminate short-term capital flows and allow the basic balance, in conjunction with intervention, to determine the exchange rate. Relieved of the need to cope with massive short-term capital flows, interest rate policy would be freer to address domestic objectives, and exchange rates would presumably be more stable.

The Tobin tax proposal presumes that the failure of private short-term capital flows to finance current accounts adds to exchange market instability and to the need to intervene. Although capital flows have largely failed to play a financing role, they have forced major changes in real exchange rates whenever government policies failed to aim for cyclical

28. James Tobin, "A Proposal for International Monetary Reform," Cowles Foundation Discussion Paper 506 (Yale University, October 1978), p. 3.
29. Ibid.

coordination and a dampening of external imbalances. Thus capital flows definitely promoted current account adjustment, although possibly exaggerating exchange rate instability.

It is not certain in what way the Tobin tax would work to stabilize exchange rates. There would be less incentive to move capital internationally in response to small yield differentials; but then the basic balance and the extent of central bank intervention would govern the exchange rate. Rather than leaning against the wind, central banks would have to take a view of exchange rates and become rate setters. Would they want to maintain nominal exchange rates or would they adjust real exchange rates in response to current account imbalances?

There is a second, and perhaps more serious, objection to the proposal. Suppose a country does not have the reserves to finance a transitory current account imbalance and thus wishes to use interest rate policy to attract capital. Clearly such a country would now have to increase interest rates by more than it would in the absence of the tax. The country would suffer the burden of financing the deficit and the Tobin tax. There is, of course, an alternative. The country could bring about a sufficiently large depreciation that the expectation of future depreciation would be reduced or eliminated; then with unchanged interest rates there would be a sufficient expected yield differential to attract capital inflows. But again, the country would be paying for the "sand in the wheels."[30]

The welfare economics of the Tobin proposal is not without question. From the standpoint of utility maximization, the choice of an optimal portfolio ranks on a par with the ability to choose one's preferred diet. To the extent that the portfolio cannot be efficiently diversified solely from home securities—and this would surely be the case for small countries—the tax is as disturbing an intervention as a tariff.

Once the principle of free capital flows is accepted, there remains the issue of how to live with them. Capital flows should operate in a stabiliz-

---

30. While I argue against the Tobin tax in its worldwide application, I do think there is a forceful case for the tax in isolated instances. I particularly note the example of the United Kingdom, where the differential adjustment speed of interest rates and inflation, in response to the stabilization policy, has led to a vast real appreciation. A real interest equalization tax is warranted to repel capital inflows and thus maintain a more nearly constant real exchange rate in the adjustment of prices to lower inflation. For a further discussion see Nissan Liviatan, "Neutral Monetary Policy and the Capital Import Tax" (Hebrew University, October 1979); and Dornbusch, *Open Economy Macroeconomics*, chap. 12.

ing manner to finance transitory current account imbalances while allowing real exchange rate changes to cope with medium-term adjustment in the current account balance. It is, in fact, not possible to identify what part of a current account balance it is appropriate to finance and what part requires adjustment. The proper policy rule for stabilizing real exchange rates when confronted with short-term and financial disturbances, without affecting the medium-term adjustment of real rates, is the following: a country with a growing current account deficit (particularly one that occurs in the process of unsynchronized cyclical movements) would both raise its real interest rate and intervene by leaning against the wind. The analysis of the present paper shows that only half the rule has, in fact, been pursued: intervention policy has leaned against the wind, but interest rate policy has been the opposite of what is recommended here.

What are the policy choices that are likely to induce more stable capital flows? It is easy to identify three different areas for reform. The first concerns policies to ease the adjustment process of an international portfolio shift from dollars to marks. That process is under way, and failure to recognize the portfolio substitution will lead to unnecessary variability in exchange rates and changes in the real exchange rates. Portfolio substitution implies a major problem for stabilization policy because its dynamics are not clear. Using sterilized intervention to cope with portfolio shifts has been an appropriate pragmatic response. Two alternatives are reshuffling more directly the currency denomination of the existing stocks of outside assets, and issuing indexed debt.

The second reform is to use monetary policy deliberately to induce stabilizing capital flows. When unanticipated, transitory disturbances arise in the current account, interest rates should be adjusted to avoid excessive real exchange rate movements. That, of course, will leave less room for domestic activism or will force the question of creating a better policy mix for domestic objectives.

The third, and perhaps the most important reform, draws on the evidence that exchange rate movements largely reflect adjustments to unanticipated current account and cyclical disturbances. This suggests that efforts to create a more predictable policy environment may well make a contribution to stabilizing exchange rates.

# Floating Exchange Rates
# and the Need for Surveillance

Jacques R. Artus and Andrew D. Crockett

## Introduction

The move from fixed to more flexible exchange rates among major currencies has been rapid and widespread in recent years, but it has fallen far short of a complete shift to a freely floating exchange-rate regime. Rate-management policies continue to play a major role. Such policies may involve intervention by the central bank in the foreign-exchange market, official or quasi-official borrowing or lending, various forms of controls on foreign transactions and payments, monetary-policy measures, and statements by public officials on the appropriateness of prevailing rates. The major issue at the moment is the extent to which national authorities should use such management policies rather than rely on the free play of market forces for the determination of their exchange rate. This issue is central to the definition of the duties and responsibilities of national authorities and of the International Monetary Fund in a world of floating exchange rates.

Two major questions in the current debate relate to the likely behavior of exchange rates in the absence of rate management and to the role of exchange-rate flexibility in the international adjustment process. An important argument for rate management is that the free play of market forces would lead in the short run to an inappropriate rate, i.e., a rate that has been unduly influenced by temporary factors of a cyclical or speculative nature and therefore diverges significantly from some longer-run equilibrium value corresponding to underlying economic conditions.[1] A second argument for rate management is that exchange-rate flexibility is not a very effective means of reducing or eliminating payments dis-

[1] An implicit assumption here is that market forces cannot produce an inappropriate rate in the longer run. This means that political and economic preferences reflected in permanent measures affecting payments flows (e.g., tariffs, capital controls, and fiscal incentives) are taken as given in the determination of the longer-run equilibrium rate. Also, in what follows, the long-run equilibrium rate should be seen as an analytical concept referring to the rate that would clear the exchange market in the absence of temporary factors and once any adjustment lags have worked themselves out, given foreseeable underlying price and economic conditions. Since such conditions are uncertain and may change rapidly, the long-run equilibrium rate is obviously not a precise value and will change whenever the foreseeable conditions that it reflects are modified.

Reprinted from "Floating Exchange Rates and the Need for Surveillance—Essays in International Finance No. 127, May 1978," pp. 1–38, with permission of Princeton University Press, Princeton, Copyright 1978.

equilibrium and may have harmful consequences for domestic economic objectives, in particular the objective of price stability. Exchange-rate movements, in this argument, lead only to offsetting local-currency price changes and result in a vicious circle of depreciation/inflation or appreciation/deflation.

Even if market forces often led to an inappropriate rate or if exchange-rate flexibility was not very effective in bringing about international adjustment, there still might not be a case for an active policy. Such a policy might be ineffective, the potential for policy errors might be large, the cost of the policy measures might be high, and finally the welfare costs of the inappropriate rate might be low.

These issues have obvious implications for the development of effective international surveillance over countries' exchange-rate policies. If the free play of market forces can be presumed to result in exchange rates that contribute to the smooth working of the international adjustment process, the interest of the international community can concentrate on cases involving deliberate rate management; the concern will be whether policy measures affecting the exchange rate are justifiable in the context of a country's overall economic strategy, and whether such measures place undue burden on other members. If, on the other hand, such a presumption about market-determined rates cannot be made, the international community will have a legitimate interest also in the policies of countries that do not pursue an active exchange-rate policy, and ultimately an assessment of whether an active or inactive policy is justified will have to be made.

This essay begins by discussing a number of important issues related to the short-term variability of exchange rates: the role played by exchange-rate risk and the risk preferences of market participants, the effects of various kinds of monetary measures, the role of exchange-rate expectations, and the costs of short-run exchange-rate variations. We conclude that the free play of market forces may lead to inappropriate rates in periods of unstable underlying economic conditions, and that such rates, if they persist, can impose significant economic costs. Inappropriate rates are less likely to result, however, if the underlying economic conditions are stable and if well-developed capital markets exist without strict capital controls.

We go on to discuss issues related to the role of exchange rates in eliminating payments disequilibria. While recognizing that there is solid empirical evidence for large feedbacks from exchange-rate movements to domestic wages and prices, we conclude that there is most often no

2

realistic alternative to the use of exchange-rate flexibility for the elimination of protracted imbalances. While exchange-rate flexibility is a necessary part of the adjustment process, however, it is not a substitute for domestic stabilization policies.

In the final section of the essay, we consider the implications of these judgments for the desirability of various forms of rate-management policies and for the development of effective international surveillance over countries' exchange-rate policies. We conclude that no simple set of rules can adequately cover the various exchange-market circumstances that are likely to arise. There will be occasions when rate management is desirable to counter short-term disturbances in the exchange markets and promote smooth balance-of-payments adjustment. There are risks, however, that deliberate management of the rate will be used to meet short-term economic goals at the expense of underlying adjustment. Effective surveillance of exchange-rate policies cannot therefore rely on a set of mechanical guidelines but must take account of the special circumstances of each individual case. In building experience with operating the new, more flexible system, the International Monetary Fund can be expected to accumulate a body of "case law" that will enable member countries to form a clearer impression of the kind of behavior most conducive to effective international adjustment.

## Issues Related to the Short-Run Variability of Exchange Rates

During the first few years of floating exchange rates among the major industrialized countries, short-run exchange-rate movements have been far greater than the corresponding movements in domestic price levels. This is illustrated in Figure 1 by comparing deviations of exchange rates and local-currency consumer price indices from their corresponding thirteen-month centered moving averages. (In this chart, the U.S. dollar and the U.S. domestic price level are used as reference points.) This experience suggests two questions: (1) What are the reasons for the short-run variability of exchange rates? And (2) are there significant economic costs when exchange rates fluctuate widely over the short run? Answering these questions requires, first, a theoretical digression on the nature of the process of exchange-rate determination. This digression provides the context for a discussion of the effects of exchange-rate risk, monetary-policy measures, and speculative excesses on the short-run variability of the exchange rate. The concluding part of the section assesses the costs of exchange-rate variability.

3

## FIGURE 1

### SHORT-RUN VARIABILITY IN EXCHANGE RATES AND RELATIVE CONSUMER PRICES

...... Deviations of exchange rates in terms of U.S. dollars from a 13-month centered moving average.

___ Deviation of the country consumer price index relative to the U.S. consumer price index from a 13-month centered moving average.

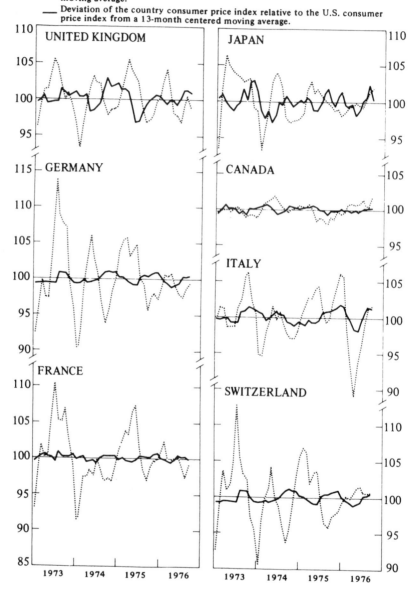

## A Simple Analytical Framework

In recent years, a broad measure of support has developed, at least among academic economists, for the asset-market theory of exchange-rate determination. The essence of this approach is to view the exchange rate as an asset price—the relative price at which the stock of money, bonds, and other financial and real assets of a country will be willingly held by domestic and foreign asset holders. The asset-market approach does not, of course, neglect the requirement that exchange rates must balance the current demand for a currency with the current supply (i.e., flow equilibrium), but it highlights the role played in the determination of current (flow) demand and supply by factors affecting the relative desired stocks of domestic and foreign assets.

The asset-market approach has its main relevance for countries with well-developed capital and money markets, where exchange controls are free enough to permit substantial arbitrage between domestic and foreign assets. In countries where the possibility of such arbitrage is limited or nonexistent, the exchange rate is determined by supply and demand in goods markets and by the amount of intervention undertaken by the authorities. For other countries, the asset-market approach has two major advantages. First, it draws attention to the multiplicity of factors affecting the exchange-rate-determination process. Asset holders continuously adjust or attempt to adjust the composition of their portfolios to reflect expected rates of return, adjusted for degrees of risk, on various domestic and foreign assets. Variability in the factors that influence expected rates of return or relative risk will tend to result in variability in exchange rates.

A second advantage of the asset-market approach is that it draws attention to the predominant role of financial factors in the short run. New unexpected financial developments leading to desired portfolio readjustments occur continuously, and they may at times cause sharp exchange-rate changes. In contrast, changes in current-account positions (apart from their effect on expectations) play a more subdued role in the short run, both because prices in goods markets usually change more gradually and because longer lags operate in the adjustment of trade and invisible flows to price changes. Also, expectational factors play a more limited role in determining trade and service flows. Thus, over any short period the potential demand for a currency resulting from changes in desired *stocks* of financial assets will be large relative to the *flow* demand arising from current balance-of-payments transactions and will be more important in determining the short-run equilibrium value of the exchange rate.

5

Relative prices of goods and of factors of production are dominant factors in exchange-rate determination in the long run, but the usefulness of the asset-market approach is much greater when the focus of the analysis is on the short run. A brief review of this approach will provide an analytical framework for discussing the various issues related to the short-run variability of exchange rates. (A more extensive review of this approach and its implications for the analysis of exchange-rate variability can be found in Schadler, 1977, and Dornbusch and Krugman, 1976.)

The foreign demand ($A'$) for assets denominated in the currency of a given country, net of the demand for foreign assets by the country's residents, can be assumed to be determined by the expected yield on that country's assets relative to the expected yield on assets denominated in other currencies.[2] The expected relative yield reflects the interest-rate differential and the expected exchange-rate change. The elasticity of demand for domestic-currency assets with respect to their expected relative yield will depend on the degree of substitutability between assets denominated in the domestic currency and assets denominated in foreign currencies. Exchange-rate risks and the risk preferences of speculators, which are discussed in more detail below, will be the major determinants of this substitutability.

In terms of flow demand, the change in $A'$ occurring during a given period will be related to changes in the interest-rate differential and in the expected exchange-rate appreciation or depreciation. At the same time, the change in $A'$, which is nothing other than the net balance on private capital flows, will have to be equal ex post to the sum of the balance on current transactions, official capital flows, and the net amount of intervention by the monetary authorities in the exchange market.

We have dwelt at some length on the analytical framework of the asset-market approach because it is the key to understanding the potential for exchange markets to produce in the short run exchange rates that are inconsistent with effective adjustment in the longer run. The factors that establish a stock equilibrium in financial markets are not necessarily consistent with those that would produce continuing flow equilibrium in goods markets. In particular, stock equilibrium may be influenced by expectational and risk-aversion factors not relevant to transactions on current account. And because the lags that govern the response of current-account flows to changes in their underlying determinants are generally presumed to be longer than those for the capital account, the exchange rate may move to a level which, while clearing the foreign-

---

[2] Foreign currencies are all grouped together here for purposes of exposition.

6

exchange market in the short term, is not consistent with continuing equilibrium in the longer term. In such cases, it may be desirable for the authorities to take action which keeps the exchange rate at (or moves it to) a level consistent with longer-term equilibrium. The following sections therefore examine those factors which may give rise to differences between the equilibrium short-term rate that is dominated by asset-market developments and the rate that would result in payments equilibrium over a longer-term time horizon.

## Exchange-Rate Risk

The degree of exchange-rate risk and the risk preferences of participants in the foreign-exchange markets influence significantly the behavior of exchange rates in the short term. Financial assets denominated in different currencies and carrying the same yield would be perfect substitutes if speculators were risk neutral.[3] In this case, temporary disturbances in the goods markets or temporary changes in "autonomous" capital flows would be rapidly offset by speculative capital flows, with little or no cost in terms of exchange-rate variations.[4] Even substantial capital flows resulting from the speculative activity of a significant number of misinformed speculators would be offset by the action of the better-informed speculators.

Even if there were variations in the degree of risk attached to particular currencies, say because of changes in the underlying economic and political conditions in the issuing countries, these would not in themselves be a source of exchange-rate change if speculators were risk neutral. Under such circumstances, it would also be impossible for the authorities to influence their exchange rate by intervening in the foreign-exchange market while offsetting the effect of this policy on the monetary base. Perfect asset substitution implies that excess money creation in one country immediately creates an incentive for capital outflows and pushes the exchange rate down to a point where the higher money stock is willingly held. Thus, the authorities can have a target for the money

[3] Risk neutrality means that speculators would make decisions on whether to acquire or dispose of assets denominated in certain currencies entirely on the basis of the mean values of their expected relative yields (interest plus capital-value change). They would not be influenced by any change in the overall degree of risk (i.e., variance of the expected yield) of their portfolio that such decisions might cause either because of differences in the degrees of risk attached to assets denominated in different currencies or because of diversification considerations.

[4] Examples of changes in "autonomous" capital flows would include, in the present context, bulky overseas direct investments related, say, to the discovery of a new oil field, a large bond issue in the Eurodollar market, or a loan to a foreign government.

7

supply or for the exchange rate, but not both. Imperfect asset substitution opens the way for exchange-rate variations caused by temporary changes in trade flows or autonomous capital flows, or by the action of misinformed speculators; at the same time, it permits, within limits, an independent monetary policy and provides justification for an active policy of intervention by the central bank in the foreign-exchange market.

Early arguments in favor of a flexible-exchange-rate regime (see, e.g., Friedman, 1953, and Sohmen, 1961) assumed that benign speculators with a firm view of where the equilibrium value of the exchange rate lay and unlimited supplies of funds would ensure the short-run stability of the exchange rate. A few years' experience with floating exchange rates has, however, led to a re-examination of the importance of exchange-rate risk and its effect on the degree of substitution between various currencies.

Essentially, currency risk is related to the perceived likelihood that the exchange rate will vary in an unpredictable way during the period a currency is held. One reason for such uncertainty is to be found in the "thinness" of exchange markets for many currencies, a characteristic that is often the result of exchange restrictions. Even with large markets, currencies of countries with unstable underlying political or economic conditions are likely to be considered particularly risky. Variations in their rate may be dominated by political factors rather than relative prices or other predictable economic variables, and the risk of exchange controls is always present. Since the depth of exchange markets and surrounding economic and political conditions vary among countries, it seems likely that perceived risk will differ considerably between currencies.

As to the risk preferences of banks and other large institutional market participants, there is a growing body of evidence (see, e.g., McKinnon, forthcoming) which suggests that they are very much risk averters as far as foreign-exchange operations are concerned. Strict legal and regulatory constraints have been imposed on the speculative activity of commercial banks in many countries. In addition to such constraints, difficulties encountered by a number of financial institutions as a result of their foreign-currency dealings have led banks to shy away from taking substantial net open positions in foreign currencies. Oil exporters and multinational corporations tend also to have a "defensive" policy; they are mainly searching for a stable haven for their funds. (Corporations or other transactors with operations in several countries may, of course, have a need to maintain cash balances and other financial assets denominated in various currencies to finance their worldwide operations.

8

This in itself does not involve risk. The multinational corporation takes risks when it speculates by reallocating its portfolio away from the equilibrium position corresponding to its normal disbursement needs. An "open position" must be understood as a deviation from the portfolio allocation corresponding to normal disbursement needs.)

Perhaps a more important reason why risk aversion can lead to exchange-rate variability is that the willingness of transactors to take additional risks is likely to decrease with their exposure in particular currencies. A small expectation of gain may be sufficient for a multinational corporation to undertake a minor reallocation in its portfolio of liquid assets, but the profit prospects might have to be much greater and more certain to induce it to accept the risk of a large open position. Except in the case where the multinational corporation is betting against a fixed rate that is clearly out of line with underlying factors, it is unlikely that it would be willing to accept a large open position. The degree of currency substitutability may thus be quite low in many cases.

Another aspect of currency substitution is that the smooth working of the adjustment mechanism requires market transactors to take open positions in "weak" currencies. It is countries with weak balances of payments that need to attract a net capital inflow, and the providers of these funds naturally ask themselves what consequences balance-of-payments adjustment will have for the value of their investment. Although an exchange-rate depreciation may appear sufficient to restore external balance, foreign holders may still be reluctant to acquire the currency if they fear that the depreciation might set off a round of domestic inflation or might be perceived as inadequate by other participants in the foreign-exchange market. Further, countries with weak balances of payments also often happen to be countries with unstable underlying economic and political conditions. The cumulative influence of all these factors may lead many transactors to consider investment in assets denominated in certain currencies as "unsuitable" in the same way that they exclude from consideration certain low-quality shares or bonds regardless of yield.

Risk aversion may therefore cause exchange rates to depart from what would be an appropriate level for longer-run equilibrium. This would be particularly likely when there is a large disturbance in trade flows. When such a disturbance occurs (for example, the oil-price increase of late 1973), it is not to be expected that current-account positions will quickly adapt to a balance financed by sustainable capital flows. Additional capital flows are needed during a transitional period while adjustment is taking place. This means that foreign residents must be prepared to take open positions in the currencies of deficit countries. If this is difficult to do

9

because markets are thin or unattractive or because of the risks involved in any open position, there is a danger that exchange rates will depreciate by more than is necessary to secure the needed adjustment in the current account.

In the case of the oil-price increase of late 1973, countries with large oil deficits and small financial markets might not have been able to attract capital on a sufficient scale to maintain an exchange-rate appropriate to longer-run adjustment. In such a situation, intervention by central banks financed from reserves or from compensatory official borrowing can have a key role to play while gradual adjustment takes place in goods markets. More generally, countries with narrow financial markets may find it necessary to intervene in the exchange market to offset the effects of cyclical and other temporary variations in their export receipts and import payments.

### Monetary Policies and Exchange Rates

Another way in which exchange rates can be pushed away from their longer-run equilibrium value is through the response of exchange markets to unexpected monetary measures, and the resultant shifts in interest differentials. It is, of course, not surprising that changes in interest differentials should have an influence on exchange rates. What is less well understood is why relatively small monetary disturbances have in recent years sometimes been accompanied by disproportionately large exchange-rate changes. (For an empirical study of the effects of monetary disturbances on the U.S. dollar/deutsche mark rate, see Artus, 1976.) An explanation that is sometimes advanced to explain these disproportionately large exchange-rate movements is based on the bandwagon hypothesis: the change in the interest-rate differential caused by the monetary shock leads initially to a small change in the exchange rate; speculators project further exchange-rate movements in the same direction and act on these projections; their action brings about further exchange-rate movements. This explanation, as it is usually presented, implies irrational market behavior. There is, however, an alternative explanation that attributes greater rationality to speculators and is intuitively more appealing. It focuses on the relation between monetary measures and expected price changes, and it also draws attention to the effects of monetary measures on nominal interest rates and on portfolio composition.

For analytical purposes, a discretionary change in monetary policy, defined as a discretionary change in the money stock, can be viewed as affecting speculators' expectations in one of three possible ways: it may

10

be seen as a transitory development, perhaps of a countercyclical nature, that will subsequently be reversed; it may be seen as a one-step change in the money stock that will not be reversed but will not change the future rate of growth of the monetary aggregates; or it may be seen as presaging a more permanent change in monetary growth rates.

If speculators view a change in monetary policy as a temporary measure that will later be reversed, they will have no reason to change their expectations as to the long-run value of the exchange rate. If we ignore for the moment the effect of the policy change on interest differentials, there is no reason to expect such a development to lead to a significant change in the current exchange rate. On the other hand, changes in monetary policies that are viewed by market participants as probably leading to permanent changes in prices can be expected to have a very different effect on the exchange rate. If there is no reason to expect that a 5 per cent increase in the money supply will be reversed at a later stage, market participants will be likely to consider that the long-run equilibrium value of the exchange rate has fallen by about 5 per cent. This fall will reflect the widespread belief that in the long run both the quantity theory of money and the purchasing-power-parity theory hold. Under these circumstances, the spot rate will fall by something close to 5 per cent after the implementation or announcement of the policy change. If the policy change is interpreted by market participants as a sign that the authorities are "giving up" in the fight against inflation and that further accommodating increases in the money supply will take place, the rate may fall by even more.

The cases considered above are illustrative only and are therefore somewhat unrealistic. The argument, however, strongly suggests that it is the effect of monetary-policy measures on expectations that determines the size of the exchange-rate change in the short run. And expectations are more likely to be volatile when domestic underlying economic conditions are unstable, leading to uncertainty about the long-run policy intentions of the authorities. During periods of rapid inflation, monetary-policy changes are likely to be interpreted mainly in relation to the fight against inflation. In such circumstances, policy measures are much less likely to be viewed as purely temporary events without any long-run implications as to prices. In countries with stable conditions, on the other hand, there is less reason to expect countercyclical measures to cause significant exchange-rate movements. Price and exchange-rate expectations would be more stable under such conditions, and in any event monetary-policy changes would probably be smaller.

The effects of movements in the money supply on prices do not come

11

through at once. In the short term, changes in interest rates are more likely to equilibrate the supply and demand for money (although this effect itself depends on how far the inflationary consequences of monetary changes are anticipated). Any change in the interest rate will affect the relative attractiveness of domestic bonds versus foreign bonds; therefore, the spot rate will initially have to move away from the expected longer-run equilibrium level by an amount sufficient to offset the interest-rate differential between domestic and foreign bonds. (For a discussion of this effect, see Dornbusch, 1976.) For example, if the 5 per cent increase in money supply considered above leads to a fall in domestic (real) interest rates in addition to the 5 per cent fall in the longer-run equilib- rium value of the spot exchange rate, then the spot rate would fall initially by 5 per cent *plus* an amount that offsets the lower interest rate.[5] The "overshooting" effect is a manifestation of the phenomenon of differences in the speed of adjustment to equilibrium—in this case between the money market and the goods market. The more rapidly domestic prices adjust to the higher money supply and attendant fall in the exchange rate, the more transitory will be the effects of monetary-policy changes on interest rates and the shorter will be the period during which the actual exchange rate differs from its longer-run equilibrium value.

For many years, one of the main attractions of a flexible-exchange-rate system has been assumed to lie in the independence it gives to national monetary authorities. In a formal sense, this occurs because, in the ab- sence of exchange-market intervention, the rate of growth of the monetary aggregates is approximately equal to the domestic credit creation that is directly under the control of the monetary authorities. Hence, the thrust of monetary policy cannot be offset by inflows or outflows of reserves.

The kind of monetary independence allowed for by a flexible-exchange- rate system has, however, been found in recent years to be less beneficial than expected, at least for high-inflation countries, and also possibly to have adverse consequences for the international community at large. Considering first the long-run aspects, it is certainly true that a flexible- exchange-rate system permits a country to "choose" a long-run inflation rate that diverges from the world rate of inflation, since it allows the country to control the rate of growth of its monetary aggregates. If

---

[5] This effect will normally be small. For example, a 1 percentage point fall in the interest-rate differential (at an annual rate) on three-month deposits might cause a fall of about one-fourth of 1 per cent in the exchange rate if the fall in the rate was expected to last three months; the fall of the exchange rate would be about 1 per cent if the fall in the interest rate was expected to last one year, 2 per cent if the fall in the interest rate was expected to last two years, etc.

12

countries faced a long-run tradeoff between output and inflation and had diverging preferences for inflation rates, the possibility for each country to choose its inflation rate would be an important advantage for all. It is now recognized, however, that in most countries such a tradeoff does not really exist in the long run; any advantage in terms of real output from having a higher-than-average rate of inflation is likely to be short-term in nature. Thus, the only long-run advantage arising from monetary independence may be that countries that are willing and able to maintain lower-than-average rates of inflation do not see their efforts frustrated by inflows of reserves pushing up the money supply.

In the short run, monetary independence may tend to increase the effectiveness of discretionary policy measures, but sometimes to the detriment of other countries. The possibility of using monetary-policy measures to affect real domestic demand is enhanced by the greater degree of control the monetary authorities have over their monetary aggregates. Of course, under a flexible-exchange-rate regime, changes in monetary policy are more quickly translated into price-level effects via the inflationary or deflationary consequences of exchange-rate movements. To the extent that prices adapt more quickly under floating rates to changes in monetary policy, real interest rates are likely to move by less in response to changes in the money supply. Even under a floating-exchange-rate regime, however, price adjustment will not be instantaneous. More important, monetary-policy measures may be quite effective in the short run in achieving real output and price goals in a floating-rate regime because of their effects on the exchange rate. It does seem to be the case that, in the short run, exchange rates are quite sensitive to interest-rate differentials created by differences in monetary policies among countries. It may therefore be tempting for countries to use this mechanism to a greater extent than is warranted in the context of balance-of-payments adjustment needs. A loosening of domestic monetary policy may drive the exchange rate down to levels where trading partners have a legitimate fear that the country concerned is exporting unemployment. Conversely, if a shift in policy toward restraint results in an exaggerated upward movement in the rate, trading partners may be concerned about the inflationary consequences for them.

These fears point to the desirability of pursuing stable monetary policies clearly related to the medium-term needs of the country concerned. They also lend support to the suggestion, adopted by a number of major countries, that market participants should receive guidance (through the establishment of medium-term monetary objectives) on how to adapt

13

their expectations in the light of current developments in the monetary aggregates.

## The Danger of Speculative Excesses

So far, we have considered ways in which market participants, acting perfectly rationally given the information available to them and their own preferences regarding risk, can move exchange rates away from their longer-run equilibrium levels. But what of the possibility of irrational speculative behavior, plain and simple?

Attitudes toward speculators have always played a key role in the debate on the merits of flexible-exchange-rate systems. At one extreme, speculators have been viewed as overreacting to most news, and as contributing to a "bandwagon" effect that causes exchange rates to fluctuate widely around their underlying equilibrium value. At the other extreme, speculation has been regarded as a stabilizing force, with speculators instantaneously processing and correctly interpreting all new developments bearing on the appropriate value of the exchange rates. A more balanced view would admit that speculative activity can be either stabilizing or destabilizing, depending on the circumstances. The task then becomes one of identifying the circumstances under which speculation might be destabilizing and considering how such speculation might be offset or its consequences limited.

The possibility of speculative excess in free markets is documented by a wide range of historical experience, from the Dutch tulip-bulb craze and the South Sea Bubble down to the Wall Street boom and bust of 1928-29. Speculative excesses are, however, much less likely to occur in the foreign-exchange market than in markets for individual commodities. A currency is only a right to purchase certain goods at market prices, and the value of a currency in terms of foreign exchange is not likely to become completely divorced from its relative purchasing power. Nevertheless, the degree of constraint imposed by arbitrage opportunities in the goods market should not itself be exaggerated. Large and protracted disturbances from purchasing-power parity can and do occur, as demonstrated by the experience of the French franc in the 1920s (see Rogers, 1929).

Speculative disturbances are more likely to occur as prospective developments in the relative purchasing power of currencies become more uncertain. In such circumstances, the desire of speculators to hold particular currencies will be affected not only by the current purchasing power of those currencies but also by perceptions of how this purchasing power will change in the future. This makes the foreign-exchange market

14

more like a stock market, where price fluctuations are due more to changing expectations of future profitability than to current profits. It follows that the possibility of speculation leading to wide fluctuations in exchange rates is less likely when underlying price developments are reasonably stable than when there is considerable uncertainty in this regard.

## The Costs of Exchange-Rate Variability

In assessing the advantages of policies designed to limit movements that might otherwise take place in the exchange rate, attention must be paid to the costs and benefits that are likely to ensue. The case for firm intervention clearly becomes stronger the greater are the potential costs of variability relative to the benefits. The benefits of exchange-rate flexibility have been shown to accrue in the form of a somewhat greater monetary independence and a somewhat greater effectiveness of monetary policies. The costs are reviewed below.

Some of the costs of exchange-rate variability arise because floating exchange rates tend to move, at least in the short run, by amounts different from underlying cost-price relationships. This means that international transactions take place with a somewhat greater risk of exchange-rate changes during the period between contract and settlement of a transaction. Such short-term uncertainty is likely to be a relatively small added cost in international trade, however, since short-term exchange-rate variability is not normally great, and, in any case, hedging mechanisms exist (e.g., through forward markets) that enable traders to protect themselves against such risks at relatively modest cost.

Departures from longer-run equilibrium rates are potentially more serious when they continue for more than a few months. The costs of having a pattern of exchange rates that is not conducive to effective balance-of-payments adjustment are of three types: (1) the static welfare loss from a suboptimal allocation of world resources; (2) the adjustment costs that inevitably ensue when the disequilibrium is eventually eliminated; (3) the political frictions that result when countries' exchange-rate objectives clash.

The static welfare loss from a suboptimal allocation of a given volume of world resources is perhaps the least serious consequence of a pattern of exchange rates that diverges from longer-run equilibrium. So long as exchange-rate fluctuations are no greater than they have been, on average, in recent years, only a small amount of trade is likely to be displaced from "efficient" channels.[6] And the welfare loss of such a misallocation

[6] It is assumed that the appropriate equilibrium exchange rate has lain somewhere within the band of fluctuation of the actual rate.

15

is reduced by the fact that only trade flows that are marginal (in their contribution to welfare) are likely to be affected. Somewhat more significant may be the dynamic consequences of frequent sustained deviations from longer-run equilibrium rates. Such deviations may increase uncertainty and lead to a shift of investment away from the more exposed traded-goods sectors. The consequences may be slower growth of foreign trade, a weaker competitive climate, and decreased incentives for growth of productivity. The influences of uncertainty are very hard to quantify, however.

More serious are the adjustment costs of correcting a disequilibrium. These costs are apt to become greater the longer a disequilibrium has persisted. Effective balance-of-payments adjustment means shifting domestic resources from (or to) producing for domestic consumption into (or out of) exports or import substitution. This normally will involve a change in the relative profitability of industries, with the likely result of redundancies, and even bankruptcies, in certain sectors. Resources cannot be moved costlessly between industries, of course; there is likely to be transitional unemployment of both labor and capital, and perhaps also inflation as the industries that need to expand, relative to the others, bid for the necessary resources.

It is important to note that these adjustment costs fall not only on countries whose exchange rates have gotten out of line with longer-run equilibrium but also on their competitors and trading partners. For example, when the overvalued exchange rate of a country eventually moves to a more appropriate level, this country will inevitably increase its share of world markets at the expense of other producers of similar goods. These other countries then have to adapt, as best they may, to weakness of export demand. In all this, the social and political costs of unemployment and structural change have to be considered along with the direct economic costs in terms of loss of production.

Last, and perhaps most important, are the political consequences of a world in which there is no mechanism for reconciling incompatibilities in exchange-rate policies. These costs are by their nature the least quantifiable. In judging them, it is necessary to balance the potential costs of such conflict (illustrated in extreme form by the events of the 1930s) against the unlikelihood of such a degeneration of international economic relations in a world where cooperation is the order of the day.

In addition to the misallocation and adjustment costs involved in disequilibrium rates, there may be a cost in terms of domestic price stability. This arises from the inflationary and deflationary impulses transmitted through the external sector when the exchange rate moves. In circum-

16

stances where inflation is a particular worry and there is thought to be downward price rigidity, an additional problem is a possible ratchet effect on prices. Downward price rigidity may result in prices going up more in response to exchange-rate depreciation than they come down when the exchange rate appreciates. If such an effect exists it would be a powerful argument for stabilizing the exchange rate against disturbances that might turn out to be temporary. The empirical evidence for such a phenomenon is not conclusive, however (see, e.g., Goldstein, 1977). The ratchet effect may be limited to countries where the authorities do not effectively control the money supply, possibly because the economic, social, and political costs entailed in achieving control are believed, either correctly or incorrectly, to be too high. In these countries, an asset-market-related depreciation of the exchange rate may indeed be a source of inflation, and the exchange-rate depreciation may be self-validating.

## The Role of Exchange Rates in Eliminating Payments Disequilibria

At the beginning of the generalized-floating period, discussion of the role of the exchange rate in the international adjustment process was overshadowed by concern for the consequences of exchange-rate varia-bility for matters such as inflation and the growth of foreign trade. In-deed, many observers seemed ready to assume that exchange-rate flexi-bility had solved the external-adjustment problem once and for all. Balance-of-payments developments in various countries put an end to this attitude rather quickly. As Marina v. N. Whitman remarked at the 1975 meeting of the American Economic Association:

> The fundamental question to which we seek an answer from recent ex-perience is, of course, whether greater flexibility of exchange rates is indeed an effective means of reducing or eliminating payments disequilibrium, thus alleviating the burden of adjustment which must otherwise be borne either by internal measures or by direct restrictions on international transactions (Whitman, 1975).

This was the crucial question during the pegged-exchange-rate period, and it remains the crucial question today. Many have come to believe that exchange-rate depreciations lead simply to higher inflation rates without any significant benefit for the current account (see, e.g., Economistes Belges de Langue Française, 1973). Certainly the evidence in Figure 2 indicates that during 1973-76 countries with depreciating currencies have had relatively high inflation rates and rather weak current-balance per-formances in spite of relatively large depreciations, while countries with appreciating currencies have had relatively low inflation rates and rather

17

# FIGURE 2

## Effective Exchange Rate, Relative Prices, and Current Balance

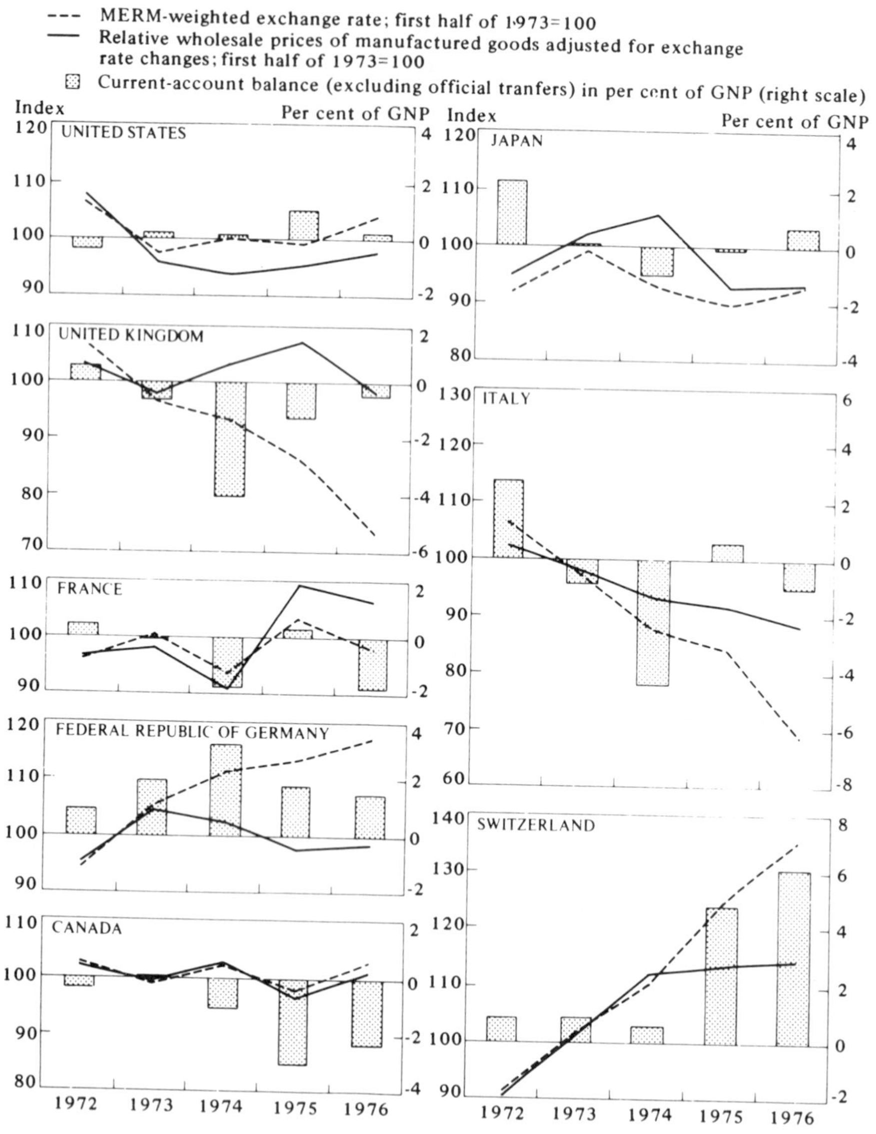

--- MERM-weighted exchange rate; first half of 1973=100

— Relative wholesale prices of manufactured goods adjusted for exchange rate changes; first half of 1973=100

▣ Current-account balance (excluding official tranfers) in per cent of GNP (right scale)

strong current-balance performances in spite of large effective apprecia-
tions. One cannot, of course, conclude from this evidence alone that a
currency depreciation is likely to push a country into a "vicious circle":
Figure 2 does not indicate whether domestic sources of inflation have
been the causal factor for the exchange-rate depreciation, or vice versa.
It does indicate, however, that on the whole the period of flexible exchange rates starting in 1973 has been characterized by a certain divergence between countries with current-account surpluses and countries
with current-account deficits rather than by a move toward a more balanced position, as might have been expected.

This section reviews, first, the "protracted" nature of external imbalances. It goes on to analyze the various forms of the vicious-circle hypothesis, and the effect of exchange-rate flexibility on the behavior of the
authorities and of private economic agents as regards the inflationary
process. The aim is to review the theoretical and empirical underpinnings
of the vicious-circle hypothesis and to make a balanced judgment as to
the appropriate role of the exchange rate in the external-adjustment
process.

## The Protracted Nature of External Imbalances

There is considerable empirical evidence that relative-price changes
have a strong effect on the structure of demand and supply in goods
markets and, in particular, on the foreign-trade performance of a
country. However, economists have become more conscious over recent
years of the slowness of the response to relative-price changes, and of the
importance of nonprice factors in the determination of trade performance. Over a given period, relative-price changes may be significant and
in the right direction but they may easily fall short of the changes that are
necessary to offset developments in nonprice factors affecting competitiveness, and this may leave observers with the impression that price
factors are of marginal importance.

Technological innovations and the marketing of new products, the
right product mix at the right time, and a reputation for quality, reliable
delivery schedules, and good after-sales service have been found to explain in large part the trade performances of countries such as the United
States in the 1950s and early 1960s and the Federal Republic of Germany
in the 1960s and 1970s. (The experience of the United States in the 1950s
was reviewed by Kindleberger, 1958, and de Vries, 1956, while Kindleberger, 1976, reviewed the German experience.) Successful trade performance over the years also creates a certain orientation of the economy
toward exports markets. Thus, it is unrealistic to expect a sudden relative-

19

price change, induced by an inflationary burst or an exchange-rate change, to affect rapidly a pattern of production and demand that reflects such "structural" factors. First, economic agents will have to be sure that the relative-price change is going to last before they undertake the large adjustment costs that are likely to be involved in any reallocation of factors of production or in any change in the allocation of demand among products and supplying countries. Second, even when economic agents are persuaded that the relative-price change is going to last, it may take years, in particular on the supply side, before the effects of the decisions they take are fully felt.

One of the consequences of the slow speed of adjustment in goods markets is the now well-known J-curve effect. In the period immediately following an exchange-rate change, the terms of trade tend to turn against a depreciating country and in favor of an appreciating country, and this may more than offset the volume effect of the exchange-rate change on the balance of trade. Figure 3 shows significant terms-of-trade effects from exchange-rate changes for four countries that have experienced substantial rate changes over the last four years.[7] Combined with the empirical evidence that the volume effects on trade flows of relative price changes take place only with a lag, this suggests that the initial effects of exchange-rate changes on the current account may be perverse. Such perverse effects could result in excessive exchange-rate movements, in particular if developments in the current account affect speculative capital movements in the same direction.

Economic agents are, of course, likely to resist any attempt to eliminate a protracted payments imbalance that has become reflected in the structure of the domestic economy. The longer the imbalance has existed, the more entrenched are the social groups that benefit from it. Resistance to adjustment will first take the form of political pressure against the exchange-rate change. Once the exchange-rate change has occurred, resistance will take the form of an attempt by social groups that benefited from the imbalance to "pass along" the price effects of the exchange-rate change to other social groups. This can be the beginning of a vicious circle of exchange-rate changes and offsetting domestic price movements.

### The Vicious-Circle Argument

A better understanding of the vicious-circle argument can be obtained by considering a country which, for whatever reason, needs to reduce

---

[7] To obtain a precise estimate of the J-curve effect, one would need to adjust the prices of the trade flows for factors other than the exchange-rate change. The strong terms-of-trade effects are clearly apparent from the chart, however.

20

# FIGURE 3

## EFFECTIVE EXCHANGE RATE AND THE TERMS OF TRADE

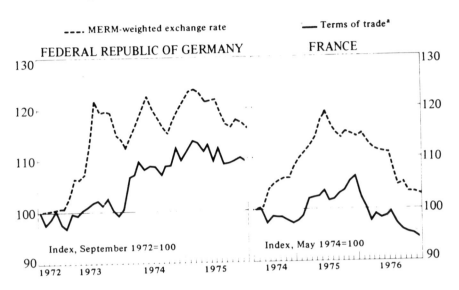

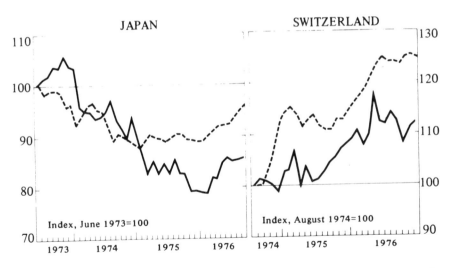

NOTE: This chart considers four particular cases in which a sharp change in the effective exchange rate has occurred.

* Export price index divided by import price index, scaled by a trade-weighted average of the terms of trade for the United Kingdom, France, the Federal Republic of Germany, Italy, and Japan.

its current-account deficit. To do so, it will have to reduce its absorption of goods relative to domestic output. It can do this in two ways: (1) it can reduce its domestic demand in terms of domestic currency, or (2) it can let its exchange rate depreciate and its price level increase in terms of domestic currency. If the policy used reduces the real wage rate, the adjustment can take place without a fall in output (an increase in unemployment); if the real wage rate cannot be reduced, the adjustment can be achieved only with a fall in output. Thus the relative effectiveness of the exchange rate as a policy instrument depends in this case on its relative effectiveness in reducing the real wage rate.

The vicious-circle argument, in its extreme form, assumes that a depreciation of the exchange rate is automatically offset by an increase in the money wage rate. Thus, the authorities are left with the choice of letting the money supply rise to accommodate the wage-price increase, in which case they are back to square one, but after experiencing a temporary increase in inflation,[8] or of refusing to accommodate the wage-price increase, in which case they arrive at the same unemployment rate they would have had if they had chosen to reduce nominal domestic demand rather than to let the exchange-rate depreciate. Under the first scenario, the exchange-rate depreciation will lead to further price increases and further depreciations of the rate in an endless succession. Under the second, it has the same consequences as the use of a deflationary policy, but without the added advantage of a temporary fall in the rate of inflation that a deflationary policy is likely to yield.

The issue behind the vicious-circle argument turns on the effectiveness of different policy instruments in bringing about the ultimate reduction in real wages that is essential for adjustment, while minimizing the harmful consequences for such other objectives as the maintenance of reasonable price stability and full employment. Fellner (1975), for example, argues that the exchange rate is more effective in reducing the real wage rate, because the concern of various labor unions with *relative* real wage positions makes it easier for the authorities to cut real wage rates by an increase in consumer prices induced by a change in the exchange rate. Such a change hits every labor group more or less equally and simultaneously, whereas a deflationary policy tends to affect labor contracts

---

[8] The change in the rate of inflation is temporary because the policy change considered here is only a once-and-for-all change in nominal demand (relative to trend). Temporary changes can, of course, lead to permanent changes if, for example, they lead to changes in expectations and accompanying changes in the rate of growth of the money supply.

22

one at a time when they are renewed. However, in many industrial countries (France, Italy, Belgium, Japan, etc.) most labor contracts are renewed at the same time, and the bargaining process involves a few *national* labor unions. Another argument for using the exchange rate is based on the hypothesis that employees may be influenced by money illusion. This argument has been weakened by the probable erosion of money illusion in periods of high inflation and by the fact that the use of indexation clauses in labor contracts is becoming increasingly widespread (see Braun, 1976).

A strong argument against the vicious-circle view and its policy implication that exchange-rate adjustments should not be relied on is that the alternative policy strategy to reduce real wage rates, a restrictive demand-management policy, seems in many cases even less effective. Is it realistic, for example, to expect countries that in the past have been unable to maintain their inflation rates in line with those of their trading partners to succeed in bringing their inflation rates substantially *below* those of their trading partners so as to gradually restore their competitiveness? Thus, the case for using the exchange rate to bring about a readjustment in real wage rates rests less on its efficiency in doing so than on the evidence that other available instruments are even less efficient. Further, while the effect of an exchange-rate change on real wage rates may not be large, the available empirical evidence shows that it is not insignificant, at least if the proper supporting policies are implemented at the time of the exchange-rate change. The evidence presented in Figure 3 above, for example, of continuing terms-of-trade effects for countries with significant exchange-rate movements is consistent with the hypothesis that exchange-rate changes are not immediately offset by changes in domestic costs. Viewed in this light, the relevant question is how the authorities can increase the effectiveness of an exchange-rate change. Here, deflationary measures, temporary nullification of indexation clauses, etc., have a crucial role to play.

## External Constraints on Inflationary Practices

Concern has often been expressed that the wider use of exchange-rate flexibility as a means of achieving balance-of-payments adjustment will weaken the discipline to adopt policies that promote price stability. Under fixed exchange rates, a country that inflates at a rate higher than that of its trading partners will tend to suffer a deterioration in its balance of payments and a loss of international reserves. If the fixed exchange rate is to be maintained, the country will ultimately have to take action to bring

23

its inflation rate into line with that of its trading partners. The usefulness of such a discipline was recognized in the 1970 report of the IMF Executive Directors, which stated:

> . . . the need to defend a fixed exchange rate against depreciation may promote willingness to impose unpopular domestic restraints; and where the attempt to defend a parity is ultimately unsuccessful, the psychological shock of a devaluation may promote broad support for the adoption of the necessary associated measures to curtail domestic demand (International Monetary Fund, 1970, p. 32).

A commitment to fixed exchange rates may also act as a limitation on cost-push pressures in the domestic economy. If price setters, either in the labor or goods markets, are aware that the exploitation of market power will lead to a loss of competitiveness and therefore to a reduction in demand, they may be induced to moderate wage or price increases. Such a result is, of course, more likely in economies that are rather open to external competition; in addition, the authorities' commitment to an exchange-rate target must be perceived as strong enough to outweigh countervailing pressures to maintain full employment.

The disciplinary pressures that operate on high-inflation countries under fixed exchange rates have, however, a counterpart in the pressures that operate on countries that have low inflation rates and a tendency toward balance-of-payments surpluses. To the extent that current-account surpluses directly raise the pressure of domestic demand, countries with low rates of price increases will be likely under a system of fixed exchange rates to experience an increase in inflationary pressures.

Which kind of disciplinary pressure will be the stronger depends on a number of factors, including the availability of conditional and unconditional liquidity to finance balance-of-payments deficits. If international liquidity is scarce, or available only on strict conditions, a stronger discipline may be exerted on deficit countries, since the consequences of continued deficits will be perceived as more serious than the consequences of prolonged surpluses. On the other hand, if balance-of-payments finance is readily available deficit countries may be much more willing to accept deficits, and it will be the surplus countries that tend to take prompter action to stem the outflow of the real resources that their surpluses represent.

As Emminger (1973) has pointed out, where finance for payments deficits is relatively freely available, flexible exchange rates may well exercise more effective discipline over inflationary policies. Under fixed exchange rates, a substantial part of the price and output consequences of a high-inflation policy can be exported abroad through a balance-of-pay-

24

ments deficit. With flexible rates, however, such a policy is more likely to result in a depreciation of the exchange rate and, in turn, an increase in domestic prices in the high-inflation country. Thus, the cost of a high-inflation policy—declining purchasing power of domestic incomes over foreign goods—will be more easily and quickly noticed by the public under flexible rates. Assuming that inflation is unpopular and that governments are responsive to public opinion, the foregoing implies in turn that inflationary pressures will be restrained more quickly under flexible rates than under fixed rates when there is the option of financing them by reserve depletion. Even under flexible rates, however, it is still possible for high-inflation countries to use exchange-rate policy to avoid, for a while, the discipline of public opinion. Since governments are always seeking the least painful way to bring inflation down, there is a danger that countries with floating rates will try to resist exchange-rate depreciation through intervention and other ways of influencing the exchange rate.

Crockett and Goldstein (1976) have argued that neither *a priori* reasoning nor empirical evidence is conclusive in showing that anti-inflationary discipline is stronger under fixed than under flexible rates. Rather, inflationary disturbances can disrupt the smoothness with which the adjustment mechanism works under either exchange-rate regime if there are not sufficient pressures to adopt timely adjustment measures. Such a result is likely to occur if balance-of-payments finance is either too easy or too difficult to obtain. Too easy access to finance is likely to deter governments from taking effective adjustment action either through domestic policies or through permitting exchange-rate movement. A shortage of liquidity, on the other hand, can prompt overhasty external adjustment, with possible adverse consequences for the stability of domestic incomes and prices.

## Assessing Exchange-Rate Policies

### The Case for Intervention

The analysis of the two preceding sections suggests that the case for intervention (or any other form of active exchange-rate policy) is rather strong. Asset-market-related disturbances have been shown to be a considerable source of exchange-rate disturbance, at least in certain cases. They may lead to disorderly conditions in the foreign-exchange market, wide fluctuations in exchange rates, and even, in certain circumstances, to an exchange-rate level that interferes with smooth external adjustment

25

in the longer run. For certain countries, a case can also be made that a freely floating exchange rate is not necessarily conducive to effective external adjustment.

In practice, however, the case for intervention is not so strong. While the free play of market forces may have undesirable effects, in most cases it may be difficult to implement a rate-management policy that would not hinder but help the working of the external adjustment process in the longer run.

The desirability of intervening to prevent disorderly market conditions is widely accepted. It is not always easy, of course, to determine in advance whether a particular disturbance in foreign-exchange markets is a temporary aberration or the beginning of an underlying trend. This would be the case, for example, where a sudden movement in the rate is prompted by political fears. If the political situation that gave rise to the exchange-rate pressures disappears, it would have been appropriate for the authorities concerned to resist the exchange-market consequences. But if the change in political climate is permanent and gives rise to a permanent reappraisal of what constitutes an appropriate exchange rate, a rapid movement in the exchange rate to its new equilibrium might have been more conducive to orderly exchange-market conditions. The important point is that it is exchange-rate movements that are temporary and reversible, not just rapid, that are a cause for concern.

The need to intervene to preserve orderly exchange markets normally arises suddenly, and thus there is no alternative to relying on the judgment of the authorities regarding the suitability of intervention in particular cases. *Ex post*, however, it is relatively easy to reach a judgment on whether such intervention has been warranted. Smoothing intervention will normally cancel out over a short period; if it does not, this can be considered *prima facie* evidence that the authorities have misjudged the nature of exchange-market conditions. Conversely, sharp exchange-rate movements around a stable exchange-rate level (or trend), when unaccompanied by official intervention, would tend to suggest that an opportunity for stabilizing intervention has been missed.

The need for intervention to preserve orderly markets is likely to be less where exchange markets have reasonable depth and breadth. In such cases, individual large transactions can be absorbed without a significant impact on the rate, and a change in sentiment on the part of some speculators is not likely to carry the exchange rate very far before countervailing expectations of other speculators are encountered. It is where exchange markets are narrow (as in the case of many smaller countries or of countries imposing exchange restrictions) or where expectations are par-

26

ticularly volatile (perhaps because of political uncertainties) that the case for smoothing intervention is likely to be strongest.

Much more difficult to appraise is the case for sustained management of the exchange rate, whether by intervention or other means. There are basically two reasons why the authorities may wish to have a rate that is different from the one that would emerge in the absence of intervention.

The first reason is that the short-run market-clearing rate may differ substantially from the longer-run equilibrium rate. A number of causes of such a divergence were examined in the section on short-run variability of exchange rates. Apart from the possibility of irrational market behavior, the most important cause was the risk-averse behavior of private market participants. Faced with the need to assess the longer-term significance of disturbances in countries' balances of payments and monetary policies, private wealth owners seek to maximize the return on their assets and to minimize their risk. In acting to minimize risk, speculators are likely to require an additional incentive to hold open positions in weak currencies. This may mean that speculators will acquire such currencies only when they become "overdepreciated" and there is a significant expectation of capital gain.

The second reason why countries may wish to intervene to influence the exchange rate is that relying on exchange-rate adjustment may complicate the task of using domestic instruments of adjustment to achieve a viable balance of payments. The extreme version of this argument is the vicious-circle hypothesis. Although this hypothesis may not be acceptable in its pure form, there may be truth in the implication that subjecting private and public decision makers to an external constraint (in the form of an exchange-rate target) will make domestic stabilization policies more effective. Since longer-term adjustment requires effective action on domestic demand, it is arguable that the main emphasis of adjustment policies should be domestic. It must be recognized that this argument for intervention implies an acceptance of a "wrong" rate in the short term to improve the prospects of achieving a more durable "right" rate in the longer term. It therefore depends crucially on backing up exchange-rate intervention with a suitable package of measures to restore domestic stability.

Ultimately, the major weakness in the case for intervention is that it is so difficult to recognize cases when intervention is really called for. Further, national authorities are often unwilling to accept significant exchange-rate movements even when they are justified by underlying economic developments. Market forces may not always be "right," but the experience of the past decade indicates clearly that they have been

27

right more often than the authorities. Perhaps the strongest temptation to pursue an inappropriate exchange-rate policy results from viewing as temporary what eventually turns out to be a permanent change in exchange-market conditions. This temptation is likely to be particularly strong when the costs, both political and economic, of adjusting to the change are perceived to be high and when there is a considerable degree of uncertainty concerning future developments affecting the balance of payments. Exchange-rate changes affect the relative profitability of production of traded and nontraded goods. Switching resources between sectors imposes adjustment costs and may also involve transitional unemployment of factors of production. Since there may also be strong economic interests that benefit from the existing structure of production, governments may be unwilling to accept the political consequences of permitting an exchange-rate realignment.

Another temptation, which is particularly prevalent when countries are attempting to bring down excessive rates of price inflation, is to use the exchange rate as part of incomes policy. It is well-known that the rate of increase in prices, particularly consumer prices, is an important element in determining wage settlements. For this reason, governments that attach high priority to reducing the rate of increase in wages are often prepared to accept measures that distort the allocation of resources in order to retard the rise in the consumer price index. This can take place through domestic subsidies, but only at the cost of increasing the budget deficit and complicating the financing of the central government. A simpler mechanism, but one that has much the same result, is to subsidize imports by holding up the exchange rate. The attraction of such a policy is enhanced when the immediate goals of implementing an effective incomes policy bulk larger in the government's overall economic strategy than the more medium-term constraint of attaining external balance.

The costs of maintaining an overvalued rate for incomes-policy purposes should not be underestimated, however. Quite apart from the distorting effect on the allocation of production between traded and nontraded sectors, any beneficial consequences on the rate of price increase in the short term are achieved at the cost of adverse consequences later on, when the rate has to be allowed to move to an equilibrium level. Only if it can be argued convincingly that the consequences of favorable effects on inflation in the short term are likely to be stronger than the consequences of unfavorable effects later will there be a case for using the exchange rate as an anti-inflationary instrument. Even so, care must be taken that any advantages of an active exchange-rate policy for the country adopting

28

it are not purchased at the expense of offsetting disadvantages for trading partners.

Although, in present circumstances, the greater risk seems to be that countries will take action that has the effect of overvaluing their exchange rate, a more long-standing concern has been the use of competitive depreciation to improve domestic employment. The temptation to use exchange policy in this way would presumably become greater if the main objective of economic policy shifted from fighting inflation to promoting employment. Also, growth in output generated by rising exports (as against domestic demand) has the attraction that overall productivity is often thought to be higher in the export sector. Furthermore, in circumstances where finance for balance-of-payments deficits is not readily available, exchange-rate depreciation may appear preferable to domestic measures as a means of promoting employment, since it is less likely to run into an external constraint. The danger of such a policy, of course, is that it promotes domestic employment objectives at the expense of employment in other countries and thus risks retaliation and a chain reaction of beggar-thy-neighbor policies.

*The Need for a Case-by-Case Approach to Surveillance*

All countries are subject to disturbances in their balance of payments that may be external or internal in origin. The interest of the international community lies in ensuring that the combination of measures chosen by a country is such as to enable timely and effective adjustment to take place, consistent with its other obligations to achieve a satisfactory employment level with reasonable price stability and to avoid actions harmful to other countries.

The need to protect the international community from potentially harmful actions by individual countries is one of the basic *raisons d'être* of the International Monetary Fund. The exchange-rate system of the original Bretton Woods Articles of Agreement sought to achieve this goal through the establishment of par values for currencies that could be changed only with international concurrence. When this system broke down in 1971 and finally collapsed in 1973, alternative means for avoiding harmful and inconsistent exchange-rate policies were sought.

In June 1974, the Fund adopted a set of nonbinding "Guidelines for the Management of Floating Exchange Rates." These guidelines reflected a widespread acceptance that the behavior of governments with respect to exchange rates was a matter for international consultation and surveillance. The main features of the guidelines were to provide encourage-

29

ment (1) to smooth out very short-term fluctuations in market rates; (2) to offer a measure of resistance to market tendencies in the slightly longer run, and in particular to rapid movements in the rate; and (3) to try to estimate, if possible, a medium-term "norm" for currencies' exchange rates, with a view to resisting movements that appeared to deviate substantially from that norm. In addition, it was recognized that intervention policies should take account of countries' reserve positions and of the effect of intervention on the exchange rate of the currency being used in intervention.

The themes of the guidelines are reflected in a number of academic contributions on the subject. In an earlier essay in this series, Ethier and Bloomfield (1975) suggested the establishment of "reference rates," which would work in much the same way as the "norms" of the guidelines. In other words, countries would not be under any obligation to defend a particular rate, but they would be prevented from intervening against their currency when it was below the reference rate and from intervening in favor of their currency when it was above the reference rate. Tosini (1977) picks up another element of the guidelines in suggesting that countries should always be encouraged to "lean against the wind" so as to moderate exchange-rate movements that would result from market forces. Her rationale for this prescription is that, in practice, rate fluctuations have proved to be excessive and that any action to dampen such swings would tend to lead to a smoother evolution of exchange rates. In a more elaborate analysis of different situations in which rate management might be considered, Mikesell and Goldstein (1975) examine a variety of circumstances in which intervention might be desirable and consider possible rules to protect the interests of the international community. Broadly speaking, their rules relate to permissible changes in official reserves between reporting periods.

The objective of formulating fairly explicit behavior rules for intervention runs into a number of theoretical and practical difficulties, however. Resistance to rapid movements in the exchange rate (or leaning against the wind) is optimal only so long as market forces are moving rates excessively in one direction. There seems to be no valid reason, however, to assume that large and sudden exchange-rate changes are necessarily inappropriate. If, for example, a major policy shift results in a substantial change in the assessment by market participants of the longer-run equilibrium exchange rate for a currency, it may be less disruptive to allow the rate to move quickly to a new value than to slow this movement down through intervention.

30

The possibility of using changes in official reserves as a standard for permissible intervention is also unlikely to be acceptable in practice. Quite apart from the difficulty of agreeing on country-by-country limits for reserve changes, it is by now widely recognized that official intervention affecting reserve holdings is only one among several ways of managing the exchange rate.

The establishment of an internationally agreed set of "norms" for exchange rates runs into severe practical difficulties, and although such norms were provided for in the IMF's 1974 guidelines, no attempt was made to establish them. Apart from the fact that the move toward greater exchange-rate flexibility has by now gone so far that the establishment of exchange-rate norms can no longer be considered politically feasible, it is obvious that the necessary conditions for the working of such a system, namely relatively stable underlying economic conditions in the major countries and some harmonization of national economic policies, are not satisfied at present.

Partly as a result of the perceived difficulties in applying rules with even the limited degree of precision involved in the 1974 IMF guidelines, the new Articles of Agreement express the obligations of Fund members in the exchange-rate field in more general terms. The relevant provision of the new Article IV is that members shall "avoid manipulating exchange rates or the international monetary system in order to prevent effective balance of payments adjustment or to gain an unfair competitive advantage over other members."

At the time this provision was drafted, late in 1975, it was decided that this general rule would be further codified later on through the adoption of "specific principles" for the guidance of all members in their exchange-rate policies. In fact, however, the decision that was eventually reached by the IMF in April 1977 (and which is reproduced here in an Appendix) was to avoid introducing too much precision in defining members' obligations and to concentrate instead on establishing procedures that would enable continuous and effective surveillance to take place. In other words, the Fund opted for a case-by-case approach to its surveillance responsibilities that would not be based on a specific set of ground rules. However, the procedures that were adopted in 1977 do provide illustrative cases of developments that would be *prima facie* evidence of a need for international consultation and review. For the most part, these developments relate to policies pursued by countries that have the effect of influencing balances of payments and reserve flows. But it is also suggested that the Fund would have an interest in "behavior of the exchange rate that appears to be unrelated to underlying economic and financial condi-

31

tions . . . ," implying that intervention to counter such exchange-rate developments is to be regarded as desirable.

These new procedures provide the necessary framework for Fund surveillance activities. They take into account that both misguided rate-management policies and the free play of market forces may lead to an "undesirable" exchange rate, but it is clear that misguided rate policies are expected to be the major source of problems.

Another advantage of these procedures is that they make allowance for the divergent interests and preferences of member countries. For a variety of reasons, countries attach different degrees of importance to the objective of exchange-rate stability. In part, these diverging preferences stem from differences among countries in the importance of the exchange rate as a price that influences the level and distribution of national output. For countries where the external sector is relatively small, the impact of external disturbances on the domestic economy is likely to be correspondingly reduced and the main objective will be to ensure that the external-adjustment mechanism does not interfere with the setting of domestic policy instruments. For countries with a larger external sector, however, the exchange rate is likely to be a price that has more pervasive effects on the level and distribution of national output. For this reason, such countries would suffer greater costs if, for whatever reason, market forces tended to move the exchange rate to a level judged incompatible with domestic objectives or inconsistent with medium-term balance-of-payments equilibrium.

The importance of the exchange rate as a disciplinary instrument may also vary between countries. In large, fairly closed economies, external developments will exercise relatively less influence over the setting of domestic policy instruments and over the wage-price formation process. For more open economies (for example those of the smaller snake members) prices and wages are likely to be much more responsive to external developments, so that exchange-rate flexibility will be less effective as an adjustment mechanism relative to domestic measures. Also, the degree to which wages and prices are indexed (formally or informally) will have important implications for the relative efficacy of different adjustment techniques.

Another possible reason for divergences in exchange-market policies is that exchange-market structures may vary between currencies. Not all countries have exchange markets with the depth and breadth of the major industrial countries, where a wide range of traders, arbitrageurs, and speculators participate. In particular, many smaller countries have a restricted market for their currencies, often centralized in the central bank.

32

Many countries also have capital controls, reducing the scope for arbitrageurs and speculators to smooth disturbances in the supply and demand for foreign exchange arising from current-account transactions. Furthermore, the current account itself may be somewhat unresponsive to exchange-rate variations, particularly where more direct means of controlling imports are employed and where the elasticity of supply of exports is low—as it may be for many primary producing countries.

These differences between countries are important in assessing the need for an active exchange-rate policy in particular cases. While a freely floating exchange rate can ensure a continuous equilibrium between the demand and supply for a country's currency in the foreign-exchange market, other measures may take considerably longer to achieve their intended effect on external payments flows. It has been argued above that exchange-rate flexibility, if unsupported by appropriate measures to influence domestic demand, is likely to be inadequate to achieve durable adjustment. To the extent that these supporting measures affect payments flows with a lag (particularly on current account), there will be a temporary disequilibrium in the balance on current and long-term capital account that has to be covered by short-term financing. If, however, the authorities do not intervene and there is weakness in private speculative demand, the exchange-rate movement needed to induce accommodating flows of private short-term capital may be excessive.

Ultimately, the exchange-rate policy of the authorities should reflect both their assessment of whether there is a large difference between the short-run market clearing rate and the longer-run equilibrium rate for their currency and their views on the appropriate role of the exchange rate in their overall economic strategy. To judge the appropriateness of the authorities' exchange-rate policy, there is no alternative to a full assessment of the implications for the balance of payments of the overall economic strategy adopted by the authorities. This involves an appraisal of the implications for the current account of existing domestic policies and prospective developments in the world economy, combined with an assessment of whether sustainable capital flows are likely to be forthcoming to finance whatever surplus or deficit emerges on current account.

These various points lead us to conclude that the only international monetary system consistent with existing economic and political realities is the present system, where the major countries rely heavily on market forces but intervene whenever these forces tend to push their exchange rates beyond what they assess to be appropriate. This mixed system, however, is bound to lead to frequent conflicts among countries with divergent interests, and surveillance by the International Monetary Fund

33

is needed to avoid harmful and inconsistent exchange-rate policies and to moderate international conflicts. Because of the real divergence in interests among countries and the many uncertainties inherent in the appraisal of exchange-rate policies—in particular, the difficulty of assessing the appropriateness of an exchange rate for the longer term—such surveillance cannot be based on a single objective indicator or even on any precise set of rules. Thus, in arriving at a judgment as to whether a country's exchange-rate policies constitute an unwarranted hindrance to the proper working of the international adjustment process, the Fund must make a comprehensive appraisal of these policies. These points have been taken into account in the new Article IV and in the Fund decision of April 1977. Whether this new system works satisfactorily depends primarily on whether the major countries show sufficient willingness to give it a fair chance.

## References

Artus, Jacques R., "Exchange Rate Stability and Managed Floating: The Experience of the Federal Republic of Germany," *IMF Staff Papers*, 23 (July 1976), pp. 312-333.

Braun, Anne W. R., "Indexation of Wages and Salaries in Developed Economies," *IMF Staff Papers*, 23 (March 1976), pp. 226-271.

Crockett, Andrew D., and M. Goldstein, "Inflation under Fixed and Flexible Exchange Rates," *IMF Staff Papers*, 23 (November 1976), pp. 509-544.

De Vries, Tom, "De Theorie van het Comparatieve Voordeel en het Dollartekort," *De Economist*, 104 (January 1956), pp. 1-39.

Dornbusch, Rudiger, "Expectations and Exchange Rate Dynamics," *Journal of Political Economy*, 84 (December 1976), pp. 1161-1176.

Dornbusch, Rudiger, and Paul Krugman, *Flexible Exchange Rates in the Short Run*, Brookings Papers on Economic Activity No. 3, 1976, pp. 537-575.

Economistes Belges de Langue Française, *Economies ouvertes face aux mutations internationales—Rapport du 2e Congrès des Economistes Belges de Langue Française*, Brussels, Centre Interuniversitaire de Formation Permanente, 1977.

Emminger, Otmar, *Inflation and the International Monetary System*, Basle, Per Jacobsson Foundation, June 16, 1973.

Ethier, Wilfred, and Arthur I. Bloomfield, *Managing the Managed Float*, Essays in International Finance No. 112, Princeton, N.J., Princeton University, International Finance Section, 1975.

Fellner, William J., "The Payments Adjustment Process and the Exchange Rate Regime: What Have We Learned?—Comments," *American Economic Review*, 55 (May 1975), pp. 148-151.

34

Friedman, Milton, "The Case for Flexible Exchange Rates," *Essays in Positive Economics*, Chicago, University of Chicago Press, 1953, pp. 157-201.

Goldstein, M., "Downward Price Inflexibility, Ratchet Effects, and the Inflationary Impact of Import Price Changes: Some Empirical Tests," unpublished, Washington, D.C., International Monetary Fund, DM/77/34, April 1977.

International Monetary Fund, *The Role of Exchange Rates in the Adjustment of International Payments*, Washington, D.C., 1970.

Kindleberger, Charles P., "The Dollar Shortage Revisited," *American Economic Review*, 68 (June 1958), pp. 388-395.

————, "Germany's Persistent Balance-of-Payment Disequilibrium Revisited," *Banca Nazionale del Lavoro Quarterly Review*, 29 (June 1976), pp. 135-164.

McKinnon, Ronald I., "Instability in Floating Foreign Exchange Rates: A Qualified Monetary Interpretation," *Money in International Exchange: The Convertible Currency System*, New York, Oxford University Press, forthcoming.

Mikesell, Raymond F., and Henry N. Goldstein, *Rules for a Floating-Rate Regime*, Essays in International Finance No. 109, Princeton, N.J., Princeton University, International Finance Section, 1975.

Rogers, James H., *The Process of Inflation in France, 1914-1927*, New York, Columbia University Press, 1929.

Schadler, Susan, "Sources of Exchange Rate Variability: Theory and Empirical Evidence," *IMF Staff Papers*, 24 (July 1977), pp. 253-296.

Sohmen, Egon, *Flexible Exchange Rates*, Chicago, University of Chicago Press, 1961.

Tosini, Paula, *Leaning Against the Wind: A Standard for Managed Floating*, Essays in International Finance No. 126, Princeton, N.J., Princeton University, International Finance Section, 1977.

Whitman, Marina v. N., "The Payments Adjustment Process and the Exchange Rate Regime: What Have We Learned?" *American Economic Review*, 65 (May 1975), pp. 133-146.

35

# APPENDIX

## Surveillance over Exchange Rate Policies*

*General Principles*

Article IV, Section 3(a), provides that "The Fund shall oversee the international monetary system in order to ensure its effective operation, and shall oversee the compliance of each member with its obligations under Section 1 of this Article." Article IV, Section 3(b), provides that in order to fulfill its functions under 3(a), "the Fund shall exercise firm surveillance over the exchange rate policies of members, and shall adopt specific principles for the guidance of all members with respect to those policies." Article IV, Section 3(b), also provides that "The principles adopted by the Fund shall be consistent with cooperative arrangements by which members maintain the value of their currencies in relation to the value of the currency or currencies of other members, as well as with other exchange arrangements of a member's choice consistent with the purposes of the Fund and Section 1 of this Article. These principles shall respect the domestic social and political policies of members, and in applying these principles the Fund shall pay due regard to the circumstances of members." In addition, Article IV, Section 3(b), requires that "Each member shall provide the Fund with the information necessary for such surveillance, and, when requested by the Fund, shall consult with it on the member's exchange rate policies."

The principles and procedures set out below, which apply to all members whatever their exchange arrangements and whatever their balance of payments position, are adopted by the Fund in order to perform its functions under Section 3(b). They are not necessarily comprehensive and are subject to reconsideration in the light of experience. They do not deal directly with the Fund's responsibilities referred to in Section 3(a), although it is recognized that there is a close relationship between domestic and international economic policies. This relationship is emphasized in Article IV which includes the following provision: "Recognizing . . . that a principal objective [of the international monetary system] is the continuing development of the orderly underlying conditions that are necessary for financial and economic stability, each member undertakes to collaborate with the Fund and other members to assure orderly exchange arrangements and to promote a stable system of exchange rates."

*Principles for the Guidance of Members' Exchange Rate Policies*

A. A member shall avoid manipulating exchange rates or the international monetary system in order to prevent effective balance of payments adjustment or to gain an unfair competitive advantage over other members.

B. A member should intervene in the exchange market if necessary to counter disorderly conditions which may be characterized inter alia by disruptive short-term movements in the exchange value of its currency.

* Reproduced from *Annual Report of the Executive Directors for the Fiscal Year Ended April 30, 1977*, Washington, D.C., International Monetary Fund, Appendix II, pp. 107-109.

36

C. Members should take into account in their intervention policies the interests of other members, including those of the countries in whose currencies they intervene.

*Principles of Fund Surveillance over Exchange Rate Policies*

1. The surveillance of exchange rate policies shall be adapted to the needs of international adjustment as they develop. The functioning of the international adjustment process shall be kept under review by the Executive Board and Interim Committee and the assessment of its operation shall be taken into account in the implementation of the principles set forth below.

2. In its surveillance of the observance by members of the principles set forth above, the Fund shall consider the following developments as among those which might indicate the need for discussion with a member:

(i) protracted large-scale intervention in one direction in the exchange market;

(ii) an unsustainable level of official or quasi-official borrowing, or excessive and prolonged short-term official or quasi-official lending, for balance of payments purposes;

(iii) (a) the introduction, substantial intensification, or prolonged maintenance, for balance of payments purposes, of restrictions on, or incentives for, current transactions or payments, or

(b) the introduction or substantial modification for balance of payments purposes of restrictions on, or incentives for, the inflow or outflow of capital;

(iv) the pursuit, for balance of payments purposes, of monetary and other domestic financial policies that provide abnormal encouragement or discouragement to capital flows; and

(v) behavior of the exchange rate that appears to be unrelated to underlying economic and financial conditions including factors affecting competitiveness and long-term capital movements.

3. The Fund's appraisal of a member's exchange rate policies shall be based on an evaluation of the developments in the member's balance of payments against the background of its reserve position and its external indebtedness. This appraisal shall be made within the framework of a comprehensive analysis of the general economic situation and economic policy strategy of the member and shall recognize that domestic as well as external policies can contribute to timely adjustment of the balance of payments. The appraisal shall take into account the extent to which the policies of the member, including its exchange rate policies, serve the objectives of the continuing development of the orderly underlying conditions that are necessary for financial stability, the promotion of sustained sound economic growth, and reasonable levels of employment.

*Procedures for Surveillance*

I. Each member shall notify the Fund in appropriate detail within thirty days after the Second Amendment becomes effective of the exchange arrangements it intends to apply in fulfillment of its obligations under Article IV, Section 1. Each member shall also notify the Fund promptly of any changes in its exchange arrangements.

37

II. Members shall consult with the Fund regularly under Article IV. The consultations under Article IV shall comprehend the regular consultations under Articles VIII and XIV. In principle such consultations shall take place annually, and shall include consideration of the observance by members of the principles set forth above as well as of a member's obligations under Article IV, Section 1. Not later than three months after the termination of discussions beween the member and the staff, the Executive Board shall reach conclusions and thereby complete the consultation under Article IV.

III. Broad developments in exchange rates will be reviewed periodically by the Executive Board, inter alia in discussions of the international adjustment process within the framework of the World Economic Outlook. The Fund will continue to conduct special consultations in preparing for these discussions.

IV. The Managing Director shall maintain close contact with members in connection with their exchange arrangements and exchange policies, and will be prepared to discuss on the initiative of a member important changes that it contemplates in its exchange arrangements or its exchange rate policies.

V. If, in the interval between Article IV consultations, the Managing Director, taking into account any views that may have been expressed by other members, considers that a member's exchange rate policies may not be in accord with the exchange rate principles, he shall raise the matter informally and confidentially with the member, and shall conclude promptly whether there is a question of the observance of the principles. If he concludes that there is such a question, he shall initiate and conduct on a confidential basis a discussion with the member under Article IV, Section 3(b). As soon as possible after the completion of such a discussion, and in any event not later than four months after its initiation, the Managing Director shall report to the Executive Board on the results of the discussion. If, however, the Managing Director is satisfied that the principles are being observed, he shall informally advise all Executive Directors, and the staff shall report on the discussion in the context of the next Article IV consultation; but the Managing Director shall not place the matter on the agenda of the Executive Board unless the member requests that this procedure be followed.

VI. The Executive Directors shall review annually the general implementation of the Fund's surveillance over members' exchange rate policies.

38

# USE OF INTERNATIONAL MONETARY FUND'S RESOURCES TO MEET BALANCE OF PAYMENTS NEEDS

INTERNATIONAL MONETARY FUND

Members of the International Monetary Fund (IMF) may draw on its financial resources to meet their balance of payments needs. They may use the reserve tranche and, under tranche policies, the four credit tranches. In addition, there are three permanent facilities for specific purposes—the facility for compensatory financing of export fluctuations (established in 1963 and liberalized in 1975 and 1979), the buffer stock financing facility (established in 1969), and the extended facility (established in 1974).

Furthermore, members may make use of temporary facilities established by the fund with borrowed resources. For 1974 and 1975, for example, following the sharp rise in oil prices, the fund provided assistance under a temporary oil facility designed to help members meet the increased cost of imports of petroleum and petroleum products. In 1978, a supplementary financing facility was established with borrowed resources amounting to special drawing rights (SDRs) of 7.784 billion from 13 member countries or their institutions and the Swiss National Bank. In March 1981, a policy of enlarged access to the fund's resources was adopted, whereby after all supplementary financing had been committed and additional borrowing arrangements had been concluded, the fund would continue to provide assistance on a scale similar to that under the supplementary financing facility.

This chapter describes the policies and principles governing members' access to the fund's general resources under the tranche policies and the permanent facilities. For any drawing, a member is required to represent to the fund that the desired purchase is needed because of its balance of payments or reserve position or developments in its reserves.

## USE OF RESOURCES

When members draw on the fund, they use their own currency to purchase the currencies of other member countries or SDRs held by the General Resources Account. Thus, a drawing results in an increase in the fund's holdings of the purchasing member's currency and a corresponding decrease in the fund's holdings of other currencies or SDRs that are sold. Thus, a drawing results in an increase in the fund's holdings of the purchasing member's currency and a corresponding decrease in the fund's holdings of other currencies or SDRs that are sold. Within a prescribed time, or earlier when its balance of payments and reserve position improves, a member must reverse the transaction, except a reserve tranche purchase, by buying back its own currency with SDRs or currencies specified by the fund.

Reprinted from "How Members Use Fund's Resources to Meet Balance of Payments Needs, IMF Survey, Supplement on the Fund, May 1981," pp. 6–10, with permission of the IMF, Copyright 1981.

Usually, repurchases are required to be made within 3–5 years after the date of purchase; but under the extended facility, the period for repurchases is within 4–10 years, under the oil facility within 3–7 years, and under the supplementary financing facility and the enlarged access policy within 3½–7 years. In addition, a member is expected normally to repurchase as its balance of payments and reserve position improves, and the fund may convert this expectation into an obligation.

## RESERVE TRANCHE

If the fund's holdings of a member's currency are less than its quota, the difference is called the reserve tranche. A member using the credit tranches or the extended fund facility has the option either to use or to retain a reserve tranche position. Purchases in the reserve tranche are subject to balance of payments need but not to prior challenge, to economic policy conditions, or to repurchase requirements.

## CREDIT TRANCHE

Further purchases are made in four credit tranches each of 25% of the member's quota. In the past, the total of purchases under credit tranche policies was normally limited to 100% of the member's quota, an amount that would raise the fund's holdings of the member's currency to 200% of its quota. However, in response to the structural and deep-rooted nature of the payments imbalance currently confronting many members, the fund is placing greater emphasis on programs involving adjustment periods of longer duration—2 or 3 years—and provision for a larger access above these limits, as detailed in the following section on Enlarged Access Policy.

All requests for the use of the fund's resources other than use of the reserve tranche are examined by the fund to determine whether the proposed use would be consistent with the provisions of the articles and with fund policies. Under credit tranche policies, use must be in support of economic measures designed to overcome a member's balance of payments difficulties. Prior to submission to the fund of a request to purchase, a member discusses with fund staff its adjustment program, includng fiscal, monetary, exchange rate, and trade and payments policies for the program period, which is usually the next 12 months but which may be extended up to 3 years in situations that call for adjustment efforts to be spread over a longer period. The criteria used by the fund in determining whether its assistance should be made available are more liberal when the request is in the first credit tranche (fund holdings of a member's currency rising from 100% but not above 125% of the member's quota) than when it is in the higher credit tranches (i.e., when the fund's holdings following the drawing exceed 125% of quota).

In the case of a request for a first credit tranche drawing, the member is expected to show that it is making reasonable efforts to overcome its difficulties. In practice, this criterion has often meant that, when differences of judgment arise, the member receives the benefit of the doubt. Use of the first credit tranche may be requested either in the form of a direct purchase or under a standby arrangement. A member requesting a direct purchase expects to draw the full amount after approval of the request; under a standby arrangement, a member may request drawings in installments during the period of the standby arrangement.

Requests for purchases in the higher credit tranches require substantial justification. Such purchases are always made under standby arrangements. The stabilization program presented by a member requesting assistance in an upper tranche is comprehensive. The amount available under a standby arrangement in the upper credit tranches is phased to be available in portions at specified intervals during the standby period, and the member's right to draw is always subject to the observance of certain key policy indicators described in the program or to a further review of the situation. The indicators, also called performance criteria, cover credit policy and policy on trade and payments restrictions; they may also extend to the financing requirements of the government, contracting or use of short-term and medium-term foreign debt, and changes in external reserves. Performance criteria allow both the member and the fund to assess progress in the implementation of policies during the standby arrangement. Failure to meet the criteria signals the need to examine whether further measures are necessary to achieve the objectives of the program. In this case, the member consults with the fund in order to reach understandings on needed changes.

## COMPENSATORY FINANCING

The compensatory financing facility is designed to extend the fund's financial support to member countries—particularly primary producing countries—encountering payments difficulties produced by temporary shortfalls in export proceeds. Members having a balance-of-payments need may draw on the fund under this facility to compensate for such shortfalls, if the fund is satisfied that the problem that caused the shortfall is a short-term one and is largely attributable to circumstances beyond the control of the member and that the member will cooperate with the fund in an effort to solve its balance-of-payments difficulties. The export shortfall for the latest 12-month period preceding a drawing request is established against an estimate of the medium-term trend of the member's exports, which is centered on the shortfall period and based on a judgmental forecast of export prospects. Members may draw up to 100% of quota. Requests for drawings beyond 50% of quota are met only if the fund is satisfied that the member has been cooperating in an effort to solve its balance-of-payments problems. Export shortfalls may include, at the option of the member, receipts from travel and workers' remittances.

Drawings under the compensatory financing facility—as well as those under the buffer stock financing and oil facilities—are additional to those which members may make in the reserve tranche and under tranche policies. An effect of this treatment is that the reserve tranche is retained as a reserve asset that can be mobilized by the member with minimum delay. This treatment also means that the member may use the fund's resources in an amount that could increase the fund's holdings of its currency beyond 200% of quota, subject to a waiver of that limit. The provisions for repurchases applying to drawings under the compensatory financing facility are generally the same as those for drawings under tranche policies.

## BUFFER STOCK FINANCING

The purpose of the buffer stock financing facility is to assist in the financing of members' contributions to international buffer stocks of primary products when members having balance-of-payments difficulties participate in such arrangements under commodity agreements that meet appropriate criteria, such as the principles laid down by the United Nations.

Drawings for buffer stock financing may be made up to the equivalent of 50% of quota. As under the compensatory financing facility, the fund is prepared to waive the 200% limit on fund holdings of the member's currency. The member is also expected to cooperate with the fund in an effort to solve its balance-of-payments difficulties. The fund has, to date, authorized the use of fund resources in connection with tin, cocoa, and sugar buffer stocks, but drawings have been made only with respect to tin and sugar. Repurchase provisions are broadly the same as for credit tranche purchases.

## EXTENDED FACILITY

Under the extended facility, the fund may provide assistance to members to meet their balance-of-payments deficits for longer periods and in amounts larger in relation to quotas than under the credit tranche policies. For example, a member might apply for assistance under the facility if it has serious payments imbalances relating to structural maladjustments in production, trade, and prices and if it intends to implement a comprehensive set of corrective policies for 2 or 3 years. Or, use of the facility might be indicated by an inherently weak balance-of-payments position that prevents the pursuit of an active development policy.

In requesting an extended arrangement, a member is expected to present a program setting forth the objectives and policies for the whole period of the extended arrangement, as well as a detailed statement of the policies and measures that it will follow in each 12-month period to meet the objectives of the program. Purchases are phased and made subject to performance clauses relating to implementation of key policy measures. Drawings under extended arrangements may take place over periods of up to 3 years. Purchases outstanding under the extended facility may not exceed 140% of the member's quota (nor raise fund holdings of a member's currency above 265% of the member's quota, excluding holdings relating to compensatory financing, buffer stock financing, and the oil facility). Purchases under the extended facility are additional to those a member may make under the first credit tranche. Repurchases under the extended facility must be made in 12 equal installments within 4–10 years after each purchase.

## SUPPLEMENTARY FINANCING FACILITY

The purpose of this facility is to enable the fund to provide supplementary financing in conjunction with the use of the fund's ordinary resources to all members of the fund facing serious payments imbalances that are large in relation to their quotas. The supplementary financing facility is used only in support of economic programs under standby arrangements reaching into the upper credit tranches or under extended arrangements. Such drawings are subject to the relevant policies of the fund, including those on conditionality, phasing, and performance criteria. The period of such standby arrangements will normally exceed 1 year and may extend up to 3 years in appropriate cases.

The fund may approve a standby or extended arrangement that provides for supplementary financing within 3 years from February 23, 1979.

The amounts available to a member qualifying to use the supplementary financing facility under a standby or an extended arrangement are apportioned between ordinary resources and supplementary financing in prescribed proportions, as shown in Table I. If a member has already used all or part of its credit tranches, the standby

TABLE I

Amounts Available Under Fund's Supplementary Financing Facility/Enlarged Access Policy in Percent of Quota[a,b]

| | Use of ordinary resources | Supplementary financing[b] enlarged access policy[c] |
|---|---|---|
| Regular tranches | 100 | 102.5[b] |
| First credit tranche | 25 | 12.5 |
| Second credit tranche | 25 | 30.0 |
| Third credit tranche | 25 | 30.0 |
| Fourth credit tranche | 25 | 30.0 |
| Extended facility | 140 | 140.0[b] |

[a]Data from IMF Treasurer's Department

[b]In special circumstances, resources beyond these limits could be made available from the supplementary financing facility.

[c]Following the full commitment of supplementary financing, additional resources beyond these limits could be made available under the enlarged access policy in accordance with guidelines adopted by the fund from time to time.

arrangement will include the amount of supplementary financing that would have been available under the decision on supplementary financing if the earlier use of the credit tranches had been made under the decision. A similar principle will apply with respect to extended arrangements. In special circumstances, when the magnitude and the nature of the member's need for financing from the fund are such as to justify additional amounts, and when access to the credit tranches or to the extended facility, as shown in Table II–VI has been exhausted, additional amounts may be provided by the fund, and these additional purchases are made wholly under the supplementary financing facility.

## ENLARGED ACCESS POLICY

Following the full commitment of supplementary financing in April 1981 and the conclusion of some borrowing arrangements in support, the fund now provides assistance under its enlarged access policy on the same terms as previously under the supplementary financing facility. The amount of assistance under this policy is determined according to guidelines adopted by the fund from time to time. Present guidelines specify limits of 150% of quota annually or 450% over a 3-year period; at the same time, a limit of 600% of quota, net of scheduled repurchases, applies on the cumulative use of fund resources. These limits, which may be exceeded in exceptional circumstances, exclude drawings under the compensatory and buffer stock financing facilities or outstanding drawings under the oil facility. The fund is discussing additional borrowing arrangements in support of its enlarged access policy. The fund may approve a standby or extended arrangement that provides for enlarged access until the Eighth General Review of Quotas becomes effective.

## CHARGES FOR USE OF RESOURCES

The fund applies charges for the use of its resources, except for reserve tranche purchases. A service charge of 0.5% is payable on purchases other than reserve tranche purchases. In addition, the fund levies charges on balances of members' currencies resulting from purchases. The rate of charge on purchases in the credit tranches and under the extended fund facility, the compensatory financing facility, and the buffer stock financing facility is determined at the beginning of each finan-

TABLE II

Exchange Rate Arrangements, September 30, 1981

| Member (currency) | Exchange rate maintained against[a,e] | Member (currency) | Exchange rate maintained against[a,e] |
|---|---|---|---|
| Afghanistan (Afghani) | | Brazil (cruzeiro) Burma (kyat) | SDR |
| Algeria (dinar)[b] | bskt | Burundi (franc) | $ |
| Argentina (peso)[b] | | Cameroon (franc) | F |
| Australia (dollar) | | Canada (dollar) | |
| Austria (schilling) | bskt | Cape Verde | |
| | | (escudo) | bskt |
| Bahamas (dollar)[b] | $ | Central African | |
| Bahrain (dinar) | | Republic (franc) | F |
| Bangladesh (taka)[b] | bskt | | |
| Barbados (dollar)[b] | $ | Chad (franc) | F |
| Belgium (franc)[b] | | Chile (peso) | $ |
| | | China, People's | |
| Benin (franc) | F | Republic | |
| Bhutan (ngultrum) | Re | (renminbi)[b] | bskt |
| Bolivia (peso) | | Colombia (peso) | |
| Botswana (pula) | bskt | Comoros (franc) | F |

TABLE II—Continued

| Member (currency) | Exchange rate maintained against[a,e] | Member (currency) | Exchange rate maintained against[a,e] |
|---|---|---|---|
| Congo (franc) | F | Kenya (shilling)[b] | SDR |
| Costa Rica (colón)[b] | $ | Korea (won) | |
| Cyprus (pound) | bskt | Kuwait (dinar) | bskt |
| Denmark (krone) | | Leo People's Dem. Rep. (kip)[b] | $ |
| Djibouti (franc) | $ | Lebanon (pound)[c] | |
| Dominica (East Caribbean dollar) | $ | Lesotho (maloti)[b] | R |
| | | Liberia (dollar) | $ |
| Dominican Rep. (peso)[b] | $ | Libya (dinar) | $ |
| Ecuador (sucre)[b] | $ | Luxembourg (franc)[b] | |
| Egypt (pound)[b] | $ | Madagascar (franc) | F |
| El Salvador (colón) | $ | Malawi (kwacha) | SDR |
| Equatorial Guinea (ekwele)[b] | P | Malaysia (ringgit) | bskt |
| Ethiopia (birr)[b] | $ | Maldives (rufiyaa)[b] | |
| Fiji (dollar) | bskt | Mali (franc) | F |
| Finland (markka) | bskt | Malta (pound) | bskt |
| France (franc) | | Mauritania (ouguiya) | bskt |
| Gabon (franc) | F | Mauritius (rupee)[b] | SDR |
| Gambia, The (dalasi) | £ | Mexico (peso) | |
| Germany, Fed. Rep. (deutsche mark) | | Morocco (dirham) | |
| | | Nepal (rupee)[b] | $ |
| Ghana (cedi)[b] | | Netherlands (guilder) | |
| Greece (drachma)[b] | | New Zealand (dollar) | |
| Grenada (East Carribean dollar)[b] | $ | Nicaragua (córdoba)[b] | $ |
| | | Niger (franc) | F |
| Guatemala (quetzal) | $ | Nigeria (naira) | |
| Guinea (syli) | SDR | Norway (krone) | bskt |
| Guinea-Bissau (peso) | SDR | Oman (rial Omani) | $ |
| Guyana (dollar) | | Pakistan (rupee) | $ |
| Haiti (gourde) | $ | Panama (balboa) | $ |
| Honduras (lempira) | $ | Papua New Guinea (kina) | bskt |
| Iceland (króna)[b] | | Paraguay (guaraní)[b] | $ |
| India (rupee) | | Peru (sol) | |
| Indonesia (rupiah) | | Philippines (peso) | |
| Iran (rial)[b] | SDR | Portugal (escudo) | |
| Iraq (dinar) | $ | Qatar (riyal) | |
| Ireland (pound) | | Romania (leu)[b] | $ |
| Israel (shekel) | | Rwanda (franc) | $ |
| Italy (lira)[b] | | St. Lucia (East Caribbean dollar) | $ |
| Ivory Coast (franc) | F | | |
| Jamaica (dollar)[b] | $ | St. Vincent (East Caribbean dollar) | $ |
| Japan (yen) | | | |
| Jordan (dinar) | SDR | São Tomé and Principe (dobra) | SDR |
| Kampuchea, Democratic (riel)[b,c,d] | | | |

(continued)

TABLE II—Continued

| Member (currency) | Exchange rate maintained against[a,e] | Member (currency) | Exchange rate maintained against[a,e] |
|---|---|---|---|
| Saudi Arabia (riyal) | | Turkey (lira)[b] | |
| Senegal (franc) | F | Uganda (shilling)[c] | |
| Seychelles (rupee) | SDR | United Arab | |
| Sierra Leone | | Emirates (dirham) | |
| (leone) | SDR | United Kingdom | |
| Singapore (dollar) | bskt | (pound) | |
| Solomon Islands | | United States | |
| (dollar) | bskt | (dollar) | |
| Somalia (shilling)[b] | $ | Upper Volta (franc) | F |
| South Africa (rand)[b] | | Uruguay | |
| Spain (peseta) | | (new peso) | |
| Sri Lanka (rupee) | | Vanuatu (vatu) | SDR |
| Sudan (pound)[b] | $ | Venezuela | |
| Suriname (guilder) | $ | (bolívar)[b] | $ |
| Swaziland | | Viet Nam (dong)[b] | SDR |
| (lilangeni)[b] | R | Western Samoa | |
| Sweden (krona) | bskt | (tala) | |
| Syrian Arab Rep. | | Yemen Arab Rep. | |
| (pound)[b] | $ | (rial) | $ |
| Tanzania (shilling) | bskt | Yemen, People's | |
| Thailand (baht) | | Dem. Rep. (dinar) | $ |
| Togo (franc) | F | Yugoslavia (dinar) | |
| Trinidad and | | Zaïre (zaïre) | SDR |
| Tobago (dollar) | $ | Zambia (kwacha) | SDR |
| Tunisia (dinar) | bskt | Zimbabwe (dollar) | bskt |

[a]Symbols: $, U.S. dollar; £, pound sterling; F, French franc; P, Spanish peseta; R, South African rand; SDR, special drawing right; bskt, currency basket other than SDR; Re, Indian rupee. Data: IMF Treasurer's and Exchange and Trade Relations Departments

[b]Member maintains multiple currency practices and/or dual exchange market. A description of the member's exchange system as of December 31, 1980 is given in the *Annual Report on Exchange Arrangements and Exchange Restrictions, 1981*.

[c]Exchange rate data not available.

[d]Information on exchange arrangements not available.

[e]Blanks indicate those members that describe their exchange rate arrangements as floating independently or as adjusting according to a set of indicators, and certain other members whose exchange arrangements are not otherwise described in this table. Belgium, Denmark, France, Germany, Ireland, Italy, Luxembourg, and the Netherlands are participating in the exchange rate and intervention mechanism of the European Monetary System and maintain maximum margins of 2.25 per cent (in the case of the Italian lira 6 per cent) for exchange rates in transactions in the official markets between their currencies and those of the other countries in this group. No announced margins are observed for other currencies.

## TABLE III

Exchange Arrangements of Fund Members, June 30, 1981[a]

| Type of exchange arrangement | Industrial countries | Oil-exporting countries | Developing countries — Non-oil developing countries — Net oil exporters | Major exporters of manufactures | Other developing countries | Total |
|---|---|---|---|---|---|---|
| Currency pegged to | | | | | | |
| U.S. dollar | — | 4 | 4 | — | 30 | 38 |
| French franc | — | — | 2 | — | 12 | 14 |
| Other currencies | — | — | — | — | 4 | 4 |
| SDR | — | 1 | — | — | 13 | 14 |
| Other composite | 4 | 2 | 2 | 1 | 13 | 22 |
| Adjusted according to a set of indicators | — | — | 1 | 2 | 1 | 4 |
| Cooperative exchange arrangements | 8 | — | — | — | — | — |
| Other | 8 | 5 | 3 | 6 | 14 | 36 |
| Total | 20 | 12 | 12 | 9 | 87 | 140[b] |

[a] Data from IMF *Annual Report 1981*.
[b] Excluding Democratic Kampuchea, for which no current information is available.

## TABLE IV

Financial Facilities of the Fund and Their Conditionality

Reserve tranche
  Condition—balance-of-payments need.
Tranche policies
  First credit tranche
    Program representing reasonable efforts to overcome balance of payments difficulties; performance criteria and installments not used.
  Higher credit tranches
    Transactions requiring that member give substantial justification of its efforts to overcome balance-of-payments difficulties; resources normally provided in the form of standby arrangements that include performance criteria and drawings in installments.
Extended facility
  Medium-term program for up to 3 years to overcome structural balance-of-payments; maladjustments; detailed statement of policies and measures for first and subsequent 12-month periods; resources provided in the form of extended arrangements that include performance criteria and drawings in installments.
Compensatory financing facility
  Existence of temporary export shortfall for reasons largely beyond the member's control; member cooperates with fund in an effort to find appropriate solutions for any balance of payments difficulties.
Buffer stock financing facility
  Existence of an international buffer stock accepted as suitable by fund; member expected to cooperate with fund as in the case of compensatory financing.
Supplementary financing facility/enlarged access policy
  For use in support of programs under standby arrangements reading into the upper credit tranches or beyond, or under extended arrangements, subject to relevant policies on conditionality, phasing, and performance criteria.

TABLE V

Possible Cumulative Purchases in Percent of Quota[a]

| | Tranche policy | Extended facility |
|---|---|---|
| Credit tranches | | |
| 4 x 25 | 100.00[b] | — |
| 1 x 25 | — | 25.0 |
| Extended facility | — | 140.0 |
| Supplementary financing/ enlarged access policy[c] | | |
| (1 × 12.5 plus 3 × 30) | 102.5 | — |
| With extended facility | — | 140.0 |
| Subtotal | 202.5 | 305.0 |
| Compensatory financing | 100.0 | 100.0 |
| Buffer stock financing | 50.0 | 50.0 |
| Cumulative total[d] | 352.5 | 455.0 |

[a]Amounts are exclusive of available reserve tranche positions. Data from IFM Treasurer's Department.

[b]This limit may be waived by a decision of the Executive Board.

[c]The total amounts that a member could purchase under a standby or extended arrangement could exceed these limits in special circumstances, that is, when the magnitude and the nature of the member's need for financing from the fund are such as to justify additional amounts. Any additional amounts that may be made available in special circumstances would be met wholly with supplementary financing and, following the full commitment of supplementary financing, with borrowed resources under the enlarged access policy.

[d]In addition, some members have used oil facility drawings. The average use by those members was equal to 75% of quota.

cial year on the basis of the estimated income and expense of the fund during the year and a target amount of net income, which for the financial year beginning May 1, 1981, is the balance between income and expense and for each subsequent year will be 3% of the fund's reserves at the beginning of the year. The rate of charge calculated for the financial year 1982 is 6.25% a year. Charges on balances in excess of 200% of members' quotas resulting from purchases under standby arrangements or in excess of 140% of quota under extended arrangements granted between April 21, 1978, and February 23, 1979, when the supplementary financing facility became effective, are the average yield to constant 5-year maturity of United States government securities in New York over the 6 months preceding the determination of the rate of charge, rounded upward to the nearest ¼ of 1%, plus ¼ of 1% per annum.

There are separate schedules of charges for the use of the supplementary financing facility and of borrowed resources under the enlarged access policy, under which charges are equal to the cost of borrowing by the fund plus a margin of 0.2% a year. The margin charged on the use of supplementary financing is increased to 0.325% after the first 3½ years. The rate of interest paid by the fund on the amounts made available to it under the supplementary financing facility is the average yield for each half of a calendar year on United States government securities with a

TABLE VI

Fund Charges on Transactions in Percent a Year[a]

Credit tranches, compensatory financing facility, and buffer stock financing facility

|  | July 1, 1974, through March 31, 1977 | April 1, 1977, through April 30, 1981 | May 1, 1981 |
|---|---|---|---|
| Service charge | 0.5 | 0.5 | 0.5 |
| Up to 1 year | 4.0 | 4.375 | 6.25 |
| 1–2 years | 4.5 | 4.875 | 6.25 |
| 2–3 years | 5.0 | 5.375 | 6.25 |
| 3–4 years | 5.5 | 5.875 | 6.25 |
| 4–5 years | 6.0 | 6.375 | 6.25 |

Extended facility

|  |  |  |  |
|---|---|---|---|
| Service charge | 0.5 | 0.5 | 0.5 |
| Up to 5 years | b | b | 6.25 |
| 5–6 years | 6.5 | 6.875 | 6.25 |
| 6–7 years | 6.5 | 6.875 | ·6.25 |
| 7–8 years | 6.5 | 6.875 | 6.25 |

Oil facility

|  | For 1974 | For 1975 |
|---|---|---|
| Service charge | 0.5 | 0.5 |
| Up to 1 year | 6.875 | 7.625 |
| 1–2 years | 6.875 | 7.625 |
| 2–3 years | 6.875 | 7.625 |
| 3–4 years | 7.000 | 7.750 |
| 4–5 years | 7.125 | 7.875 |
| 5–6 years | 7.125 | 7.875 |
| 6–7 years | 7.125 | 7.875 |

Supplementary financing facility

| Service charge | 0.5 |
|---|---|
| Up to 3½ years | Rate of interest paid by the fund plus 0.2% |
| Over 3½ years | Rate of interest paid by the fund plus 0.325% |

Enlarged access policy

| Service charge | 0.5 |
|---|---|
| Periodic charge | Net cost of borrowing by the fund plus 0.2% |

[a]Except for service charge, which is payable once on a transaction and is stated as a percentage of the amount of the transaction. Data from IMF Treasurer's Department.

[b]Same as for regular tranches.

maturity of 5 years, rounded upward to the nearest ¼ of 1%. Under the policy on enlarged access, the cost of borrowing will be determined by taking into account the income from the investment of any undisbursed proceeds of borrowing and any net gain or loss from exchange valuation adjustments of the currency balances representing the undisbursed proceeds, in addition to interest and other costs paid or accrued to lenders.

## REMUNERATION ON CREDITOR POSITIONS

As drawings are made on the fund, the fund's holdings of the currencies purchased are reduced. When the fund's holdings of a member's currency are reduced below the "norm," the member acquires a creditor position in the fund on which it earns remuneration (i.e., interest). The "norm" is not a uniform percentage of quota: for each member it is the sum of 75% of its quota prior to the Second Amendment of the Articles of Agreement plus the amounts of any subsequent quota increases.

Prior to the Second Amendment, a member's creditor position was the amount by which the fund's holdings of its currency were below 75% of quota. Now it is the amount by which the fund's holdings are below the "norm." As quotas are increased in the future, the "norm" will, over time, gradually be raised toward 100%.

The fund pays remuneration on creditor positions at a rate determined by a formula based on short-term market interest rates in the United States, the Federal Republic of Germany, the United Kingdom, France, and Japan. A weighted average of daily interest rates in the five countries is calculated for the 3-week period ending two business days before the start of each calendar quarter. Since May 1, 1981, the rate of remuneration for a calendar quarter is 85% of the interest rate on the SDR, which, in turn, is 100% of the combined market rate.

# Part IX

## Perspectives and Implications

Klein

# SOME ECONOMIC SCENARIOS
# FOR THE 1980's

Nobel Memorial Lecture, 8 December, 1980

by

## LAWRENCE R. KLEIN

University of Pennsylvania, Philadelphia, Pennsylvania 19104, USA

At the beginning of a decade it is tempting to look ahead for the next ten years. In addition to end-of-decade targets, there is considerable interest, at the present time, in end-of-century targets. Analysis of multi-decade developments depends on an even longer view, and I shall focus my attention on the medium-term outlook for one decade as much as possible.

This analysis will proceed through the medium of two econometric models, one for the United States and one for the World as a whole. I shall refer to available simulations of the Wharton Model of the United States for a single (large) country appraisal. The U.S. weight in the total for all OECD countries is more than one-third of aggregate production. Any sizeable action by the U.S. is, therefore, reflected in the totals.

Other countries are going in their own chosen directions, and it will be useful to try pull them all together in world model simulations from the equation system of project LINK. The LINK system is an amalgamation of econometric models from 17 OECD industrial countries, eight socialist countries, and four regional models of developing countries.[1]

## BASE CASE – UNITED STATES

First, let us consider a baseline simulation for the United States. There is general recognition that something large (a "sea change") has come over the leading countries in the OECD area. In the case of the United States, real GNP growth, from the end of World War II until the end of the 1960's, averaged just under 4%. The baseline projection shows a distinct tendency for the economy to hover in the neighborhood of 3% growth. Slower growth, more inflation, high interest costs, an elevated rate of unemployment and balance-of-payments problems are manifest in the long sequence of tables generated by the Wharton Annual Model. Some annual growth rates, recorded at five year intervals, are listed in Table 1.

[1] This is the present country/regional make-up of the LINK system. In some versions of system simulations – set up a year or so ago – there are four fewer OECD country models and one fewer centrally planned model.

This is a pattern familiar not only in the United States but in other industrial countries. The changed economic profile between the post World War II recovery/expansion period (1945−1970) and the period since 1970, through the end of the decade, is a result of some profound changes in the underlying economic environment. They are related to such major events as

   (i)   energy supply-demand imbalance and a shift from inexpensive to dear prices;

  (ii)   pressure on available food supplies and a shift towards higher food prices;

 (iii)   accelerated inflation;

 (iv)   declining productivity growth;

Table 1. Five Year Average Annual Percentage Growth Rates

Five Years Ending:

|  | 1960 | 1965 | 1970 | 1975 | 1980[1] | 1985[1] | 1990[1] |
|---|---|---|---|---|---|---|---|
| Real GNP | 2.4 | 4.7 | 3.0 | 2.3 | 3.3 | 3.0 | 3.0 |
| GNP Deflator | 2.4 | 1.6 | 4.2 | 6.8 | 7.3 | 8.0 | 7.6 |
| Nominal GNP | 4.9 | 6.3 | 7.4 | 9.2 | 10.8 | 11.3 | 10.8 |
| Real Consumption | 2.8 | 4.3 | 3.7 | 3.0 | 3.6 | 2.7 | 3.0 |
|   Durables | 0.1 | 6.9 | 3.9 | 4.9 | 3.5 | 3.1 | 2.6 |
|   Nondurables | 2.4 | 3.2 | 3.0 | 1.6 | 2.8 | 1.4 | 2.0 |
|   Services | 4.1 | 4.6 | 4.3 | 3.6 | 4.3 | 3.4 | 3.8 |
| Total Real Investment | 0.2 | 7.3 | 0.6 | −1.6 | 5.3 | 5.9 | 3.8 |
|   Nonresidential | 1.5 | 7.7 | 2.8 | 0.6 | 4.8 | 3.2 | 4.5 |
|   Residential | −0.1 | 4.3 | −1.3 | −0.8 | 2.2 | 7.5 | 2.2 |
| Real Trade Flows |  |  |  |  |  |  |  |
|   Imports | 5.5 | 6.2 | 9.9 | 0.5 | 8.8 | 3.1 | 3.7 |
|   Exports | 5.1 | 6.5 | 6.4 | 6.0 | 7.7 | 3.9 | 3.3 |
| Real Government |  |  |  |  |  |  |  |
|   Spending | 2.8 | 3.9 | 3.6 | 1.0 | 1.4 | 1.9 | 2.2 |
|   Federal | 0.9 | 2.1 | 2.0 | −2.7 | 2.0 | 3.3 | 2.0 |
|   State and Local | 5.1 | 5.9 | 5.0 | 3.6 | 1.0 | 0.9 | 2.3 |
| Employment | 1.1 | 1.6 | 2.0 | 1.5 | 2.7 | 1.4 | 1.3 |
| Civilian Labor Force | 1.4 | 1.3 | 2.1 | 2.3 | 2.5 | 1.3 | 1.1 |

[1] Forecast values: Wharton EFA. Real values in prices of 1972

  (v)   rapid expansion of the labor force;

 (vi)   increasing attention paid to problems of quality of life.

These issues started to appear in the late 1960's, many in the wake of the Vietnam war, and prevailed during the 1970's which proved to be a turbulent decade for the U. S. economy. Averages for the past decade, after smoothing of cyclical movements show changed trends in growth, inflation, unemployment rates, interest rates, internal deficits, and external deficits. It is also a period in which the U. S. economy became highly internationalized; i. e., increasingly subject to pressures of international events, less self-contained, and not at all insulated. The differences between the 1970's and earlier decades are matters of recorded history. The average trends estimated from the decade 1971−1980 govern the projections for the 1980's and 1990's. There is no indication that the trends of the 1970's were aberrations and that we are likely to return to the heady days of earlier postwar decades. The reasons for this changed performance are contained in the six points listed above, but in this essay, I want to look at the problem through the medium of econometric model simulation, rather than point-by-point analysis of the six items.[2]

Table 2. Selected U. S. Economic Indicators Projections to 1990.

WHARTON ANNUAL AND INDUSTRY FORECASTING MODEL
PRE-MEETING CONTROL SOLUTION - OCT 1980

SELECTED INDICATORS

| I T E M | 1980 | 1981 | 1982 | 1983 | 1984 | 1985 | 1986 | 1987 | 1988 | 1989 | 1990 |
|---|---|---|---|---|---|---|---|---|---|---|---|
| GROSS NATIONAL PRODUCT (CUR $) | 2559.2 | 2659.0 | 3212.8 | 3552.3 | 3939.6 | 4362.5 | 4859.2 | 5412.7 | 5983.2 | 6614.8 | 7286.2 |
| % CHANGE | 8.0 | 11.7 | 12.4 | 10.6 | 10.9 | 10.7 | 11.4 | 11.4 | 10.5 | 10.6 | 10.1 |
| GROSS NATIONAL PRODUCT (72 $) | 1416.5 | 1448.8 | 1511.1 | 1553.6 | 1606.9 | 1645.3 | 1696.5 | 1745.2 | 1795.5 | 1852.9 | 1903.2 |
| % CHANGE | -1.0 | 2.2 | 4.3 | 2.8 | 3.4 | 2.4 | 3.1 | 2.9 | 2.9 | 3.2 | 2.7 |
| GROSS NAT. PROD. DEFL. (1972=100.0) | 180.8 | 197.3 | 212.6 | 228.7 | 245.2 | 265.1 | 286.4 | 310.2 | 333.2 | 357.0 | 382.6 |
| % CHANGE | 9.2 | 9.3 | 7.7 | 7.6 | 7.2 | 8.1 | 8.0 | 8.3 | 7.4 | 7.1 | 7.2 |
| POPULATION (MILLIONS) | 222.51 | 224.57 | 226.72 | 228.69 | 231.08 | 233.27 | 235.46 | 237.63 | 239.77 | 241.87 | 243.93 |
| % CHANGE | 0.9 | 0.9 | 1.0 | 1.0 | 1.0 | 0.9 | 0.9 | 0.9 | 0.9 | 0.9 | 0.9 |
| LABOR FORCE (MILLIONS) | 104.43 | 106.65 | 108.16 | 109.41 | 110.68 | 112.00 | 113.42 | 114.79 | 116.04 | 117.16 | 118.19 |
| % CHANGE | 2.0 | 1.6 | 1.4 | 1.2 | 1.2 | 1.2 | 1.3 | 1.2 | 1.1 | 1.0 | 0.9 |
| PARTICIPATION RATE | 63.9 | 64.1 | 64.3 | 64.4 | 64.5 | 64.6 | 64.8 | 64.9 | 65.1 | 65.3 | 65.4 |
| % CHANGE | -0.2 | 0.3 | 0.3 | 0.1 | 0.2 | 0.2 | 0.3 | 0.2 | 0.3 | 0.3 | 0.3 |
| EMPLOYMENT (MILLIONS) | 97.05 | 97.93 | 99.74 | 101.23 | 102.80 | 103.91 | 105.37 | 106.84 | 108.20 | 109.59 | 110.91 |
| % CHANGE | 0.1 | 0.9 | 1.9 | 1.4 | 1.6 | 1.1 | 1.4 | 1.4 | 1.3 | 1.3 | 1.2 |
| WAGE RATE PER WEEK, ALL INDUSTRIES | 313.8 | 340.1 | 384.9 | 416.2 | 453.4 | 496.0 | 542.3 | 594.9 | 650.4 | 709.8 | 777.6 |
| % CHANGE | 8.4 | 10.4 | 10.6 | 8.7 | 8.4 | 9.4 | 9.3 | 9.7 | 9.3 | 9.1 | 9.5 |
| PRODUCTIVITY - ALL INDUSTRIES | 14.600 | 14.794 | 15.143 | 15.348 | 15.631 | 15.835 | 16.099 | 16.334 | 16.595 | 16.908 | 17.160 |
| % CHANGE | -1.1 | 1.3 | 2.4 | 1.4 | 1.8 | 1.3 | 1.7 | 1.5 | 1.6 | 1.9 | 1.5 |
| PRODUCTIVITY - ALL MANUFACTURING | 7.927 | 8.112 | 8.354 | 8.563 | 8.823 | 9.018 | 9.287 | 9.565 | 9.845 | 10.158 | 10.448 |
| % CHANGE | -1.3 | 2.3 | 3.0 | 2.5 | 3.0 | 2.2 | 3.0 | 3.0 | 2.9 | 3.2 | 2.8 |
| REAL PER CAPITA GNP (THOU 72 $) | 6.368 | 6.451 | 6.665 | 6.788 | 6.954 | 7.053 | 7.205 | 7.344 | 7.489 | 7.661 | 7.802 |
| % CHANGE | -1.9 | 1.3 | 3.3 | 1.8 | 2.5 | 1.4 | 2.2 | 1.9 | 2.0 | 2.3 | 1.8 |
| REAL PER CAP DISP INC (THOU 472 $) | 4.450 | 4.502 | 4.603 | 4.688 | 4.810 | 4.870 | 4.986 | 5.071 | 5.171 | 5.290 | 5.411 |
| % CHANGE | -1.3 | 1.2 | 2.2 | 1.9 | 2.6 | 1.3 | 2.4 | 1.7 | 2.0 | 2.3 | 2.3 |
| CORPORATE PROFITS BEFORE TAXES | 230.5 | 236.5 | 245.1 | 276.2 | 321.8 | 343.2 | 451.2 | 531.3 | 586.9 | 659.9 | 724.0 |
| % CHANGE | -2.0 | 2.6 | 5.3 | 10.9 | 16.5 | 19.1 | 17.7 | 17.5 | 10.7 | 12.4 | 9.7 |
| BOND RATE (%) | 12.33 | 12.28 | 11.43 | 10.98 | 10.73 | 10.29 | 10.32 | 10.66 | 9.96 | 9.63 | 9.54 |
| PRIME COMMERCIAL PAPER RATE (%) | 11.23 | 11.17 | 9.66 | 9.01 | 9.29 | 8.73 | 8.87 | 9.29 | 8.66 | 8.23 | 8.17 |
| MONEY SUPPLY | 1124.2 | 1221.2 | 1324.9 | 1326.9 | 1539.9 | 1734.6 | 1961.8 | 2190.4 | 2453.5 | 2712.0 | 3020.2 | 3359.6 |
| % CHANGE | 8.5 | 8.7 | 8.6 | 10.0 | 12.6 | 13.1 | 12.0 | 11.6 | 10.7 | 11.4 | 11.2 |
| UNEMPLOYMENT RATE (%) | 7.51 | 8.17 | 7.74 | 7.48 | 7.12 | 7.23 | 7.09 | 6.92 | 6.76 | 6.46 | 6.17 |
| SAVINGS RATE (%) | 4.24 | 4.79 | 4.90 | 5.23 | 5.58 | 5.36 | 5.61 | 5.44 | 5.51 | 5.80 | 6.04 |
| SURPLUS OR DEFICIT, FEDERAL (CUR $) | -51.1 | -58.3 | -37.7 | -28.3 | -23.5 | -7.7 | -9.8 | -0.7 | -5.0 | 3.3 | -1.1 |
| SURPLUS OF DEF, STATE & LOC (CUR $) | 23.2 | 31.6 | 42.6 | 38.9 | 39.4 | 37.9 | 37.0 | 36.5 | 41.3 | 45.5 | 44.4 |
| COMPEN. TO EMPLOYEES TO NAT. INCOME | 77.0 | 77.1 | 77.0 | 76.6 | 75.8 | 75.5 | 75.0 | 74.9 | 74.9 | 74.8 | 75.3 |
| PROFITS TO NATIONAL INCOME | 11.2 | 10.3 | 9.6 | 9.6 | 10.1 | 10.4 | 11.4 | 12.0 | 12.0 | 12.2 | 12.2 |

A trend projection of a large scale econometric model has a special interpretation. In the initial two or three years of such an extrapolation, an attempt is made to introduce as much specific business-cycle content as possible by moving principal policy magnitudes along specified short-run courses that interpret budget commitments, tax statutes, behavior of monetary authorities, and various economic regulations. This portion of the extrapolation may properly be labelled as a multi-dimensional forecast. From that point forward, major inputs are placed on recent medium-term trend paths. A set of exoge-

[2] The Wharton Model projections for the United States, reported here, were prepared by Vijaya Duggal, Gene Guill, George Schink, and Yacov Sheinin.

*Perspectives and Implications*

nous inputs are sought, by trial and error, that generate a balanced growth path for the economy. By balanced growth, I refer to several established long run characteristics that are used to constrain the solution. These are:

(a) equality between the real growth rate and real interest rate;
(b) a stable savings ratio;
(c) a stable wage share of GNP;
(d) a stable velocity ratio;
(e) tolerable deficits, internal and external.

It is not easy, but it is generally possible, to find a set of input values which, together with initial conditions, generate a model solution with these properties. There is no guarantee that such a solution, determined from a model of some 1000 or more interrelated equations, is unique, but there is no indication that a very different one exists that also meets these enumerated conditions.

It was evident at the beginning of the 1970's — as early as 1970, in fact — that if we were to try to bring projected solutions of the Wharton Model closer to established long run trends for growth and unemployment that internal pressures would be built up that would unbalance the solution for the economy. Inflation would pick up, the domestic deficit would grow abnormally large, and the net foreign balance would move into serious deficit. It did not seem possible to start from prevailing initial conditions and end up with a solution to the model that moved on a higher growth path, conforming to history and satisfying the constraints imposed on the long run extrapolation. Generally speaking, the model would produce higher rates of inflation and large domestic and foreign deficits. Further development of feedbacks to capital flows and dollar exchange rates were not explicitly developed.

The trial and error simulation procedure gives the following indications:
  (i)   the long term growth rate has fallen by about one percentage point;
  (ii)  the inflation rate has been raised by about five percentage points;
  (iii) the current account balance is barely maintained;
  (iv)  productivity growth is resumed, but at a rate lower by about one percentage point;
  (v)   nominal interest rates are generally higher than in the past;
  (vi)  domestic fiscal balance is eventually attained.

At the very beginning of a new economic situation, determined to a great extent by adverse external circumstances, we should expect to find an immediate decline in the growth rate, but should the production path of the economy be shifted downwards, once and for all by a level amount, and then revert to the former growth rate, or should the growth rate itself, be lowered? Equilibrium growth theory and intuition suggest that after the initial growth decline the economy should return to the old growth rate. The level of production should be shifted downwards, but the rate of expansion should recover to the old position. Large scale econometric models do not seem to produce that result, at least over the period of one decade. There appears to be a downshift of the entire growth rate; thus, the United States are now expected to grow at about 3 % instead of 4 %, and that is a familiar pair of numbers often cited to describe expectations in a number of individual countries of Western Europe. For

Japan, the downshift in long term equilibrium growth is from about 10 percent to 5 %. This is an interesting finding that pervades many econometric modeling exercises for different countries and, as we shall see, for the world, too.

If the economy of the United States is stimulated toward recovery of the higher growth path of the 1950's and 1960's, a gap in trade and payments appears. But if the economy is allowed to proceed along the more moderate path of 3 %, the current account stays close to balance with only slight deterioration in spite of a continuing increase (assumed) in the real price of imported oil. There is some tendency towards energy conservation, but the value of oil imports is expected to grow significantly, year by year. Mainstays of the American current balance are growing agricultural exports and an impressive positive balance for services, or invisible accounts. Among the latter, the most important growth item is investment income. Many U. S. firms unsettled the balance of payments when they invested capital abroad in earlier formative years. But eventually, they made good on their investments, which was always the intention. U. S. based multinational enterprises now enjoy good income from abroad. In many cases, foreign income is much more favourable than domestic income.

Two other developments also contribute to net investment income from invisibles; high interest rates abroad, especially in the Euro-dollar market, enable U. S. corporate treasurers to realize good earnings from short term investments of working capital. High oil prices, which hurt our balance in the visible, merchandise sector, are offset by high earnings of U. S. multinational oil companies.

The U.S. economy is fundamentally beset by "fiscal drag". When the economy is operating in the neighborhood of full employment, present tax and revenue statutes are capable of generating very large receipts, generally large enough to cover all reasonable expenditures, extrapolated along historical growth paths. There will be some fresh tax cuts, and these are, indeed, factored into the baseline projection. But that is not enough to prevent overall expansion, by large sums, of revenues for the account of central government. Although we seldom realize balanced internal budgets, after the year is over, we do project them in baseline simulations. Major disturbances that bring forth new outlays and hold back the expansion of the personal income base cause internal accounts to fall into deficit positions much more frequently than is expected when a decade projection is made.

It is not solely energy considerations, such as the shift to relatively higher energy prices, that have caused the new slower profile of economic expansion in the United States, but energy is a key factor in these aspects of economic change. It is not possible to appreciate fully the new dimensions of the modern economy without devoting a great deal of attention to the role of energy. Accordingly, the Wharton Model, among other econometric interpretations of the United States, has incorporated a great deal of energy detail. It is evident from the accompanying table that progress is expected in energy conservation, a natural component of economic efficiency. Inefficiency in the use of an expensive scarce resource such as energy should eventually result in its more

*Perspectives and Implications*

careful use. The baseline projection for the American economy shows a steady downward trend in the energy (BTU) to GNP ratio, from 52.48 (Thou BTU/1972$) in 1980 to 44.57 in 1990. Were it not for energy conservation, in response to a relative price shift, the problem of bringing external trade accounts into current balance would be much more difficult, with added pressure on the dollar and thus on domestic inflation; therefore, energy use in response to the laws of economics forms an important component of this entire look into the future.

It is not only the free working of the market economy that brings about increasing energy efficiency but also legislative mandates on the fuel efficiency of the automobile fleet. The steady improvement of the statistics on average miles-per-gallon is clearly evident in Table 3. By meeting these standards from the side of the fleet supply, consumers and producers are implicitly contributing to the improvement in the energy to GNP ratio. These institutional considerations are part of the exogenous input into the baseline case.

Table 3. Energy and Related U.S. projections to 1990

| | 1980 | 1981 | 1982 | 1983 | 1984 | 1985 | 1986 | 1987 | 1988 | 19 |
|---|---|---|---|---|---|---|---|---|---|---|
| Gasoline and oil consumption (bill $72) | 24.0 | 23.3 | 23.7 | 23.0 | 22.3 | 21.8 | 21.4 | 21.1 | 20.8 | 20. |
| Miles per gallon (new) | 17.19 | 18.84 | 20.47 | 22.02 | 22.78 | 23.18 | 23.39 | 23.59 | 23.80 | 24. |
| Miles per gallon (all) | 13.75 | 14.23 | 14.88 | 15.70 | 16.64 | 17.63 | 18.62 | 19.60 | 20.51 | 21. |
| Crude oil imports (m b) | 2017 | 1956 | 2138 | 2030 | 1996 | 1952 | 1921 | 1898 | 1872 | 18 |
| Import price ($/b) | 32.25 | 38.46 | 44.99 | 50.39 | 54.92 | 59.87 | 65.25 | 71.13 | 77.53 | 84. |
| Energy consumption (quad BTU) | 74.35 | 75.24 | 77.10 | 77.57 | 78.41 | 78.91 | 80.11 | 81.30 | 82.41 | 83. |
| Energy GNP ratio thous. BTU/1972$) | 52.48 | 51.93 | 51.02 | 49.93 | 48.80 | 47.96 | 47.22 | 46.58 | 45.90 | 45. |

The projected decline in energy use per unit of production is not simply a "hope" built into the solution of the Wharton Model; it is, in fact, a continuation of an existing trend that has been apparent but too little appreciated since 1973. The energy-GNP ratio fell from 60.41 to 54.50 over the period 1973–79.

In searching for a set of economic policies that give rise to the balanced solution, termed the baseline case, I have been mindful of contemporary politics. Since the Kennedy-Johnson years, the federal administration in the United States has been conservative, undoubtedly becoming more conservative with the passage of time and with mounting frustration in dealing with inflation. The fiscal and monetary policies of the baseline case are appropriately constrained to be conservative also. They continue basic downward trends in public expenditures as a percent of GNP and keep taxes high enough to generate an eventual domestic budget balance. The growth of money supply is prudent. In the long run there is a tendency for this model to conform to the quantity theory of money, i.e., nominal GNP and money supply expand at the same rate of change.[3] This is shown by a tendency toward steady velocity of

[3] See L.R. Klein, "Money in a General Equilibrium System: Empirical Aspects of the Quantity Theory", *Economie Appliquée*, XXXI (1–2, 1978), 5–14.

circulation. This is a conservative monetary policy, to go hand-in-hand with an assumed conservative fiscal policy.

Table 4. Estimated M4 Velocity in U.S. Baseline Projections

| 1980 | 2.28 | 1984 | 2.27 | 1988 | 2.21 |
|------|------|------|------|------|------|
| 1981 | 2.34 | 1985 | 2.22 | 1980 | 2.19 |
| 1982 | 2.42 | 1986 | 2.21 | 1990 | 2.17 |
| 1983 | 2.31 | 1987 | 2.21 |      |      |

## BASE CASE—THE WORLD ECONOMY[4]

In many respects, the economic evolution of the United States over the next decade should indicate a general pattern for most developed industrial economies. To be sure, every country will have its own special situation, but the principal simulation results—moderate growth, less inflation, and overall balance—should prevail for several if not all industrial market economies. Next let us consider the world as a whole, not just the group of industrial countries which comprise the OECD, but the centrally planned and developing countries as well. Interest centers on their interaction and the way the world economy evolves.

During the 1960's economic development was rapid. Among industrialized countries, Japan's growth was unusually high, exceeding 10% annually. The growth rate of all industrialized countries averaged 5.1% over the decade but fell to only 3.2% during the greater part of the 1970's as a result of business cycle swings. The centrally planned economies turned in some individual good performances, but the cultural revolution in China, internal upheavals in Czechoslovakia, and difficulties elsewhere held their growth rate to 4.9%, just under the OECD average. The socialist countries picked up considerably in the 1970's but now face the same problems as the market economies in the period ahead.

For the developing countries, the results are very mixed depending on country classification. According to World Bank estimates, low income countries grew at rates significantly under 4 % in both the 1960's and 1970's. Performance was close to 6 % in the middle income grouping, and even higher for Persian Gulf Oil exporters. These tabulations cut off notably in 1978, just prior to the revolution in Iran, which has disrupted economic activity for some time to come.

---

[4] Members of the research team of Project LINK contributed markedly to the results reported in this section. They are Victor Filatov, Shahrokh Fardoust, Yuzo Kumasaka, Michael Papaioannou, and Baudouin Velge.

Table 5. Some World Historical Statistics

| | GDP Growth | | Inflation rate | | Export Growth | | Import Growth | |
|---|---|---|---|---|---|---|---|---|
| | 1960−70 | 1970−78 | 1960−70 | 1970−78 | 1960−70 | 1970−78 | 1960−70 | 1970−7 |
| Low Income Countries | 3.9 | 3.6 | 3.0 | 10.6 | 5.0 | −0.8 | 5.0 | 3.2 |
| Middle Income Countries | 6.0 | 5.7 | 3.1 | 13.1 | 5.5 | 5.2 | 6.8 | 5.8 |
| Industrial Countries | 5.1 | 3.2 | 4.2 | 9.4 | 8.7 | 5.7 | 9.4 | 5.1 |
| Persian Gulf Oil Exporters | 13.0 | 6.0 | 1.2 | 22.2 | 9.5 | −1.2 | 11.1 | 21.1 |
| Centrally Planned | 4.9 | 5.6 | − | − | − | − | − | − |

Source: World Development Report, 1980
World Bank

Inflation rates were modest prior to the economic dislocations of the past decade, with single-digit rates well under 5 % customary in the non-socialist world. There were some significant exceptions in the developing world. After the large increments in food and fuel prices during the early 1970's, and the absorption of the legacy of Vietnam, prices took off to new heights. The average, 1970−78, was just below 10 % for the industrialized countries, but the situation has worsened considerably in the most recent years. This is one of the bleakest aspects of the future outlook.

There are no satisfactory price reports from the centrally planned economies. Very recently, they have shown a series of once-for-all price changes, but their opening of their borders to trade on a significantly larger scale means that they will have to absorb a large degree of imported inflation. Where appropriate price indexes are available, they indicate price increases comparable to those in the West.

Another dimension in the world economy is the growth and pattern of world trade. The decade of the 1960's was a "golden era" in trade development. Both exports and imports grew faster than did aggregate production. As recession hit the world economy in the 1970's, trade growth also receded, but it remained significantly above the growth in production. On a world scale, the growth in trade volume was about 50 % faster than production growth.

For most of the historical period since the end of World War II, the fixed parity system of the Bretton Woods Agreement took care of adjustments in trade balances, while developing a thriving multilateral system of trade. The build up of large surpluses by countries like Japan and Germany and the relative weakness of the United States, United Kingdom and a few other key countries brought the downfall of this system at the end of the 1960's or beginning of the 1970's. The managed floating system was being given a chance to operate, when the world was shocked by the oil embargo of 1973, followed by high energy pricing by OPEC. Now there are large surpluses and deficits among countries, subject to a great deal of turnover from year to year, as regards who is in surplus and who is in deficit. Overshadowing the short run adjustments among various OECD members is the very large balance of oil exporting nations. After the first buildup of surplus balances by OPEC in 1974−75, the excess funds were recirculated throughout the world economy through inflation, dollar devaluation, and OPEC's high propensity to import.

This situation has been halted, and a large surplus for oil exporting countries is presently matched by a deficit for oil importing countries in both the developed and developing world.

The basic assumption for world model projections into the 1980's is that oil production will be more moderate; price increases will be maintained above western inflation rates; and the surplus of oil exporting countries will be used for the development of the non-oil sectors of their economies or invested throughout the world.

By using the initial conditions of recent world economic history, an assumption about the course of oil prices, and extrapolated trends of major exogenous variables, we can compute a baseline projection of the world economy as a whole. The interrelated system of national and regional econometric models that constitute project LINK is the statistical medium through which this calculated projection is made.[5]

Each component model of the LINK system is put through a trend extrapolation exercise analogous to that described above for the Wharton Model of the United States, the main difference being that the models of project LINK, including the U.S. component, do not have the large detailed input-output and energy sectors that are present in the version of the Wharton Model that is being used for these longer term analyses. In the U.S. case, the projected American economy of the LINK model is monitored by the known results of the annual Wharton Model.

The main advantage of using the integrated LINK system for this medium term projection is to develop the growth patterns of world trade and inflation as part of the outcome of the calculation rather than as assumed inputs. For the individual assessment of growth patterns in each separate country or region, assumed values for world trade and import prices must be established in advance.

The base case projection for the world economy bears some close resemblances to the results discussed already for the U.S. case, since most parts of the world are experiencing the same kinds of economic pressures and converging towards a similar response and outcome.

[5] R. J. Ball, ed., *The International Linkage of National Economic Models*, J. Waelbroeck, ed., *The Models of Project LINK*, and J. Sawyer, ed., *Modelling the International Transmission Mechanism*. (Amsterdam: North-Holland Publishing Co., 1973, 1976, 1979). See also B. G. Hickman and L. R. Klein, "A Decade of Research by Project LINK", ITEMS (New York: Social Science Research Council), vol. 33, (December, 1979) 49−56.

Table 6. World Summary Measures of Growth and Inflation 1980–1990 Baseline (annual percentage changes)*

| Country Grouping | 1980 | 1981 | 1982 | 1983 | 1984 | 1985 | 1986 |
|---|---|---|---|---|---|---|---|
| Gross Domestic Product | | | | | | | |
| 13 LINK OECD | | | | | | | |
| Countries[1] | 1.3 | 2.3 | 4.2 | 3.7 | 3.4 | 3.4 | 3.2 |
| Level[2] | (2613.5) | (2672.5) | (2785.5) | (2889.0) | (2987.9) | (3087.5) | (3184.7) |
| Developing Countries | 5.0 | 5.6 | 5.3 | 5.7 | 5.4 | 5.6 | 5.4 |
| Non-Oil Exporting | 5.5 | 5.6 | 5.3 | 5.7 | 5.4 | 5.6 | 5.3 |
| Oil Exporting | 2.2 | 5.3 | 5.3 | 5.3 | 5.3 | 5.3 | 6.3 |
| Centrally Planned | | | | | | | |
| Countries[3] | 4.2 | 3.4 | 4.3 | 4.4 | 4.6 | 4.4 | 4.4 |
| World[4] | 2.2 | 3.0 | 4.3 | 4.1 | 3.9 | 3.9 | 3.7 |
| | | | | | | | |
| Private Consumption Deflator | | | | | | | |
| 13 LINK OECD | | | | | | | |
| Countries | 11.2 | 8.5 | 6.4 | 5.9 | 5.6 | 5.4 | 5.4 |
| (GDP Deflator) | (9.5) | (8.1) | (6.6) | (5.9) | (5.6) | (5.6) | (5.5) |
| Developing Countries | 25.3 | 29.1 | 20.7 | 18.1 | 16.2 | 13.0 | 11.2 |
| Non-Oil Exporting | 26.8 | 31.6 | 22.7 | 19.6 | 17.4 | 13.8 | 11.8 |
| Oil Exporting | 14.4 | 11.3 | 6.3 | 7.4 | 8.0 | 7.5 | 6.9 |
| World[5] | 13.8 | 12.3 | 9.1 | 8.2 | 7.6 | 6.8 | 6.5 |

* Weighted averages of own country/region growth rates.
[1] 13 LINK OECD countries are Australia, Austria, Belgium, Canada, Finland, France, Federal Republic of Germany, Italy, Japan, Netherlands, Sweden, United Kingdom, and the United States of America.
[2] Billions of 1970 U.S. $ at 1970 exchange rates.
[3] Includes only Eastern Europe CMEA and the U.S.S.R.
[4] World = .6565×OECD+.1494·DEVE+.1851×CMEA.
[5] World = .8145×OECD+.1855 DEVE. Inflation measures for CMEA are not avialable.
Period averages are calculated as the geometric mean of the first through last period growth rates.

On average, the industrialized countries are projected to lose one or two percentage points of growth. During the 1960's they expanded at more than 5%, but a longer stretch of time including the 1950's would reduce that estimate. In the projection, the growth rate is about 3%, the same as in the cyclical decade of the 1970's. GDP growth of the developing countries is reduced in this projection, as is that of the centrally planned economics. All told, when the figures are averaged on a world-wide basis, the resulting figure for growth is between 3.5 and 4.0% for the decade ahead. The corresponding figure was in excess of 5% for the 1960's and somewhat smaller during the 1970's.

Historically, world trade has expanded more rapidly than production, in a ratio of about 1.5. In the projection, however, the ratio falls considerably, so that world trade is expected to grow by little more than 10% above the growth rate of production. This is a new situation, with new large economies entering the world trade system on a large scale — China, the U.S.S.R., and other socialist countries — together with an awareness of an increasing degree of interrelatedness among nations. The United States is noticeably more concerned about its international economic relations, and more involved too. Countering these tendencies are efforts at import substitution, the introduction

| 1987 | 1988 | 1989 | 1990 | 1981–1985 | 1986–1990 | 1980–1990 | 1981–1990 |
|---|---|---|---|---|---|---|---|
| 2.9 | 2.9 | 2.9 | 2.7 | 3.4 | 2.9 | 3.0 | 3.2 |
| (3277.2) | (3373.1) | (3471.6) | (3564.4) | | | | |
| 5.3 | 5.3 | 5.3 | 5.3 | 5.5 | 5.3 | 5.4 | 5.4 |
| 5.2 | 5.2 | 5.2 | 5.2 | 5.6 | 5.2 | 5.4 | 5.4 |
| 6.3 | 6.3 | 6.3 | 6.3 | 5.3 | 6.3 | 5.5 | 5.8 |
| 4.6 | 4.4 | 4.6 | 4.6 | 4.2 | 4.5 | 4.3 | 4.4 |
| 3.6 | 3.5 | 3.6 | 3.4 | 3.8 | 3.6 | 3.6 | 3.7 |
| 5.4 | 5.5 | 5.4 | 5.4 | 6.4 | 5.4 | 6.4 | 5.9 |
| (5.5) | (5.5) | (5.4) | (5.6) | (6.4) | (5.5) | (6.2) | (5.9) |
| 11.1 | 10.8 | 10.5 | 10.3 | 19.3 | 11.5 | 15.9 | 15.0 |
| 11.7 | 11.4 | 11.0 | 10.8 | 20.9 | 11.3 | 17.0 | 16.0 |
| 6.9 | 6.8 | 6.8 | 6.6 | 8.1 | 6.8 | 8.1 | 7.4 |
| 6.5 | 6.5 | 6.3 | 6.3 | 8.8 | 6.5 | 8.2 | 7.6 |

of some measures of protectionism, and some attempts by oil exporting nations to restrain the growth of their output.

In this moderate growth, relatively slow trade era, it is expected that eventually anti-inflationary policies will take hold. These are promoted by the conservative economic attitudes of policy makers now prevalent in the United States. The overall inflation rate does not fall back to the very low ranges that prevailed some twenty years ago. In place of the less than 5% rates that we once enjoyed, a reduction to single digit ranges and ultimately to about 5–6% is considered a significant achievement. In the developing world, a reasonable target would be about 15%, on average.

The growth of the OPEC surplus, covered over in these tables as a result of the amalgamation of all developing countries is matched, over the decade, by the deficit of the industrial countries. There is some deficit, as well, among the socialist countries. This projection assumes that these offsetting balances are recycled through the world financial system. The actual process may be quite difficult to accomplish.

Within the OECD area, there is a great deal of shifting between surplus and deficit areas. While the U.S. goes from deficit towards balance by 1990, Japan and Germany initially move into deficit, as do France, Italy, and the United

Table 6. (continued) World trade summary

| | 1980 | 1981 | % △ | 1982 |
|---|---|---|---|---|
| **13 LINK OECD Countries[1]** | | | | |
| Exports[2] | 1036.0 | 1219.3 | 17.7 | 1420.8 |
| Imports | 1090.0 | 1275.2 | 17.0 | 1439.0 |
| Balance | −54.0 | −55.8 | | −18.2 |
| **Developing Countries** | | | | |
| Exports | 532.1 | 631.6 | 18.7 | 693.7 |
| Imports | 438.5 | 564.3 | 28.7 | 629.2 |
| Balance | 93.6 | 67.4 | | 64.5 |
| **Centrally Planned Countries[3]** | | | | |
| Exports | 138.4 | 156.7 | 13.3 | 177.2 |
| Imports | 146.7 | 168.2 | 14.6 | 186.5 |
| Balance | −8.3 | −11.4 | | −9.3 |
| **Rest of the World[4]** | | | | |
| Exports | 163.4 | 203.0 | 19.8 | 189.8 |
| Imports | 194.7 | 203.1 | 4.3 | 226.8 |
| Balance | −31.3 | −0.1 | | −37.0 |
| **World Exports** | 1869.9 | 2210.7 | 18.2 | 2481.5 |
| **World Export Price** | 3.3 | 3.8 | 14.8 | 4.1 |
| **World Exports (Real)*** | 572.2 | 589.4 | 3.0 | 612.6 |
| **World Export Price of Fuel** | 10.9 | 13.2 | 20.8 | 14.3 |
| **World Exports of Fuel (Real)*** | 40.9 | 39.6 | −3.2 | 41.3 |

* Constant dollar measures have base 1970 = 1.0

+ Figures in parentheses are annual average trade balances.

[1] 13 LINK OECD countries are Australia, Austria, Belgium, Canada, Finland, France, Federal Republic of Germany, Italy, Japan, Netherlands, Sweden, United Kingdom and the United States of America.

[2] Measures are for merchandise trade, F.O.B.

[3] Includes only Eastern Europe CMEA and the U.S.S.R.

[4] Period averages are calculated as the compound annual growth rate of the last over first year projection.

Kingdom. The Japanese situation is projected to change drastically and promptly back into surplus by mid-decade, while the German case follows a more moderate path towards balance and reaches a small surplus by 1990.

The analysis of U.S. growth prospects is applicable by analogy to the industrial countries as a whole. Restrictive policies to fight inflation, to pay for expensive oil imports, protect exchange value of the currency, and to recoup productivity losses keep the economy on a moderate path. The slowdown in the industrial world holds back the export potential of developing countries. In order to cope with adverse trade and payments deficits, restrictive policies are followed. In this environment, capital inflows for development are harder to

| %Δ | 1983 | %Δ | 1984 | %Δ | 1985 | %Δ |
|---|---|---|---|---|---|---|
| 16.5 | 1592.3 | 12.1 | 1788.9 | 12.3 | 2012.7 | 12.5 |
| 12.8 | 1622.5 | 12.7 | 1833.2 | 13.0 | 2055.5 | 12.1 |
| | −30.2 | | −44.3 | | −42.7 | |
| 9.8 | 773.5 | 11.5 | 859.0 | 11.1 | 954.7 | 11.1 |
| 11.5 | 719.9 | 14.4 | 807.2 | 12.1 | 912.9 | 13.1 |
| | 53.6 | | 51.8 | | 41.8 | |
| 13.0 | 201.1 | 13.5 | 229.7 | 14.2 | 261.3 | 13.8 |
| 10.9 | 207.3 | 11.2 | 235.3 | 13.5 | 265.9 | 13.0 |
| | −6.2 | | −5.6 | | −4.6 | |
| −6.5 | 239.7 | 26.3 | 286.9 | 19.7 | 329.2 | 14.8 |
| 11.7 | 256.9 | 13.2 | 288.9 | 12.5 | 323.7 | 12.0 |
| | −17.2 | | −2.0 | | 5.5 | |
| 12.2 | 2806.6 | 13.1 | 3164.5 | 12.8 | 3557.9 | 12.4 |
| 8.0 | 4.4 | 7.9 | 4.7 | 7.8 | 5.1 | 7.5 |
| 3.9 | 641.9 | 4.8 | 671.4 | 4.6 | 702.3 | 4.6 |
| 8.5 | 15.8 | 10.5 | 17.4 | 9.9 | 19.1 | 9.7 |
| 4.3 | 42.9 | 3.9 | 44.5 | 3.7 | 46.1 | 3.5 |

come by. High debt service burdens, in a number of cases, act as additional constraints. Conservative governments in the OECD area are less disposed than previously to grant concessionary aid.

The centrally planned economies used to consider themselves well insulated against the economic ills of the rest of the world. This is no longer the case.

The centrally planned economies, dissatisfied with the outcome of their own efforts to achieve good economic growth performance, have changed strategy and decided to import high technology from the West, as well as necessary grains to supplement their domestic agricultural supplies. This new approach has opened their economies to Western inflation because imports have been

Table 6. (continued)

|  | 1986 | % Δ | 1987 | % Δ | 1988 | % Δ |
|---|---|---|---|---|---|---|
| **13 LINK OECD Countries** | | | | | | |
| Exports | 2259.8 | 12.3 | 2522.4 | 11.6 | 2824.7 | 12.0 |
| Imports | 2295.4 | 11.7 | 2551.0 | 11.1 | 2840.3 | 11.3 |
| Balance | −35.6 | | −28.5 | | −15.6 | |
| **Developing** | | | | | | |
| Exports | 1055.8 | 10.6 | 1161.0 | 10.0 | 1273.5 | 9.7 |
| Imports | 1020.1 | 11.7 | 1134.0 | 11.2 | 1258.6 | 11.0 |
| Balance | 35.7 | | 27.0 | | 14.8 | |
| **Centrally Planned Countries** | | | | | | |
| Exports | 298.2 | 14.1 | 333.2 | 11.7 | 376.0 | 12.8 |
| Imports | 304.9 | 14.7 | 339.5 | 11.3 | 383.9 | 13.1 |
| Balance | −6.7 | | −6.2 | | −7.9 | |
| **Rest of the World** | | | | | | |
| Exports | 367.8 | 11.7 | 408.3 | 11.0 | 454.8 | 11.4 |
| Imports | 361.2 | 11.6 | 400.7 | 10.9 | 446.1 | 11.3 |
| Balance | 6.6 | | 7.7 | | 8.7 | |
| **World Exports** | 3981.6 | 11.9 | 4425.1 | 11.1 | 4929.0 | 11.4 |
| **World Export Price** | 5.4 | 7.1 | 5.8 | 7.1 | 6.2 | 6.9 |
| **World Exports (Real)** | 733.9 | 4.5 | 761.9 | 3.8 | 793.7 | 4.2 |
| **World Fuel Price** | 20.9 | 9.3 | 22.9 | 9.7 | 25.0 | 9.2 |
| **World Fuel Exports (Real)** | 47.9 | 3.9 | 49.2 | 2.7 | 50.7 | 3.1 |

reflecting rising world prices. Gold and oil sales at correspondingly rising prices have been used by the Soviet Union to finance part of their import needs, but they are fully enmeshed in world inflation accounting in balancing rising export prices.

The economies of Eastern Europe have had to cope with trade deficits and unusual borrowing in order to pay for imports, over and above their abilities to produce exports for the world markets. As their external accounts have got out of line, they have had to resort to the "stop" phase of familiar "stop-go" policies. In addition, Poland and other Eastern countries have been confronted with domestic labor unrest in an inflationary environment.

The People's Republic of China are resorting to similar trade policies, but mindful of the complications that arise when socialist countries rush headlong into an open economy format, they are taking lessons from the European experience and moderating their original trade and capital import plans. Although the Chinese are approaching this phase of development quite cautiously, they have enough pent-up growth potential at the present time to support a growth rate in excess of the average for centrally planned economies.

| 1989 | % △ | 1990 | % △ | 1981–1985 % △ | 1986–1990 % △ | 1980–1990 % △ | 1981–1990 % △ |
|---|---|---|---|---|---|---|---|
| 3172.8 | 12.3 | 3545.1 | 11.7 | 13.3 | 11.9 | 13.1 | 12.6 |
| 3174.3 | 11.8 | 3532.1 | 11.3 | 12.7 | 11.4 | 12.5 | 11.9 |
| −1.5 | | | 13.0 | $(-38.2)^+$ | $(-13.6)^+$ | $(-28.5)^+$ | $(-25.9)^+$ |
| | | | | | | | |
| 1399.8 | 9.9 | 1534.0 | 9.6 | 10.8 | 9.8 | 11.2 | 10.3 |
| 1403.2 | 11.5 | 1559.0 | 11.1 | 12.7 | 11.2 | 13.5 | 11.9 |
| −3.4 | | −25.0 | | $(55.8)^+$ | $(19.8)^+$ | $(38.3)^+$ | $(37.8)^+$ |
| | | | | | | | |
| 424.6 | 12.9 | 477.2 | 12.3 | 13.6 | 12.5 | 13.2 | 13.1 |
| 433.6 | 12.9 | 487.2 | 12.3 | 12.1 | 12.4 | 12.7 | 12.5 |
| −9.0 | | −10.0 | | $(-7.4)^+$ | $(-7.9)^+$ | $(-7.7)^+$ | $(-7.7)^+$ |
| | | | | | | | |
| 512.5 | 12.7 | 576.7 | 12.5 | 12.8 | 11.9 | 13.4 | 12.3 |
| 498.7 | 11.8 | 554.7 | 11.2 | 12.3 | 11.3 | 11.0 | 11.8 |
| 13.9 | | 22.0 | | $(-10.2)^+$ | $(11.7)^+$ | $(-2.1)^+$ | $(0.8)^+$ |
| | | | | | | | |
| 5509.7 | 11.8 | 6133.0 | 11.3 | 12.6 | 11.4 | 12.6 | 12.0 |
| 6.6 | 6.8 | 7.1 | 7.8 | 7.6 | 7.1 | 8.0 | 7.2 |
| 830.4 | 4.6 | 861.9 | 3.8 | 4.5 | 4.1 | 4.2 | 4.3 |
| 27.3 | 9.1 | 29.7 | 9.0 | 9.6 | 9.2 | 10.6 | 9.4 |
| 52.3 | 3.2 | 54.0 | 3.3 | 3.8 | 3.1 | 2.8 | 3.5 |

In the near term, China is growing at 7 percent or more. For the longer term, 6% seems to be attainable, although they could slip backward by another percentage point, or so.

A special group of nations among the developing countries are the OPEC nations or, more broadly, the oil exporting nations. They have little or no balance of payments constraint attached to their development plans for the medium term at least. Although they may be in a position to develop at a more rapid rate, they are reconsidering the experience of the past few years in which rapid exploitation of oil resources did not optimize their purchasing power for capital and other imports and created dangerous or fatal unrest in several countries. Many of these countries were not able to absorb imports efficiently at the more rapid pace. Their overseas investments have been only partially successful. As a consequence of all these problems, oil exporting nations are opting for a more moderate rate of industrialization. Both the oil and non-oil sectors of their economies will be expected to phase down to a slower growth path.

No matter where we look in the assessment of the world economy, there are fundamental reasons for expecting a slower rate of development.

Recognition of lack of abundant energy resources for the world, more particularly crude oil resources, has by itself contributed to the slowing down of the world economy. This can be seen by looking at the results of alternative simulations of the LINK model with different assumptions about energy prices. A standard procedure is to compute a baseline projection, as has already been described here, and then compare this result with an alternative projection where specific changes in external factors have been imposed on the model. In the case being examined here, the change imposed is an increase in the world oil price by 10%; i.e., the exogenous path of world oil prices, set by OPEC, is raised, year by year, to a new path that is uniformly about 10% above the baseline path.

On occasion, we have made simulations in which the oil price was kept fixed at some base year value or in which the *real* price of oil was kept fixed — by allowing the nominal price to move by the same percentage change as an accepted index of inflation, say a general price index in the OECD countries. The LINK system has a basic symmetry property. Results with lower oil prices are opposite in sign, with similar magnitude, to those with an increase in price.

The general findings can best be described by considering main elasticities of the system. These are percentage changes in principal magnitudes associated with a change in oil price, other external inputs remaining unchanged.

In the first place, fuel import demand falls by about 1.1% for a 10% increase in price. The elasticity coefficient is about 0.11. This degree of sensitivity appears in the first year of a projection and persists for a whole decade. After 10 years, if the price is higher by 10%, the trade volume is lower by about 1%.

Higher oil price discourages world economic activity and adds to world inflation. Industrial world GDP falls by about 0.5% when oil prices are initially 10% and then about 6% higher after the first year. This works out to be an elasticity coefficient of 0.06. For this same change in oil price, OECD inflation measured by the GDP implicit deflator is up by about 0.2% and consumer price rises by 0.3% to 0.4%.

When we consider that world energy prices have gone up much more than 10%, after 1973 — they quadrupled and then more than doubled again — we can see that energy issues had much to do with the present state of stagflation. It is not a simple matter of finding a multiple of the 10% change used in the elasticity calculations, because those changes were introduced in an artificial ceteris paribus situation, while many things changed in the actual world environment after oil prices first jumped. In fact, *real* oil prices did not permanently rise after the initial change in 1974–75; they did, however, after the latest change in 1979.

In any event, we can plainly see that world economic activity and world inflation are highly sensitive to world energy prices. The present slowing down of economic growth, accompanied by higher inflation, is due, in part at least, to higher energy prices.

The baseline projections made here for both the United States and the world as a whole are done in a *benign* environment; that is to say, there are no untoward major disturbances contemplated for the 1980's in this case. Since the end of World War II and the immediate readjustment period there have been three completed decades, each with its own disturbing factors that upset an otherwise benign environment. These have been:

1950's    Korean War
          "Cold" War
          Suez Canal Closing
1960's    Vietnam War
1970's    Breakdown of Bretton Woods
          Massive Harvest Failures
          Oil Embargo cum OPEC Cartel Pricing
          Iranian Revolution

These major events had enormous impact on economic performances all over the world. Within each decade there were other disturbances as well, less dramatic, yet economically significant.

In thinking about possible "futures" for the 1980's, it may be convenient to formulate baseline cases without contingency planning for such disturbances because, in many respects, the kinds of formal models that we use decompose, approximately, into a systematic (baseline) component and a disturbance component.

This property is associated with *linearity* in formal model theorizing, and it appears to be a reasonable approximation. Therefore, we proceed by first working out the base case and then superimposing disturbances on it.

Sometime during the 1980's there can very well be—according to many thinkers there *will* be—another significant interruption of delivery of oil supplies and another large harvest failure. During the 1970's there was war in the Middle East but not on the scale and duration of Korea or Vietnam. In a sense, the oil embargo and OPEC pricing listed above are economic surrogates for the Middle East War.

Will the military experience of the 1950's and 1960's be repeated during the 1980's? This is certainly a contingency. There could well be large scale cold or hot war during the coming decade. Also, the international economy has been so upset by events in the food and fuel sectors that we tend to look to those areas for the reappearance of disturbances. It is likely that a large scale economic disruption will occur during the decade, but there will probably be disturbances in surprising new areas. Shortages of basic materials other than food and fuel could develop. There could be a wave of debt defaults running throughout the developing world or among relatively poor countries of the developed world. There could be a massive dislocation in the physical environment, in atmospheric or water pollution, or urban congestion. It is worthwhile exploring in some detail the economic dimensions of a few of there disturbances.

## POSSIBLE DISTURBANCES

Cartel Pricing: For the baseline case, we have assumed that, after the immediate effects of the Iranian Revolution have been worked out, crude oil prices would rise, on average, by about 10% annually for the whole decade. This turns out to be about 3% above the inflation rate that is relevant to the oil exporting countries, namely, the export price of OECD countries. In the baseline solution, this key price grows at about 6−7% annually. Between 1980 and the attaining of the trend pattern for the rest of the decade, there is an assumed transitional period before world inflation and the growth rate of oil prices decline; thus, the 1981 real price increases by more than 3%. In addition, a disturbance did appear during 1980, continuing at the end of this year, in the form of war in the Middle East. We are witnessing an unusual event in which two members of an effective cartel are engaged in open warfare. This has significantly reduced oil supplies and provides another reason for marking up the price in the transitional period, at least.

The steady rise of 3% in the real price of oil, together with an approximate solution path for OECD export prices at 7%, amount, in a numerical sense, to the indexation of oil prices. It has often been mentioned that a stated objective of OPEC is to devise an index formula for oil pricing. The simplest of such formulas is implicit in the baseline case. Variations on this case have been worked out with either higher or lower rates of increase of real oil prices. Another route to follow is to have a multivariate indexation formula, in which the oil price is also tied to GDP growth in the developed world, to the exchange value of the dollar, and other relevant indicators.

Indexation formulas give smooth steady paths for the course of oil prices, but at least one unsteady path is worth consideration as a result of a possible disturbance. If forseeable world supplies of oil are balance against estimated world demand, there appears to be a large scale shortfall developing by mid-decade. This deficit could be made up either by having price rise steadily on a faster gradient, by having a one-time large upward step of 50−100%, or by rationing.

The fact that the OPEC Cartel has been as cohesive and long lasting as it has, surprises many economists. At the present time it even appears to be surviving open warfare between two members. We should be prepared, therefore, to experience other surprises of similar proportions. There are, of course, some unique features about OPEC that are not easily duplicated. Cartels in other fields of economic activity may not reach out to such important products from an industrial viewpoint. Cartels in diamonds, mercury, chromite or other industrial products would not have as great an impact on overall world activity. In most food lines, grains for example, large developed countries, which would be more disposed towards maintenance of a multilateral free trade system, are major export suppliers and could, therefore, inhibit or prohibit effective cartel action. Petroleum products are peculiarly concentrated, as far as surplus capacity for export is concerned, in the hands of a fairly cohesive politico-social group, dedicated to pan-Arabism or to aspirations of developing nations. If export capacity of important industrial products were concentrated in the hands of a cohesive group of nations (religious, political, social, geographical) another effective cartel could arise. Such a possible field of action does not presently appear to exist on the world scene.

The world has been economically disturbed by food, as well as by fuel. There is certainly a possibility of unforeseen supply shocks in the provision of food during the 1980's. Harvest failures have been occurring with alarming regularity as ambitious attempts are being made to upgrade diets over a large part of the world. The year 1980 initiates the decade on an insecure footing with some measurable disturbances. The embargo of American grain shipments to the U.S.S.R., as a result of the invasion of Afghanistan has been controversial, not only because of its political impact but also because of doubts about its effectiveness. When all the arguments are sifted, it does appear that the embargo has been effective in delaying the delivery of larger meat supplies to the Soviet population. When it is placed in juxtaposition to the disappointing Soviet harvest of 1980 and the food shortages in Poland, it is evident that there are quite significant stresses on the world economy. To add to the list of food supply setbacks, we can also cite the drought, resulting in a poor crop in feed grains in the United States in 1980. We are starting out the decade with upward pressure on food/agricultural prices. The downward drift in world inflation, which is an important component of the baseline economic scenario for the 1980's is temporarily being thrown off course by rises in food/agricultural prices and in energy prices. If more disturbances like these occur during the course of the decade, we could have significantly worse economic performance than in the base case.

There is an important difference between food and fuel disturbances. The fuel disturbances of the 1970's, stemming from the 1973 Middle East War, were institutionalized and made permanent by the control power of OPEC. Food/agricultural prices, however, fluctuated during the decade, since supply responses to high prices have been relatively quick in agriculture. U.S. grain supplies, in particular, were expanded on a large scale after the massive depletion of stocks by Soviet purchases in 1972/73. The responsiveness of U.S. and other suppliers tends to soften the effect of agricultural disturbances when spread over a two or three year horizon.

The next disturbance to the world economy could well be entirely unforeseen. In searching for new and different areas where contingent planning would be helpful, we may cite the possibility of simultaneous debt default. Many developing countries, some centrally planned economies, and some poorer developed countries are seriously in debt. The degree of seriousness is indicated by debt service ratios showing the extent to which trade gain can cover (or fail to cover) needs for interest payment and debt amortization. There have been several singular cases of a nation's inability to meet current debt service requirements, but they have always been met in recent years without the precipitation of a crisis. Debt rescheduling has been successful for dealing with the specific situations that have arisen—Peru, Zaire, Zambia, Poland, Turkey, to name a few. As long as such cases can be kept isolated from the routine functioning of the world financial system, a major disturbance can be averted. Commercial banks and international institutions have become alerted to the situation as a result of experiences in these singular cases. They have, accordingly, become more prudent in loan activity. This is another reason why moderation in the face of economic activity has become a characteristic of the baseline projection for the decade ahead. Although there is good reason to believe that disturbances in the form of a wave of debt defaults will not occur, such an event is by no means impossible.

## AN OPTIMISTIC CASE

While some economists may feel that the base case, itself, is optimistic—at least complacent and trouble free—there are many industrialists, policy makers, financiers, and economists who strive for a better outcome. If there is a single measure, among the many that properly describe the economy, that indicates the unsatisfactory nature of present performance and its extrapolation along the path of the baseline case, it is the poor performance of *productivity*. In Table 2, it can be seen that productivity growth during 1980, in the United States, has been *negative*, while its trend projection, at rates of change that are mainly between 1.0 and 2.0% annually, fall considerably below a previously established long run path. It used to grow at a rate in excess of 3% in the United States. Outside the U.S., across most international lines, it is also true that the productivity improvement factor has fallen, perhaps not as drastically as in the United States, but it is uniformly lower in recent years.

A central focus for policy targets that gives some promise for better economic performance is, therefore, a policy mix that attempts to enhance productivity growth. Since we are not sure of the causes that led to the productivity slowdown it is especially difficult to prescribe policies for productivity improvement. Of the possible sources of productivity decline, it is widely felt that relatively weak capital formation, in the private sector, plays a major role.

Both through general capital expansion, and through modernization, it is expected that higher rates of fixed capital formation will lead to better productivity growth. It is necessary to make firms want to invest and to use the new capital at a high capacity rate.

Capital formation is important but not the whole story, because there is still a long way to go towards revitalizing the economy even after some objectives on capital formation have been reached. In a study for the New York Stock Exchange, last year, the operators of the Wharton Model examined the resulting net gains for the United States as a consequence of raising the investment share of GNP to about 12% from a stagnating level of about 10%.[6]

In a rounded policy package, the raising of the investment and savings rates by 2 percentage points are formidable steps forward, but they apparently do not suffice to restore the rate of growth of productivity to its historical trend of

---

[6] *Building a Better Future: Economic Choices for the 1980's.* New York Stock Exchange, New York, (December, 1979).

the early 1960's. This drastic upward shift in investment is estimated to add about 0.2 to 0.5 percentage points to the overall rate of productivity expansion. This in turn is associated with an overall improvement of the GNP growth rate by about 0.5 percentage points. These are promising policies but standing by themselves, they are not enough.

It is one thing to assume in a statistical mathematical model solution that the investment ratio is to be higher (by about 2.0 percentage points) and quite another to design policies that will, indeed, raise the share of capital formation in GNP by 2.0 percentage points. The policy discussion in the United States is narrowing to the provision of tax incentives for investment through liberalized depreciation rules and investment tax credits. There is still ample room for improvement of the rate of return on capital through changes in the appropriate tax parameters. Additional policy measures concern tax benefits for R & D outlays, more federal spending for R & D, and more federal spending for basic research. In addition to these standard fiscal measures on the taxing and spending sides of the national budget, there is expectation that productivity growth will be helped by relaxation of restricitve regulations and by the promotion of worker training programs on-the-job. To complement policies designed to raise the investment share of GNP, there should be corresponding policies to raise the savings share. One possible route is to encourage private savings for pension systems, possibly through policies to make retirement pensions portable between jobs. Another way is to exempt some interest on savings accounts from income taxation. The basic issue, however, is to shift the proportions in the make-up of U.S. GNP, namely, to reduce the consumption ratio by 2 percentage points while raising the investment ratio by an equal amount. This is the same thing as saying that the savings ratio should be increased by the same amount as the increase in the investment ratio. In other words, the objective is to shift the U.S. economy from being fundamentally a high consumption to being a high investment economy.

Other countries may view the problem differently, but there should be broad agreement that capital formation has had a relatively poor recovery since the start of the cyclical upswing after 1975. World-wide, the problem is to stimulate investment, but, as in the United States, that will be only a step in the right direction; it will lead to only modest improvement. Clearly, more imaginative policy thinking will have to deal with higher productivity growth.

A feature of the baseline path is the gradual reduction of the average rate of inflation. Many economists would argue that the central economic problem in both the long and short run is to reduce the inflation rate and that many things will "fall-into-place" once inflation has been controlled and gradually reduced.

Many policies can contribute to this worthy end, but a principal line would be to tie changes in the rate of productivity growth to the inflation rate. In the long run, if the rate of inflation is to be lowered, the growth rate of productivity must be significantly increased on a lasting basis.

If the rate of return on capital can be raised, if R & D activity can be made popular again, if economic regulation is liberalized, if worker productivity can be improved through training schemes, and if the rate of inflation can be moderately but steadily decreased, then there is a chance that we can enjoy an investment boom in the 1980's that compares favorably with the great expansionary era of the 1950's and early 1960's.

The appropriate policy measures for raising capital formation and productivity growth are being looked at essentially on individual national bases. But coordinated fiscal and monetary policies offer a new dimension in which to act. If all major countries act synchronously to stimulate capital formation or to ease monetary stringency, there can be added reinforcement effects. International amplification of national fiscal multipliers is estimated to be as large as 1.25 to 1.50. Simultaneous expansions operate through the world trading system because as countries expand, they generally increase import demand. This, in turn, helps partner country export activity and feeds back again on domestic expansion in each individual country. The stronger the response, both nationally and internationally, the less the stimulus has to be in order to arrive at a specified objective. The more we can moderate the use of fiscal/monetary policy, the better is the prospect for lower inflation. By keeping inflation on a favorable path, we stand to gain much through better trade performance.

The figures in Table 7 give a rough indication of what might be expected if the federal government were to stimulate private fixed investment so that it would grow by an extra 2% annually. The policies are different among countries, but they generally consist of tax changes, support from public capital formation and support from general government spending.

The growth rate of GDP is improved over the course of this scenario by about 0.5 percentage points at the beginning, but gradually the investment stimulus tends to wear off by mid-decade. Similarly, there are gains in reducing inflation, again by 1.0 or 2.0 percentage pints. A major contributing factor to the inflation gains is the improvement in productivity (real output per worker hour). It, too, performs better at the outset than at the end in 1985.

Table 7. Coordinated Investment Stimulus 13 LINK Members in OECD 1979–1985
Differences in Percentage Growth Rates

|                     | 1979 | 1980 | 1981 | 1982 | 1983 | 1984 | 1985 |
|---------------------|------|------|------|------|------|------|------|
| GDP                 | 0.3  | 0.6  | 0.5  | 0.2  | 0.2  | 0.2  | 0.1  |
| Inflation           |      |      |      |      |      |      |      |
| (consumer prices)   | 0.0  | −0.2 | −.1  | 0.0  | 0.0  | 0.0  | 0.0  |
| Productivity gains  | 0.3  | 0.4  | 0.3  | 0.0  | 0.0  | −0.3 | −0.1 |

*Edmund S. Phelps.* Studies in Macroeconomic Theory, Volume 1: *Employment and Inflation.* Volume 2: *Redistribution and Growth.*

*Marc Nerlove, David M. Grether, and José L. Carvalho.* Analysis of Economic Time Series: *A Synthesis*

*Thomas J. Sargent.* Macroeconomic Theory

*Jerry Green and José Alexander Scheinkman (Eds.).* General Equilibrium, Growth and Trade: *Essays in Honor of Lionel McKenzie*

*Michael J. Boskin (Ed.).* Economics and Human Welfare: *Essays in Honor of Tibor Scitovsky*

*Carlos Daganzo.* Multinomial Probit: *The Theory and Its Application to Demand Forecasting*

*L. R. Klein, M. Nerlove, and S. C. Tsiang (Eds.).* Quantitative Economics and Development: *Essays in Memory of Ta-Chung Liu*

*Giorgio P. Szegö.* Portfolio Theory: *With Application to Bank Asset Management*

*M June Flanders and Assaf Razin (Eds.)* Development in an Inflationary World

*Thomas G. Cowing and Rodney E. Stevenson (Eds.).* Productivity Measurement in Regulated Industries

*Robert J. Barro (Ed.).* Money, Expectations, and Business Cycles: *Essays in Macroeconomics*

*Ryuzo Sato.* Theory of Technical Change and Economic Invariance: *Application of Lie Groups*

*Iosif A. Krass and Shawkat M. Hammoudeh.* The Theory of Positional Games: *With Applications in Economics*

*Giorgio Szegö (Ed.).* New Quantitative Techniques for Economic Analysis

*John M. Letiche (Ed.).* International Economic Policies and Their Theoretical Foundations: A Source Book

In preparation

*Victor A. Canto, Douglas H. Joines, and Arthur B. Laffer (Eds.).* Foundation of Supply Side Economics

*Murray C. Kemp.* Production Sets